AF602133

The Haptophyte Algae

The Systematics Association
Special Volume No. 51

The Haptophyte Algae

Edited by

J. C. GREEN
Plymouth Marine Laboratory,
UK

B. S. C. LEADBEATER
School of Biological Sciences,
The University of Birmingham,
UK

Published for the SYSTEMATICS ASSOCIATION by
CLARENDON PRESS · OXFORD

This book has been printed digitally and produced in a standard specification in order to ensure its continuing availability

OXFORD
UNIVERSITY PRESS

Great Clarendon Street, Oxford OX2 6DP

Oxford University Press is a department of the University of Oxford.
It furthers the University's objective of excellence in research, scholarship, and education by publishing worldwide in

Oxford New York

Auckland Cape Town Dar es Salaam Hong Kong Karachi
Kuala Lumpur Madrid Melbourne Mexico City Nairobi
New Delhi Shanghai Taipei Toronto
With offices in
Argentina Austria Brazil Chile Czech Republic France Greece
Guatemala Hungary Italy Japan South Korea Poland Portugal
Singapore Switzerland Thailand Turkey Ukraine Vietnam

Published in the United States
by Oxford University Press Inc., New York

Reprinted 2011

ISBN 978-0-19-857772-0

Preface

In the last half century or so, particularly since the introduction of the electron microscope as a routine instrument in biological work, there has been a dramatic increase in our knowledge of the diversity of micro-organisms, including the microalgae. Several new major groups have been recognized, one of the earliest being the class Haptophyceae (= Prymnesiophycene), now in a separate division, the Haptophyta. Representatives of the Haptophyta include the coccolithophorids, unicellular algae with an investment of calcified plates (coccoliths), together with many unmineralized forms, usually flagellates. The marine coccolithophorids have long been marine organisms to students of the phytoplankton, and also to micropalaeontolosts and sedimentologists who have found their calcified remains to be valuable in dating rocks and sediments from prehistoric to recent times. However, apart from species of certain genera, such as *Phaeocystis* and *Prymnesium*, the unmineralized forms have remained of rather esoteric interest until the occurrence in the last few years, particularly in Scandinavian waters, of dense blooms of toxic species of *Chrysochromulina*, and the suggestion from oceanographers that marine phytoplankton, particularly members of the Haptophyta, may affect climate through the release of volatile sulfur compounds.

There has thus been an upsurge of interest in the Haptophyta leading to many recent publications on all aspects of their biology, and the Editors judged that the time was right, therefore, to review as many aspects of current work on the Haptophyta as possible, including their taxonomy, structure, ecology, biochemistry, origin, and evolution. To this end a number of potential contributors were approached; all responded enthusiastically to the challenge and the Editors are happy to present here the fruits of their collective labours.

A word of explanation is needed on the nomenclature and taxonomy of the group. This subject is covered in detail in Chapter 1, but it is worth noting that both the descriptive class name Haptophyceae and the typified name Prymnesiophyceae are valid and are used throughout the individual volume depending on the preferences of authors. At a higher taxonomic level the valid division name is Haptophyta, though the name Prymnesiophyta also occurs in the literature. However, the latter name has never been validly published and the Editors have, therefore, used the valid term Haptophyta as the standard name of the division throughout the volume. The informal terms 'haptophyte' and 'prymnesiophyte' are also in use but have no taxonomic status. The latter term has been avoided as its homophony implies acceptance of the invalid division name Prymnesiophyceae. With respect to the taxonomy at levels below that of class, there are a number of schemes currently available (see Chapter 1) and the editors have therefore not attempted to impose a standard system on authors.

The Editors are grateful to Dr D. M. John, Editor-in-Chief of the Systematics Association, and the Staff of the Oxford University Press for their help and advice in the preparation of this Volume. We are also indebted to the following who have provided critical reviews of various chapters: G. T. Boalch, D. J. Cooper, P. A. Cranwell, W. Gieskes, C. M. Happey-Wood, B. R. Heimdal, G. M. Kennaway, I. Laing, S. Mann, N. J. P. Owens, J. Priddle, J. A. Raven, F. E Round, M. Whitfield, and D. M. Williams.

Plymouth J. C. G.
Birmingham B. S. C. L.
Autumn 1993

Contents

Contributors

G. L. A. BARKER
Department of Biology, University of Bristol, Bristol, BS8 1UG, UK

M. BAUMANN
Alfred Wegener Institute for Polar and Marine Research, Am Handelshafen, 12, 27570 Bremerhaven, Germany

C. BILLARD
Laboratoire de Biologie et Biotechnologies Marines, Université de Caen, 14032-Caen, France

PAUL R. BOWN
Department of Geology, University College, Gower St., London WC1E 6BT, UK

M. R. BROWN
CSIRO Division of Fisheries, Marine Laboratories, GPO Box 1538, Hobart, Tasmania 7001, Australia

COLIN BROWNLEE
Marine Biological Association, Citadel Hill, Plymouth, PL1 2PB, UK

GEERT-JAN BRUMMER
NIOZ, PO Box 59, 1790 AB den Burg, The Netherlands

KURT R. BUCK
Monterey Bay Aquarium Research Institute, 160 Central Avenue, Pacific Grove, California 93950, USA

JACKIE A. BURNETT
Department of Geology, University College, Gower St., London WC1E 6BT, UK

T. CAVALIER-SMITH
Department of Botany, The University of British Columbia, Vancouver, B.C. V6T 1Z4, Canada

FRANCISCO P. CHAVEZ
Monterey Bay Aquarium Research Institute, 160 Central Avenue, Pacific Grove, California 93950, USA

MAUREEN H. CONTE
Department of Chemistry and Geochemistry, Woods Hale Oceanographic Institute, Woods Hole, MA 02543, USA.

L. F. DONG
School of Biological Sciences, University College of Swansea, Singleton Park, Swansea SA2 8PP, UK

GEOFFREY EGLINTON
Biogeochemistry Research Centre, School of Chemistry, University of Bristol, Bristol BS8 1TS, UK

P. R. VAN EMBURG
Department of Biochemistry, Gorlaeus Laboratory, Leiden University, Post Box 9502, 2300 RA Leiden, The Netherlands

J. C. GREEN
Plymouth Marine Laboratory, Citadel Hill, Plymouth PL1 2PB, UK

ROGER HARRIS
Plymouth Marine Laboratory, Prospect Place, West Hoe, Plymouth PL1 3DH, UK

P. K. HAYES
Department of Biology, University of Bristol, Bristol, BS8 1UG, UK

BERIT RIDDERVOLD HEIMDAL
Department of Fisheries and Marine Biology, University of Bergen, 5020 Bergen, Norway

JAN E. VAN HINTE
Faculty of Earth Sciences, Free University, PO Box 7161, 1007 MC Amsterdam, The Netherlands

T. HORI
Institute of Biological Sciences, University of Tsukuba, Tsukuba, Ibaraki, 305 Japan

ISAO INOUYE
Institute of Biological Sciences, University of Tsukuba, Tsukuba, Ibaraki, 305 Japan

S. W. JEFFREY
CSIRO Division of Fisheries, Marine Laboratories, GPO Box 1538, Hobart, Tasmania 7001, Australia

HARRIET L. J. JONES
School of Biological Sciences, The University of Birmingham, Edgbaston, Birmingham B15 2TT, UK

R. W. JORDAN
Department of Earth Sciences, Faculty of Science, Yamagata University, Yamagata, 990 Japan

MASANOBU KAWACHI
Institute of Biological Sciences, University of Tsukuba, Tsukuba, Ibaraki, 305 Japan

CHRISTIANE LANCELOT
Université Libre de Bruxelles, Groupe de Microbiologie des Milieux Aquatiques, Campus de la Plaine, CP 221, Boulevard du Triomphe, 1050 Bruxelles, Belgium

M. LANGE
Alfred Wegener Institute for Polar and Marine Research, Am Handelshafen, 12, 27570 Bremerhaven, Germany

B. S. C. LEADBEATER
School of Biological Sciences, The University of Birmingham, Edgbaston, Birmingham B15 2TT, UK

PETER S. LISS
School of Environmental Sciences, University of East Anglia, Norwich NR4 7TJ, UK

GILLIAN MALIN
School of Environmental Sciences, University of East Anglia, Norwich NR4 7TJ, UK

HARVEY J. MARCHANT
Australian Antarctic Division, Channel Highway, Kingston, Tasmania 7050, Australia

L. K. MEDLIN
Alfred Wegener Institute for Polar and Marine Research, Am Handelshafen, 12, 27570 Bremerhaven, Germany

M. J. MERRETT
School of Biological Sciences, University College of Swansea, Singleton Park, Swansea SA2 8PP, UK

ØJVIND MOESTRUP
Department of Phycology, Botanical Institute, University of Copenhagen, Ø. Farimagsgade 2D, DK-1353 Copenhagen K, Denmark

N. NIMER
School of Biological Sciences, University College of Swansea, Singleton Park, Swansea SA2 8PP, UK

VÉRONIQUE ROUSSEAU
Université Libre de Bruxelles, Groupe de Microbiologie des Milieux Aquatiques, Campus de la Plaine, CP 221, Boulevard du Triomphe, 1050 Bruxelles, Belgium

KOZO TAKAHASHI
Department of Marine Sciences and Technology, School of Engineering, Hokkaido Tokai University, Minamisawa 5-1-1-1, Minami-ku, Sapporo, Japan 005

HELGE A. THOMSEN
Department of Phycology, Botanical Institute, University of Copenhagen, Ø. Farimagsgade 2D, DK-1353 Copenhagen K, Denmark

SUZANNE M. TURNER
School of Environmental Sciences, University of East Anglia, Norwich NR4 7TJ, UK

MARCEL VELDHUIS
NIOZ, PO Box 59, 1790 AB den Burg, The Netherlands

JOHN K. VOLKMAN
CSIRO Division of Oceanography, Marine Laboratories, GPO Box 1538, Hobart, Tasmania 7001, Australia

J. P. M. DE VRIND
Department of Biochemistry, Gorlaeus Laboratory, Leiden University, Post Box 9502, 2300 RA Leiden, The Netherlands

E. W. DE VRIND-DE JONG
Department of Biochemistry, Gorlaeus Laboratory, Leiden University, Post Box 9502, 2300 RA Leiden, The Netherlands

PETER WESTBROEK
Department of Chemistry, University of Leiden, PO Box 9502, 2300 RA Leiden, The Netherlands

S. W. WRIGHT
Australian Antarctic Division, Channel Highway, Kingston, Tasmania 7050, Australia

JEREMY R. YOUNG
Palaeontology Department, The Natural History Museum, Cromwell Road, London SW7 5BD, UK

1. Systematic history and taxonomy

J. C. GREEN
Plymouth Marine Laboratory, Citadel Hill, Plymouth, UK

and R. W. JORDAN*
Institute of Oceanographic Sciences, Deacon Laboratory, Wormley, Godalming, UK
**Present address: Department of Earth Sciences, Yamagata University, Yamagata, Japan*

Abstract

Representatives of the algal division Haptophyta occur predominantly in marine habitats, relatively few being known from freshwater or from terrestrial situations. Certain species, some of them toxic to animal life, are known to form extensive blooms. The history of systematic studies on the Haptophyta is reviewed briefly and some of the taxonomic difficulties arising from a reliance on relatively few morphological features, particularly at levels below that of class, are discussed. Recent advances in ultrastructural, biochemical, and genetic studies are outlined in relation to some of the outstanding problems.

Introduction

The algal division Haptophyta includes a diversity of organisms known principally from marine environments, but with a few freshwater and terrestrial representatives (Stein 1878; Lackey 1939; Parke *et al.* 1961; Green 1973; Green and Parke 1975; Nicholls 1978; Kling 1981; Wujek and Gardiner 1985). Representatives include unicellular motile and non-motile forms, the latter sometimes occurring as palmellae, gelatinous colonies or filaments. The most familiar representatives of the Haptophyta are probably the coccolithophorids, unicellular organisms, motile or non-motile, with an investiture of calcified plates (coccoliths). Coccolithophorids may form extensive blooms, detectable by remote sensing techniques (Holligan *et al.* 1993) due to the reflective nature of the coccoliths. However, blooms of toxic planktonic haptophytes in

The Haptophyte Algae (ed. J. C. Green and B. S. C. Leadbeater), Systematics Association Special Volume No. 51, pp. 1–21. Clarendon Press, Oxford, 1994.

Scandinavian waters, with the associated mass mortality of marine life and severe economic losses, have recently brought attention to other groups, particularly species of *Chrysochromulina* (Underdahl *et al.* 1989; Tangen 1991), although localized outbreaks of fish mortality caused by species of *Prymnesium* had been recorded previously from other parts of the world (for references, see Moestrup, Chapter 14).

Most known members of the Haptophyta are photosynthetic, but at least one non-photosynthetic coccolithophorid is known (*Balaniger balticus*; Thomsen and Oates 1978). A number of non-photosynthetic freshwater organisms have been included in the Haptophyta by some authors (e.g. Starmach 1985) on the basis of their flagellar arrangement, but it is now known that a number of these organisms have affinities elsewhere (for references, see Jordan and Green 1994).

The feature most readily associated with the Haptophyta is the haptonema, a filiform organelle whose length may be many times that of the cell carrying it. For example, the haptonema of *Chrysochromulina strobilus* is over 100 μm in length, the diameter of the cell body itself being only about 8.0 μm (Parke *et al.* 1959). In a number of other genera, however, the haptonema may be very reduced. In species of *Isochrysis* and *Chrysotila* it is very short (1–2 μm; Green and Parke 1975; Green and Pienaar 1977), and in *Dicrateria* and *Imantonia* the haptonema is detectable only in electron micrographs as, at most, a small proboscis between the flagella (Green and Pienaar 1977).

Many members of the Haptophyta also carry unmineralized scales on the cell-body. The scales in their simplest form consist of a two-layered plate, each layer composed of fibrils arranged in a radial pattern. Such scales often occur alone or as an underlayer beneath the coccoliths (where present) and, in most coccolithophorids the coccoliths themselves are formed on a base-plate scale (for a review, see Leadbeater, Chapter 2). Haptonemata, coccoliths, and two-layered fibrillar scales are found only in the Haptophyta, and taxonomic studies of the group have been largely concerned with the interpretation and assessment of the importance of these structures.

Historical background

The coccolithophorids

The earliest records of evidence of organisms we now know as members of the Haptophyta were those of Ehrenberg (1836) who reported the presence of microscopic platelets in chalk deposits (Table 1.1). However, he considered that these were formed by chemical precipitation rather than by biological activity. Later, Huxley (1858) reported the presence of

Table 1.1 Some key dates in studies of the division Haptophyta

1836	Ehrenberg discovers crystallites in chalk
1858	Huxley reports the presence of 'coccoliths' in deep-sea deposits
1861	Wallich discovers coccospheres
1878	Stein describes a freshwater coccolithophorid (but does not realize what it is)
1898	Murray and Blackman confirm the algal origin of coccoliths and coccospheres
1900	Scherffel describes a haptonema ('third flagellum') in *Phaeocystis*
1902	Lohmann's first major classification of the coccolithophorids
1910	Pascher introduces the order Isochrysidales for chrysophycean flagellates with two equal flagella
1911	Büttner describes *Wysotskia*, another flagellate with 'three flagella', probably a species of *Prymnesium*
1920	Description of *Prymnesium* by Massart and first report (Liebert and Deerns) of an icthyotoxic bloom, now attributable to a species of *Prymnesium*
1926	Conrad proposes that chrysophycean organisms with 'three flagella' should be included in a new family
1930	Schiller's major monograph on the coccolithophorids
1937	Carter describes more species of *Prymnesium*
1939	Lackey describes the first species of *Chrysochromulina* from the USA
1952	Butcher describes *Pavlova gyrans*
	First TEM illustrations of coccoliths
1955	First TEM illustrations of *Chrysochromulina* and definition of the haptonema by Parke and Manton
1960	Parke and Adams report both holococcolithophorid and heterococcolithophorid stages in the life-history of *Coccolithus pelagicus*
1962	Introduction of the class name Haptophyceae by Christensen
1972	First SEM illustrations of coccoliths
1972/1976	Hibberd introduces the concept of a separate division for the Haptophyceae and introduces the typified class name Prymnesiophyceae
1986/1989	Cavalier-Smith splits the Prymnesiophyceae into two subclasses, then into two classes

such 'coccoliths' (since they reminded him of the alga *Protococcus*) within deep-sea deposits. Wallich (1860) and Sorby (1861) confirmed the presence of coccoliths in surface sediments and chalk, but also noted that they were associated with hollow spheres (coccospheres; Wallich 1861) which both workers interpreted as being related to foraminifera, either representing part of the life cycle (Wallich 1861) or as independent but related organisms (Sorby 1861).

During the 1870s, a number of investigators appeared to disregard the coccolith literature and resurrected Ehrenberg's theory that coccoliths were formed as a result of precipitation (e.g. Barrois 1876). However, there were others who maintained that coccoliths were of algal origin, possibly representing single cells of calcareous algae such as *Melobesia* (Carter 1871). Wyville-Thomson (1874) thought that they were the armature of minute algae, but it was not until the report of a chromatophore inside a coccosphere, and the illustration of a dividing cell (Murray and Blackman 1898), that the true nature of coccolithophorids as unicellular algae, each with an investment of coccoliths, was fully appreciated.

The haptonema

The earliest record of a haptonema is that of Scherffel (1900). In his detailed description of swarmers of *Phaeocystis globosa,* he noted that the cells had two flagella, approximately the same length as the cells. However, after fixing the swarmers with osmium tetroxide, he observed a short rod-like body associated with the flagella. He interpreted this as an 'ancillary flagellum' and noted that he was unaware of any other flagellate with this structure.

Later, Büttner (1911) described a flagellate (*Wysotskia gladiociliata*) with cells having a short 'third flagellum'. He thought that this must be a stage of division, but the organism has since been identified as probably a species of *Prymnesium* (*P. gladiociliatum*; Carter 1937; Jordan and Green 1994). The type species of *Prymnesium* (*P. saltans*) was described by Massart (1920) who noted that the cell could attach itself to the surface of the slide or coverslip by the short 'third flagellum'. Conrad (1926) later observed *P. saltans* in brackish waters in Belgium, formalized the name, and realizing the importance of this species, created a new family, Prymnesiacées, within the Chrysomonadales Pascher.

Also in 1920, Liebert and Deerns reported the occurrence of a flagellate which they named informally 'the chryso-monadine from Workum', but which, from their excellent drawings is unmistakeably a species of *Prymnesium.* It occurred in high concentrations (0.5×10^6 cells ml^{-1}) and was associated with extensive fish mortality. This event seems to be, therefore, the first of its kind recorded in which the responsible agent can be identified as a member of the Prymnesiophyceae.

N. Carter (1937) described further species of *Prymesium* from a saline pool on the Isle of Wight (UK), and two years later Lackey (1939) described from America the first species of *Chrysochromulina* (*C. parva*). Again, it was thought that the cell carried three flagella, though in this case the ancillary flagellum was very long, 5–8 times the diameter of the cell in length, i.e. 25–40 μm.

It is now well known that the advent of the electron microscope as a routine instrument for the study of microorganisms, combined with developments in laboratory culture techniques, facilitated the study of haptophyte algae from the early 1950s, beginning with a series of elegant studies by Parke and Manton on species of *Chrysochromulina* isolated from the waters around Plymouth. Parke *et al.* (1955) demonstrated that the 'third flagellum' was not a flagellum after all, but a novel organelle which they named the haptonema. They also demonstrated the presence of a variety of unmineralized scales, sometimes very elaborate in structure, forming an investment on the cell body (Leadbeater, Chapter 2). Subsequently, reduced haptonemata were recognized in other genera, particularly in the coccolithophorid *Pleurochrysis scherffelii* (Stosch 1958), demonstrating the affinities of the coccolithophorids with *Prymnesium* and *Chrysochromulina.*

Taxonomy

Nomenclature

Recent attempts to classify the haptophyte algae show a broad measure of agreement (Table 1.2). However, a variety of names has arisen in the literature and those of the higher taxa are summarized in Table 1.3. The names are either typified, i.e. based on the existing genus, or descriptive, reflecting a physical attribute common to members of the group. It should be noted that, at the division level, the Chromophyta is conceptually a much wider division than Haptophyta (Christensen 1962, 1989), and that the name Prymnesiophyta, although in common use, is not valid.

At the class level, both the typified name Prymnesiophyceae and the descriptive name Haptophyceae are valid. The advantage of using the typified name is that should the taxon be split, which happens more frequently at class and lower taxonomic levels, it is obvious which new group takes the existing name. Cavalier-Smith (1993; see also below and Chapter 22) has recently split the class, but has, however, introduced two new names, one typified (Pavlovea), the other descriptive (Patelliferea).

Higher rank taxonomy

In spite of the apparent anomalies in their structure, e.g. more or less equal flagella and the presence of the third appendage, haptophyte algae were for many years included in the Chrysophyceae, principally on the basis of their generally yellow-brown pigmentation. Papenfuss (1955) provides an example of the earlier view by including the coccolithophorids in the order Isochrysidales (chrysophytes with two equal flagella;

Table 1.2 Recent taxonomic schemes for the haptophyte algae

	Christensen (1962)	Parke and Dixon (1976)	Tappan (1980)	Christensen (1980)
Division	**Chromophyta**	**Haptophyta**	**Haptophyta**	**Haptophyta**
Class	**Haptophyceae**	**Haptophyceae**	**Coccolithophyceae**	**Prymnesiophyceae**
Subclass				
Order	**Isochrysidales**	**Isochrysidales**	**Isochrysidales**	
Families	Isochrysidaceae	Isochrysidaceae	Gephyrocapsaceae	
	(Derepyxidaceae)	Gephyrocapsaceae	Isochrysidaceae	
	Diacronemataceae	Hymenomonadaceae	Hymenomonadaceae	
		Ochrosphaeraceae	Ochrosphaeraceae	
			(Thoracosphaeraceae)	
Order	**Prymnesiales**	**Prymnesiales**	**Prymnesiales**	**Prymnesiales**
Families	Prymnesiaceae	Prymnesiaceae	Prymnesiaceae	Prymnesiaceae
	Coccolithophoraceae	Phaeocystaceae	Phaeocystaceae	Phaeocystaceae
	Phaeocystaceae			Chrysotilaceae
				Gloeothamniaceae
				Pavlovaceae
				Coccolithophorids included in **3 families:**
				Thoracosphaeraceae
				Braarudosphaeraceae
				Ceratolithaceae
Order		**Coccosphaerales**	Coccolithophorids distributed between **6 orders** and **36 families**	
Families		Zygosphaeraceae		
		Calciosoleniaceae		
		Pontosphaeraceae		
		Halopappaceae		
		Helicosphaeraceae		
		Coccolithaceae		
		Rhabdosphaeraceae		
		Braarudosphaeraceae		
Class				
Subclass				
Order		**Pavlovales**	**Pavlovales**	
Families		Pavlovaceae	Pavlovaceae	

Table 1.2 (*cont.*)

Cavalier-Smith (1990)[1]	Chrétiennot-Dinet (1990)	Chrétiennot-Dinet *et al.* (1993)	Cavalier-Smith (1993)	Jordan and Green (1994)
Haptomonada			**Haptophyta**	**Haptophyta**
Haptophyceae	**Prymnesiophyceae**	**Prymnesiophyceae**	**Patelliferea**	**Prymnesiophyceae**
Prymnesidae				**Prymnesiophycidae**
	Isochrysidales	**Isochrysidales**	**Isochrysidales**	
	Isochrysidaceae	Isochrysidaceae		
	Prymnesiales	**Prymnesiales**	**Prymnesiales**	**Prymnesiales**
	Prymnesiaceae	Prymnesiaceae		Isochrysidaceae
	Phaeocystaceae	Phaeocystaceae		Noëlaerhabdaceae
				Prymnesiaceae
				Phaeocystaceae
	Coccosphaerales	**Coccolithophorales**	**Coccosphaerales**	Coccolithophorids
	Coccolithophorids	Coccolithophorids		distributed between
	distributed between	distributed between		**12 families**
	13 families	**14 families**		
			Pavlovea	
Pavlovidae				**Pavlovophycidae**
	Pavlovales	**Pavlovales**	**Pavlovales**	**Pavlovales**
	Pavlovaceae	Pavlovaceae		Pavlovaceae

[1]See, Cavalier-Smith (1989).

Table 1.3 Names of major taxa used in haptophyte systematics

Level	Name	Introduced	Validated	Descriptive (D) *or* Typified (T)
Division	Chromophyta	Christensen 1962	Christensen 1990[1]	D
	Haptophyta	Hibberd 1972	Cavalier-Smith 1986	D
	Haptomonada	Cavalier-Smith 1990[2]	Cavalier-Smith 1986	D
	Prymnesiophyta	Hibberd 1976	Never validly published	T
Class	Coccolithophyceae	Rothmaler 1951	Not valid (Silva 1980)	D
	Haptophyceae	Christensen 1962	Silva 1980	D
	Prymnesiophyceae	Hibberd 1976	Hibberd 1976	T
	Patelliferea[3]	Cavalier-Smith 1993	Cavalier-Smith 1993	D
	Pavlovea[3]	Cavalier-Smith 1993	Cavalier-Smith 1993	T
Subclass	Prymnesidae[3]	Cavalier-Smith 1986	Cavalier-Smith 1986	T
	Pavlovidae[3]	Cavalier-Smith 1986	Cavalier-Smith 1986	T
	Prymnesiophycidae[3]	Jordan and Green 1994	Cavalier-Smith 1986	T
	Pavlovophycidae[3]	Jordan and Green 1994	Cavalier-Smith 1986	T

[1] See Christensen (1989). [2] See Cavalier-Smith (1989).

[3] Cavalier-Smith (1986) divided the class Prymnesiophyceae by erecting two subclasses, the Pavlovidae, to include *Pavlova* and related genera, and the Prymnesidae, to include all other prymnesiophycean algae. The two subclasses were later both raised to the rank of class, the Prymnesidae to the Patelliferea and the Pavlovidae to the Pavlovea (Cavalier-Smith 1993). The names Haptophyceae and Prymnesiophyceae therefore become redundant under Cavalier-Smith's classification (Cavalier-Smith 1993, see also Chapter 22). Jordan and Green (1994) have retained Cavalier-Smith's two subclasses, but have used the conventional botanical suffixes for the names of algal subclasses.

Pascher 1910), and those organisms believed to have three flagella in an new order, the Prymnesiales. However, Christensen (1962), accepting the importance of the haptonema and other characteristics of these aberrant chrysophytes, erected a new class, the Haptophyceae, included in the division Chromophyta (Table 1.2). Subsequently, Hibberd (1976) introduced the typified class name Prymnesiophyceae and concluded that the Prymnesiophyceae demonstrated features justifying their separation from other chromophyte algae in a separate division, the Prymnesiophyta. He thus also introduced a typified name for the division, though the descriptive name Haptophyta was already in the literature (Hibberd 1972; Table 1.3).

Some of the more important morphological features of the Haptophyta are shown in Table 1.4. They have been summarized in several earlier publications (e.g. Hibberd 1976; Cavalier-Smith 1989) and only a few

Table 1.4 Principal morphological features characteristic of members of the division Haptophyta

Chloroplasts with no girdle lamella
Non-heterokont; tubular hairs never found
Flagella usually 2, equal or subequal; if unequal, no tubular hairs (Pavlovophycidae have fibrous hairs and knob-scales on longer flagellum)*
Eyespots found only in Pavlovophycidae, usually associated with invagination of plasmalemma; not associated with flagellar swelling**
Haptonema (or trace) present
Unmineralized body scales basically a fibrillar 2-layered plate; many species with calcified coccoliths. Silicification rare.

* Possibly modified hairs rather than scales (see Cavalier-Smith, Chapter 22).

**Except in *Diacronema vlkianum* (Green and Hibberd 1977).

examples will be given here. The most obvious difference is the structure and arrangement of the flagellar apparatus (for reviews, see Leadbeater 1989; Andersen 1991). A characteristic feature of most chromophytes is the presence of heterokont flagella, two unequal flagella, one of them sometimes present only as a basal body, with the longer, usually anterior, flagellum carrying an array of tubular flagellar hairs. In contrast, members of the Prymnesiophyceae typically have two equal or subequal flagella which never carry tubular hairs. Unequal flagella are found only in the Pavlovales, and here the longer flagellum carries an array of fine fibrillar hairs and small knob-like bodies. These bodies have been interpreted as modified scales, but it is now suggested (Cavalier-Smith, Chapter 22) that they are probably modified hairs. A swelling on the shorter flagellum intimately associated with an eyespot, characteristic of several heterokont groups, is never found in haptophyte algae. Eyespots in the Haptophyta are found only the Pavlovales, where they are often associated with a tubular invagination of the cell membrane close to the flagellar apparatus (Green 1980).

Other distinctive features of the Haptophyta, scales, coccoliths, and haptonemata, have already been mentioned briefly. The cell body of most members of the class, other than members of the Pavlovales, is covered either with more or less ornate unmineralized scales, or scales and calcified coccoliths (for reviews, see Leadbeater and Green 1993; Leadbeater, Chapter 2). The basic scale is a two-layered plate with a radial arrangement of fibrils on the proximal face and a spiral or concentric pattern on the distal face, and although scales are found in other chromophyte algae, they are not constructed in such a uniform manner. Calcified scales (coccoliths) are exclusive to the Haptophyta, but silicification, which occurs frequently in the Chrysophyta and in the diatoms, is recorded in the Haptophyta only in cysts of *Prymnesium* (Pienaar 1980).

The most distinctive feature of the Haptophyta, however, is the haptonema, an organelle apparently unique to this group of organisms (see above). It is intimately associated with the flagella by means of a number of fibrous connections, and though it is unlikely that the haptonema evolved from a flagellum (Cavalier-Smith 1986; Chapter 22), there can be no doubt that flagellar movement and the action of the haptonema are linked in some way in species where haptonematal coiling can be demonstrated (Gregson *et al.* 1993; Green and Hori, Chapter 3; Inouye and Kawachi, Chapter 4).

At taxonomic levels above that of division there is less agreement on the status and position of the Haptophyta. Christensen (1980) included the class Haptophyceae in the division Chromophyta in a general group, the Contophora, encompassing those algae known to have flagellate cells at least some time in their life-cycle. Others have included the haptophytes in more restricted groups. For example, Cavalier-Smith (1986, 1989) included them in a subkingdom Chromophyta in a kingdom Chromista, but Leedale (1974), in an even more restrictive scheme, suggested that the haptophytes (= prymnesiophytes) could be conceived as comprising a kingdom in a multi-kingdom evolutionary scheme. The advantage of the more restrictive approach is that it helps to overcome problems of polyphyly inherent in larger groupings of organisms (e.g. Lee *et al.* 1985; Margulis *et al.* 1990).

Lower rank taxonomy

Below class level, there is considerable diversity of opinion on the taxonomy of the Prymnesiophyceae (Haptophyta), both in the number of orders and the number of families (Table 1.2).

An appreciation of the importance of the presence of two more or less equal flagella in some 'chrysophytes' led to the erection of the order Isochrysidales by Pascher (1910). Conrad (1926) erected the family Prymnesiaceae (as the Prymnésiacées) to include organisms with two equal flagella and an ancillary 'third flagellum', and Papenfuss (1955) based his order Prymnesiales on this family. These orders were retained by Christensen (1962) and have been used by many, though not all, taxonomists in recent years (Table 1.5). The most significant differences lie between members of the Pavlovales and the other groups, and have formed the basis of the recent separation of the Pavlovales at either subclass (Cavalier-Smith 1986, 1989) or class level (Cavalier-Smith 1993). At the level of family, genus, and species haptophyte taxonomy depends heavily on details of morphology, particularly that of scales and coccoliths, and some of the problems inherent in this approach can be illustrated with reference to coccolithophorid systematics.

The first major attempt at classification of the coccolithophorids by Lohmann (1902) placed them in a family, the Coccolithophoridae, with two subfamilies depending on the presence or absence of a flagellar field. Genera were differentiated by coccolith shape. Lemmermann (1908) subsequently included the coccolithophorids in a class, with Lohmann's two subfamilies raised to orders characterized by the presence or absence of coccolith perforations. As the number of species and genera increased, so did the complexity of the classification system (e.g. Schiller 1930; Deflandre 1952).

The advent of the electron microscope permitted the observation of coccoliths in greater detail, first with the TEM (Braarud *et al.* 1952) and later with the SEM (Borsetti and Cati 1972). The increase in resolution and magnification led to a more detailed terminology and more precision in the definition of different coccolith types (Braarud *et al.* 1955; Farinacci 1971). Thus, coccolith morphology remains the most important criterion

Table 1.5 Principal morphological features characteristic of the orders of the Prymnesiophyceae (division: Haptophyta)

Subclass: Prymnesiophycidae[1]			Subclass: Pavlovophycidae[1]
Isochrysidales[2,3]	**Prymnesiales**[2,3,4]	**Coccosphaerales**[2] or **Coccolithophorales**[3]	**Pavlovales**[4]
Unicellular, motile or non-motile, sometimes filamentous	Unicellular, usually motile, sometimes non-motile. Cells free, occasionally colonial.	Unicellular, motile or non-motile	Unicellular or forming palmelloid masses
Flagella 2, equal or subequal One flagellum sometimes autofluorescent	Flagella 2, equal or subequal One flagellum sometimes autofluorescent	Flagella 2, equal or subequal	Flagella unequal, shorter may be reduced, longer with investment of fibrous hairs, and knob-scales. Always heterodynamic.
Haptonema reduced or absent	Haptonema usually conspicuous, often well-developed and coiling	Haptonema conspicuous or reduced	Haptonema short, non-coiling

Table 1.5 (*cont.*)

	Subclass: Prymnesiophycidae[1]		Subclass: Pavlovophycidae[1]
Isochrysidales[2,3]	**Prymnesiales[2,3,4]**	**Coccosphaerales[2]** or **Coccolithophorales[3]**	**Pavlovales[4]**
Body-scales small unmineralized (occasionally absent)[5]	Body-scales various, one or more layers, and often several types of scale, per cell	Calcified coccoliths present at some stage during life cycle; non-mineralized scales also present	Body-scales absent
	Flagellar root system based on 3–4 microtubular roots[6]		Flagellar root system unique to group, based on two microtubular roots[6]
	Mitotic spindle axis straight; no fibrous root MTOC[7]		Mitotic spindle axis V-shaped; fibrous root MTOC[7]

[1]e.g. Jordan and Green (1994); [2]e.g. Parke and Dixon (1976); [3]e.g. Chrétiennot-Dinet (1990).
[4]Christensen (1980) and Jordan and Green (1994) include all members of the Prymnesiophyceae in only two orders, the Prymnesiales and the Pavlovales.
[5]Parke and Dixon (1976) include some coccolithophorids in the Isochrysidales on the basis of the reduced or absent haptonema. There is some justification for including *Emiliania huxleyi* here on the grounds that *E. huxleyi* and species of *Isochrysis* and *Chrysotila* share a number of unusual lipids (see Conte *et al.*, Chapter 19).
[6]See Green and Hori, Chapter 3; [7]See Hori and Green, Chapter 5.

in coccolith identification and taxonomy with families characterized by one or more coccolith types, genera depending on combinations of different coccolith types or special structures, and species delimitation depending on variations in the structure of the particular type of coccolith characteristic of the genus. However, the use of the electron microscope has introduced a number of new problems. For example, we now have very detailed information about the form and structure of coccoliths, but we have become so dependent on the EM that observations using light microscopy are rarely made. Many newly described species of both coccolithophorids and unmineralized flagellates (e.g. Estep *et al.* 1984) have, therefore, been erected from electron microscope information only, with little or no information on the habit of the live cell, cell shape, flagellar apparatus, or even *in situ* coccolith or scale distribution. A further problem exists in that many members of the Haptophyta have proved to be difficult to maintain in culture, making access to live material even more difficult. In this context, problems of taxonomy also arise from the lack of information on the life cycles of most members of the group. The majority of species are known only as unicellular, monophasic organisms though there is accumulating evidence for complex life cycles and sexuality (see Billard, Chapter 9).

Over the last decade or so, a number of aspects of structure and biochemistry have been investigated with a view to obtaining further information of value in assessing haptophyte relationships. The structure of the flagellar root system has also received considerable attention (for references, see Green and Hori, Chapter 3). It has been shown that some coccolithophorids have complex root systems with sheets and 'crystalline' bundles of microtubules (MTs), but comparable root systems containing bundles or large sheets of MTs have also been reported in a number of non-coccolithophorid species. Similarly, relatively simple root systems also occur in both coccolithophorid and non-coccolithophorid genera. Only members of the Pavlovales seem to display any degree of uniformity, with a root system that appears to be characteristic of this particular assemblage.

There have also been a number of studies of nuclear division in the Haptophyta (see Hori and Green, Chapter 5). There is a degree of uniformity in the general pattern of nuclear division in the Prymnesiales; a straight spindle axis, lack of any type of obvious pole structure, and formation of the new nuclear envelope on the poleward face of the mass of daughter chromatin. However, there are differences of detail, but the information we have at present is insufficient to allow detailed taxonomic decisions to be made. It is again the Pavlovales which are clearly distinct from the other orders; the spindle in *Pavlova* has a V-shaped axis, there are well-defined fibrous pole structures, and the nuclear envelope is persistent for much of the nuclear division.

Recent biochemical studies have yielded a considerable body of new information on chlorophylls and accessory pigments (for references, see Jeffrey and Wright, Chapter 6), and unusual long-chain unsaturated lipids (Conte *et al.*, Chapter 19). With respect to pigments, members of the Prymnesiophyceae may be grouped into four principal pigment types, with different combinations of chlorophylls from the chlorophyll *c* group, fucoxanthin, 19′-hexanoyloxyfucoxanthin, and 19′-butanoyloxyfucoxanthin, but there is no clear-cut pattern correlating with existing taxonomic concepts of the class. Similarly, it has been shown by Eglinton and coworkers (Conte *et al.*, Chapter 19) that unusual long-chain $C_{37\text{-}39}$ polyunsaturated methyl and ethyl ketones and esters occur in the coccolithophorid *Emiliania huxleyi*, and in species of *Isochrysis* and *Chrysotila*, supporting their inclusion together in the Isochrysidales. However, these compounds have not been found in other members of the Haptophyta and their presence in *Emiliania huxleyi* serves to emphasize the differences between this species and other coccolithophorids.

Conclusions

In summary, although there is much recent information on the ultrastructure and biochemistry of the Haptophyta it is not yet possible to erect a taxonomic system which we can confidently feel reflects the relationships between the different genera. The relatively new techniques of molecular genetics will no doubt yield a wealth of new information in the same way that the advent of electron microscopy opened up new fields in morphological and structural studies (Manhart and McCourt 1992). Most of the work to date on the chlorophyll *c* containing algae has been concerned primarily with phylogeny at the level of major taxa (e.g. Williams 1991; Battacharya *et al.* 1992), but, recently, such techniques have been applied to taxonomic problems at the generic level in *Phaeocystis* and *Emiliania* (Medlin *et al.*, Chapter 21).

The taxonomy of the Haptophyta, therefore, still depends heavily on a few morphological characters. Of course such features will always be valuable, especially in identification, but at present there are problems in defining a taxonomic scheme, especially at the ordinal level, which can take into account the variation and apparent anomalies encountered in several of the present schemes. We have recently (Jordan and Green 1994) adopted the approach of Cavalier-Smith (1986) in recognizing the fundamental differences between *Pavlova* and related genera, and other members of the Haptophyta, by separating them at the level of subclass. However, within the subclass Haptophyta we have followed Christensen (1980) by including all taxa in a single order, the Prymnesiales (Table 1.5). Perhaps this reflects the attitude of Schwarz (1894), who place all coccolithophorids in one species because he claimed that it was impossible to distinguish differences between them! However, it does avoid problems such as reconciling the inclusion of some coccolithophorids in typically 'non-coccolithophorid' orders, based on the presence of unmineralized scales in the motile stage and the absence of a haptonema, and the occurrence of complex compound microtubular root systems in both coccolithophorid and non-coccolithophorid species. Nevertheless, there remains an urgent need for further information at all levels before we can understand fully the relationships of the haptophyte algae to each other, and to other major algal and protistan groups.

References

Andersen, R. A. (1991). The cytoskeleton of chromophyte algae. *Protoplasma*, **164**, 143–59.

Barrois, C. (1876). Mémoire sur l'embryologie de quelques éponges de la Manche. *Annales des Sciences Naturelles (Zoologie)*, **Series 6**, **Part 3**, 1–84.

Battacharya, D., Medlin, L., Wainwright, P. O., Aritzia, E. V., Bibeau, C., Stickel, S. K., and Sogin, M. L. (1992). Algae containing chlorophylls *a* + *c* are paraphyletic: molecular evolutionary analysis of the Chromophyta. *Evolution*, **46**, 1801–17.

Borsetti, A. M. and Cati, F. (1972). Il nannoplankton calcareo vivente nel Tirreno centromeridionale. *Giornale di Geologia*, Ser. 2a, **38**, 395–452.

Braarud, T, Gaarder, K. R., Markali, J., and Nordli, E. (1952). Coccolithophorids studied in the electron microscope. Observations on *Coccolithus huxleyi* and *Syracosphaera carterae*. *Nytt Magasin for Botanikk*, **1**, 129–34.

Braarud, T., Deflandre, P., Halldal, P., and Kamptner, E. (1955). Terminology, nomenclature, and systematics of the Coccolithophoridae. *Micropalaeontology*, **1**, 157–9.

Butcher, R. W. (1952). Contribution to our knowledge of the smaller marine algae. *Journal of the Marine Biological Association of the United Kingdom*, **31**, 175–92.

Büttner, J. (1911). Die forbigen Flagellaten des Kieler Hafens. *Wissenschaftliche Meeresuntersuchungen, Abteilung Kiel.*, **12**, 119–33.

Carter, H. J. (1871). On '*Melobesia unicellularis*', better known as the coccolith. *Annals and Magazine of Natural History*, Series 4, **7**, 184–9.

Carter, N. (1937). New or interesting algae from brackish waters. *Archiv für Protistenkunde*, **90**, 1–68.

Cavalier-Smith, T. (1986). The kingdom Chromista: Origin and systematics. *Progress in Phycological Research*, **4**, 309–47.

Cavalier-Smith, T. (1989). The kingdom Chromista. In *The chromophyte algae: problems and perspectives*, Systematics Association Special Volume No. 38, (ed. J. C. Green, B. S. C. Leadbeater, and W. L. Diver), pp. 381–407. Clarendon Press, Oxford. (Actual year of publication, 1990).

Cavalier-Smith, T. (1993). Kingdom Protozoa and its 18 phyla. *Microbiological Reviews*, **57**, 953–94.

Chrétiennot-Dinet, M.-J. (1990). *Atlas du phytoplancton marin*, Vol. 3. Éditions du CNRS, Paris.

Chrétiennot-Dinet, M.-J., Sournia, A., Ricard, M., and Billard, C. (1993). A classification of the marine phytoplankton of the world from class to genus. *Phycologia*, **32**, 159–79.

Christensen, T. (1962). Alger. In *Botanik* Bd. 2, *Systematisk Botanik*, Nr. 2, (ed. T. W. Böcher, M. Lange, and T. Sørensen), pp. 1–178. Munksgaard, Copenhagen.

Christensen, T. (1980). *Algae*. AiO Tryk as, Odense.

Christensen, T. (1989). The Chromophyta, past and present. In: *The chromophyte algae: problems and perspectives*, Systematics Association Special Volume No. 38, (ed. J. C. Green, B. S. C. Leadbeater, and W. L. Diver), pp. 1–12. Clarendon Press, Oxford. (Actual year of publication, 1990).

Conrad, W. (1926). Recherches sur les flagellates de nos eaux saumâtres. II. Chrysomonadines. *Archiv für Protistenkunde*, **56**, 167–231.

Deflandre, G. (1952). Classe des Coccolithophoridés (Coccolithophoridae Lohmann, 1902). In: *Traité de Zoologie*, Vol. 1, (ed. P.-P. Grassé), pp. 439–470. Masson et Cie., Paris.

Ehrenberg, C. G. (1836). Bermerkungen über feste mikroskopische, anorganische Formen in den erdigen und derben Mineralien. *Bericht über die Verhandlungen der Königlich Preussichen Akademie der Wissenschaften*, Berlin, **1836**, 84–5.

Estep, K. W., Davis, P. G., Hargraves, P. E., and Sieburth, J. McN. (1984). Chloroplast containing microflagellates in natural populations of North Atlantic nanoplankton, their identification and distribution, including a description of five new species of *Chrysochromulina* (Prymnesiophyceae). *Protistologica*, **20**, 613–34.

Farinacci, A. (1971). Round table on calcareous nannoplankton, Roma, September 23–28, 1970. In *Proceedings of the II planktonic conference, Roma 1970*, (ed. A. Farinacci), II, pp. 1341–60. Edizioni Tecnoscienza, Rome.

Green, J. C. (1973). Studies in the fine structure and taxonomy of flagellates in the genus *Pavlova*. II. A freshwater representative, *Pavlova granifera* (Mack) comb. nov. *British Phycological Journal*, **8**, 1–12.

Green, J. C. (1980). The fine structure of *Pavlova pinguis* Green and a preliminary survey of the order Pavlovales (Prymnesiophyceae). *British Phycological Journal*, **15**, 151–91.

Green, J. C. and Hibberd, D. J. (1977). The ultrastructure and taxonomy of *Diacronema vlkianum* (Prymnesiophyceae) with special reference to the haptonema and flagellar apparatus. *Journal of the Marine Biological Association of the United Kingdom*, **57**, 1125–36.

Green, J. C. and Parke, M. (1975). New observations upon members of the genus *Chrysotila*, with remarks upon their relationships within the Haptophyceae. *Journal of the Marine Biological Association of the United Kingdom*, **55**, 109–21.

Green, J. C. and Pienaar, R. N. (1977). The taxonomy of the Order Isochrysidales (Prymnesiophyceae) with special reference to the genera *Isochrysis* Parke, *Dicrateria* Parke and *Imantonia* Reynolds. *Journal of the Marine Biological Association of the United Kingdom*, **57**, 7–17.

Gregson, A. J., Green, J. C., and Leadbeater, B. S. C. (1993). Structure and physiology of the haptonema in *Chrysochromulina* (Prymnesiophyceae). I. Fine structure of the flagellar/haptonematal root system in *C. acantha* and *C. simplex*. *Journal of Phycology*, **29**, 674–86.

Hibberd, D. J. (1972). Chrysophyta: definition and interpretation. *British Phycological Journal*, **7**, 281.

Hibberd, D. J. (1976). The ultrastructure and taxonomy of the Chrysophyceae and Prymnesiophyceae (Haptophyceae): a survey with some new observations on the ultrastructure of the Chrysophyceae. *Botanical Journal of the Linnean Society*, **72**, 55–80.

Holligan, P. M., Fernandez, E., Aiken, J., Balch, W. M., Boyd, P., Burkill, P. H., *et al.* (1993). A biogeochemical study of the coccolithophore, *Emiliania huxleyi*, in the North Atlantic. *Global Biogeochemical Cycles*, **7**, 879–900.

Huxley, T. H. (1858). Appendix A. In *Deep-sea soundings in the North Atlantic Ocean between Ireland and Newfoundland* (J. Dayman), pp. 63–8. Her Majesty's Stationery Office, London.

Jordan, R. W. and Green, J. C. (1994). A check-list of the extant Haptophyta of the world. *Journal of the Marine Biological Association of the United Kingdom*, **74**, 149–74.

Kling, H. J. (1981). *Chrysochromulina laurentiana*: an electron microscopic study of a new species of Prymnesiophyceae from Canadian Shield lakes. *Nordic Journal of Botany*, **1**, 551–5.

Lackey, J. B. (1939). Notes on plankton flagellates from the Scioto River. *Lloydia*, **2**, 128–43.

Leadbeater, B. S. C. (1989). The phylogenetic significance of flagellar hairs in the Chromophyta. In: *The chromophyte algae: problems and perspectives*, Systematic Association Special Volume No. 38, (ed. J. C. Green, B. S. C. Leadbeater, and W. L. Diver), pp. 145–65. Clarendon Press, Oxford.

Leadbeater, B. S. C. and Green, J. C. (1993). Cell coverings in microalgae. In *Ultrastructure of microalgae* (ed. T. Berner), pp. 71–98. CRC Press, Boca Raton, Florida.

Lee, J. J., Hutner, S. H., and Bovee, E. C. (ed.) (1985). *An illustrated guide to the protozoa*. Society of Protozoologists, Lawrence, Kansas.

Leedale, G. F. (1974). How many are the kingdoms of organisms? *Taxon*, **23**, 261–70.

Lemmermann, E. (1908). Flagellatae, Chlorophyceae, Coccosphaerales und Silicoflagellatae. *Nordisches plankton*, Vol. 21, (ed. K. Brandt and C. Apstein), pp. 1–40. Lipsius and Tischer, Kiel and Leipzig.

Liebert, F. and Deerns, W. M. (1920). Onderzoek naar de oorzaak van een vischsterfte in den polder Workumer-Nieuwland, nabij Workum. *Verhandelingen en Rapporten uitgegeven door de Rijksinstituten voor Visscherijonderzoek*, **1**, 81–93.

Lohmann, H. (1902). Die Coccolithophoridae, eine Monographie der Coccolithen bildenden flagellaten, zugleich ein Beitrag zur Kenntnis des Mittelmeerauftriebs. *Archiv für Protistenkunde*, **1**, 89–165.

Manhart, J. R. and McCourt, R. M. (1992). Molecular data and species concepts in the algae. *Journal of Phycology*, **28**, 730–7.

Margulis, L., Corliss, J. O., Melkonian, M., and Chapman, D. J. (ed.) (1990). *Handbook of Protoctista*. Jones and Bartlett Publishers, Boston.

Massart (1920). Recherches sur les organismes inférieur. VIII. Sur la motilité des flagellates. *Bulletin de l'Academie Royale de Belgique, Classe des Sciences*, Série 5, **6**, 116–41.

Murray, G. and Blackman, V. H. (1898). On the nature of the coccospheres and rhabdospheres. *Philosophical Transactions of the Royal Society, Series B*, **190**, 427–41.

Nicholls, K. H. (1978). *Chrysochromulina breviturrita* sp. nov., a new freshwater member of the Prymnesiophyceae. *Journal of Phycology*, **14**, 499–505.

Papenfuss, G. F. (1955). Classification of the algae. In: *A century of progress in the natural sciences*, 1853–1953, pp. 115–224. California Academy of Sciences, San Francisco.

Parke, M. and Adams, I. (1960). The motile (*Crystallolithus hyalinus* Gaarder and Markali) and non-motile phases in the life history of *Coccolithus pelagicus* Schiller. *Journal of the Marine Biological Association of the United Kingdom*, **39**, 263–74.

Parke, M. and Dixon, P. S. (1976). Check-list of British marine algae — third revision. *Journal of the Marine Biological Association of the United Kingdom*, **56**, 527–94.

Parke, M., Manton, I. and Clarke, B. (1955). Studies on marine flagellates. II. Three new species of *Chrysochromulina*. *Journal of the Marine Biological Association of the United Kingdom*, **34**, 579–604.

Parke, M., Manton, I. and Clarke, B. (1959). Studies on marine flagellates. V. Morphology and microanatomy of *Chrysochromulina strobilus* sp. nov. *Journal of the Marine Biological Association of the United Kingdom*, **38**, 169–88.

Parke, M., Lund, J. W. G., and Manton, I. (1961). Observations on the biology and fine structure of the type species of *Chrysochromulina* (*C. parva*. Lackey) in the English Lake District. *Archiv für Mikrobiologie*, **42**, 333–52.

Pascher, A. (1910). Chrysomonaden aus dem Herschberger Grossteiche. *Monographien und Abhandlungen zur Internationale Revue der gesamten Hydrobiologie und Hydrographie*, **1**, 1–66.

Pienaar, R. N. (1980). Observations on the structure and composition of the cyst in *Prymnesium* (Prymnesiophyceae). *Electron Microscopy Society of South Africa — Proceedings*, **10**, 73–4.

Rothmaler, W. (1951). Die Abteilung und Klassen der Pflanzen. *Feddes Repertorium Specierum Novarum Regni Vegetabilis*, **54**, 256–66.

Scherffel, A. (1900). *Phaeocystis globosa* nov. spec. nebst einigen Betrachtungen über die Phylogenie niederer, unsbesondere brauner Organismen. *Wissenschaftliche Meeresuntersuchungen Abteilung Helgoland*, **4**, 1–29.

Schiller, J. (1930). Coccolithineae. In: *Kryptogamen-Flora von Deutschland, Österreich und der Schweiz*, Vol. 10, (ed. L. Rabenhorst), pp. 89–267. Akademische Verlagsgesellschaft, Leipzig.

Schwarz, E. H. L. (1894). Coccoliths. *Annals of Natural History*, **14**, 341–6.

Silva, P. C. (1980). Names of classes and families of living algae. *Regnum Vegetabile.* **103**, 1–156.

Sorby, H. C. (1861). On the organic origin of the so-called 'crystalloids' of the chalk. *Annals and Magazine of Natural History*, Ser. 3, **8**, 193–200.

Starmach, K. (1985). Chrysophyceae and Haptophyceae. In *Süsswasserflora von Mitteleuropa*, Vol. 1, (ed. H. Ettl, J. Gerloff, H. Heynig, and D. Mollenhauer). Gustav Fischer, Stuttgart.

Stein, F. R. von (1878). *Der Organismus der Infusionsthiere. III. Abtheilung Die Naturgeschichte der Flagellaten oder Geisselinfusorien* 1. *Hälfte.* Wilhelm Engelmann, Leipzig.

Stosch, H. A. von (1958). Der Geisselapparat einer Coccolithophoride. *Naturwissenschaften*, **45**, 140–1.

Tangen, K. (1991). Serious fish kills due to algae in Norway. *Red Tide Newsletter* (*Sherkin Island*), **4**, 9–10.

Tappan, H. (1980). *The paleobiology of plant protists.* W. H. Freeman, San Francisco.

Thomsen, H. A. and Oates, K. (1978). *Balaniger balticus* gen. et sp. nov. (Prymnesiophyceae) from Danish coastal waters. *Journal of the Marine Biological Association of the United Kingdom*, **58**, 773–9.

Underdahl, B., Skulberg, O. M., Dahl, E., and Aune, T. (1989). Disastrous blooms of *Chrysochromulina polylepis* (Prymnesiophyceae) in Norwegian coastal waters 1988 — mortality in marine biota. *Ambio*, **18**, 265–70.

Wallich, G. C. (1860). Results of soundings in the North Atlantic. *Annals of Natural History*, **6**, 457–8.

Wallich, G. C. (1861). Remarks on some novel phases of organic life and on the boring powers of minute annelids, at great depths in the sea. *Annals of Natural History*, Ser. 3, **8**, 52–8.

Williams, D. M. (1991). Cladistic methods and chromophyte phylogeny. *BioSystems*, **25**, 101–12.

Wujek, D. E. and Gardiner, W. E. (1985). Chrysophyceae (Mallomonadaceae) from Florida. II. New species of *Paraphysomonas* and the prymnesiophyte *Chrysochromulina. Florida Scientist*, **48**, 59–63.

Wyville-Thomson, C. (1874) Preliminary notes on the nature of the seabottom procured by the soundings of *HMS Challenger* during her cruises in the 'Southern Sea' in the early part of the year 1874. *Proceedings of the Royal Society of London, Series B*, **23**, 32–49.

2. Cell coverings

B. S. C. LEADBEATER

School of Biological Sciences, The University of Birmingham, Birmingham, UK

Abstract

The unit of cell covering in most members of the Isochrysidales, Prymnesiales, and Coccolithophorales is a two-layered microfibrillar scale. The proximal layer consists of a quadriradial pattern of microfibrils. The distal layer is more variable with spiral or interwoven patterns of microfibrils being common. Scales are usually arranged on the cell surface in one or more layers, and in species of *Chrysochromulina* the distal surface of outerlayer scales is often modified to form a spine, cup, or cylinder. Scale form and substructure is therefore an important taxonomic character at the species level. The coccoliths of some coccolithophorids are based on the characteristic microfibrillar scale. In some species, such as *Pleurochrysis carterae*, the coccolith baseplate is a two-layered scale whereas in others the scale component may be only represented by a few curved radial microfibrils. In *Calyptrosphaera*, and the crystallolith-bearing cells of *Coccolithus pelagicus*, the covering of scales and crystalloliths is bounded distally by a continuous investment.

Pleurochrysis scales contain four distinctive components: (i) the proximal radial microfibrils, the main constituent of which is a sulfated polysaccharide; (ii) the distal spiral microfibrils of cellulose and protein; (iii) a glycopeptide covering to the spiral microfibrils; and (iv) an amorphous matrix of acidic polysaccharide. Scales are produced within cisternae of the Golgi apparatus. Radial microfibrils are produced within a cisterna first, followed by the deposition of the spiral microfibrils and then the amorphous matrix.

Members of the Pavlovales do not possess microfibrillar scales. *Pavlova* species may have a covering of 'knob' scales on the outer surface of the plasma membrane.

Introduction

The majority of haptophytes have some form of covering, even if only a layer of mucilage, external to the plasma membrane. In most members of

The Haptophyte Algae (ed. J. C. Green and B. S. C. Leadbeater), Systematics Association Special Volume No. 51, pp. 23–46. Clarendon Press, Oxford, 1994.

the Isochrysidales, Prymnesiales, and Coccolithophorales the unit of covering is a two layered microfibrillar scale. The general substructure of haptophyte scales is distinctive and usually permits easy positive recognition of haptophyte cells in mixed nanoplankton collections (Leadbeater, 1972*b*; Thomsen *et al.*, Chapter 10). Scale morphology and substructure, which can be relatively easily observed in shadowcast whole mounts for electron microscopy, are characters used extensively in haptophyte taxonomy at the species level.

Mineralization of haptophyte cell coverings, especially with calcium carbonate, is common. In particular, the coccolithophorids, which constitute a large group of mainly marine species within the orders Isochrysidales and Coccolithophorales, carry a covering of calcified structures, the coccoliths. In some species an organic microfibrillar scale provides the baseplate for the calcified portion of the coccolith. Biogenesis of coccoliths, like scale development, is an intracellular activity and only when complete are the individual structures extruded on to the surface of the protoplast. Some benthic haptophytes, such as *Chrysotila*, also undergo mineralization but in this case calcification occurs externally within the mucilage sheath.

Members of the Pavlovales do not possess microfibrillar scales at any stage in their life cycle. However, motile cells of *Pavlova* may possess an outer covering of scattered 'knob-scales' which are superficially similar to those encountered on the flagella. The available evidence suggests that these structures are produced intracellularly like those that cover the long flagellum.

The aim of this chapter is to review information currently available on the form, substructure, chemical composition, and development of haptophyte scales, and to discuss the way in which scales are combined to form continuous cell coverings. Consideration will be given to the role that scales may play in the construction of some coccoliths, the range of coverings in the Pavlovales, and the occurrence of extracellular calcification in *Chrysotila*. For further detailed information on the crystalline substructure and intracellular development of coccoliths, Chapters 7 and 8 should be consulted.

Microfibrillar scales

Scale substructure

The basic haptophyte scale is a discoid or elliptical plate comprising two layers of microfibrils (Fig. 2.1). The microfibrils of the proximal layer, that is those closest to the plasma membrane, are arranged more or less radially in four quadrants. The number of microfibrils in a quadrant varies from 5–6 in *Chrysotila lamellosa* (as *Ruttnera spectabilis*; Green and Parke 1974), 20–30 in *Pleurochrysis scherffelii* (Brown and Romanovicz 1976), to more than 40 in *Chrysochromulina mactra* and *Crystallolithus hyalinus* (Manton and Leedale 1963; Manton 1972*a*). Whilst, in general, the number of microfibrils per quadrant is higher in larger scales, some large plate scales, such as those of *Chrysochromulina discophora* and *C. vexillifera,* have an intermediate number (*c.* 30) of relatively widely spaced microfibrils (Manton 1983; Manton and Oates 1983*a*). The proximal scale surface is usually flat without any embellishments and the radial organization of microfibrils, which is the most highly conserved of scale features, is the only microfibrillar layer present in the scales of *Phaeocystis pouchetii, Chrysotila lamellosa,* and *Imantonia rotunda* where they are equally visible from both sides of the scale (Parke *et al.* 1971; Green and Parke 1974; Green and Pienaar 1977). In *Chrysochromulina fragilis* the large plate scales are very thin and featureless apart from a regular arrangement of curved radial ridges extending inwards for a short distance from the periphery of the scale (Leadbeater 1972*a*).

The distal layer of a scale also usually contains microfibrils, but their arrangement in different species is much less consistent than those of the proximal layer. In vegetative cells of *Pleurochrysis scherffelii* a band of 8–10 microfibrils is arranged in a spiral of 4–5 gyres on the distal side of the scale (Brown and Romanovicz 1976). Many other patterns occur, for instance in *Chrysochromulina parva* between 2–5 widely spaced microfibrils undergo about 1.5 turns (Parke *et al.* 1962) and in *C. tenuispina,* which apparently has one layer of spirally arranged microfibrils sandwiched between two layers of radial microfibrils, the five spiral microfibrils are widely spaced and undergo about 1.5 gyres (Manton 1978*b*). In *Pleurochrysis carterae* the microfibrils on the distal surface of the circular underlayer scales form an irregularly looped meshwork (Manton and Leedale 1969; Billard, Chapter 9) whilst in *Navisolenia aprilei* they form an interwoven meshwork (Leadbeater and Morton 1973). In *Hymenomonas roseola* the distal layer comprises a peripheral band of approximately concentric microfibrils and a central region of irregularly interwoven microfibrils (Manton and Peterfi 1969), whereas in *Chrysochromulina microcylindra, C. megacylindra,* and *C. mactra* the distal layer consists only of a peripheral band of concentric microfibrils (Leadbeater 1972*a*; Manton 1972*a, b*). It is not uncommon for the distal surface to contain a substantial deposit of smooth, amorphous material which partially or completely obscures any distal layer microfibrils that might be present e.g. *Chrysochromulina latilepis, C. herdlensis, C. mantoniae,* and *C. chiton* (Manton 1967*b*, 1982; Leadbeater 1972*a*).

In contrast to the proximal scale surface, the distal surface is frequently elaborated with various projections and embellishments. These are particularly apparent on the distal surface of the outermost scales on a cell. At the edge of individual scales there may be a thread-like marginal thickening as in *Chrysochromulina latilepis* (Manton 1982). Marginal thickenings are also present on the distal surface of the plate scales of *Chrysochromulina strobilus, C. camella, C. cymbium*, and *C. acantha* (Leadbeater and Manton 1969*a, b*, 1971). More substantial peripheral rims, which may be inflexed or upstanding, are present on the large rounded and the small elongated scales respectively of *Chrysochromulina polylepis* (Manton and Parke 1962), and are a feature of many haptophyte scales. More substantial projections from the distal surface include a wide variety of spines most of which have buttress-like supports of one kind or another, for example the spine scales of *Chrysochromulina alifera, C. ephippium, C. pringsheimii, C. acantha, C. hirta*, and *C. latilepis* (Parke *et al.* 1956; Parke and Manton, 1962; Leadbeater and Manton 1971; Manton 1978*a*, 1982). In some species, such as *C. pringsheimii* and *C. mantoniae*, the buttresses are extended as ridges along the entire length of the spine. The spines of *C. parkeae* are probably hollow with periodic striations (thickenings) which maintain the spine's rigidity (Green and Leadbeater 1972). The length of spines may be very short (0.5 μm) as in *C. ephippium* or long (20–30 μm) as in *C. pringsheimii, C. latilepis*, and *C. parkeae*. Cylinder-shaped projections are present on the outer scales of *Chrysochromulina ericina, C. megacylindra* (Fig. 2.18), *C. microcylindra, C. mactra, C. pachycylindra*, and *C. cyathophora* (Manton and Leedale 1961; Leadbeater 1972*a*; Manton 1972*a*; Manton *et al.* 1981*a, b*). In *C. ericina* and *C. mactra* the cylinder appears to have been achieved by hypertrophy of the distal surface of the scale (Manton and Leedale 1961; Manton 1972*a*). However, as Manton *et al.* (1981*b*) point out, although there is a superficial similarity in appearance between the cylinders of the species listed above, the cylindrical structure has been achieved in several different ways and possession of cylinder scales does not necessarily imply a close phyletic relationship between the species concerned. Cup-shaped outer scales are a feature of *Chrysochromulina strobilus, C. cymbium*, and *C. camella* (Fig. 2.10), but in these species there are sufficient other similarities in scale substructure to conclude that there is a close phyletic relationship between them (Leadbeater and Manton 1969*a, b*).

Both the size and shape of scales vary considerably between species. Sizes range from 0.10 × 0.13 μm for the small scales of *Phaeocystis pouchetii* to 5.0 × 3.5 μm for the large rimmed scales of *Chrysochromulina parkeae* (Parke *et al.* 1971; Green and Leadbeater 1972). The base plates of placoliths of *Coccolithus pelagicus* are 7.5 × 5.0 μm (Manton and Leedale 1969). The form of scales varies considerably; plate scales may be circular, elliptical, or rhomboidal. Some scales, such as the smallest plate scales of *C. parkeae* are 'waisted' at the ends (Green and Leadbeater 1972) and the large elongated scales of *C. polylepis* may have a terminal forked projection (Manton and Parke 1962).

Calcified microfibrillar scales

A number of coccolithophorids bear coccoliths that are based on the characteristic prymnesiophycean microfibrillar scale. Motile, coccolith-bearing cells of *Pleurochrysis carterae* possess two layers of scales, the outer layer of which is calcified (Fig. 2.3). The coccolith base plate bears a pattern of radial microfibrils arranged in quadrants on its proximal surface whilst the distal surface is more or less amorphous. In decalcified coccoliths a distal organic rim is apparent in which the calcium carbonate is located giving the scale its classic cricolith appearance (Manton and Leedale 1969; Leadbeater 1970). A comparable substructure is found in the coccoliths of *Pleurochrysis scherffelii, Ochrosphaera neapolitana, Jomonlithus littoralis,* and *Umbilicosphaera hulburtiana* (Leadbeater 1971; Inouye and Chihara 1983; Billard, Chapter 9).

Coccolithus pelagicus, one of the few coccolithophorids for which the life cycle has been studied in culture, has a placolith bearing phase (heterococcolithophorid) and a crystallolith bearing phase (holococcolithophorid) within one cycle (Parke and Adams 1960), and both forms of coccolith are based on microfibrillar scales. The relatively massive placoliths are each based on a large oval scale which in shadowcast whole mounts appears patternless on both surfaces except for a slight marginal thickening and a transverse central ridge or crease. However, in section, a coarse pattern of radial microfibrils is revealed (Manton and Leedale 1969). The holococcolithophorid or '*Crystallolithus hyalinus*' stage of *Coccolithus pelagicus* produces oval crystalloliths each of which comprises a pattern of rhombohedral crystals attached to the distal surface of a microfibrillar scale (Fig. 2.4). The proximal surface contains the characteristic radial microfibrils arranged in quadrants and the distal surface has an approximately concentric pattern of microfibrils (Manton and Leedale 1963). Individual crystals are surrounded by a densely staining layer presumed to be organic in composition (Rowson *et al.* 1986). A similar substructure is seen in the crystalloliths of *Calyptrosphaera sphaeroidea* (Klaveness 1973) with the difference that the crystals are arranged to form a crown that projects outwards from the scale. Scales, sometimes very thin and with few microfibrils, form the coccolith base plates of a number of species such as *Syracosphaera pulchra* and *Rhabdosphaera stylifera* (Leadbeater and Morton 1973). In both species the coccolith scale takes the form of a diaphanous film with a regular peripheral arrangement of curved radial microfibrils. Other genera in which organic microfibrils are present within coccoliths include *Calciopappus, Ophiaster, Papposhaera, Pappomonas* (Fig. 2.2) *Turrisphaera,* and *Wigwamma* (Manton and Sutherland 1975; Manton *et al.* 1976*a, b,* 1977; Manton and Oates 1983*b*). However, whilst all coccoliths possess an organic component, not all are based on

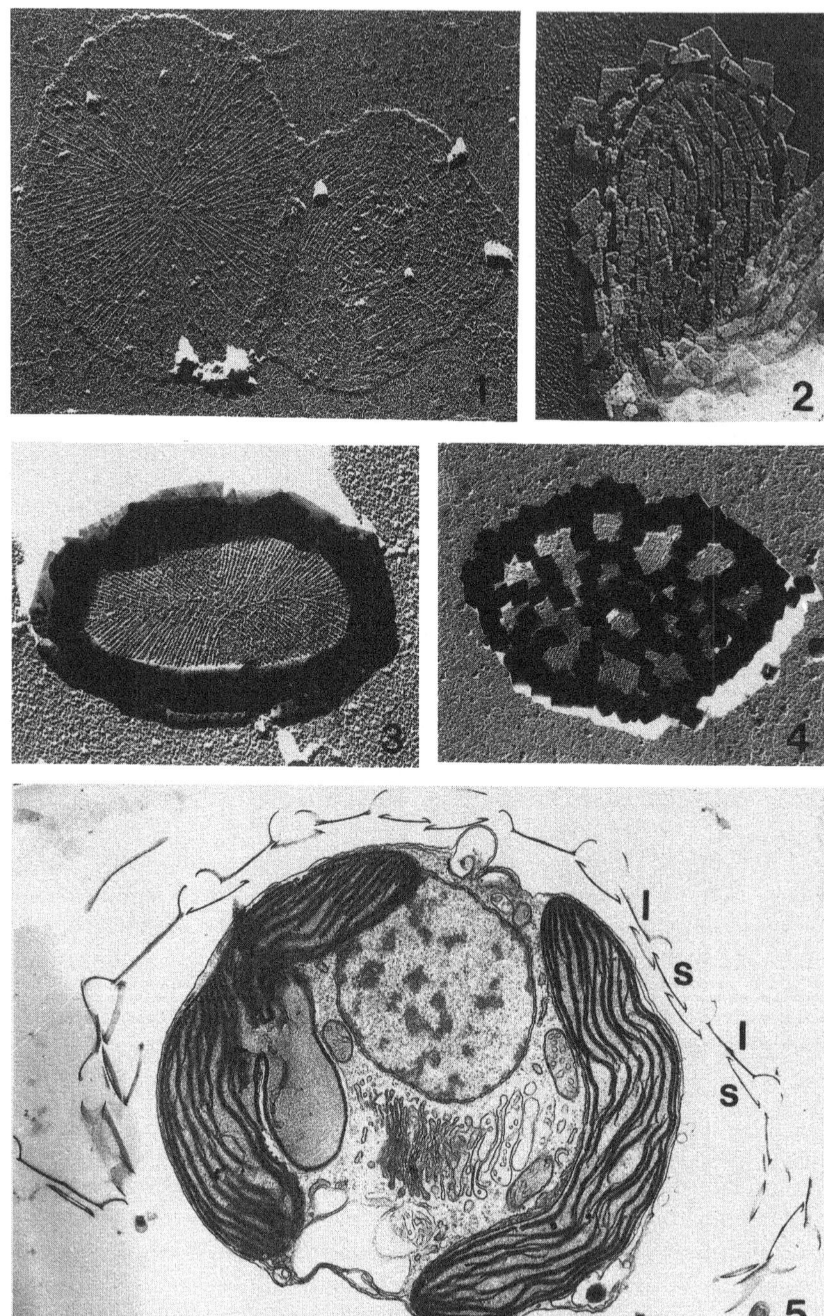
1
2
3
4
5
l
s
l
s

Figs 2.1–2.5 **(1–4)** Shadowcast whole mounts of unmineralized and mineralized scales. (**1**) *Apistonema* sp.: radial microfibrils denote proximal surface visible on left hand scale; spiral microfibrils denote distal surface visible on right hand scale. × 35 000, from Leadbeater (1970). (**2**) *Pappomonas flabellifera*: calcified plate scale showing fine radial microfibrils. × 46 000. (**3, 4**) *Pleurochrysis carterae* and the '*Crystallolithus hyalinus*' stage of *Coccolithus pelagicus* respectively: coccoliths showing radial microfibrils of underlying scale. × 30 000. (**5**) *Chrysochromulina chiton*: section of cell showing arrangement of scales within scale case (l = large scale, s = small scale). × 10 000, from Manton (1967*b*).

microfibrillar scales, as has been adequately demonstrated for *Emiliania huxleyi* (de Vrind-de Jong *et al.*, Chapter 8).

Organization of scales and coccoliths on the cell surface

There are many variations in the number of scale layers and different scale types incorporated into a covering. The most straightforward arrangement is a single layer of identical scales which may be precisely imbricated with regular overlaps as in '*Apistonema*' swarmers (Leadbeater 1970). Many species have two evenly arranged layers of scales. In *Phaeocystis pouchetii,* a layer of small circular scales with upstanding rims overlies a layer of smaller oval scales with strongly inflexed rims (Parke *et al.* 1971). There is no overlap between scales belonging to the same layer and, in detached cases, scales of the outer layer are present in close-packed arrangement. The majority of *Chrysochromulina* species possess an underlayer of plates and an outer layer of scales with distal elaborations such as spines, cups, or cylinders. In some species, neighbouring plate scales adjacent to the plasmalemma may overlap each other by about one-quarter of the width of a scale (Manton 1982). In *C. camella* the outer cup scales are close packed (Fig. 2.10) giving an overall hexagonal symmetry when seen in tangential section (Leadbeater and Manton 1969*b*). There may be relatively minor variations in the size and arrangement of scales on different parts of the cell, for instance the plate scales adjacent to the flagellar pole in *Hymenomonas roseola, Pleurochrysis carterae,* and *Chrysochromulina bergenensis* (Fig. 2.15) are smaller than the remainder of plate scales elsewhere on the cell (Manton and Peterfi 1969). In *Chrysochromulina bergenensis, C. chiton* (Figs 2.5 and 2.6), *C. mantoniae,* and *Navisolenia aprilei* the underlayer scales are not evenly distributed but are located beneath the adjacent boundaries of the larger outer layer scales (Manton 1967*b*; Leadbeater 1972*a*; Leadbeater and Morton 1973; Manton 1982). It is not uncommon for species with spine scales, such as *Chrysochromulina pringsheimii* and *C. parkeae,* to bear longer spines at the poles and shorter spines or plate scales without spines on the remainder of the cell (Parke and Manton 1962; Green and Leadbeater 1972). Some species, such as *C. parkeae,* have four or more different scale types which are precisely located on different regions of the cell surface (Green and Leadbeater 1972).

In *Pleurochrysis scherffelii* and *Apistonema* sp., which have filamentous stages in their life cycle, a quasi-continuous cell wall is achieved by the close packing of many layers of overlapping rimless plates (Figs 2.20 and 2.21; Brown *et al.* 1970; Leadbeater 1970; Leadbeater 1971; Brown and Romanovicz 1976). In some *Apistonema* filaments the outer scales may have a surface deposit of dark staining granules (Fig. 2.21).

In two holococcolithophorids, the '*Crystallolithus hyalinus*' phase of *Coccolithus pelagicus* and *Calyptrosphaera sphaeroidea,* the covering of crystalloliths which overlie layers of plate scales are enclosed within a continuous investment described by Manton and Leedale (1963) as a 'skin' (Klaveness 1973; Rowson *et al.* 1986). A continuous outer investment is also present on the surface of *Chrysochromulina bergenensis,* and here an underlying deposit of columnar bodies and subspherical dense particles or droplets is located between the lower scales and the plasma membrane (Fig. 2.15; Manton and Leadbeater 1974). In *C. herdlensis* dense staining columnar deposits are present on the outer surface of the large plate scales (Fig. 2.7).

A number of authors have observed fibrillar or columnar material either between the plasma membrane and overlying scales, or between individual layers of scales which may explain how scales are held in place on the cell surface (Fig. 2.7). One of the best examples is to be found in *Hymenomonas roseola* where densely staining columnar material is located between the plasma membrane and the most proximal layer of scales (Manton and Peterfi 1969). Similar material is present between the plasma membrane and overlying scales in *Coccolithus pelagicus* (Manton and Leedale 1969) and *Chrysochromulina herdlensis* (Fig. 2.7). Fibrillar material, presumed to be pectic in nature, is present between overlapping scales in *Pleurochrysis scherffelii* (Brown *et al.* 1970), and a layer of fibrils of unknown composition overlies the cup scales of *Chrysochromulina camella* (Fig. 2.10; Leadbeater and Manton 1969*b*). The scale cases of some species, such as *Chrysochromulina pringsheimii,* can be removed intact from cells without disrupting the arrangement of scales (Parke and Manton 1962), whereas other scale cases disrupt immediately on treatment. The use of glutaraldehyde fixation during preparation for EM frequently disrupts scale cases leaving the protoplast without a covering.

Some coccolithophorids, such as *Coccolithus pelagicus* and *Syracosphaera pulchra,* possess one or more layers of microfibrillar scales between the plasma membrane and the outer layer of coccoliths (Manton and Leedale 1963; Leadbeater and Morton 1973). Coccoliths are usually close-packed on the cell surface and may interdigitate with each other; in *Helicosphaera* sp. the imbrication is regular (Chrétiennot-Dinet 1990), whereas there is no obvious regular pattern in *Emiliania huxleyi* (Medlin *et al.*, Chapter 21). Coccoliths on a single cell may be of similar size and morphology, but in many species there may be several distinct coccolith forms and sizes. Spines are often confined to one or both poles of the cell, for instance the spines of *Calyptrolithina* are at the flagellar pole only, whereas in *Acanthoica* they are at both ends of the cell (Chrétiennot-Dinet 1990).

Chemical composition of scales

Scale composition of two haptophytes, *Pleurochrysis scherffelii* and *Chrysochromulina chiton*, has been studied in detail using standard physical and chemical techniques including histochemical staining in conjunction with electron microscopy (Brown *et al.* 1970, 1973; Allen and Northcote 1975; Brown and Romanovicz 1976; Romanovicz and Brown 1976). Scale cases of *Pleurochrysis* react with the periodic acid Schiff's test denoting the presence of carbohydrate, and they take up cationic dyes such as ruthenium red and alcian blue indicating the existence of a pectin-like acidic polysaccharide. With modification of the staining solutions and cell pretreatment it has been established that both sulphate and carboxyl groups are present in the acidic polysaccharide fraction of the *Pleurochrysis* wall (Romanovicz and Brown 1976).

Using a serial extraction protocol, Romanovicz and Brown (1976) were able to obtain four scale fractions which were then subjected to protein, carbohydrate and, after hydrolysis, sugar analyses. Scale material remaining after each stage of treatment was negatively stained and observed with electron microscopy. From these preparations the dimensions of the microfibrils were obtained. The first stage, involving treatment of isolated scales with water adjusted to pH 5.0 with 0.1N trifluoracetic acid (TFA) at 37 °C, solubilized an amorphous subcomponent permitting clear observation of the remaining radial and spiral microfibrils. The second stage involved treatment with 0.5N TFA at 37 °C for 3 hours which solubilized the radial microfibrils. The third stage, involving treatment with 2N TFA at 121 °C for 3 hours, removed the coating material closely bound to the spiral microfibrils. The fourth stage, which involved treatment with 6N HCl at 100 °C for 24 hours solubilized the spiral microfibrils. The results of the chemical treatment of the different fractions obtained using this protocol are summarized in Table 2.1.

Since the dimensions of radial microfibrils could only be obtained whilst they were still attached to the spiral microfibrils it was only possible to measure one dimension. When negatively stained with uranyl acetate, a dark central band was observed extending the entire length of the

Table 2.1 Chemical analysis of *Pleurochrysis* scale components. After Romanovicz and Brown (1976)

Scale subcomponent	Protein %	Carbohydrate %	Principal monosaccharides
Amorphous material	6	94	galactose, glucose, fucose
Radial microfibrils	7	93	galactose, arabinose, fucose
Coating of spiral microfibrils	9	91	galactose, fucose, mannose
Spiral microfibrils	37	63	glucose

microfibril. Isolated spiral microfibrils, obtained in the third fraction, were rectangular in cross section and also possessed a central band. A summary of measurements is presented in Table 2.2. By combining the results of physical and chemical analyses with electron microscopical observations it has been possible to build up a reasonably detailed picture of the substructural composition of scales (Romanovicz and Brown 1976).

Brown *et al.* (1970) were the first to suggest that the occurrence of glucose in their scale hydrolysates indicated the existence of cellulose in prymnesiophycean scales. This has since been confirmed by chemical analysis (Herth *et al.* 1972), and by data obtained from infrared absorption and x-ray diffraction analyses carried out on isolated spiral microfibrils (Romanovicz and Brown 1976). In addition to cellulose, the spiral microfibrils contain 37% protein which may be located around a core of crystalline cellulose, or possibly four protofibrils of crystalline cellulose may be stabilized by a central peptide moiety. The glycopeptide coating of

Table 2.2 *Pleurochrysis* scale microfibril dimensions from negatively stained whole mounts for TEM. After Romanovicz and Brown (1976)

Microfibril	Total width Å	Central Band Å
Radial microfibrils	60	24
Spiral microfibrils narrow axis	37	11
Spiral microfibrils broad axis	49	16

the spiral microfibrils, which contains hydroxyproline as the most abundant amino acid residue, appears to be responsible for maintaining the spiral configuration of the cellulosic microfibrils (Romanovicz and Brown 1976). It is possible that the hydroxyproline-containing peptide is analagous to the peptide 'extensin' found in other cell walls (Romanovicz 1981). The non-cellulosic components comprising the radial microfibrils and the amorphous material, are similar in solubility and staining properties. The presence of protein in the short polypetides would account for the carboxyl groups indicated by the alcian blue staining. The amorphous, sulfated, pectin-like polysaccharide almost certainly consolidates the overall pattern of microfibrils in scales. In an investigation of the antigenic properties of *Pleurochrysis* scales, Kavookjian and Brown (1978) obtained rabbit antibodies that reacted with the amorphous coating material and a previously unknown component located at the periphery on both sides of all scales. They referred to this latter component as the 'amorphous rim modification substance', and suggested that its function may be to maintain the integrity of the spiral microfibrillar organization. The component was resistant to treatment with water, adjusted to pH 5.0 at 37 °C, which dissolved most of the amorphous material but, as with the radial microfibrils, it was soluble in 0.5N TFA at 37 °C for 3 hours.

Allen and Northcote (1975) found that the scales of *Chrysochromulina chiton* contained 32.5% carbohydrate and 65% protein, and that the microfibrils contained a β1-4 glucan indicating the existence of cellulose. However, their protocol did not distinguish between the different types of microfibrils present. The scale matrix of *C. chiton* comprised a neutral polysaccharide which was composed of galactose, glucose, rhamnose, and two acidic polysaccharides containing uronic acids, glucose, ribose, and galactose.

Scale production

Manton and Parke (1962) were the first to observe scale profiles within the Golgi apparatus of an haptophyte cell, in this case *Chrysochromulina polylepis*. In the same publication they suggested that there might be a diurnal rhythm in scale production since more scale profiles were found within the Golgi apparatus of cells fixed during the light period than in those fixed during the dark period.

The Golgi apparatus of most haptophyte cells is located close to the flagellar bases. A Golgi apparatus active in producing scales in *Chrysochromulina, Coccolithus,* or *Pleurochrysis* is usually large comprising 20–30 cisternae (Figs 2.9 and 2.14; Manton 1966, 1967*a*; Brown *et al.* 1970). The *cis*, or proximal, face of the Golgi apparatus is usually adjacent

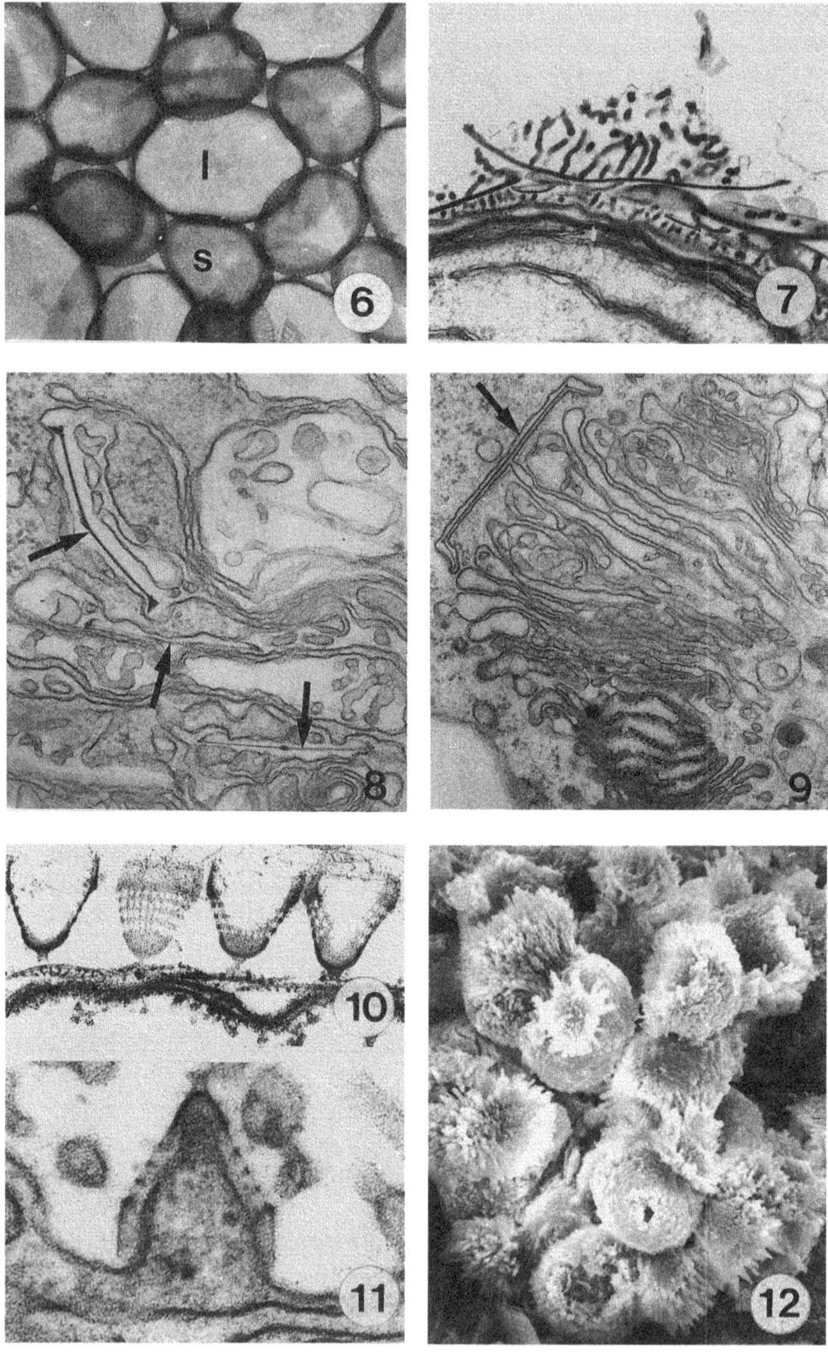
l
s
6
7
8
9
10
11
12

Figs 2.6–2.12 (**6**) *Chrysochromulina chiton*: part of scale case showing arrangement of small scales (s) around a large scale (L). × 15 000. (**7**) *C. herdlensis*: showing plate scales with columnar material on distal surface and shorter columnar material between lowest layer of scales and plasma membrane. × 30 000, from Manton and Leadbeater (1974). (**8**) *C. chiton*: part of Golgi apparatus showing scales in cisternae (*arrows*). × 40 000, from Manton (1967*b*). (**9**) *C. chiton*: Golgi apparatus showing large scale in peripheral cisternal extension perpendicular to long axis of Golgi stack; note 'peculiar' cisternae towards bottom of figure. × 25 000, from Manton (1967*b*). (**10, 11**) *C. camella* from Leadbeater and Manton (1969*b*). (**10**) Vertical section of scale covering showing plate and cup shaped scales with overlying layer of fibrils. × 30 000. (**11**) Cup-scale developing in Golgi cisterna; note plug of cytoplasm extending within cup. × 85 000. (**12**) *Chrysotila lamellosa*: spherulites of $CaCO_3$ surrounding individual cells. × 600.

to one or more flattened cisternae of ER and the accompanying small vesicles (Fig. 2.5). Proximal Golgi cisternae are usually flattened with no visible contents. In the middle portion of the Golgi stack, cisternae may be distended centrally in a regular fashion and contain densely staining material on the inner surface of the cisternal membrane (Figs 2.9 and 2.15; Manton 1967*a*; 1972*b*). Manton (1967*a*) referred to these dilations as 'peculiar Golgi', and found that in *Chrysochromulina chiton* they stained most intensely in cells that were nearing the end of their light period when kept in a 16h light: 8h dark regime. Brown *et al.* (1970) observed similar dilations in *Pleurochrysis* and, since their appearance always preceded the development of scale microfibrils, they referred to the dilations as 'reaction centres'. Golgi cisternae in the middle of the stack and towards the plasma membrane contain developing scales (Fig. 2.8). Large plate scales usually occur singly within an individual cisterna but if scales are small, as in *Prymnesium parvum* or *Chrysochromulina bergenensis*, then more than one scale may be present within a cisterna (Fig. 2.15; Manton 1966). In species with large scales, such as *C. chiton* (Fig. 2.9), or cylinder-shaped scales, such as *C. megacylindra* (Fig. 2.19), *C. microcylindra*, and *C. mactra*, the scale is produced within a lateral extension of a cisterna on the outside, and at right angles to the long axis, of the Golgi stack (Manton 1967*a*, 1972*a*, *b*; Leadbeater 1972*a*). In *C. microcylindra* a cisterna containing a plate scale may also contain a peripheral extension with a developing cylinder scale (Leadbeater 1972*a*). In *C. microcylindra*, *C. megacylindra*, and *C. mactra* (Leadbeater 1972*a*; Manton 1972*a*, *b*), and also species with cup-shaped scales, a plug of cytoplasm fills the internal cavity of the developing scale and may serve as a mould around which the scale is shaped (Fig. 2.11; Leadbeater and Manton 1969*a*, *b*).

Brown and co-workers (Brown *et al.* 1970; Brown and Romanovicz 1976; Brown and Willison 1977) have proposed a well defined sequence of events for the intracellular biogenesis of scales which accords with the concept of cell surface-orientated membrane flow. A thorough description of the proposed sequence can be found in Brown and Romanovicz (1976). The earliest detectable scale product that can be recognized within a Golgi cisterna is a pool of radial microfibril precursor. This material has a granular structure, stains densely, and is probably equivalent to the material observed in the 'peculiar Golgi' of *Chrysochromulina chiton* (Manton 1967*a*). This precursor pool then undergoes crystallization to form folded (double row configuration) microfibrillar rods and subsequently the polysaccharide of the microfibrils becomes sulfated. Golgi cisternae, containing double rows of parallel microfibrillar rods, are Z-shaped and the transformation of microfibrils into the single layer, quadriradial arrangement involves unfolding of the microfibrils, which is probably achieved by the pulling out and flattening of the Z-shaped cisternae. Next the spiral microfibrils are added beginning at the periphery of the flat plate of radial microfibrils and progressing centripetally. Finally, the pectinlike amorphous subcomponents are added, again centripetally. Once completed, the scale is extruded, usually adjacent to the base of the flagellar apparatus, and becomes part of the cell covering.

Development of coccoliths based on microfibrillar scales

Manton and Peterfi (1969) and Manton and Leedale (1969) demonstrated that the coccoliths of *Hymenomonas roseola, Pleurochrysis carterae,* and *Coccolithus pelagicus* were produced in the Golgi apparatus in a similar manner to the scales of unmineralized haptophytes. This is not surprising seeing that the coccolith baseplate in these species is a microfibrillar scale. The large coccoliths of *Coccolithus pelagicus* develop within T-shaped extensions of individual Golgi cisternae perpendicular to the long axis of the stack (Figs 2.13 and 2.14), in a similar manner to large plate scale development in *Chrysochromulina chiton* (Manton and Leedale 1969). In *Pleurochrysis carterae* the Golgi apparatus consists of 20 or more cisternae and its overall appearance including the presence of 'peculiar dilations' is

Figs 2.13–2.17 (**13, 14**) *Coccolithus pelagicus*: developing coccolith baseplate (**13**) and coccolith (**14**) in Golgi cisternae; note cisterna containing developing coccolith in 14 is at right angles to long axis of Golgi stack. × 20 000 and × 15 000 respectively. (**15**) *Chrysochromulina bergenensis*: section to show cell covering with outer continuous investment; note small scales at anterior end of cell (s) and within Golgi cisternae (*arrows*). × 22 500, from Manton and Leadbeater (1974). (**16, 17**) *Pleurochrysis scherffelii*: two stages in the development of coccoliths in Golgi cisternae; *arrows* denote coccolithosomes. × 60 000 and × 40 000 respectively, from Leadbeater (1971).

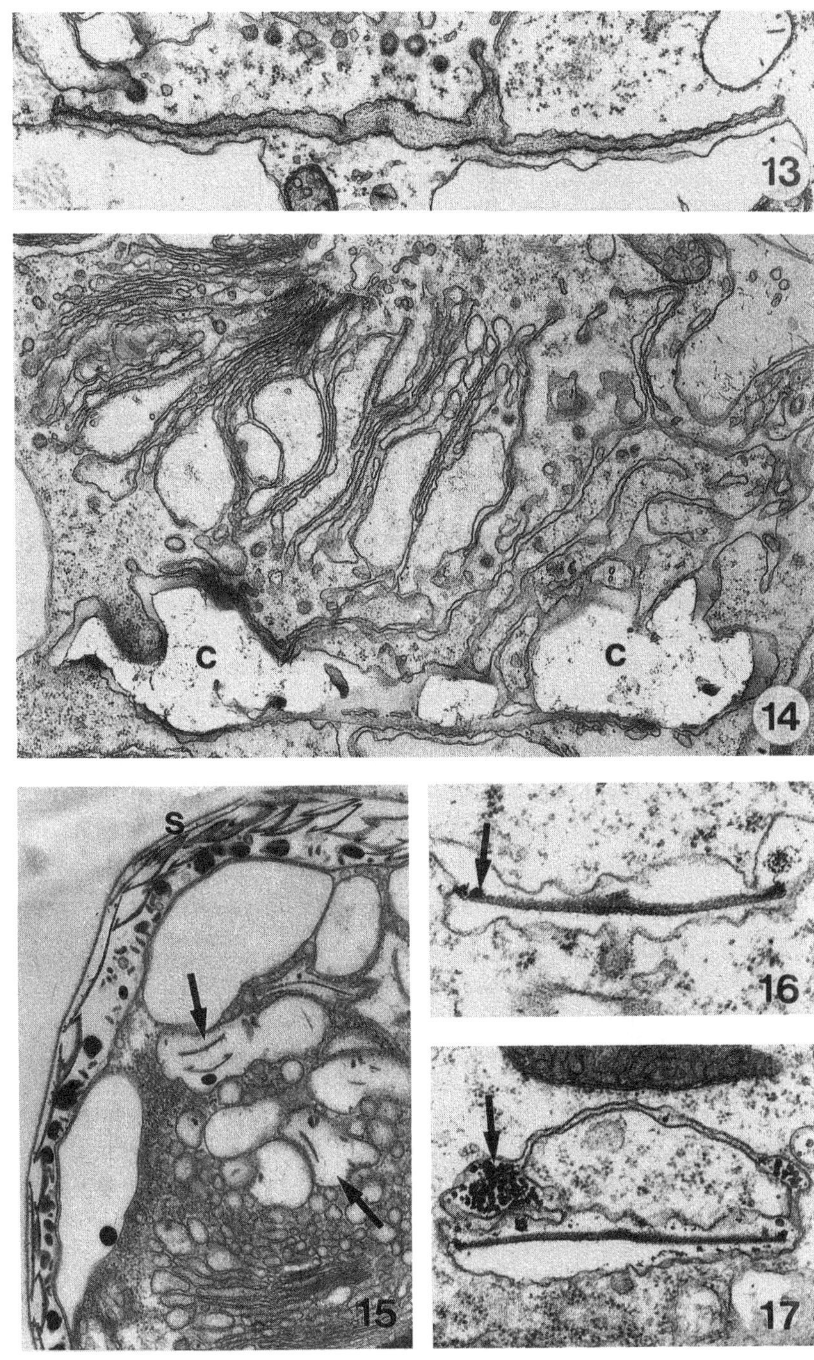
13
14
c
c
s
15
16
17

typical of that for other prymnesiophycean cells (Manton and Leedale 1969; Pienaar 1971; Leadbeater 1970, 1971; Outka and Williams 1971; Wal *et al.* 1983). Wal *et al.* (1983) recognized seven distinct stages in the development of coccoliths. The formation of the microfibrillar baseplate is completed before calcification takes place. On completion of the baseplate, the Golgi cisterna becomes associated with vesicles containing densely staining granules called coccolithosomes (Outka and Williams 1971; Gayral and Fresnel 1983; Fresnel and Billard 1991). These vesicles are usually at some distance from the Golgi stack and contact between them and the Golgi apparatus is established by long meandering membrane-bounded cisternae (Fig. 2.17; Leadbeater 1971). The coccolithosomes eventually appear within baseplate-containing Golgi cisternae and become aggregated at the periphery of the distal surface of the baseplate (Figs 2.16 and 2.17). Westbroek *et al.* (1985) consider that coccolithosomes provide both the organic matrix and the calcium for the calcium carbonate crystals (see Chapter 8). When the coccolith is fully calcified it is extruded to the surface of the protoplast close to the flagellar pole.

The development of crystalloliths in the '*Crystallolithus hyalinus*' stage of *Coccolithus pelagicus* remains somewhat enigmatic. Developing crystallolith baseplates are present in Golgi cisternae, but calcium carbonate crystal deposition has never been observed within Golgi cisternae. Rowson *et al.* (1986) came to the conclusion that crystal deposition must occur outside the plasma membrane within the space enclosed by the continuous outer envelope or 'skin'. Crystals have not been observed within the Golgi cisternae of *Calyptrosphaera sphaeroidea* (Klaveness 1973) nor were they seen on the crystallolith baseplates within the Golgi cisternae of the '*Crystallolithus hyalinus*' cells observed by Manton and Leedale (1963).

Extracellular calcification in *Chrysotila*

Chrysotila lamellosa is a benthic prymnesiophycean species that produces motile *Isochrysis*-like scale-bearing swarmers (Green and Parke 1975). Cells may become non-motile and divide rapidly to form rafts and blocks of cells. Each cell becomes invested with a thick layer of hyaline mucilage which, as it enlarges, becomes lamellated and disrupts the regular organization of cells. Calcification occurs extracellularly in association with the mucilage sheath. The first appearance of calcium carbonate is in the form of small flakes, dumbells, rods, or cruciform structures (Green and Course 1983). Further growth involves the development of bundles of acicular crystals and subsequent accretion of $CaCO_3$ produces spherical bodies which may be 100 μm or more in diameter. SEM examination reveals that the calcified structures have a rugose surface (Fig. 2.12). Extracellular calcification in *Chrysotila* would seem to be good example of 'biologically induced' mineralization in contrast to the 'biologically controlled' intracellular deposition of $CaCO_3$ in coccolithophorids (Green and Course 1983).

Cell coverings of the Pavlovales

Microfibrillar scales have not been observed on any member of the Pavlovales. Most species of *Pavlova* lack a distinctive covering although non-motile cells may be invested with layers of mucilage (Green 1980). A cisterna of ER is located immediately below the plasma membrane, and the layer of cytoplasm between the two membranes may be so thin that a five-layered periplast is formed (Fig. 2.22). In some species 'knob-scales', superficially similar to those on the long flagellum (Fig. 2.23), are located on the outer surface of the plasma membrane. They are usually spherical although they may be clavulate or clavate (Green 1980). Knob-scales are formed within the Golgi apparatus but little is known of how they come to be deposited on the plasma membrane.

Scale coverings and heteromorphic life cycles

The biphasic life cycle of *Coccolithus pelagicus* with placolith and crystallolith bearing stages has already been mentioned. Recently Thomsen *et al.* (1991) have demonstrated that some planktonic coccolithophorids, such as species of *Turrisphaera* and *Papposphaera*, are in fact two phases in a single life cycle. *Pleurochrysis*, *Hymenomonas* and *Apistonema* also have multiphasic life cycles, each phase bearing scales of different morphology and with a different pattern of microfibrils. Whether or not haptophyte life cycles involve an alternation between sexual and asexual phases with haploid and diploid generations is still a matter of conjecture. However, Billard (Chapter 9) shows that there is a consistency in the morphology and microfibril arrangement in the underlayer scales of specific phases in the life cycles of different haptophytes with bi- or multiphasic life cycles. Billard's work is a start in bringing order to a diverse series of observations but, clearly, further study is necessary, especially in establishing unequivocally the existence of sexual reproduction in haptophytes.

Functions of cell coverings

At present the function of scales and coccoliths is not clear (Green *et al.* 1990). It has been suggested that a coccolith layer may serve as a light scattering device, thus protecting the cell from excessive irradiation, but Paasche (1964) working with *Emiliania huxleyi* found no evidence for this. Sikes and Wilbur (1982) found that a covering of coccoliths imparted

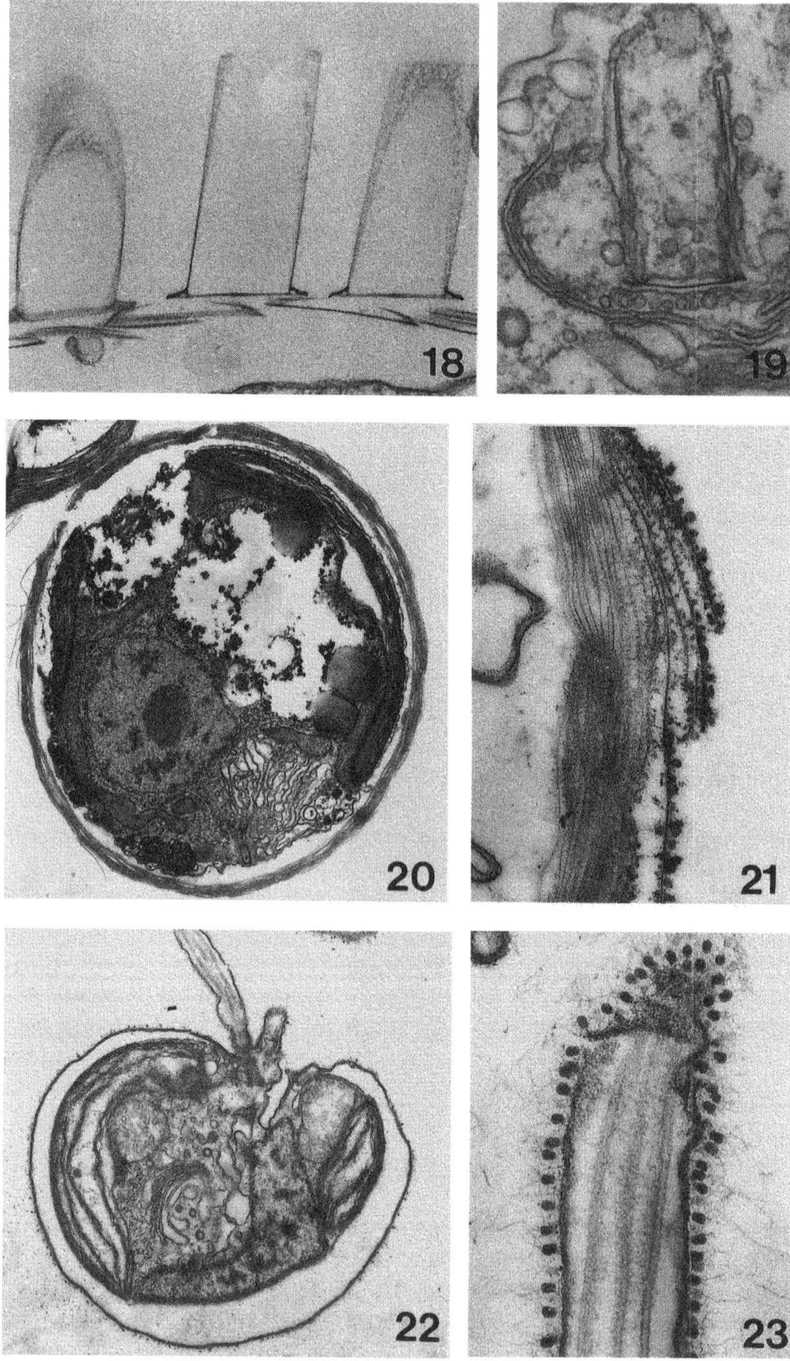
18
19
20
21
22
23

Figs 2.18–2.23 (**18, 19**) *Chrysochromulina megacylindra*: row of cylinder scales on surface of cell (**18**) and cylinder scale within Golgi cisterna (**19**). × 25 000, from Manton (1972*b*). (**20**) *Pleurochrysis scherffelii*: vegetative cell with thick cell wall made up of many layers of scales. × 8000. (**21**) *Apistonema* sp.: cell wall made up of many layers of scales; outer scales covered by deposit of dense staining granules. × 30 000, from Leadbeater (1970). (**22**) *Pavlova pinguis*: showing outer loose 'covering' comprising complex of plasma membrane and ER membrane. × 12 500. (**23**) *Pavlova* sp.: LS of long flagellum showing covering of 'knob' scales. × 50 000.

enhanced tolerance to low salinity to the underlying cell, but the mechanism by which this is achieved is not known. It is probable that the stabilization of a 'boundary layer' by a cell covering will modify the movement of molecules in the microenvironment outside the plasma membrane, but the significance of this has not been investigated (Manton, 1986). In the mixotrophic haptophyte *Chrysochromulina spinifera*, food particles apparently adhere to spines on the cell surface. These immobilized particles are collected up periodically by the haptonema and transported to the posterior end of the cell where they are ingested (Iuouye and Kawachi Chapter 4, this volume). Finally, Estep and McIntyre (1989) have suggested that the scaly coverings of prymnesiophytes may protect cells from toxins released into the environment by other planktonic cells.

Acknowledgements

I am grateful to Professor L. V. Evans and the University of Leeds for granting permission for the use of negatives lodged in the Professor Irene Manton Archive. I also wish to thank Dr J. C. Green for supplying the photograph of *Chrysotila lamellosa*; Mr Steve Price for help with photography and preparation of the manuscript and to Dr Jackie Eccleston-Parry and Dr Harriet Jones for reading and commenting on the text during preparation. The following are acknowledged for permission to reproduce illustrations: *Annals of Botany*, Figs 16, 17; *Archives of Microbiology*, Figs 10, 11; *British Phycological Journal*, (Figs 1, 18, 19, 21; *Journal of Cell Science*,) Figs 5, 8, 9; *Journal of the Marine Biological Association UK*, Figs 13, 14; *Royal Danish Academy of Sciences and Letters*, Figs 7, 15.

References

Allen, D. M. and Northcote, D. H. (1975). The scales of *Chrysochromulina chiton*. *Protoplasma*, **83**, 389–412.

Brown, R. M. and Romanovicz, D. K. (1976). Biogenesis and structure of Golgi-derived cellulosic scales in *Pleurochysis*. I. Role of the endomembrane system in scale assembly and exocytosis. *Applied Polymer Symposium*, **28**, 537–85.

Brown, R. M., Franke, W. W., Kleinig, H., Falk, H., and Sitte, P. (1970). Scale formation in chrysophycean algae. I. Cellulosic and non-cellulosic wall components made by the Golgi apparatus. *Journal of Cell Biology*, **45**, 246–71.

Brown, R. M., Herth, W., Franke, W. W., and Romanovicz, D. K. (1973). The role of the Golgi apparatus in the biosynthesis and secretion of a cellulosic glycoprotein in *Pleurochrysis:* a model system for the synthesis of structural polysaccharides. In *Biogenesis of plant cell wall polysaccharides,* (ed. F. Loewus), pp. 207–57. Academic Press, New York.

Brown, R. M. and Willison, J. H. M. (1977). Golgi apparatus and plasma membrane involvement in secretion and cell surface deposition with special emphasis on cellulose biogenesis. In *International cell biology 1976–1977,* (ed. B. R. Brinkley and K. R. Porter), pp. 267–83. Rockefeller University Press, New York.

Chrétiennot-Dinet, M.-J. (1990). *Atlas du phytoplankton marin,* Vol. 3. Éditions du CNRS, Paris.

Estep, K. W. and McIntyre, F. (1989). Taxonomy, life cycle, distribution and dasmotropohy of *Chrysochromulina:* a theory accounting for scales, haptonema, muciferous bodies and toxicity. *Marine Ecology Progress Series,* **57**, 11–21.

Fresnel, J. and Billard, C. (1991). *Pleurochrysis placolithoides* sp. nov. (Prymnesiophyceae), a new marine coccolithophorid with remarks on the status of cricolith-bearing species. *British Phycological Journal,* **26**, 67–80.

Gayral, P. and Fresnel, J. (1983). Description, sexualité et cycle de développement d'une nouvelle Coccolithophoracée (Prymnesiophyceae): *Pleurochrysis pseudoroscoffensis* sp. nov. *Protistologica,* **19**, 245–61.

Green, J. C. (1980). The fine structure of *Pavlova pinguis* Green and a preliminary survey of the order Pavlovales (Prymnesiophyceae). *British Phycological Journal,* **15**, 151–91.

Green, J. C. and Course, P. A. (1983). Extracellular calcification in *Chrysotila lamellosa* (Prymnesiophyceae). *British Phycological Journal,* **18**, 367–82.

Green, J. C. and Leadbeater, B. S. C. (1972). *Chrysochromulina parkeae* sp. nov. (Haptophyceae), a new species recorded from S. W. England and Norway. *Journal of the Marine Biological Association of the United Kingdom,* **52**, 469–74.

Green, J. C. and Parke, M. (1974). A reinvestigation by light and electron microscopy of *Ruttnera spectabilis* Geitler (Haptophyceae), with special reference to the fine structure of zoids. *Journal of the Marine Biological Association of the United Kingdom,* **54**, 539–50.

Green, J. C. and Parke, M. (1975). New observations upon members of the genus *Chrysotila* Anand, with remarks upon their relationships within the Haptophyceae. *Journal of the Marine Biological Association of the United Kingdom,* **55**, 109–21.

Green J. C. and Pienaar, R. N. (1977). The taxonomy of the order Isochrysidales (Prymnesiophyceae) with special reference to the genera *Isochrysis* Parke, *Dicrateria* Parke and *Imantonia* Reynolds. *Journal of the Marine Biological Association of the United Kingdom,* **57**, 7–17.

Green, J. C., Perch-Nielsen, K., and Westbroek, P. (1990). Phylum Prymnesiophyta. In *Handbook of Protoctista,* (ed. L. Margulis, J. O. Corliss, M. Melkonian, and D. J. Chapman), pp. 293–317. Jones & Bartlett, Boston.

Herth, W., Franke, W. W., Stadler, J., Bittiger, H., Keilich, G., and Brown, R. M. (1972). Further characterisation of the alkali-stable material from the scales of *Pleurochrysis scherffelii*: a cellulosic glycoprotein. *Planta*, **105**, 79–92.

Inouye, I. and Chihara, M. (1983). Ultrastructure and taxonomy of *Jomonlithus littoralis* gen. et sp. nov. (Class Prymnesiophyceae), a coccolithophorid from the North West Pacific. *Botanical Magazine* Tokyo, **96**, 365–76.

Kavookjian, A. M. A. and Brown, R. M. (1978). Antigenic properties of the Golgi-derived scales of *Pleurochrysis scherffelii*. *Cytobios*, **23**, 45–70.

Klaveness, D. (1973). The microanatomy of *Calyptrosphaera sphaeroidea* with some supplementary observations on the motile stage of *Coccolithus pelagicus*. *Norwegian Journal of Botany*, **20**, 151–62.

Leadbeater, B. S. C. (1970). Preliminary observations on differences of scale morphology at various stages in the life cycle of '*Apistonema-Syracosphaera*' *sensu* von Stosch. *British Phycological Journal*, **5**, 57–61.

Leadbeater, B. S. C. (1971). Observations on the life history of the haptophycean alga *Pleurochrysis scherffelii* with special reference to the microanatomy of the different types of motile cells. *Annals of Botany*, **35**, 429–39.

Leadbeater, B. S. C. (1972*a*). Fine structural observations on six new species of *Chrysochromulina* (Haptophyceae) from Norway with preliminary observations on scale production in *C. microcylindra* sp. nov. *Sarsia*, **49**, 65–80.

Leadbeater, B. S. C. (1972*b*). Identification, by means of electron microscopy, of flagellate nanoplankton from the coast of Norway. *Sarsia*, **49**, 107–24.

Leadbeater, B. S. C. and Manton, I. (1969*a*). New observations on the fine structure of *Chrysochromulina strobilus* Parke and Manton with special reference to some unusual features of the haptonema and scales. *Archiv für Mikrobiologie*, **66**, 105–20.

Leadbeater, B. S. C. and Manton, I. (1969*b*). *Chrysochromulina camella* sp. nov. and *C. cymbium* sp. nov., two new relatives of *C. strobilus* Parke and Manton. *Archiv für Mikrobiologie*, **68**, 116–32.

Leadbeater, B. S. C. and Manton, I. (1971). Fine structure and light microscopy of a new species of *Chrysochromulina* (*C. acantha*). *Archiv für Mikrobiologie*, **78**, 58–69.

Leadbeater, B. S. C. and Morton, C. (1973). Ultrastructural observations on the external morphology of some members of the Haptophyceae from the coast of Jugoslavia. *Nova Hedwigia*, **24**, 207–33.

Manton, I. (1966). Observations on scale production in *Prymnesium parvum*. *Journal of Cell Science*, **1**, 375–80.

Manton, I. (1967*a*). Further observations on the fine structure of *Chrysochromulina chiton* with special reference to the haptonema, 'peculiar' Golgi structure and scale production. *Journal of Cell Science*, **2**, 265–72.

Manton, I. (1967*b*). Further observations on scale formation in *Chrysochromulina chiton*. *Journal of Cell Science*, **2**, 411–18.

Manton, I. (1972*a*). Preliminary observations on *Chrysochromulina mactra* sp. nov. *British Phycological Journal*, **7**, 21–35.

Manton, I. (1972*b*). Observations on the biology and microanatomy of *Chrysochromulina megacylindra* Leadbeater. *British Phycological Journal*, **7**, 235–48.

Manton, I. (1978*a*). *Chrysochromulina hirta* sp. nov., a widely distributed species with unusual spines. *British Phycological Journal*, **13**, 3–14.

Manton, I. (1978*b*). *Chrysochromulina tenuispina* sp. nov. from Arctic Canada. *British Phycological Journal*, **13**, 227–34.

Manton, I. (1982). *Chrysochromulina latilepis* sp. nov. (Prymnesiophyceae = Haptophyceae) from the Galapagos Islands, with preliminary comparisons with relevant taxa from South Africa. *Botanica Marina*, **25**, 163–9.

Manton, I. (1983). Nanoplankton from the Galapagos Islands: *Chrysochromulina discophora* sp. nov. (Haptophyceae = Prymnesiophyceae). Another species with exceptionally large scales. *Botanica Marina*, **26**, 15–22.

Manton, I. (1986). Functional parallels between calcified and uncalcified periplasts. In *Biomineralization in lower plants and animals*, (ed. B. S. C. Leadbeater and R. Riding), pp. 157–72. Clarendon Press, Oxford.

Manton, I. and Leadbeater, B. S. C. (1974). Fine-structural observations on six species of *Chrysochromulina* from wild Danish marine nanoplankton including a description of *C. campanulifera* sp. nov. and a preliminary summary of the nanoplankton as a whole. *Det Kongelige Danske Videnskabernes Selskab Biologiske Skrifter*, **20**, 1–26.

Manton, I. and Leedale, G. F. (1961). Further observations on the fine structure of *Chrysochromulina ericina* Parke and Manton. *Journal of the Marine Biological Association of the United Kingdom*, **41**, 145–55.

Manton, I. and Leedale, G. F. (1963). Observations on the microanatomy of *Crystallolithus hyalinus* Gaarder and Markali. *Archiv für Mikrobiologie*, **47**, 115–36.

Manton, I. and Leedale, G. F. (1969). Observations on the microanatomy of *Coccolithus pelagicus* and *Cricosphaera carterae*, with special reference to the origin and nature of coccoliths and scales. *Journal of the Marine Biological Association of the United Kingdom*, **49**, 1–16.

Manton, I. and Oates, K. (1983*a*). Nanoplankton from the Galapagos Islands: *Chrysochromulina vexillifera* (Haptophyceae = Prymnesiophyceae) a species with semivestigial body spines. *Botanica Marina*, **26**, 517–25.

Manton, I. and Oates, K. (1983*b*). Nanoplankton from the Galapagos islands: two genera of spectacular coccolithophorids (*Ophiaster* and *Calciopappus*), with special emphasis on unmineralized periplast components. *Philosophical Transactions of the Royal Society of London*, Series B, **300**, 435–62.

Manton, I. and Parke, M. (1962). Preliminary observations on scales and their mode of origin in *Chrysochromulina polylepis* sp. nov. *Journal of the Marine Biological Association of the United Kingdom*, **42**, 565–78.

Manton, I. and Peterfi, L. S. (1969). Observations on the fine structure of coccoliths, scales and the protoplast of a freshwater coccolithophorid, *Hymenomonas roseola* Stein, with supplementary observations on the protoplast of *Cricosphaera carterae*. *Proceedings of the Royal Society of London*, Series B, **172**, 1–15.

Manton, I. and Sutherland, J. (1975). Further observations on the genus *Pappomonas* Manton et Oates with special reference to *P. virgulosa* sp. nov. from West Greenland. *British Phycological Journal*, **10**, 377–85.

Manton, I., Sutherland, J., and McCully, M. (1976*a*). Fine structural observations on coccolithophorids from South Alaska in the genera *Papposhaera* Tangen and *Pappomonas* Manton and Oates. *British Phycological Journal*, **11**, 225–38.

Manton, I., Sutherland, J., and Oates, K. (1976*b*). Arctic coccolithophorids: two species of *Turrisphaera* gen. nov. from West Greenland, Alaska and the North West Passage. *Proceedings of the Royal Society of London*, Series B, **194**, 179–94.

Manton, I., Sutherland, J., and Oates, K. (1977). Arctic coccolithophorids: *Wigwamma arctica* gen. et sp. nov. from Greenland and Artic Canada, *W. annulifera* sp. nov. from South Africa and S. Alaska and *Calciarcus alaskensis* gen. et sp. nov. from S. Alaska. *Proceedings of the Royal Society of London*, Series B, **197**, 145–68.

Manton, I., Oates, K., and Course, P. A. (1981*a*). Cylinder-scales in marine flagellates from the genus *Chrysochromulina* (Haptophyceae = Prymnesiophyceae) with a description of *C. pachycylindra* sp. nov. *Journal of the Marine Biological Association of the United Kingdom*, **61**, 17–26.

Manton, I., Oates, K., and Sutherland, J. (1981*b*). Cylinder-scales in marine flagellates from the genus *Chrysochromulina* (Haptophyceae = Prymnesiophyceae). Further observations on *C. microcylindra* Leadbeater and *C. cyathophora* Thomsen. *Journal of the Marine Biological Association of the United Kingdom*, **61**, 27–33.

Outka, D. E. and Williams, D. C. (1971). Sequential coccolith morphogenesis in *Hymenomonas carterae*. *Journal of Protozoology*, **18**, 285–97.

Paasche, E. (1964). A tracer study of the inorganic carbon uptake during coccolith formation and photosynthesis in the coccolithophorid *Coccolithus huxleyi*. *Physiologia Plantarum*, Supplement III, 1–82.

Parke, M. and Adams, I. (1960). The motile (*Crystallolithus hyalinus* Gaarder and Markali) and non-motile phases in the life history of *Coccolithus pelagicus* (Wallich) Schiller. *Journal of the Marine Biological Association of the United Kingdom*, **39**, 263–74.

Parke, M. and Manton, I. (1962). Studies on marine flagellates. VI. *Chrysochromulina pringsheinii* sp. nov. *Journal of the Marine Biological Association of the United Kingdom*, **42**, 391–404.

Parke, M., Manton, I., and Clarke, B. (1956). Studies on marine flagellates. III. Three further species of *Chrysochromulina*. *Journal of the Marine Biological Association of the United Kingdom*, **35**, 387–414.

Parke, M., Lund, J. W. G., and Manton, I. (1962). Observations on the biology and fine structure of the type species of *Chrysochromulina* (*C. parva* Lackey) in the English Lake District. *Archiv für Mikrobiologie*, **42**, 333–52.

Parke, M., Green, J. C., and Manton, I. (1971). Observations on the fine structure of zoids of the genus *Phaeocystis* (Haptophyceae). *Journal of the Marine Biological Association of the United Kingdom*, **51**, 927–41.

Pienaar, R. N. (1971). Coccolith production in *Hymenomonas carterae*. *Protoplasma*, **73**, 217–24.

Romanovicz, D. K. (1981). Scale formation in flagellates. In *Cytomorphogenesis in plants*, (ed. O. Kiermayer), pp. 27–62. Springer-Verlag, Heidelberg.

Romanovicz, D. K. and Brown, R. M. (1976). Biogenesis and structure of Golgi-derived cellulosic scales in *Pleurochrysis*. II. Scale composition and supramolecular structure. *Applied Polymer Symposium*, **28**, 587–610.

Rowson, J. D., Leadbeater, B. S. C., and Green, J. C. (1986). Calcium carbonate deposition in the motile (*Crystallolithus*) phase of *Coccolithus pelagicus* (Prymnesiophyceae). *British Phycological Journal*, **21**, 359–70.

Sikes, C. S. and Wilbur, K. M. (1982). Functions of coccolith formation. *Limnology and Oceanography*, **27**, 18–26.

Thomsen, H. A., Østergaard, J. B., and Hansen, L. E. (1991). Heteromorphic life-histories in arctic coccolithophorids (Prymnesiophyceae). *Journal of Phycology*, **27**, 634–42.

Wal, P. van der, de Jong, E. W., Westbroek, P., de Bruijn, W. C., and Mulder-Stapel, A. A. (1983). Polysaccharide localization, coccolith formation and Golgi dynamics in the coccolithophorid *Hymenomonas carterae*. *Journal of Ultrastructural Research*, **85**, 139–58.

Westbroek, P., Wal, P. van der, Emburg, P. R. van, de Vrind-de Jong, E. W., and Bruijn, W. C. de. (1985). Calcification in the coccolithophorids *Emiliania huxleyi* and *Pleurochrysis carterae*. I. Ultrastructural aspects. In *Biomineralization in lower plants and animals*, (ed. B. S. C. Leadbeater and R. Riding), pp. 189–203. Clarendon Press, Oxford.

3. Flagella and flagellar roots

J. C. GREEN
Plymouth Marine Laboratory, Citadel Hill, Plymouth, UK

and T. HORI
Institute of Biological Sciences, University of Tsukuba, Japan

Abstract

Members of the Prymnesiophyceae (division: Haptophyta) typically have two flagella, equal or unequal but similar in form in the subclass Prymnesiophycidae, unequal and dissimilar in the subclass Pavlovophycidae. The basal apparatus in all members of the Prymnesiophyceae typically includes two basal bodies, the haptonematal base, an array of microtubular and fibrous roots, and interconnections between the bases of the appendages and other organelles. In the Prymnesiophycidae there are basically three or four microtubular roots, one having its proximal termination close to the left flagellar base and the haptonema, one terminating ventrally between the basal bodies, and one or two associated with the right basal body. In the Pavlovophycidae there are two microtubular roots associated with the basal body of the left (shorter) flagellum. At present, information on the structure of the basal apparatus supports the separation of the Pavlovophycidae and Prymnesiophycidae as major taxa, but is of little value in the determination of relationships between species.

Introduction

In recent years there have been several analyses of the flagellar root structure in chromophyte algae and the reader is referred to Preisig (1989), Andersen (1991) and Inouye (1993), for recent reviews. With reference to the Prymnesiophyceae (division: Haptophyta), there is published information on some 30 species (Table 3.1), though in a number of cases the information is incidental to the main theme of the paper and the data are by no means complete. In this chapter we shall first describe the range of structure in the basal apparatus in the subclass Prymnesiophycidae comprising the majority of known prymnesiophycean species in one order, the Prymnesiales, and then summarize briefly the

The Haptophyte Algae (ed. J. C. Green and B. S. C. Leadbeater), Systematics Association Special Volume No. 51, pp. 47–71. Clarendon Press, Oxford, 1994.

Table 3.1 Presence (1) or absence (0) of flagellar apparatus characters in members of the Prymnesiophycidae, based on Table 1 of Roberts and Mills (1992) with additional data. For further explanation, see notes at foot of Table.

Species	R1	CR1	F1	R2	CR2	R3	F3	R4	Suppl.	References
Chrysochromulina acantha Leadbeater & Manton	1	0	0	0	0	1[a]	0	1[b]	1[c]	Gregson *et al.* (1993)
C. apheles Moestrup & Thomsen	1	0	0	1	0	1	0	1		Moestrup and Thomsen (1986)
C. chiton Parke & Manton	1	?	?	?	?	?	?	?	1?[d]	Manton (1968)
C. simplex Estep *et al.*	1	0	0	0	0	1[a]	0	?		Gregson *et al.* (1993)
C. aff. *simplex*	1	?	?	?	?	?	?	?		Birkhead and Pienaar (1990)
Chrysochromulina sp.	1	1	0	1	0	1	0	1		Pienaar (personal communication)
Coccolithus pelagicus (Wallich) Schiller	1	1	?	?	?	?	?	?		Klaveness (1973)
Cruciplacolithus neohelis McIntyre & Bé	1	1?	1?	1	1?	0	0	0		Fresnel (1986)
Emiliania huxleyi (Lohmann) Hay & Mohler (as *Coccolithus huxleyi* (Lohmann) Kamptner)	1	?	?	?	?	1?	?	?		Klaveness (1972)
Hymenomonas coronata Mills	1	0	1	1	1	1	1	0		Roberts and Mills (1992)
H. globosa (Magne) Gayral & Fresnel	1	1	?	1?	?	1?	?	?		Gayral and Fresnel (1976)
H. roseola Stein	1	1	1	1	1	?	?	?		Manton and Peterfi (1969)
H. lacuna Pienaar	1	1	?	1	1	?	?	?		Pienaar (1976)

Table 3.1 (*cont.*)

Species	R1	CR1	F1	R2	CR2	R3	F3	R4	Suppl.	References
Imantonia rotunda Reynolds	1	1[e]	0	1	0	1	0	0	1[f,g]	Green and Hori (1986)
Isochrysis galbana Parke	1	1	1	1	0	1	0	0		Hori and Green (1991)
Jomonlithus littoralis Inouye & Chihara	1	1	1	1	1	1	0	0		Inouye and Chihara (1983)
Ochrosphaera neapolitana Schussnig (as *O. verrucosa* Schussnig)	1	1	?	1	1	?	?	?		Inouye and Chihara (1980)
Platychrysis pienaarii Gayral & Fresnel	1	1?	?	1	?	1	?	?		Gayral and Fresnel (1983*a*)
P. pigra Geitler	1	1?	?	?	?	1	?	1?		Chrétiennot (1973)
P. simplex Gayral & Fresnel	1	1?	?	1	?	1	?	?		Gayral and Fresnel (1983*a*)
Pleurochrysis carterae (Braarud & Fagerland) Christensen (as *Cricosphaera carterae* (Braarud & Fagerland) Braarud)	1	1	1	1	1	1[h]	0	1[h]		Manton and Peterfi (1969); Beech and Wetherbee (1988)
P. carterae var. *dentata* Johansen & Doucette	1	1	1?	1	1	?	?	?		Johansen *et al.* (1988)
P. placolithoides Fresnel & Billard	1	1	1	1	1	1	?	?		Fresnel and Billard (1991)
P. pseudoroscoffensis Gayral & Fresnel	1	1	?	?	?	?	?	?		Gayral and Fresnel (1983*b*)
P. roscoffensis (Dangeard) Fresnel & Billard (as *Cricosphaera roscoffensis* (Dangeard) Gayral & Fresnel)	1	1	?	1	1	1	?	?		Gayral and Fresnel (1976)

Table 3.1 (*cont.*)

Species	R1	CR1	F1	R2	CR2	R3	F3	R4	Suppl.	References
Pleurochrysis sp.	1	1	1	1	1	1[h]	0	1[h]		Inouye and Pienaar (1985)
Prymnesium parvum N. Carter	1	?	?	1	?	1	?	?		Manton (1964*b*, 1966, 1968); Manton and Leedale (1963)
P. patelliferum Green *et al.*	1	0	1	1	0	1	0	0	1[j]	Green and Hori (1990)
Prymnesium sp.	1	1	0	0	0	1	0	1	?1[d]	Birkhead and Pienaar (1991); Pienaar (personal communication)
Syracosphaera pulchra Lohmann	1	0	0	1	0	1[h]	1	1[h]		Inouye and Pienaar (1988)
Umbilicosphaera sibogae var. *foliosa* (Kamptner) Okada & McIntyre	1	1	0	1	1	1[h]	0	1[h]		Inouye and Pienaar (1984)
Symbiont of Acantharia	1	1?	?	1	?	1	?	1		Febvre and Febvre-Chevalier (1979)

In a number of cases the information has been inferred from micrographs published in the references cited, and, where there is doubt, this is indicated by (1?). No information is indicated by (?) alone. R1–4 = microtubular roots 1–4; CR1, 2 = accessory compound components of R1 and R2; F1, 3 = fibrous roots associated with R1 and R3; Suppl. = additonal roots of uncertain homology.

[a] As R2 in Gregson *et al.* (1993);
[b] As a component of R2 in Gregson *et al.* (1993);
[c] As R3 in Gregson *et al.* (1993);
[d] Possible MT root extending from the base of the haptonema;
[e] As R5 in Green and Hori (1986);
[f] R6 in Green and Hori (1986); an additonal root, R5 in Green and Hori (1986), may be a component of a compound R1 (see text, p. 55);
[g] As R4 in Green and Hori (1986);
[h] As a component of a bifurcate R3 in original description (see references);
[j] As R4 in Green and Hori (1986).

information on the Pavlovophycidae (Pavlovales). For further discussion on the systematics of the Prymnesiophyceae, the reader is referred to Green and Jordan (Chapter 1).

Terminology

Members of the Prymnesiophycidae usually have two flagella, equal or unequal in length, but similar in form, with no scales or tubular hairs. In many early papers on the fine structure of prymnesiophytes there was no attempt to identify particular flagella or basal bodies, but recently they have been referred to by several authors as left or right with respect to the haptonematal base, the open side of the C-shaped arc of haptonematal microtubules facing the left basal body (e.g. Inouye and Pienaar 1984, 1985; Fresnel 1986; Beech and Wetherbee 1988; Green and Hori 1990; Hori and Green 1991), and this terminology will be followed here.

Recent research on flagellar transformation during the cell cycle has led to the concept of mature and immature flagella, the immature flagellum (No. 2) changing to a mature flagellum (No. 1) during mitosis and then remaining unchanged through subsequent generations (Beech *et al.* 1991). Beech *et al.* (1988) have studied flagellar transformation in *Pleurochrysis carterae* (Prymnesiophycidae) and have shown that the left basal body remains unchanged in structure and position during mitosis. It therefore represents the mature (No. 1) structure, but for convenience and ease of description, we shall continue here to refer to the basal bodies as 'left' and 'right' respectively.

Basal apparatus of the Prymnesiophycidae

Microtubular roots

The components of the basal apparatus receiving most attention have been the microtubular roots (see Table 3.1). In a number of species, one or more of these roots are composed of broad sheets and bundles of microtubules (MTs) which frequently appear in sections and have attracted comment since the earliest microanatomical studies on members of the Prymnesiophyceae (e.g. Figs 14, 15 in Manton and Leedale 1963).

Although there is considerable variation in the detailed structure of the microtubular roots, it is possible to detect consistency in the overall arrangement (Andersen 1989, 1991; Preisig 1989; Fig. 3.1A). There is always a root associated with the left flagellar base and this is referred to as R1. Other roots are labelled in a clockwise direction (viewed anteriorly) and are variable in number though there is usually a root (R3) associated with the right flagellar base.

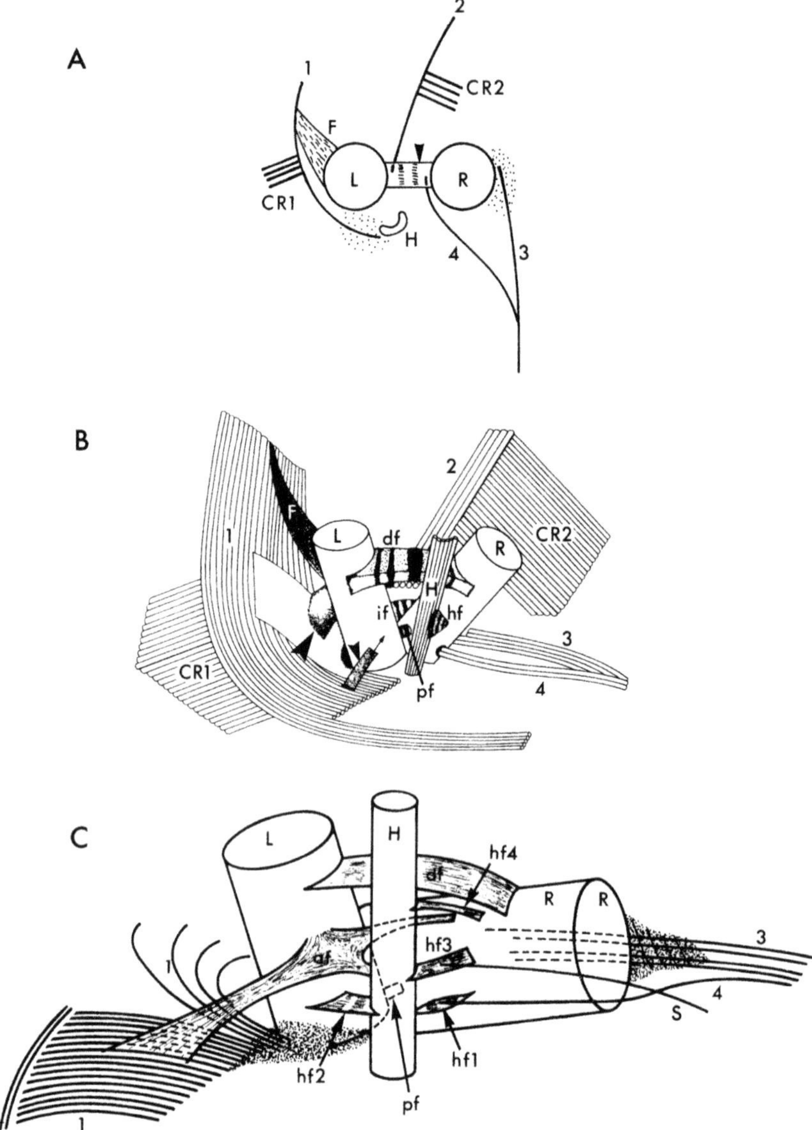

Fig. 3.1 Diagrammatic representations of the architecture of the basal apparatus in members of the Prymnesiophycidae. (A) Generalized arrangement showing the positions of the principle microtubular roots (1–4), compound components (CR1, CR2) of roots 1 and 2, a fibrous root (F) associated with root 1, the haptonematal base (H), the left and right basal bodies (L and R), and the distal connecting fibre (*arrowhead*). (B, C) Schematic reconstructions of the basal apparatus of *Pleurochrysis* sp. (B) and *Chrysochromulina acantha* (C). Labelling as in A with, in addition, distal,

Root R1 Root R1 seems to be present in all species (Table 3.1) and is often the most conspicuous component of the flagellar root system. The main component is a sheet of MTs, occasionally as many as 20–30 in number (Fig. 3.1B, C; Inouye and Pienaar 1985; Beech and Wetherbee 1988; Green and Hori 1990; Fresnel and Billard 1991), or as few as two to five (Green and Hori 1986; Moestrup and Thomsen 1986). The sheet of MTs originates close to the base of the haptonema, often in association with amorphous densely staining material (Inouye and Pienaar 1984; Green and Hori 1986, 1990; Birkhead and Pienaar 1991; Hori and Green 1991; Gregson *et al.* 1993; see also illustrations in, for example, Chrétiennot 1973; Febvre and Fevre-Chevalier 1979; Gayral and Fresnel 1983*a*, *b*; Inouye and Chihara 1983; Fresnel 1986; Moestrup and Thomsen 1986; Inouye and Pienaar 1988; Roberts and Mills 1992). It then follows a course round the left side of the left basal body either as a coherent band, at least initially (Inouye and Pienaar 1984, 1985; Beech and Wetherbee 1988), or with the MTs gradually splaying apart as in *Prymnesium patelliferum* (Green and Hori 1990), or with a select number of MTs breaking away from the sheet as in some species of *Chrysochromulina* (Moestrup and Thomsen 1986; Gregson *et al.* 1993; Fig. 3.1C). In all cases, root R1 is closely associated with a mitochondrial profile and the chloroplast lying on the left side of the cell, often running over the surface of the organelles in question for at least part of its course. It has been suggested (Moestrup and Thomsen 1986) that this association is concerned with the transfer of ATP to the flagellar apparatus.

A particular development of the R1 sheet of microtubules has been noted in certain coccolithophorids. In *Pleurochrysis pseudoroscoffensis*, *Pl. carterae*, and *Pl. placolithoides* (Gayral and Fresnel 1983*b*; Beech and Wetherbee 1988; Fresnel and Billard 1991), part of root R1 extends deep into the cell in a cytoplasmic tongue which is partly surrounded by an invagination of the peripheral cisterna lying immediately beneath the plasmalemma. Gayral and Fresnel (1983*b*) and Beech and Wetherbee (1988) discuss the possibility that the cytoplasmic tongue may be concerned with the production and deployment of scales. Fresnel and Billard

intermediate, and proximal fibres linking the basal bodies (df. if, pf), and fibres linking the haptonematal base and the basal bodies (hf1, hf2, etc.). In B, a plate of material on root 1 is linked (*large arrowhead*) to the left basal body, and an accessory fibre (*smaller arrowhead*), cut away in order not to obscure other detail, links root 1 to the distal fibre, as indicated by the small arrow. In C, an accessory fibre (af) connects root 1, the haptonematal base, and the basal bodies, and a supplementary root (S) has its origin between the basal bodies. (B) after Inouye and Pienaar (1985); (C) after Gregson *et al.* (1993).

(1991) ascribe a contractile function to this part of root R1 and suggest that it may be responsible for the contractile movement observed in both *Pl. pseudoroscoffensis* and *Pl. placolithoides.*

The outstanding feature of root R1 in many species is the presence of an accessory bundles of MTs branching off from the main sheet of MTs (Fig. 3.1B) and these sometimes massive close-packed 'crystalline' bundles often attracted the attention of early workers (Manton and Peterfi 1969). They have been reported since in several species (Table 3.1), especially in the coccolithophorids, and it was thought that such compound roots were a feature of these particular organisms. However, compound roots have been demonstrated in some non-coccolithophorid species, for example, *Isochrysis galbana* (Hori and Green 1991), a species of *Prymnesium* (Birkhead and Pienaar 1991), a species of *Chrysochromulina* (Pienaar, personal communication), and possibly in the unnamed symbiont of acantharian actinopods (Febvre and Febvre-Chevalier 1979). In addition, Chrétiennot (1973) reported crystalline bundles in *Platychrysis pigra* which may well be components of a compound root system, but this awaits confirmation. The compound R1 root system of *I. galbana* is interesting in that there are two accessory bundles of MTs both of which converge to meet the sheet of MTs near its origin close to the left basal body. The sheet of MTs and one of the bundles are closely associated with a mitochondrial profile and with the chloroplast, whilst the other bundle component passes into the cell close to the Golgi body.

A double compound root has not been recorded elsewhere, but examination of Plate VB in Pienaar (1976) shows two groups of MTs converging on a root composed of a sheet of MTs close to the flagellar apparatus of *Hymenomonas lacuna*, suggesting that roots of this type may be more common than is apparent at present. However, it is now known that some coccolithophorids do not have compound R1 roots. *Syracosphaera pulchra* and *Hymenomonas coronata*, for example, have simple R1 roots composed of a sheet of only 10 or 8–10 MTs (Inouye and Pienaar 1988; Roberts and Mills 1992), and preliminary analysis of the root system of *Cruciplacolithus neohelis* has shown that root R1 does not appear to be a compound structure (Fresnel 1986).

A frequent feature of root R1 is the presence of a plaque of densely staining material on the face of the sheet of microtubules facing the left basal body. Such material has been either reported or illustrated in *Pleurochrysis roscoffensis* and *Hymenomonas globosa* (Gayral and Fresnel 1976), *Jomonllithus littoralis* (Inouye and Chihara 1983), *Pleurochrysis pseudoroscoffensis* (Gayral and Fresnel 1983*b*), *Pleurochrysis* sp. (Inouye and Pienaar 1985; Fig. 3.1B), *Chrysochromulina apheles* (Moestrup and Thomsen 1986). *Pl. carterae* (Beech and Wetherbee 1988), *Syracosphaera pulchra* (Inouye and Pienaar 1988), and *Isochrysis galbana* (Hori and

Green 1991). Such a structure may prove to be a component of most R1 roots and it has been suggested by Moestrup and Thomsen (1986) that this is a clear indication of the homology of these particular roots associated with the left basal body. Whether the electron dense plaque has a general function is not known, but it has been shown that in *Pleurochrysis* sp. (Inouye and Pienaar 1985), *Pl. carterae* (Beech and Wetherbee 1988), and *Syracosphaera pulchra* (Inouye and Pienaar 1988) that a fibrous strand extends from the plaque, anchoring the root to the left basal body (see below).

It may be noted here that an additional root arising close to the origin of R1 has been described (as R5) by Green and Hori (1986) in *Imantonia rotunda*. This is composed of a single MT which extends for a short distance towards the adjacent chloroplast, and it was suggested by Green and Hori (1986) that R1 and R5 together in *I. rotunda* represented a much reduced compound root.

Root R2 Root R2 has its origin between the basal bodies (Fig. 3.1A, B), often in association with a striated fibre (the distal connecting fibre) linking the basal bodies, and it leaves the flagellar apparatus in a ventral direction (terminology of Beech and Wetherbee 1988).

In some coccolithophorids, root R2 is also known to be a compound structure (Table 3.1; Fig. 3.1A, B). In *Hymenomonas coronata*, root R1 is not compound, but the R2 complex is well developed with a bundle of several dozen MTs extending towards the right side of the cell from a ribbon of 3–5 MTs. An additional pair of MTs branch off from the ribbon in the same region as the major bundle and pass towards the left side of the cell (Roberts and Mills 1992).

Not all coccolithophorids have compound R2 roots. In *Syracosphaera pulchra* (Inouye and Pienaar 1988), and possibly *Cruciplacolithus neohelis* (Fresnel 1986), root R2 consists only of a ribbon of four MTs arranged in an arc. In the non-coccolithophorid species, the R2 root may be present or absent. In *Imantonia rotunda* the R2 root composed of three MTs arises between the flagellar bases beneath the distal connecting fibre (Green and Hori 1986). In *Isochrysis galbana*, root R2 consists of a single MT having its origin between the flagellar basal bodies, but close to a proximal connecting band (Hori and Green 1991). After leaving the flagellar apparatus it passes round the left basal body and follows a course deep into the cell, passing between the accessory bundles of MTs of root R1 (see above). A simple R2 root of three MTs has also been described in *Prymnesium patelliferum* (Green and Hori 1990) arising between the flagellar basal bodies, closer to the left one, and initially running for a short distance parallel to it before bending over at the level of the distal connecting fibre and passing away from the flagellar apparatus in a ventral direction. Of the two species of *Chrysochromulina* which have been analysed in detail, only *C. apheles* has a root identified as R2, a single MT terminating at its proximal end on the surface of the distal connecting fibre and passing away from the flagellar apparatus towards the left chloroplast (Moestrup and Thomsen 1986). Gregson *et al.* (1993) have not found a root corresponding to R2 in *C. acantha* (Fig. 3.1C), their root R2 (in Gregson *et al.* 1993) being homologous with the root R3 of other members of the Prymnesiophycidae (see below).

Examination of micrographs in various publications indicates the presence of a root in the R2 position in a number of other species. These include *Prymnesium parvum* (Manton 1966), *Cricosphaera roscoffensis*, and *Hymenomonas globosa* (Gayral and Fresnel 1976), *Hymenomonas lacuna* (Pienaar 1976), the symbiont of Acantharia (Febvre and Febvre-Chevalier 1979), *Platychrysis pienaarii*, and *Pl. simplex* (Gayral and Fresnel 1983*a*). It is interesting to note that the root which we interpret as R2 in *Pl. pienaarii* (see Fig. 30 in Gayral and Fresnel 1983*a*) bifurcates sharply a short distance from the flagellar apparatus, a feature not reported in other species.

Roots R3 and R4 In all members of the Prymnesiophycidae for which detailed information is available, a microtubular root is associated with the right basal body, but it is always composed of relatively few MTs and is never compound in structure. In several species the root appears to be bifurcate at the proximal end, with one component originating on the outer (right) face of the basal body and another arising on the left side of the basal body towards the haptonema. Both sets of MTs pass away from the flagellar apparatus in a dorsal direction and come together to form a single microtubular strand, sometimes with the addition of extra MTs (Inouye and Pienaar 1984, 1985, 1988; Beech and Wetherbee, 1988; Gregson *et al.* 1993; Fig. 3.1A–C). There is a similar arrangement in *C. apheles*, but in this species the two components have been identified as separate roots (Moestrup and Thomsen 1986). In this context, Inouye (1993) considers that the two components of the bifurcate R3 root should be considered as separate roots (R3 and R4; Fig. 3.1A–C) as they terminate proximally at separate MT organizing centres (MTOCs). This has also been suggested by Gregson *et al.* (1993) and such an interpretation is, therefore, accepted here, the root having its origin between the basal bodies being denoted R4.

The R3/R4 root complex is not always a bifurcate structure. In *Imantonia rotunda* (Green and Hori 1986), *Prymnesium patelliferum* (Green and Hori 1990), and *Isochrysis galbana* (Hori and Green 1991) root R3 is composed only of MTs originating on the right side of the basal body. In *Imantonia rotunda* there is a single pair of MTs, whilst in *Isochrysis galbana* there is one pair proximally, but with the addition of a second pair to form a 2 + 2 root. In *Prymnesium patelliferum* the arrangement is unusual with the root originating as a pair of MTs close to the proximal end of the basal body, the addition of a further pair which run adjacent to them for a very short distance only (i.e. 2 + 2 for one or two sections), followed by the addition of a single MT as the root leaves the vicinity of the basal bodies, thus forming a 2 + 1 final arrangement. In *Hymenomonas coronata*, the root associated with the right basal body consisted of a band of three or four MTs originating on its ventral side and passing to the right side of the cell (Roberts and Mills 1992).

As in the case of root R2, examination of published micrographs suggests that root R3 may present in a number of other species including *Prymnesium parvum* (Manton 1964*b*, 1966), *Emiliania huxleyi* (Klaveness 1972), *Platychrysis pigra* (Chrétiennot 1973), *Pleurochrysis roscoffensis* and *Hymenomonas globosa* (Gayral and Fresenel 1976), and *H. lacuna* (Pienaar 1976), and the symbiont of Acantharia (Febvre and Febvre-Chevalier 1979). Only in *Cruciplacolithus neohelis* has root R3 been reported as definitely absent (Fresnel 1986).

The MTs of root R3 at their proximal ends are usually embedded in densely staining material which presumably acts as an MTOC. This material often appears amorphous or, occasionally, finely fibrillar (e.g. in *Chrysochromulina acantha*; Gregson *et al.* 1993; Fig. 3.1C).

Other microtubular roots In *Imantonia rotunda* (Green and Hori 1986), *Prymnesium patelliferum* (Green and Hori 1990), and *Chrysochromulina acantha* (Gregson *et al.* 1993) a microtubular root has been described arising dorsally between the basal bodies (Fig. 3.1C). In *I. rotunda* (as R4 in Green and Hori 1986), it consists of a single MT which terminates proximally between the distal and intermediate fibres joining the basal bodies, rather closer to the left basal body. It runs parallel to the basal body for a short distance before taking up a position close to R3. In *Prymnesium patelliferum*, a single MT root (R4 in Green and Hori 1990) has its proximal origin close to the proximal bands connecting the basal bodies. It also initially follows a course approximately parallel to the basal bodies before turning in a dorsal direction and acquiring a second MT to become a two-stranded root. A single MT root in *Chrysochromulina acantha* (as R3 in Gregson *et al.* 1993) originates close to a striated fibre linking the haptonematal base and the right basal body (Fig. 3.1C), and is joined to one of the basal body triplets by a fine fibre. However, in contrast to the 'R4' root in *I. rotunda* and *P. patelliferum*, it is seen in only a few sections and is probably, therefore, very short (Gregson *et al.* 1993). It is possible that the R4 root of *Imantonia rotunda*, which, distally, runs close to R3, may be homologous with the R4 roots described above (p. 00). However, the diverse positions of the proximal terminations of these additional dorsal roots suggests that they may not all be homologous either with each other or with any of the standard roots described. It is probably premature, therefore, to give them a uniform designation.

In *Imantonia rotunda* there is an additional root of approximately three MTs associated with the right basal body, but which runs in a ventral direction near R2. No comparable root has been described in any other member of the Prymnesiophycidae. Finally, Birkhead and Pienaar (1991; also personal communication) have noted in a species of *Prymnesium* a possible microtubular root extending from the base of the haptonema, though this awaits confirmation. A similar root may be present also in *Chrysochromulina chiton* (see Fig. 24 in Manton 1968). Birkhead and Pienaar (1991; also personal communication) have also noted in *Prymnesium* sp. a hitherto unreported microtubular root extending towards the right side of the cell from the right basal body, and a similar root may also be present in *Platychrysis simplex* (see Fig. 30 in Gayral and Fresnel 1983*a*). The orientation of this root suggests that it may be homologous with root R3 of *Hymenomonas coronata* (Roberts and Mills 1992).

Fibrous roots

Fibrous roots have been described in a number of members of the Prymnesiophyceae, including some coccolithophorids (see Table 3.1; Fig. 3.1A, B), *Prymnesium patelliferum* (Green and Hori 1990) and *Isochrysis galbana* (Hori and Green 1991). The most conspicuous of such roots (F1) extends from the face of the left basal body away from the haptonema to the adjacent face of the sheet of MTs of root R1. In some species, the F1/R1 extends for some considerable distance into the cell in a cytoplasmic tongue, passing close to one of the chloroplasts and to numerous mitochondrial profiles (Gayral and Fresnel 1983*b*; Beech and Wetherbee 1988; Roberts and Mills, 1992).

In *Syracosphaera pulchra* and *Hymenomonas coronata*, an additional fibrous root is associated with root R3 (Inouye and Pienaar 1988; Roberts and Mills 1992). In the former species, the fibrous root F3 appears to originate on the same side of the right basal body as the microtubular root R3, but in the latter species, the fibrous root has its proximal origin on the opposite side of the right basal body from R3.

Accessory and connecting fibres

In addition to the fibrous roots, other structures have been described linking certain roots to the basal bodies or to other structures. For example, in species of *Pleurochrysis* (Inouye and Pienaar 1985; Beech and Wetherbee 1988; Fig. 3.1B) and probably in *Hymenomonas globosa* (Gayral and Fresnel 1976), a large non-striated accessory fibre connects a dense plaque of material on the face of the R1 sheet of microtubules to the left basal body, whilst a smaller accessory fibre in *Pl. carterae* links the R1 microtubules to the proximal part of the left basal body. The R1 MTs of the unnamed species of *Pleurochrysis* examined by Inouye and Pienaar (1985) are connected also to the distal fibre which extends between the flagellar bases (Fig. 3.1B). A large accessory fibre also connects root R1 to various other organelles in *Prymnesium patelliferum* and *Chrysochromulina acantha*. In *Pr. patelliferum* it is situated distally between the flagellar bases, and links R1 to the haptonematal base and to the mitochondrial profile on the right side of the cell (Green and Hori 1990), whilst in *C. acantha* the non-striate accessory fibre links the sheet of R1 MTs to both basal bodies and to the haptonematal base (Gregson *et al.* 1993; Fig. 3.1C).

Fibres connecting the flagellar basal bodies to each other and to the haptonematal base are commonly reported and are probably universal (Fig. 3.1A–C). However, there is much variation in the exact way that these structures are arranged and whether they are striate or non-striate. The information obtained from some of the more complete accounts of flagellar structure is summarized in Table 3.2.

The transition region

Moestrup (1982) and Preisig (1989) have discussed the early work on the transition region of the flagellum and only the more important features are summarized here.

The most detailed early study was that of Manton (1964*a*, 1965) on *Prymnesium parvum*. In this species, there are two widely spaced transitional plates extending between the microtubule doublets of the axoneme, the distal plate thickened where it joins the doublets, the more proximal thickened centrally. The two plates appear to extend to the flagellar membrane though the outer parts of the plate consist of less well-defined diffuse material. They appear to be joined by a fine central thread running longitudinally along the flagellum. Also in the transition region of *P. parvum*, the outer axonematal doublets are linked to form an internal star-like structure. A similar double partition transition region is illustrated by Green and Hori (1986) in *Imantonia rotunda*, but in this species the proximal partition does not appear to extend fully across the flagellum (see Fig. 18 in Green and Hori 1986). In *Isochrysis galbana*, there is only one transitional plate extending across the central part of the axoneme, but it does not extend to the flagellar periphery (Hori and Green 1991). However, distal to this structure, and thus beyond the transition region, there is a fibrous plug, approximately 100 nm in length, extending across the flagellum, through which the axonematal microtubules, including the central pair, pass. The fibrous plug appears to be composed at least in part, of helically arranged fine fibrils.

Table 3.2 The distribution of connecting fibres in the basal apparatus of members of the Prymnesiophycidae (BB = basal body; s = striate; ns = non-striate)

Species	Flagellar connecting fibres			Haptonematal fibres	Additional fibres	References
	Distal	Intermediate	Proximal			
Chrysochromulina acantha	ns	–	ns	Distal to right BB; ns Interm. to right BB; s Prox. to right BB; s Prox. to left BB; s	Distal accessory fibre interconnecting BBs and haptonematal base; ns	Gregson et al. (1993)
Chrysochromulina apheles	s	–	s	2 at similar level: s to junction of distal fibre and right BB; s directly to right BB		Moestrup and Thomsen (1986)
Chrysochromulina sp.	s	ns	double; s	Distal to right BB; s Proximal to right BB; s Also connected to BBs by distal and proximal flagellar connecting fibres	Accessory fibre attached to right BB ventrally; s	Pienaar (Personal communication)

Table 3.2 (*cont.*)

Species	Flagellar connecting fibres			Haptonematal fibres	Additional fibres	References
	Distal	Intermediate	Proximal			
Cruciplacolithus neohelis	s (position uncertain)			no information		Fresnel (1986)
Imantonia rotunda	s	double; ns	ns	no haptonema		Green and Hori (1986)
Hymenomonas coronata	As *Pleurochrysis* spp.			s to left BB s to right BB		Roberts and Mills (1992)
Isochrysis galbana	s	–	ns plus fine fibre between two triplets	–		Hori and Green (1991)
Jomonlithus littoralis	s	–	ns	1 (s) to right BB? (see Fig. 19)		Inouye and Chihara (1983)
Pleurochrysis sp.	s	s; on ventral side of BBs	s	1 only (s) to right basal body		Inouye and Pienaar (1985)
Pleurochrysis carterae	s	s	s	ns to left BB s to right BB		Beech and Wetherbee (1988)
Pleurochrysis placolithoides		?		s to right BB (see Fig. 17)		Fresnel and Billard (1991)

Table 3.2 (*cont.*)

Species	Flagellar connecting fibres			Haptonematal fibres	Additional fibres	References
	Distal	Intermediate	Proximal			
Prymnesium parvum	s	–	s; double	s to right BB		Manton (1964b)
Prymnesium patelliferum	s	ns; on ventral side of BBs	s; double	To right BB, s, distal ns, proximal		Green and Hori (1990)
Prymnesium sp.	s	ns	double; s	Distal to right BB; s Proximal to right BB; s Also connected to BBs by distal and proximal flagellar connecting fibres	Accessory fibre attached to right BB ventrally; s	Piennar (personal communication)
Syracosphaera pulchra	s	–	s	1 to left BB 1 to right BB		Inouye and Pienaar (1988)
Umbilicosphaera sibogae	s	–	s	1 to right BB; ns		Inouye and Pienaar (1984)
Symbiont of Acantharia	s			?		Febvre and Febvre-Chevalier (1979)

In *Pleurochrysis* sp., Inouye and Pienaar (1985) described up to nine tiers of electron-dense rings in the flagellar transition region (eight according to Beech and Wetherbee (1988) who suggest that the most distal is an axosome). Beech and Wetherbee (1988) have also described a series of five closely stacked tubular rings in the proximal part of the transition region of the flagella of *Pleurochrysis carterae*, and, in this species, there is particularly elaborate axosome situated distally to the rings. Beyond the axosome (and thus distal to the transition region itself) there is also helical band of at least seven gyres located between the peripheral doublets and the central pair of MTs. These structures are probably more common than present records would suggest as similar configurations having been described by Roberts and Mills (1992) in *Hymenomonas coronata*, and are illustrated in *H. roseola* (Manton and Peterfi 1969) and *Pleurochrysis pseudoroscoffensis* (Gayral and Fresnel 1983*b*). A helical band has also been described in a preliminary way in the flagellum of a species of *Prymnesium* (Birkhead and Pienaar 1991). It is situated distal to the transition region and consists of 10 or more gyres of fibrillar material, but in this species it is lies between the peripheral doublets and the flagellar membrane.

Transitional core

Inouye and Pienaar (1988) have drawn attention to electron-dense material situated in the central lumen of the basal bodies of *Syracosphaera pulchra*. Similar material has been observed in a number of other species including the coccolithophorid *Hymenomonas coronata* (Roberts and Mills 1992) and several non-coccolithophorid species (*Prymnesium parvum*, Manton 1964*a*; *Chrysochromulina chiton*, Manton 1968; the symbiont of Acantharia, Febvre and Febvre-Chevalier 1979; *C. apheles*, Moestrup and Thomsen 1986; *Platychrysis pienaarii* and *Phaeocystis pouchetii*, Inouye and Pienaar 1988; *Imantonia rotunda*, Green and Hori 1986; *Isochrysis galbana*, Hori and Green 1991).

The flagellar apparatus in the Pavlovophycidae

The flagellar apparatus in members of the Pavlovophycidae (with one order, the Pavlovales) differs in several respects from that of members of the Prymnesiophycidae. Detailed observations have been made on the basal apparatus of *Diacronema vlkianum* (Green and Hibberd 1977) and *Pavlova pinguis* (Green 1980), but incidental information is available in other publications and the reader is referred to Green (1980) and Green *et al.* (1990) for reviews.

There are two unequal flagella inserted laterally, the longer, directed forwards during swimming, often bearing an array of small knob-like

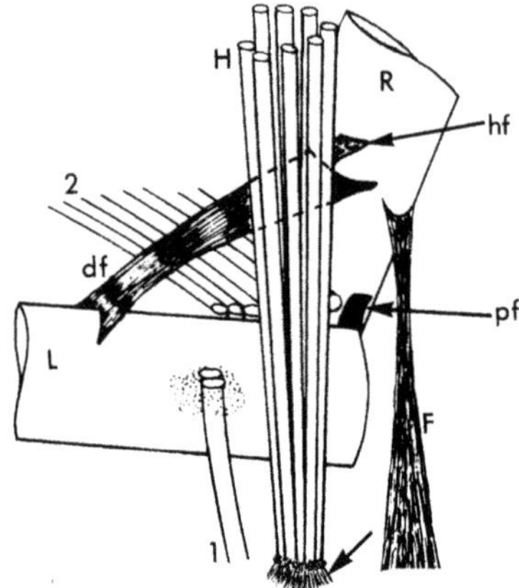

Fig. 3.2 Schematic representation of the basal apparatus of *Pavlova pinguis* showing the left (L; short flagellum) and right (R; long flagellum) basal bodies, the microtubules of the haptonematal base (H), distal (df) and proximal (pf) connecting fibres, a haptonematal fibre (hf), microtubular roots (1, 2), and a fibrous root (F) associated with the right basal body. There is an additional fibrous root associated with the haptonematal base (*arrow*). Root 1 is numbered by analogy with root 1 of the Prymnesiophycidae, having its proximal termination close to the haptonematal base and the left basal body, but the roots may not be homologous. Adapted from Green (1980).

scales (Leadbeater, Chapter 2; possibly modified hairs, see Cavalier-Smith, Chapter 22) and fine, non-tubular hairs. The shorter flagellum, sometimes with distal attenuations, carries no scales and is occasionally reduced to a vestigial structure. The basal bodies are situated in two different planes, the angle between their long axes often variable, but commonly approximately 90° (Fig. 3.2). The open side of the arc of MTs forming the base of the haptonema faces the basal body of the posteriorly directed or shorter flagellum which is probably, therefore, homologous with the left (mature) basal body of other species of the Prymnesiophyceae (see above, p. 51). The basal bodies are connected by two connecting fibres, a large distal striated fibre and a short proximal fibre. The haptonematal base is attached by a single apparently striated haptonematal fibre to the anterior (right) basal body.

The major microtubular root in *Diacronema vlkianum* (Green and Hibberd 1977) and *Pavlova pinguis* (Green 1980) is a 5 + 2 or 6 + 1 structure (R2; Fig. 3.2), attached at its proximal end by short fibrils arising between the MTs to the posterior basal body between the distal and proximal connecting bands. It follows a course away from the basal bodies on the side opposite the haptonema and can be traced running beneath the periplast. A similar root is illustrated in a number of other species (Green and Manton 1970; Green 1973, 1975, 1976; Billard 1976; Veer and Leewis 1977; Gayral and Fresnel 1979). An additional microtubular root of only two MTs (R1; Fig. 3.2) is associated with the posterior (left) basal body on the side towards the haptonema. The longitudinal axis of this root is more or less perpendicular to that of the basal body, and the root follows a course into the cell approximately parallel to that of the haptonema.

A conspicuous fibrous root is associated with the basal body of the longer flagellum in several species (Billard 1976; Veer 1976; Veer and Leewis 1977; Green 1980). It follows a course deep into the cell passing close to the nucleus. In both *Diacronema vlkianum* (Green and Hibberd 1977) and *Pavlova pinguis* (Green 1980) the haptonema also has a fibrous root. In *D. vlkianum* it radiates as an array of fibrils arising from a densely staining plate where the haptonematal MTs terminate proximally, but in *Pavlova pinguis* it forms a more compact structure which passes deep into the cell, bifurcating in the region of the nucleus. In *P. calceolata* fine filaments extend radially from the base of the anterior flagellum (Veer 1976).

The transition region of members of the Pavlovophycidae has only one transverse partition, situated approximately at the level of the junction of the flagellum and the cell body (Green 1975, 1980; Green and Hibberd 1977). The partition extends between the outer doublets of the axoneme and, in *Diacronema vlkianum*, has a triple-layered structure (Green and Hibberd 1977). There is often a thickened axosome (Green 1975, 1980; Green and Hibberd 1977). Between the outer doublets and the flagellar membrane, there may be amorphous osmiophilic material; in *Pavlova pinguis* the amorphous material may be finely fibrillar (Green 1980). In *D. vlkianum*, at the level of the partition, the doublets of the axoneme are linked by fine strands forming a polygonal pattern. Within the lumen of the basal body there may be osmiophilic material. This is best illustrated in *D. vlkianum* in which it occurs either as four blocks of material (just below the partition), or in a spoked arrangement more proximally (Green and Hibberd 1977). Similar material is illustrated in *Pavlova lutheri* (Green 1975) and may be homologous with the transitional core material of other members of the Prymnesiophyceae.

Discussion

Although there is considerable variation of root structure and arrangement, it appears that within the Prymnesiophycidae and the Pavlovophycidae, respectively, there are consistent features, and, within each subclass, it is therefore possible to make comparisons between species. Thus, in the Prymnesiophycidae, most species have microtubular roots in the R1, R2, and R3 position; R1 associated with the left basal body and the haptonema, R2 arising between the basal bodies, and R3 associated with right basal body. All the microtubular roots are associated proximally with osmiophilic amorphous or fibrous material. The function of this material is uncertain, but it may serve as an MTOC.

In some species, root R3 is bifurcate at the proximal end with R3 and R4 components terminating on either side of the right basal body (see p. 56–7). The interpretation of the components of the bifurcate root as two separate roots (Gregson *et al.* 1993; Inouye 1993) suggests a superficial resemblance between the architecture of the prymnesiophycidean root system and that of some other chromophyte classes, e.g. the Chrysophyceae and Oomycetes (see fig. 2E in Andersen 1989; fig. 3E in Andersen 1991). However, if the right basal body of the Prymnesiophycidae is correctly interpreted as the immature basal body, roots R3 and R4 are not homologous with roots R3 and R4 of the heterokont classes as illustrated by Andersen (1989, 1991), since in the latter groups they are associated with the mature non-hairy flagellum. Again, the prymnesiophycidean root R1 and the root R1 of the heterokont groups, which both may act as MTOCs for secondary bundles of MTs, cannot be homologous.

It is difficult to detect any homology between the microtubular roots of the Pavlovophycidae and members of either the Prymnesiophycidae or the heterokont groups. It is conceivable that the short 2-MT root is a relict of a secondary bundle originally associated with a root R1 such as that described in *Isochrysis galbana* by Hori and Green (1991), but this seems unlikely in view of the numerous other differences between the groups both in terms of the flagellar apparatus and other features of the cell.

The possible value of ultrastructural features of flagellar and basal body microanatomy as taxonomic and phylogenetic markers in algae has been discussed in numerous papers. With respect to the Prymnesiophyceae, Manton (1964*a*, 1965) drew attention to features of the proximal parts of the flagellum of *Prymnesium parvum* and discussed their possible phylogenetic significance in the context of the limited information then available. During the next fifteen years fine structural information on a number of prymnesiophycean species was published and the data on flagellar and flagellar root structure was summarized by Moestrup (1982), but at that time it was not possible to draw any firm systematic conclusions.

The more detailed studies of the prymnesiophycean flagellar basal apparatus published over the last decade have done little to resolve such problems. It is apparent that members of the order Pavlovales stand apart from the main body of the Prymnesiophyceae, both in the arrangement of their flagellar apparatus and other structural features of the cell and they have, therefore, been included in a separate subclass

(Pavlovophycidae) or class (for a summary, see Green and Jordan, Chapter 1). Within the Prymnesiophycidae, it appeared at first that complex root systems were characteristic of the coccolithophorids (Inouye and Pienaar 1985), but subsequently it was shown that some coccolithophorids have relatively simple root systems (Inouye and Pienaar 1988), and that some non-cocolithophorid species may have relatively complex systems (Green and Hori 1990; Hori and Green 1991). Even within one genus, there may be considerable variation. For example, *Chrysochromulina apheles* has a basal apparatus in which the maximum number of MTs in any root is 4(5) + 1 + 1. In addition, there are three relatively small connecting fibres. However, *C. acantha* and *C. simplex* have R1 roots with approximately 20 MTs and the system of connecting fibres is more complex with a number of fibres interconnecting the flagellar and haptonematal bases (Gregson *et al.* 1993). A yet more complex system has been found in an undescribed species of *Chrysochromulina* (Pienaar and Birkhead, personal communication) in which the R1 root consists of approximately 20 MTs with an additional compound component. There is also a complex system of striated and unstriated connecting fibres. The basal apparatus of this species of *Chrysochromulina* is of particular interest in that it strongly resembles that of a new species of *Prymnesium* both in the structure of the basal apparatus and in the morphology of its body scales. Within the Prymnesiophycidae, therefore, although it is possible to discern a basic architectural pattern (Andersen 1989, 1991; Preisig 1989), there is much variation in the structure of the basal apparatus both within and between genera, and we have to agree with Roberts and Mills (1992) that the lack of correlation between flagellar apparatus and other morphological features compounds existing systematic problems. The data on the structure of the flagellar apparatus must be analyzed in conjunction with other characters, morphological, biochemical, and genetic.

Acknowledgements

The authors are grateful to Professor R. N. Pienaar and Miss M. N. Birkhead for generously providing unpublished data on species of *Chrysochromulina* and *Prymnesium*.

References

Andersen, R. A. (1989). The Synurophyceae and their relationship to other golden algae. *Beiheft zur Nova Hedwigia*, **95**, 1–26.

Andersen, R. A. (1991). The cytoskeleton of chromophyte algae. *Protoplasma*, **164**, 143–59.

Beech, P. L. and Wetherbee, R. (1988). Observations on the flagellar apparatus and peripheral endoplasmic reticulum of the coccolithophorid, *Pleurochrysis carterae* (Prymnesiophyceae). *Phycologia*, **27**, 142–58.

Beech, P. L., Wetherbee, R., and Pickett-Heaps, J. D. (1988). Transformation of the flagella and associated components during cell division in the coccolithophorid *Pleurochrysis carterae. Protoplasma*, **145**, 37–46.

Beech, P. L., Heimann, K., and Melkonian, M. (1991). Development of the flagellar apparatus during the cell cycle in unicellular algae. *Protoplasma*, **164**, 23–37.

Billard, C. (1976). Sur une nouvelle espèce de *Pavlova, P. virescens* nov. sp., (Haptophyceés). *Bulletin de la Société Phycologique de France*, **21**, 18–27.

Birkhead, M. and Pienaar, R. N. (1990). The ultrastructure of *Chrysochromulina* cf. *simplex* (Prymnesiales). *Proceedings of the Electron Microscopy Society of Southern Africa*, **20**, 85–6.

Birkhead, M. and Pienaar, R. N. (1991). The flagellar apparatus of a new species of *Prymnesium. Proceedings of the Electron Microscopy Society of Southern Africa*, **21**, 73–4.

Chrétiennot, M.-J. (1973). The fine structure and toxonomy of *Platychrysis pigra* Geitler (Haptophyceae). *Journal of the Marine Biological Association of the United Kingdom*, **53**, 905–14.

Febvre, J. and Febvre-Chevalier, C. (1979). Ultrastructural study of zooxanthellae of three species of Acantharia (Protozoa: Actinopoda), with details of their taxonomic position in the Prymnesiales (Prymnesiophyceae, Hibberd, 1976). *Journal of the Marine Biological Association of the United Kingdom*, **59**, 215–26.

Fresnel, J. (1986). Nouvelles observations sur une Coccolithacée rare: *Cruciplacolithus neohelis* (McIntyre et Bé) Reinhardt (Prymnesiophyceae). *Protistiologica*, **22**, 193–204.

Fresnel, J. and Billard, C. (1991). *Pleurochrysis placolithoides* sp. nov. (Prymnesiophyceae), a new marine coccolithophorid with remarks on the status of cricolith-bearing species. *British Phycological Journal*, **26**, 67–80.

Gayral, P. and Fresnel, J. (1976). Nouvelles observations sur deux Coccolithophoracées marines: *Cricosphaera roscoffensis* (P. Dangerad) comb. nov. et *Hymenomonas globosa* (F. Magne) comb. nov. *Phycologia*, **15**, 339–55.

Gayral, P. and Fresnel, J. (1979). *Exanthemachrysis gayraliae* Lepailleur (Prymnesiophyceae, Pavlovales): Ultrastructure et discussion taxinomique. *Protistologica*, **15**, 271–82.

Gayral, P. and Fresnel, J. (1983*a*). *Platychrysis pienaarii* sp. nov. et *P. simplex* sp. nov. (Prymnesiophyceae): description et ultrastructure. *Phycologia*, **22**, 29–45.

Gayral, P. and Fresnel, J. (1983*b*). Description, sexualité et cycle de développment d'une nouvelle Coccolithophoracée (Prymnesiophyceae): *Pleurochrysis pseudoroscoffensis* sp. nov. *Protistologica*, **19**, 245–61.

Green, J. C. (1973). Studies in the fine structure and taxonomy of flagellates in the genus *Pavlova.* II. A freshwater representative, *Pavlova granifera* (Mack) comb. nov. *British Phycological Journal*, **8**, 1–12.

Green, J. C. (1975). The fine-structure and taxonomy of the haptophycean flagellate *Pavlova lutheri* (Droop) comb. nov. (= *Monochrysis lutheri* Droop). *Journal of the Marine Biological Association of the United Kingdom*, **55**, 785–93.

Green, J. C. (1976). Notes on the flagellar apparatus and taxonomy of *Pavlova mesolychnon* van der Veer, and on the status of *Pavlova* Butcher and related genera within the Haptophyceae. *Journal of the Marine Biological Association of the United Kingdom*, **56**, 595–602.

Green, J. C. (1980). The fine structure of *Pavlova pinguis* Green and a preliminary survey of the Order Pavlovales (Prymnesiophyceae). *British Phycological Journal*, **15**, 151–91.

Green, J. C. and Hibberd, D. J. (1977). The ultrastructure and taxonomy of *Diacronema vlkianum* (Prymnesiophyceae) with special reference to the haptonema and flagellar apparatus. *Journal of the Marine Biological Association of the United Kingdom*, **57**, 1125–36.

Green, J. C. and Hori, T. (1986). The ultrastructure of the flagellar root system of *Imantonia rotunda* (Prymnesiophyceae). *British Phycological Journal*, **21**, 5–18.

Green, J. C. and Hori, T. (1990). The architecture of the flagellar apparatus of *Prymnesium patellifera* (Prymnesiophyta). *Botanical Magazine* (Tokyo), **103**, 191–207.

Green, J. C. and Manton, I. (1970). Studies in the fine structure and taxonomy of flagellates in the genus *Pavlova*. I. A revision of *Pavlova gyrans*, the type species. *Journal of the Marine Biological Association of the United Kingdom*, **50**, 1113–30.

Green, J. C., Perch-Nielsen, K., and Westbroek, P. (1990). Phylum Prymnesiophyta. In *Handbook of the Protoctista*, (ed. L. Margulis *et al.*), pp. 293–317. Jones and Bartlett, Boston.

Gregson, A. J., Green, J. C., and Leadbeater, B. S. C. (1993). Structure and physiology of the haptonema in *Chrysochromulina* (Prymnesiophyceae). I. Fine structure of the flagellar/haptonematal root system in *C. acantha* and *C. simplex*. *Journal of Phycology*, **29**, 674–86.

Hori, T. and Green, J. C. (1991). The ultrastructure of the flagellar root system of *Isochrysis galbana* (Prymnesiophyta). *Journal of the Marine Biological Association of the United Kingdom*, **71**, 137–52.

Inouye, I. (1993). Flagella and flagellar apparatus of algae. In *Ultrastructure of microalgae*, (ed. T. Berner), pp. 99–133. CRC Press, Boca Raton, Florida.

Inouye, I. and Chihara, M. (1980). Laboratory culture and taxonomy of *Hymenomonas coronata* and *Ochrosphaera verrucosa* (Class Prymnesiophyceae) from the Northwest Pacific. *Botanical Magazine* (Tokyo), **93**, 195–208.

Inouye, I. and Chihara, M. (1983). Ultrastructure and taxonomy of *Jomonlithus littoralis* gen. et sp. nov. (Class Prymnesiophyceae), a coccolithophorid from the Northwest Pacific. *Botanical Magazine* (Tokyo), **96**, 365–76.

Inouye, I. and Pienaar, R. N. (1984). New observations on the coccolithophorid *Umbilicosphaera sibogae* var. *foliosa* (Prymnesiophyceae) with reference to cell covering, cell structure and flagellar apparatus. *British Phycological Journal*, **19**, 357–69.

Inouye, I. and Pienaar, R. N. (1985). Ultrastructure of the flagellar apparatus in *Pleurochrysis* (Class Prymnesiophyceae). *Protoplasma*, **125**, 24–35.

Inouye, I. and Pienaar, R. N. (1988). Light and electron microscope observations of the type species of *Syracosphaera, S. pulchra* (Prymnesiophyceae). *British Phycological Journal*, **23**, 205–17.

Johansen, J. R., Doucette, G. J., Barclay, W. R., and Bull, J. D. (1988). The morphology and ecology of *Pleurochrysis carterae* var. *dentata* var. nov. (Prymnesiophyceae), a new coccolithophorid from an inland saline pond in New Mexico, USA. *Phycologia*, **27**, 78–88.

Klaveness, D. (1972). *Coccolithus huxleyi* (Lohm.) Kamptn. II. The flagellate cell, aberrant cell types, vegetative propagation and life-cycles. *British Phycological Journal*, **7**, 309–18.

Klaveness, D. (1973). The microanatomy of *Calyptrosphaera sphaeroidea*, with some supplementary observations on the motile stage of *Coccolithus pelagicus*. *Norwegian Journal of Botany*, **20**, 151–62.

Manton, I. (1964*a*). The possible significance of some flagellar bases in plants. *Journal of the Royal Microscopical Society*, **82**, 279–85.

Manton, I. (1964*b*). Further observations on the fine structure of the haptonema in *Prymnesium parvum*. *Archiv für Mikrobiologie*, **49**, 315–30.

Manton, I. (1965). Some phyletic implications of flagellar structure in plants. *Advances in Botanical Research*, **2**, 1–34.

Manton, I. (1966). Observations on scale production in *Prymnesium parvum*. *Journal of Cell Science*, **1**, 375–80.

Manton, I. (1968). Further observations on the microanatomy of the haptonema in *Chrysochromulina chiton* and *Prymnesium parvum*. *Protoplasma*, **66**, 35–53.

Manton, I. and Leedale, G. F. (1963). Observations on the fine structure of *Prymnesium parvum* Carter. *Archiv für Mikrobiologie*, **45**, 285–303.

Manton, I. and Peterfi, L.S. (1969). Observations on the fine structure of coccoliths, scales and the protoplasts of a freshwater coccolithophorid, *Hymenomonas roseola* Stein, with supplementary observations on the protoplast of *Cricosphaera carterae*. *Proceedings of the Royal Society*, Series B, **172**, 1–15.

Moestrup, Ø. (1982). Flagellar structure in algae: a review, with new observations particularly on the Chrysophyceae, Phaeophyceae (Fucophyceae), Euglenophyceae, and *Reckertia*. *Phycologia*, **21**, 427–528.

Moestrup, Ø. and Thomsen, H. A. (1986). Ultrastructure and reconstruction of the flagellar apparatus in *Chrysochromulina apheles* sp. nov. (Prymnesiophyceae = Haptophyceae). *Canadian Journal of Botany*, **64**, 593–610.

Pienaar, R. N. (1976). The microanatomy of *Hymenomonas lacuna* sp. nov. (Haptophyceae). *Journal of the Marine Biological Association of the United Kingdom*, **56**, 1–11.

Preisig, H. R. (1989). The flagellar base ultrastructure and phylogeny of chromophytes. In *The chromophyte algae: problems and perspectives*, Systematics

Association Special Volume No. 38, (ed. J. C. Green *et al.*), pp. 167–87. Clarendon Press, Oxford.

Roberts, K. R. and Mills, J. T. (1992). The flagellar apparatus of *Hymenomonas coronata* (Prymnesiophyta). *Journal of Phycology*, **28**, 635–42.

Veer, J. van der (1976). *Pavlova calceolata* (Haptophyceae), a new species from the Tamar estuary, Cornwall, England. *Journal of the Marine Biological Association of the United Kingdom*, **56**, 21–30.

Veer, J. van der and Leewis, R. J. (1977). *Pavlova ennorea* sp. nov., a haptophycean alga with a dominant palmelloid phase, from England. *Acta Botanica Neerlandica*, **26**, 159–76.

4. The haptonema

ISAO INOUYE and MASANOBU KAWACHI
Institute of Biological Sciences, University of Tsukuba, Japan

Abstract

Morphological, behavioural, and physiological features of the haptonema are reviewed and their biological significance is discussed. The haptonematal components are arranged in an absolute configuration in relation to the flagellar apparatus and cellular organization, and the direction of bending and coiling is also fixed. Coiling is a Ca^{2+}-mediated phenomenon. The adhesive property of the haptonema is probably related to the presence of concanavalin A-binding substances and/or electric charge associated with the haptonematal surface. Haptonematal functions are provisionally proposed.

Introduction

The haptonema is a thread-like organelle that superficially resembles the flagellum. This unique organelle was known as 'the third flagellum' until Parke *et al.* (1955) recognized it as being distinct from flagella and gave it the name 'haptonema'. It has so far been found only in haptophytes.

The fundamental properties of the haptonema were recognized in the period from 1950s to early 1970s mainly by British phycologists (for bibliography, see Hibberd 1980). They found that: (1) The axoneme of the haptonema comprises several singlet microtubules, typically six or seven, which are surrounded by a cisterna of the endoplasmic reticulum (ER) and the plasmalemma; (2) The haptonema can adheres to suitable substrata and show gliding motility; (3) In some species the haptonema coils spontaneously or when physically stimulated, and on chemical fixation and cell death; (4) The haptonema can bend, but it does not beat like a flagellum. All these features were comprehensively reviewed by Hibberd (1980). Among the many early articles published the following are the most important. Manton (1964, 1968) investigated the ultrastructure of the haptonema in detail and Leadbeater (1971*a*) investigated

The Haptophyte Algae (ed. J. C. Green and B. S. C. Leadbeater), Systematics Association Special Volume No. 51, pp. 73–89. Clarendon Press, Oxford, 1994.

the behaviour of the haptonema using ciné-microscopy. Major descriptions of this organelle appearing in present-day text books of phycology are based on these pioneering works. However, investigations on this organelle have made slow progress during the past twenty years.

In this chapter, we will review morphological, behavioural, and physiological features of this unique organelle, incorporating data recently obtained in our laboratory (Kawachi and Inouye 1994; Kawachi *et al.* 1994).

Morphology

In many members of the Prymnesiales, the haptonema is well developed and functional, that is, it is capable of coiling and adhesion to external substrata. It is particularly long and conspicuous in species of the genus *Chrysochromulina*, e.g. up to 160 μm long in *C. camella* (Leadbeater 1971*a*) (Figs 4.18, 4.26). The long and functional haptonema, also, is widely distributed in the coccolithophorids.

The number of microtubules is fixed for a particular species, but there are variations between species. The usual number is six or seven (Fig. 4.1), but in an undescribed haptophyte we have recently discovered, the haptonema contains fourteen microtubules (Fig. 4.3). The ultrastructure of the haptonematal tip has been investigated by Manton (1964, 1968). It is normally flat-ended, but tapered, pointed or spatulate forms also occur (Manton 1964). Manton and Leedale (1963) and Leadbeater (1971*a*) demonstrated the presence of a swelling that moves up and down the haptonema. The morphological basis of this dynamic structure has not been investigated.

In the free part of the haptonema, the shaft, microtubules are arranged usually in a ring surrounded by three concentric membranes (Fig. 4.1), the inner two of which are of ER that is continuous with the peripheral ER of the cell. The haptonematal ER also shows a C-shaped profile in transverse section, where it is linked to the plasma membrane by 'bridges' (Manton 1964). The cytoplasmic core contains ribosomes, but there has been no report of any other constituents making up the haptonematal shaft. However, in many species, an electron dense structure is present at the centre of microtubular ring, connecting singlet microtubules to one

Figs 4.1–4.8 (**1**) Transverse section of a haptonema comprising seven singlet microtubules surrounded by three membranes (*Chrysochromulina hirta*) (× 110 000). (**2**) Electron dense structure connecting microtubules and dense dots (*arrows*) situated on the outer membrane of the ER (*Prymnesium parvum*) (× 120 000). (**3**) Haptonema of an undescribed species including fourteen electron dense microtubules (× 48 000). (**4**) Longitudinal section of a haptonema showing the basal

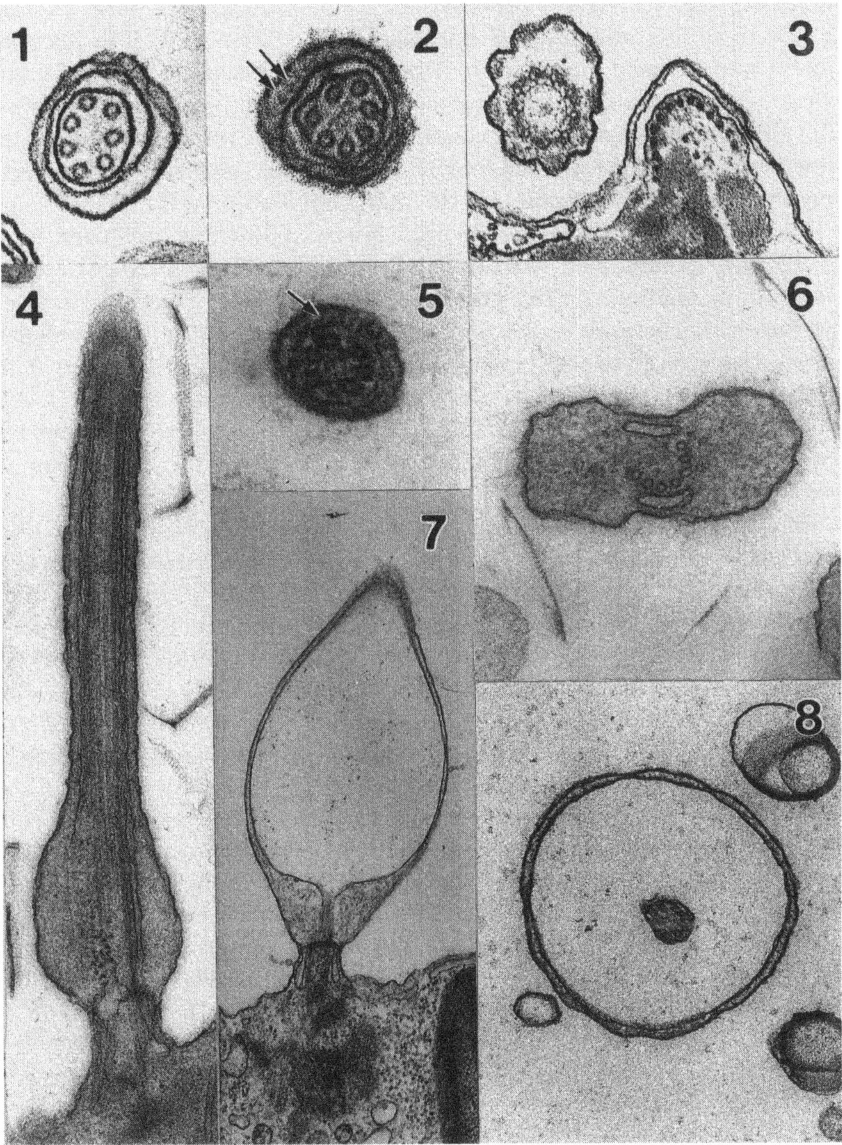

swelling and the ER that lies along the haptonematal membrane (*Chrysochromulina alifera*) (× 39 000). (**5**) Dense bars (*arrow*) connecting the ER and haptonematal membrane (*Prymnesium parvum*) (× 116 000). (**6**) Transverse section of the basal swelling of *Chrysochromulina alifera* showing an arc of microtubules and reduced ER cavity embedded in the cytoplasmic core (× 65 000). (**7, 8**) Basal swelling of *Chrysochromulina parkeae*. (**7**) Longitudinal section (× 27 000). (**8**) Transverse section (× 50 000).

another (Fig. 4.2) (see Gregson *et al.* 1993). Similar structures can be found in the haptonemata of *Chrysochromulina chiton.* (Fig. 3 in Manton 1967) and *C. apheles* (Fig. 22 in Moestrup and Thomsen 1986). In the space between the haptonematal membrane and the outer membrane of the ER, there are some structures which appear in transverse sections as dots situated on the outer ER membrane (Fig. 4.2) or dense bars connecting the outer ER membrane and the plasmamembrane (Fig. 4.5).

Before entering into the cell, the microtubules are organized in a trough-like arrangement surrounded by a ring of ER, showing a C- or arc-like arrangement in transverse section (Hibberd 1980). The configuration of these microtubules in relation to the flagellar apparatus and cellular organelles is fixed so that only one absolute configuration is present. The concave side of the microtubular trough faces one of the flagella that is referred to as the left flagellum (Inouye and Pienaar 1985). The microtubular configuration at successive vertical levels of the haptonema is shown in Fig. 4.18.

In close proximity to the base, the haptonema is often swollen (Fig. 4.4), though swelling is absent in some species, e.g. *Prymnesium parvum* (Manton 1964). The cytoplasm of this region is more electron dense than in the distal region and the microtubular configuration takes the form of an arc or half circle (Figs 4.6, and 4.18). The ER is considerably reduced and fragmented into two portions: one portion lies along the convex side of the microtubular arc and the other along the opposite side of the haptonema (Fig. 4.6). As far as we have investigated, the basal swelling is most conspicuous in *Chrysochromulina parkeae* (Figs 4.7, 4.8).

The free part of the haptonema, slightly distal to its entry into the cell, is morphologically unique (Hibberd 1980). The ER forms a finger-like process with a flat end extending towards the concave side of the microtubular arc (Figs 4.9, 4.16). The cytoplasmic core is more electron dense than that at any other level so that, in longitudinal sections, it is seen as a transverse septum connecting the flat end of the finger-like process and ER membrane of the opposite side (Fig. 4.16). This region of the haptonema is called the transition region. The haptonema is often detached around this level when it is discarded in response to physical or

Figs 4.9–4.17 (**9**) Transverse section of the transition region of the haptonema of *Prymnesium parvum* showing the finger-like process of ER (*arrow*) (× 113 000). (**10**) Transverse section of the haptonema of *Chrysochromulina alifera* at the level of entry into the cell. Note the arc-like arrangement of microtubules as in Figs 4.6 and 4.9 (× 110 000). (**11–15**) Transverse sections of the haptonema at successive levels from distal to proximal (*Chrysochromulina hirta*). Note the configurational change and addition of new microtubules (numbered 8 and 9). Dense material arising from the

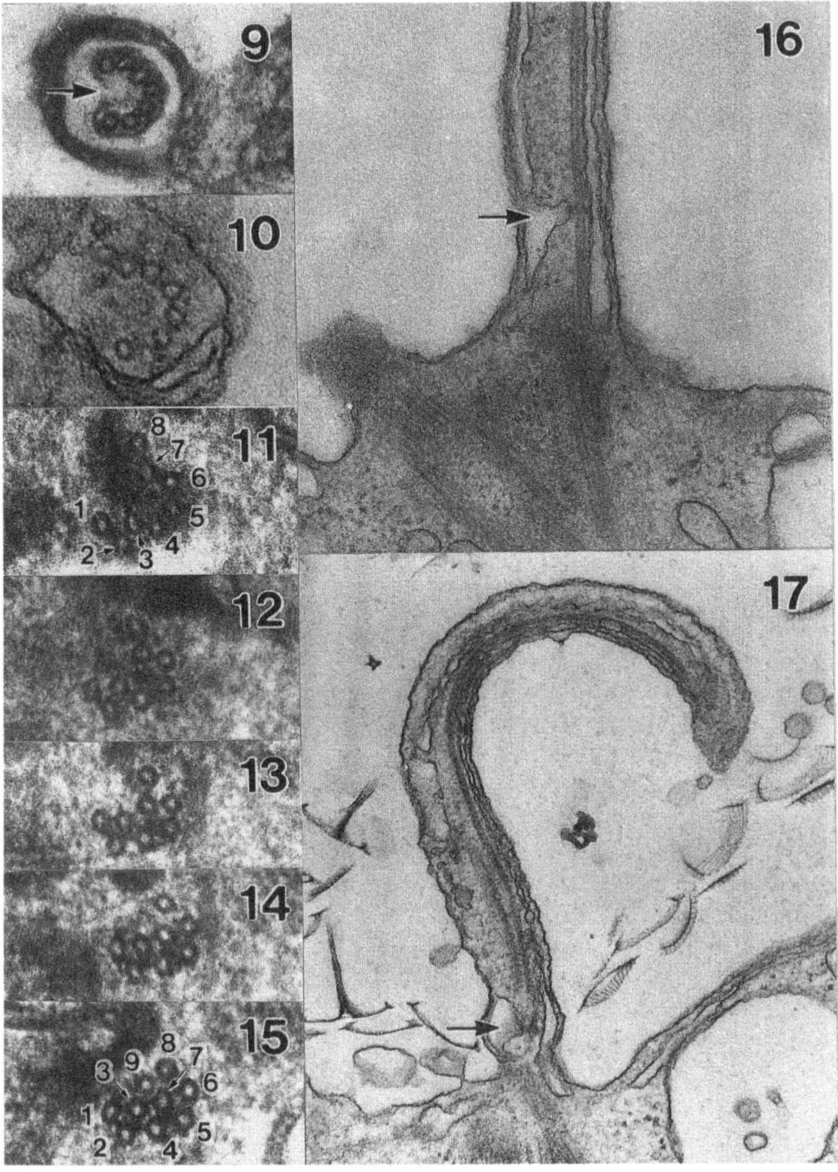

centre of the microtubular arc is seen in 11 and 12. Numbering is after Moestrup and Thomsen (1986) (× 16 000). (**16**) Longitudinal section of the transition region (*Chrysochromulina* sp.). Note finger-like process of ER (*arrow*) and the transverse septum (× 50 000). (**17**) Longitudinal section of a coiled haptonema (*Chrysochromulina alifera*). Direction of coiling is towards the convex side of the microtubular arc. *Arrow* indicates the finger-like process of ER (× 41 000).

chemical stimuli. Manton (1964) suggested that this special region acts as a 'locking device' of the extended organelle.

The haptonema enters the cell at a definite position in relation to the two flagella (see Green and Hori, Chapter 3). Microtubules extend deep into the cytosol showing configurational changes (Figs 4.10–4.15) which have been described in detail in early publications of Manton (1964), Leadbeater and Manton (1969), and recently by Moestrup and Thomsen (1986). An example of a haptonematal base, that of *Chrysochromulina hirta*, is given in Figs 4.11–4.15, and represented diagrammatically in Fig. 4.18. Two microtubules are added one by one in sequence resulting in nine at the most proximal end. The microtubular arc becomes rearranged to show a characteristic profile, usually referred to as a zig-zag (Figs 4.11, 4.12). This comprises two superposed arcs each made up of four microtubules (Fig. 4.14). Immediately above the level of the distal connecting fibre associated with the basal bodies, electron dense material is present at the centre of the microtubular arc (Figs 4.11, 4.12). This material extends over and beyond the connecting fibre (not shown). In the most proximal region nine microtubules are arranged in such a way that every three adjacent microtubules form an equilateral triangle, and seven of these are hexagonally packed (Fig. 4.15). The haptonema

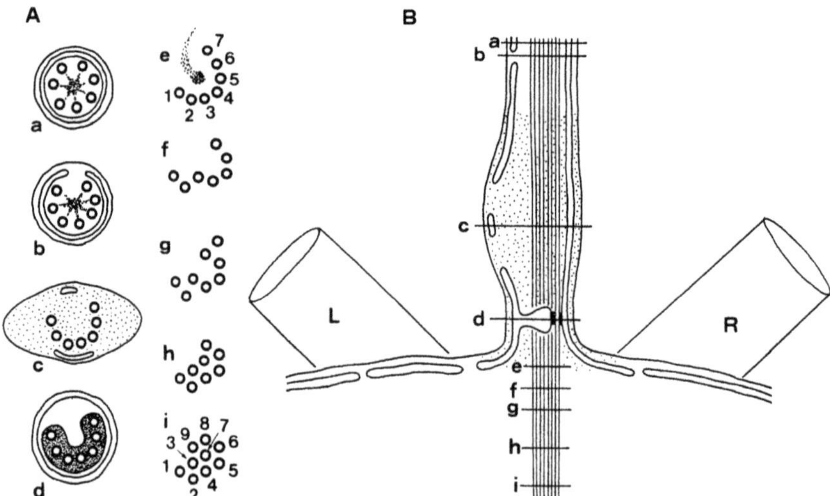

Figs 4.18 Diagrammatic representation of the structure of a typical haptonema. (A) Transverse sections of a haptonema at successive levels from distal to proximal ends. (B) Longitudinal section of a haptomena (proximal part); (a)–(i) in (A) correspond to (a)–(i) in (B). L = left flagellum; R = right flagellum. Numbering of microtubules after Moestrup and Thomsen (1986). Not to scale.

terminates near the left basal body. At various vertical levels the haptonema is linked to both basal bodies by several fibrous bands. A fibrous haptonematal root arising from the most proximal end of the haptonema and dispersing inside the cell, is known only in the Pavlovales (Green and Hibberd 1977; Green 1980).

The length of the haptonema varies at different stages of cell cycle. It shortens during replication of the flagellar-haptonematal complex so that it is very short in cells that have just completed cell division (Manton and Leedale 1963; Inouye and Pienaar 1988). The haptonema increases its length in several hours after cell division. Regeneration of the haptonema was studied experimentally (Leadbeater 1971*a*; Gregson *et al.* 1993). After excision by bubbling air, the haptonema regenerated and recovered its normal length in 12–15 h.

The haptonema is rudimentary or even entirely absent in the Isochrysidales and some coccolithophorids. *Imantonia rotunda, Dicrateria inornata,* and motile cells of *Emiliania huxleyi* lack haptonematal microtubules (Klaveness 1972; Green and Pienaar 1977). Other species have a few microtubules in the shaft; three in *Chrysotila lamellosa* and *C. stipitata* (Green and Parke 1974, 1975) and five in *Isochrysis galbana* (Hori and Green 1991). Some species of *Pleurochrysis* and *Hymenomonas* have short and bulbous haptonemata containing up to six microtubules surrounded by irregularly arranged ER (Manton and Peterfi 1969; Leadbeater 1971*b*; Inouye and Pienaar 1985). The haptonema is usually rudimentary in the Pavlovales being attenuated and containing only one to four microtubules, though the number increases inside the cell to seven or eight (e.g. Green 1976; Green and Hibberd 1977; Gayral and Fresnel 1979).

Behaviour and physiology

Adhesion and translocation of external particles

The haptonema is capable of coiling, bending, and of adhesion to external substrata. These behavioural features have been demonstrated by Leadbeater (1971*a*) with excellent ciné-photographs. However, the biological and functional significance of these movements has long been a puzzle. The functional significance of the adhering nature of the haptonema was recently demonstrated by Kawachi *et al.* (1991). In the mixotrophic species *Chrysochromulina hirta,* the haptonema plays a primary role in food capture. Prey particles adhere to the haptonema whilst the cell swims with the haptonema projecting ahead and with the flagella beating alongside the cell body (forward swimming) (Fig. 4.19). The adhering particles are translocated down to a particular point on the haptonema

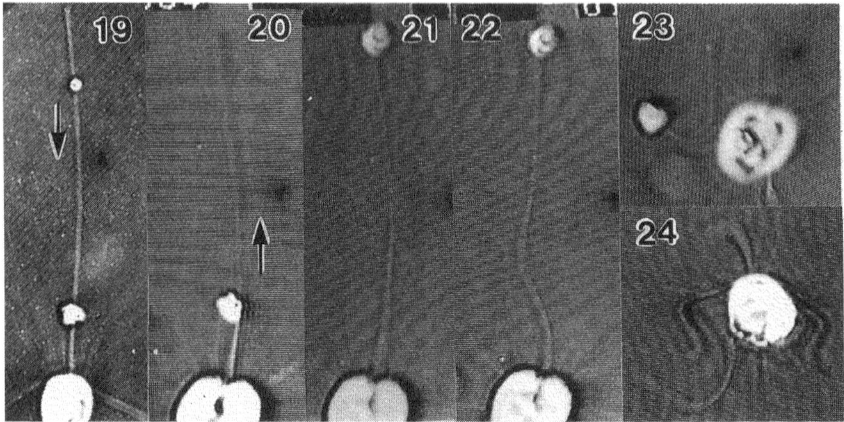

Figs 4.19–4.24 (**19–23**) High-speed video images of prey capture in *Chrysochromulina hirta*. (**19**) Adhesion and translocation of a particle and formation of aggregate at the particle aggregating centre (PAC) (× 1200). (**20, 21**) Upward translocation of an aggregate from the PAC to the tip (× 1600). (**22, 23**) Delivery of the aggregate to the cell surface by bending of the haptonema (× 1600). *Arrows* indicate the direction of translocation of particles. (**24**) Cell bending haptonema without captured particle (× 1600).

about 2 μm distal from the base, called the particle aggregating centre (PAC), and accumulate there resulting in the development of a massive aggregate (Figs 4.19, 4.20). Morphologically, the PAC probably corresponds to the basal swelling. In the aggregate, individual particles tightly adhere to one another suggesting that some cement material is secreted at the PAC. Once formed, the aggregate moves upwards and reaches the haptonematal tip (Figs 4.20, 4.21). The haptonema bends into a sigmoid shape soon after the aggregate arrives at the tip (Fig. 4.22) and eventually delivers it to the cell surface at the posterior end of the cell where it is ingested into a food vacuole (Fig. 4.23). Bending of the haptonema occurs even in cells that do not translocate aggregates of prey (Fig. 4.24). It is worth noting that the cell ceases flagellar beating just before the haptonema starts to bend, and that the flagella show sudden and vigorous activity when the aggregate is detached from the haptonema. These phenomena suggest that haptonematal behaviour is precisely regulated in an integrated functional system, the flagellar-haptonematal complex. Close functional association between the haptonema and flagella may also occur during the process of coiling (see below).

The direction of bending during aggregate transport to the cell surface is fixed in relation to cellular organization. This was confirmed using *Chrysochromulina spinifera*, another species that uses the haptonema

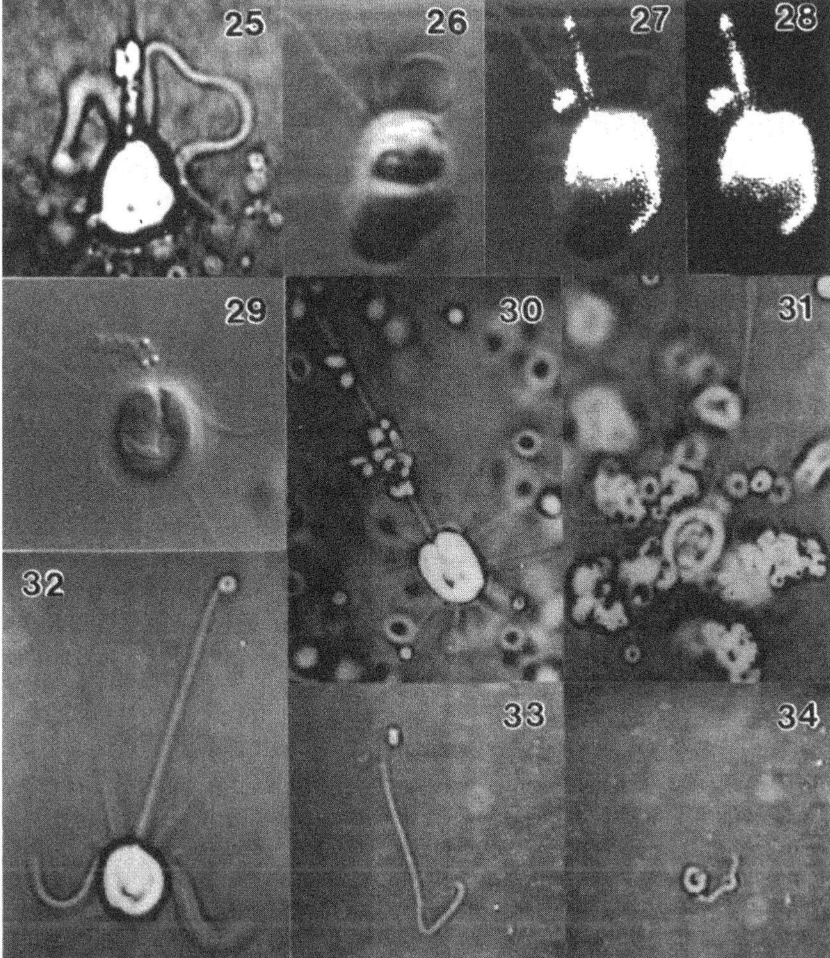

Fig. 4.25–4.34 (**25**) Video image of *Chrysochromulina spinifera*, showing two unequal flagella and captured microspheres on the haptonema (× 2000). (**26–28**) Confocal laser micrograph of *Chrysochromulina spinifera*. Fluorescence of FITC-labelled concanavalin A is localized on the haptonema. (**26**) Phase contrast; (**27**) superimposed micrograph of phase contrast and fluorescence images; (**28**) fluorescence image. (× 2500). (**29**) *Chrysochromulina hirta* with the haptonema tightly coiled. Spine scales are also seen (× 1800). (**30, 31**) Video images of *Chrysochromulina hirta*: (**30**) cell capturing bacteria with haptonema; (**31**) cell with amidine-coated latex microspheres attached to the cell body and spine scales (× 1200). (**32**) *Chrysochromulina hirta* cell treated with procaine. The haptonema is extended, but partly coiled at the tip (× 1800). (**33, 34**) Detached haptonema of *Chrysochromulina hirta* treated with procaine (**33**) and followed by A23187 (**34**). Note induction of coiling (× 1800). (Figs 4.32–4.34 from Kawachi and Inouye 1994)

as a food capturing device. This species has two unequal flagella which can be used as positional markers (Fig. 4.25). The haptonema always bends passing between the two flagella, closer to the longer flagellum (Fig. 4.39A). The longer flagellum corresponds to the left flagellum/basal body and the food vacuole is located some distance below it (Fig. 39A). The concave side of trough-like microtubules faces the left basal body and the direction of haptonematal bending is therefore interpreted as approximately toward the open side of the trough (Fig. 4.39A).

There is no doubt that the haptonematal surface membrane is a dynamic structure. However, it is presently unknown what kind of mechanisms are involved in the adhesion and translocation of prey particles and what kind of molecules might be responsible for these phenomena. The adhesive nature of the flagellar surface has been studied in *Chlamydomonas*. *Chlamydomonas* cells attach to solid or semisolid surfaces by means of their flagella and the whole cell can move on the substrate. This phenomenon is referred to as gliding motility or flagellar surface motility. Bloodgood (1990) reviewed flagellar surface motility and listed questions that should be answered in future studies. The following questions may also be applicable to the haptonema: (1) What haptonematal surface components are responsible for the mechanical force required to adhere to external particles or other substrate? (2) What is the motor molecule responsible for generating movement of external particles? (3) How is the haptonematal surface motility system regulated? (4) Why does prey particle adhesion induce translocation instead of coiling of the haptonema?

In *Chlamydomonas*, it has been revealed that glycoproteins are involved in adhesion of the flagella (Bloodgood 1990). It is possible that similar substances are involved in haptonematal adhesion and surface motility. In order to examine this possibility, we added FITC-labelled concanavalin A (Co A) that specifically binds α-D-glucose and α-D-mannose into the medium. It is interesting that, in the case of *Chrysochromulina spinifera*, fluorescence was detected only on the haptonema (Figs 4.26–4.28). It should be noted that no fluorescence was observed on the flagella of *C. spinifera*, while *Chlamydomonas* flagella showed bright fluorescence under the same conditions. Results so far obtained indicate that the haptonematal membrane contains glycoproteins or glycolipids that bind with Co A, and that this membrane is distinct from the plasma membrane of both the flagella and cell body. This suggests that the haptonematal membrane forms a distinct membrane domain on *Chrysochromulina* cells.

Another possible force that could generate adhesion is an electric charge associated with the haptonematal membrane. We investigated this possibility using electrically charged latex microspheres. When we added microspheres with a negative charge (carboxylate or albumin-coated) into the culture of *Chrysochromulina hirta*, microspheres attached to the haptonema immediately. A similar effect was also found when bacteria were used (Fig. 4.30). The cell surface of bacteria is believed to be negatively charged. In contrast, particles with a positive charge, such as amidine-coated microspheres, attached to spine scales that cover the cell body but not to the haptonema (Fig. 4.31). This suggests that the haptonematal surface of *C. hirta* has a positive charge, and again supports the suggestion that the haptonematal membrane forms a distinct domain. We therefore postulate that an electrical charge may be, at least partly, responsible for adhesion. It is generally believed that the cell membrane is negatively charged so that if our interpretation is correct, the haptonematal membrane is very unusual. More detailed physiological studies are needed to substantiate this hypothesis.

Coiling

The second prominent feature of haptonematal behaviour is coiling and uncoiling. If the haptonema is well developed and relatively long, coiling is a universal feature (Fig. 4.29), although some species coil their haptonemata more readily than others (Leadbeater 1971*a*). Manton (1964, 1968) put forward a 'turgor hypothesis' as a possible mechanism for the coiling and uncoiling of haptonemata. She postulated that coiling is induced by a change of turgor pressure in the ER cavity, and that a gradual increase of internal pressure built up to produce uncoiling of the haptonema, whereas a quick decrease results in coiling. There have been no data supporting this fascinating idea and , at present, there are no data to explain how mechanical forces induce coiling and uncoiling.

Coiling is frequently observed when cells are swimming rapidly, with flagella beating behind the cell (backward swimming). Many species undergo forward swimming as an ordinary mode of locomotion with the haptonema projecting ahead. In such species, rapid backward swimming occurs when cell collide with obstacles. On contact, cells instantly coil their haptonemata, undergo a spinning movement, reverse the direction and increase the speed of the flagellar heat. Cells swim away from obstacles in this way, implying that coiling produces an immediate avoiding response. It is postulated, therefore, that one of the functions of the haptonema is obstacle sensing, at least in species that usually swim forward with the haptonema directed ahead (see also Gregson *et al.* 1993). When spontaneously triggered, coiling is accompanied by curling of the flagella (Leadbeater 1971*a*), suggesting close functional association between the haptonema and flagella.

Coiling is an extremely rapid process, taking 1/100 s to 1/60 s (Leadbeater 1971*a*). It is even less than 1/200 s when conditions are favourable (unpublished observations). When compared with coiling,

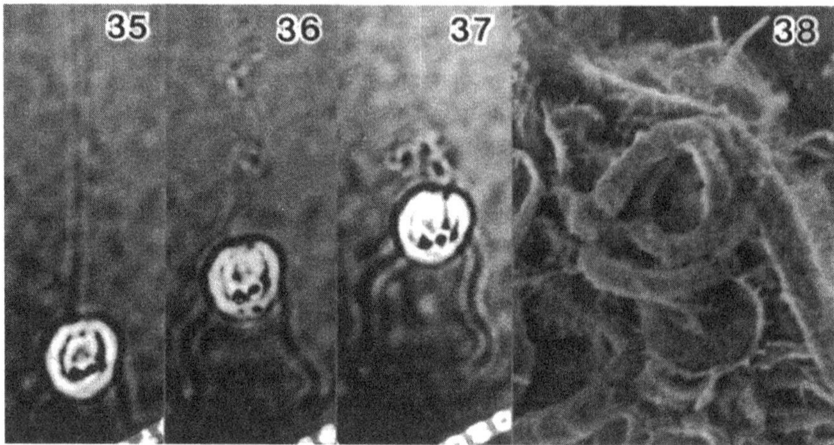

Figs 4.35–4.38 (**35–37**) Sequential images of coiling in *Chrysochromulina parva.* Note that coiling starts at several points (× 2000). (**38**) Scanning micrograph of *Chrysochromulina* sp., the haptonema coiled in a clockwise helix (× 25 000).

uncoiling is a much slower process and usually takes several seconds. It begins from the base and proceeds towards the tip. Since the haptonema coils when cells are chemically fixed or dried, it has been suggested that coiling is the low-energy state (Estep and MacIntyre 1989), and thus uncoiling is the energy-dependent form. It is possible to make haptonemata repeat coiling and uncoiling by repeating physical stimuli, such as tapping the microscope. However, repeated shock results in gradual loss of sensitivity (Leadbeater 1971*a*). Coiling starts most frequently from the distal tip and proceeds toward the base, but it may also begin at several points along the length (Figs 4.35–4.37).

The direction of coiling is fixed with respect to the configuration of the haptonematal microtubules and cell organization. Detailed observation of coiling was made by Leadbeater (1971*a*) using high-speed ciné-microscopy. He found that the helix was formed on the side towards the cell, i.e. when viewed from the ventral side, the helix is formed at the left side of the haptonema (Fig. 4.39B, C). The helix of the haptonema is formed in a precise pattern. Our observation show that clockwise helices are always formed (Figs 4.38, 4.39). The direction of coiling is thus opposite to the bending that occurs during prey transport, i.e., towards the convex side of the microtubular arc of the haptonema (Fig. 4.39A). In the helix, the microtubules are arranged to form an arc with its concave side facing towards the axis of the helix, whereas microtubules are arranged in a circle when the haptonema is in an extended condition (Gregson *et al.* 1993).

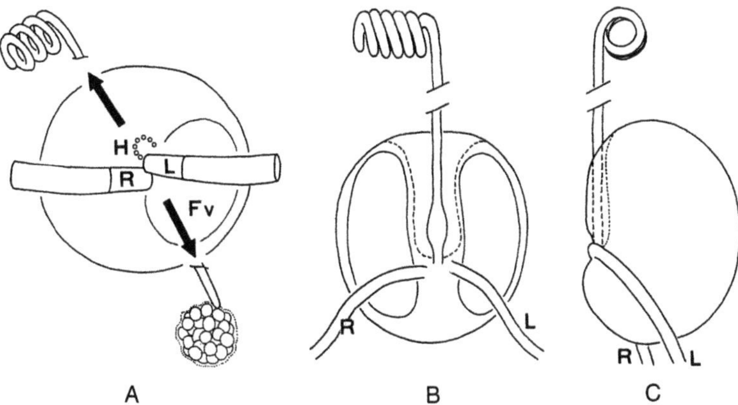

Fig. 4.39 Diagrammatic representation of haptonemata showing the directions of coiling and bending for the transport of aggregates. (**A**) *Chrysochromulina* cell with anterior flagella and haptonema. Coiling is towards the convex side of the microtubular arc of the haptonema, and bending is towards the opposite side. (**B, C**) Ventral and lateral views of *Chrysochromulina* cell with laterally inserted flagella and haptonema. Coiling occurs towards the dorsal side of the cell and into a clockwise helix. Fv = food vacuole; H = haptonema; L = left flagellum; R = right flagellum. Not to scale.

Since both food capturing and coiling are associated with particular movement of flagella as mentioned earlier, it is postulated that haptonematal behaviour is affected by the Ca^{2+} concentration in the haptonematal lumen as for the change of flagellar beat pattern of other organisms. To test this hypothesis, several experiments were designed using *Chrysochromulina hirta* (Kawachi and Inouye 1994). We first used the Ca^{2+} ionophore A23187, which generates channels in the cell membrane through which Ca^{2+} ions can pass. Since the Ca^{2+} concentration inside the cell is normally kept lower than that outside, the occurrence of such channels probably causes an influx of Ca^{2+} ions. When A23187 was added to the medium, coiling was induced in most cells (Fig. 4.29). In another experiment, we added EGTA to the medium to reduce the Ca^{2+} concentration. In such a condition, coiling was suppressed and the haptonema was kept uncoiled even if cells were physically stimulated by tapping or chemical fixation. These two experiments indicate that coiling is a Ca^{2+}-mediated phenomenon, and is caused by the influx of Ca^{2+} from outside the cell. Similar experiments were undertaken by Gregson *et al.* (1993) using EGTA and lanthanum chloride, a calcium channel blocker. They estimated that the threshold concentration of extracellular Ca^{2+} ions for haptonematal coiling to occur lies between 10^{-7}M and 10^{-6}M.

It has been found that, in muscle cells, the ER is used as a Ca^{2+} pool and Ca^{2+} ions released from it induce muscular contraction (Iino 1989). The ER that surrounds haptonematal microtubules may, therefore, be another possible source of Ca^{2+} supply. It is also known that caffeine induces Ca^{2+}-induced-Ca^{2+}-release (CICR) from the ER pool of muscle cells, and that procaine inhibits it (Iino 1989). When caffeine was added to the medium, 100% of cells coiled their haptonemata. In contrast, the coiling was suppressed by adding procaine to the medium. Haptonemata did not coil at all, or ceased coiling at an intermediate stage so that many cells with partly coiled haptonemata were observed (Fig. 4.32). These observations suggest that the ER is also involved in the coiling mechanism and may serve as a Ca^{2+} reservoir. When cells pre-treated by procaine (so that coiling of the haptonemata was inhibited by interruption of the Ca^{2+} supply) were exposed to A23187 the haptonemata coiled. Probably Ca^{2+} ion influx from the outside medium induced coiling. Figures 4.33 and 4.34 are such an example of detached haptonemata.

It is obvious from the above experiments that the haptonema is a highly Ca^{2+}-dependent organelle. Perhaps the change of direction and beating frequency of flagella in prymnesiophyte cells is also due to the influx of Ca^{2+} ions into the flagella and their basal regions, as this is one of the universal features of flagella of many flagellate organisms and gametic cells. With these observations in mind, we think coiling is the first phenomenon of an avoiding response. Gregson *et al.* (1993) reached the same conclusion.

The mechanical force which generates coiling and/or bending of the haptonema is presently unknown. No dynein-like structure is associated with microtubules. Gregson *et al.* (1993) have demonstrated structures that connect adjacent haptonematal microtubules to one another and suggested that a sliding mechanism may be involved in coiling.

Concluding remarks

There is no doubt that the haptonema is a more complex organelle in both morphological and physiological aspects than has previously been thought. Taking into account observations previously made by many phycologists and those we have made, we provisionally suggest haptonematal functions as follows: (1) The haptonema is a sensory device for detecting obstacles and potentially unsuitable external environments, and coiling is part of an avoiding response that generates rapid backward swimming; (2) It is used as a device for capturing, aggregating, and transporting prey. The adhesive property also makes it possible for cells to attach to an external substrate. There seem to be at least two different processes that form the physiological basis of these functions, i.e. a Ca^{2+}-mediated response and haptonematal surface motility. All functions seem

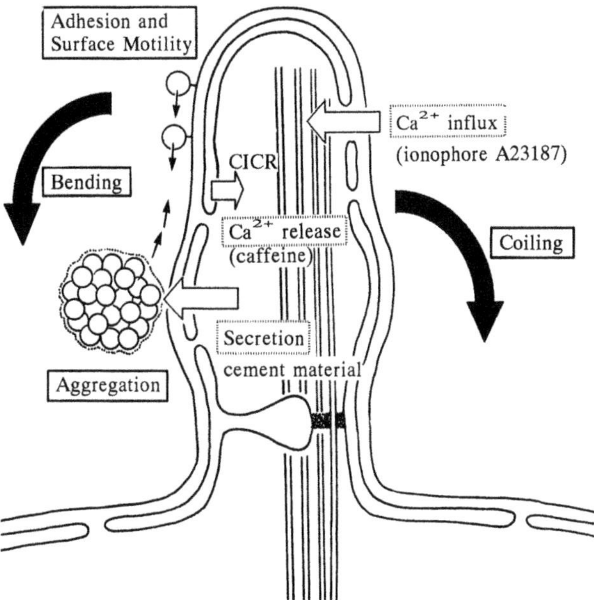

Fig. 4.40 Diagrammatic representation of various phenomena so far recognized in the haptonema (*boxed in solid lines*) and the possible physiological events (*boxed in dotted lines*) that allow these functions. Influx of Ca^{2+} ions from outside or by calcium-induced-calcium-release (CICR) from the ER may induce coiling. Aggregation of prey particles may be realized by the secretion of cement material around the basal swelling. Concanavalin A-binding substances or electrical charges on the haptonematal surface may be responsible for the adhesion of external particles or to a substrate. Motor molecules that generate surface motility and the mechanical forces for bending are presently unknown. *Small arrows* indicate the direction of surface motility. Not to scale.

to be precisely regulated in an integrated flagellar-haptonematal system. Possible haptonematal functions are summarized in Fig. 4.40.

The haptonema is an organelle unique to the Haptophyta and no homologous structure has ever been observed in other organisms. It is therefore of considerable interest to know 'What is the origin of the haptonema?' Our understanding on this organelle is, however, still at an initial stage and we have no information to answer this question (but see Cavalier-Smith, Chapter 22) . Many problems remain to be solved before we start to discuss the origin of the haptonema and its evolutionary and phylogenetic significance. Investigations into the physiological and molecular mechanisms of the flagellar-haptonematal complex may be very important in this respect.

References

Bloodgood, R. A. (1990). Gliding motility and flagellar glycoprotein dynamics in *Chlamydomonas*. In *Ciliary and flagellar membranes*, (ed. R. A. Bloodgood), pp. 91–128. Plenum Press, New York.

Estep, K. W. and MacIntyre, F. (1989). Taxonomy, life cycle, distribution and dasmotrophy of *Chrysochromulina*: a theory accounting for scales, haptonema, muciferous bodies and toxicity. *Marine Ecology Progress Series*, **57**, 11–21.

Gayral, P. and Fresnel , J. (1979). *Exanthemachrysis gayraliae* Lepailleur (Prymnesiophyceae, Pavlovales): ultrastructure et discussion taxinomique. *Protistologica*, **15**, 271–82.

Green, J. C. (1976). Notes on the flagellar apparatus and taxonomy of *Pavlova mesolychnon* van der Veer, and on the status of *Pavlova* Butcher and related genera within the Haptophyceae. *Journal of the Marine Biological Association of the United Kingdom*, **56**, 595–602.

Green, J. C. (1980). The fine structure of *Pavlova pinguis* Green and a preliminary survey of the order Pavlovales (Prymnesiophyceae). *British Phycological Journal*, **15**, 151–91.

Green, J. C. and Hibberd, D. J. (1977). The ultrastructure and taxonomy of *Diacronema vlkianum* (Prymnesiophyceae) with special reference to the haptonema and flagellar apparatus. *Journal of the Marine Biological Association of the United Kingdom*, **57**, 1125–36.

Green, J. C. and Parke, M. (1974). A re-investigation by light and electron microscopy of *Ruttnera spectabilis* Geitler (Haptophyceae), with special reference to the fine structure of zoids. *Journal of the Marine Biological Association of the United Kingdom*, **54**, 539–50.

Green, J. C. and Parke, M. (1975). New observations upon members of the genus *Chrysotila* Anand, with remarks upon their relationships within the Haptophyceae. *Journal of the Marine Biological Association of the United Kingdom*, **55**, 109–21.

Green, J. C. and Pienaar, R. N. (1977). The taxonomy of the order Isochrysidales (Prymnesiophyceae) with special reference to the genera *Isochrysis* Parke, *Dicrateria* Parke and *Imantonia* Reynolds. *Journal of the Marine Biological Association of the United Kingdom*, **57**, 7–17.

Gregson, A. J., Green, J. C., and Leadbeater, B. S. C. (1993). Structure and physiology of the haptonema in *Chrysochromulina* (Prymnesiophyceae). II. Mechanisms of haptonematal coiling and the regeneration process. *Journal of Phycology*, **29**, 686–700.

Hibberd, D. J. (1980). Prymnesiophytes (=Haptophytes). In *Developments in marine biology*, Vol. 2, *Phytoflagellates* (ed. E. R. Cox), pp. 273–317. Elsevier North Holland, New York.

Hori, T. and Green, J. C. (1991). The ultrastructure of the flagellar root system of *Isochrysis galbana* (Prymnesiophyta). *Journal of the Marine Biological Association of the United Kingdom*, **71**, 137–52.

Iino, M. (1989). Calcium-induced calcium release mechanism in guinea pig *Taenia caeci*. *Journal of General Physiology*, **94**, 363–83.

Inouye, I. and Pienaar, R. N. (1985). Ultrastructure of the flagellar apparatus on a species of *Pleurochrysis* (Class Prymnesiophyceae). *Protoplasma* **125**, 24–35.

Inouye, I. and Pienaar, R. N. (1988). Light and electron microscope observations of the type species of *Syracosphaera, S. pulchra* (Prymnesiophyceae). *British Phycological Journal*, **23**, 205–17.

Kawachi, M. and Inouye, I. (1994). Ca^{2+}-mediated induction of the coiling of the haptonema on *Chrysochromulina hirta* (Prymnesiophyta = Haptophyta). *Phycologia*, **33**, 53–7.

Kawachi, M., Inouye, I., Maeda, O., and Chihara, M. (1991). The haptonema as a food-capturing device: observations on *Chrysochromulina hirta* (Prymnesiophyceae). *Phycologia*, **30**, 563–73.

Klaveness, D. (1972). *Coccolithus huxleyi* (Lohm.) Kamptn. II. The flagellate cell, aberrant cell types, vegetative propagation and life cycles. *British Phycological Journal*, **7**, 309–18.

Leadbeater , B. S. C. (1971*a*). Observations by means of ciné photography on the behaviour of the haptonema in plankton flagellates of the class Haptophyceae. *Journal of the Marine Biological Association of the United Kingdom*, **51**, 207–17.

Leadbeater , B. S. C. (1971*b*). Observations on the life history of the haptophycean alga *Pleurochrysis scherffelii* with special reference to the microanatomy of the different types of motile cells. *Annals of Botany*, **35**, 429–39.

Leadbeater , B. S. C. and Manton, I. (1969). New observations on the fine structure of *Chrysochromulina strobilus* Parke and Manton with special reference to some unusual features of the haptonema and scales. *Achiv für Mikrobiolgie*, **66**, 105–20.

Manton, I. (1964). Further observations on the fine structure of the haptonema in *Prymnesium parvum*. *Archiv für Mikrobiolgie*, **49**, 315–30.

Manton, I. (1967). Further observations on the fine structure of *Chrysochromulina chiton* with special reference to the haptonema, 'peculiar' Golgi structure and scale production. *Journal of Cell Science*, **2**, 265–72.

Manton, I. (1968). Further observations on the microanatomy of the haptonema in *Chrysochromulina chiton* and *Prymnesium parvum*. *Protoplasma*, **66**, 35–53.

Manton, I. and Leedale, G. F. (1963). Observations on the microanatomy of *Crystallolithus hyalinus* Gaarder and Markali. *Archiv für Mikrobiolgie*, **47**, 115–36.

Manton, I. and Peterfi, L. S. (1969). Observations on the fine structure of coccoliths, scales and the protoplast of a freshwater coccolithophorid, *Hymenomonas roseola* Stein, with supplementary observations on the protoplast of *Cricosphaera carterae*. *Proceedings of the Royal Society of London*, Series B, **172**, 1–15.

Moestrup, Ø. and Thomsen, H. A. (1986). Ultrastructure and reconstruction of the flagellar apparatus in *Chrysochromulina apheles* sp. nov. (Prymnesiophyceae = Haptophyceae). *Canadian Journal of Botany*, **64**, 593–610.

Parke, M., Manton, I., and Clarke, B. (1955). Studies on marine flagellates. II. Three new species of *Chrysochromulina*. *Journal of the Marine Biological Association of the United Kingdom*, **34**, 579–609.

5. Mitosis and cell division

T. HORI
Institute of Biological Sciences, University of Tsukuba, Japan

and J. C. GREEN
Plymouth Marine Laboratory, Citadel Hill, Plymouth, UK.

Abstract

The fine structural changes during mitosis in members of the Prymnesiophyceae are reviewed with special reference to the behaviour of the periplastidial endoplasmic reticulum (PER) and the structures associated with the spindle poles. As in other chromophyte algae, the nuclear envelope (NE) and PER are confluent in interphase cells of members of the Prymnesiophyceae. In some species, PER-NE confluence is maintained during chloroplast replication, whilst in other species the PER-NE link breaks down. Disruption of the PER-NE link usually occurs at prophase, PER-NE confluence being restored at metaphase or in the later stages of mitosis. Some organelles, especially mitochondria, may play an important role in determining the positioning of the spindle poles and in extension of the anaphase spindle, since long mitochondrial profiles are well developed around each spindle pole and along the interzonal spindle at anaphase. Hitherto, structured microtubule organizing centres (MTOCs) have not been observed in the Prymnesiophyceae other than in *Pavlova*, but we report here the presence of MTOCs composed of osmiophilic amorphous material on the surface of mitochondria in cells of *Phaeocystis globosa*.

Introduction

During the last three decades, considerable progress has been made in studies of algal cell division, and the resulting information has led, for example, to a complete revision of green algal systematics and phylogeny (e.g. Mattox and Stewart 1984; Sluiman 1989). In contrast, mitosis in certain other classes is known only in one or two representatives (for reviews, see Hori 1979; Green 1989), and comparisons in terms of phylogenetic relationships are, therefore, difficult. However, there have now been several studies of mitosis in members of the Prymnesiophyceae

The Haptophyte Algae (ed. J. C. Green and B. S. C. Leadbeater), Systematics Association Special Volume No. 51, pp. 91–109. Clarendon Press, Oxford, 1994.

Table 5.1 Species from the Prymnesiophyceae in which mitosis has been described using the electron microscope. Species names in parentheses are those used in the original papers. (Taxonomy following Jordan and Green 1994.)

Species	References
Prymnesiophycidae	
Prymnesium parvum N.Carter	Manton (1964)
Chrysochromulina chiton Parke & Manton	Manton (1966); Green *et al.* 1989
Pleurochrysis carterae (Braarud & Fagerland) Christensen (= *Hymenomonas carterae*)	Stacey and Pienaar (1980)
Pleurochrysis roscoffensis (Dangeard) Fresnel & Billard (= *Cricosphaera roscoffensis* var. *haptonemofera*)	Hori and Inouye (1981)
'*Apistonema*' stage of unknown coccolithophorid	Mesquita and Santos (1983)
Isochrysis galbana Parke	Hori and Green (1985*a*)
Emiliania huxleyi (Lohmann) Hay & Mohler	Hori and Green (1985*b*)
Imantonia rotunda Reynolds	Hori and Green (1985*c*)
Phaeocystis globosa Scherffel	This study
Pavlovophycidae	
Pavlova lutheri (Droop) Green	Green and Hori (1988)
Pavlova salina (Carter) Green	Green and Hori (1988)

(division: Haptophyta) (Table 5.1) and in this paper we review the information in relation to the chromophyte algae in general.

The class Prymnesiophyceae (=Haptophyceae) is composed of two distinct groups of algae, the Prymnesiophycidae (with one order the Prymnesiales) forming one subclass, the Pavlovophycidae (with one order the Pavlovales) the other (taxonomy following Jordan and Green 1994). In the present context, the most striking difference in cell ultrastructure of the two groups is the presence (Pavlovophycidae) or absence (Prymnesiophycidae) of a fibrous flagellar root passing deep into the cell (Green and Hori 1988).

Other differences include the morphology of the flagella, the presence or absence of plate-like body scales, the presence or absence and position of the eyespot, and the overall symmetry of the cell (Green 1980).

Hitherto, reports on cell division in the algae have concentrated only on nuclear division and cytokinesis, but it must be remembered that cell division involves also replication of the cell organelles, either after cytokinesis or immediately before mitosis, i.e. either early or late in interphase. Separation of the daughter organelles then follows cytokinesis. The prymnesiophycean algae are useful in studying such processes since they are generally small unicellular organisms with a minimum number of organelles, i.e. a nucleus, one or two chloroplasts, a few mitochondria (probably part of a mitochondrial network; Beech and Wetherbee 1984), one or two Golgi bodies (dictyosomes), a stigma or eyespot (in some species only), the flagellar apparatus consisting of two basal bodies with associated microtubular and fibrous roots, and a haptonema. Other sub-organellar structures may act as markers in following the behaviour of some organelles; e.g. the pyrenoid is useful in following the course of replication of the chloroplasts during division. In algal cells, the replication of the cell organelles may be by division, duplication or *de novo* formation, and this paper offers both a review of organelle reproduction in the Prymnesiophyceae and a comparison of nuclear division and cytokinesis.

Replication of cell organelles

Division of chloroplast and pyrenoid

Chloroplasts of prymnesiophycean algae usually have pyrenoids either immersed in the chloroplast matrix (most species) or bulging into the cytoplasm from the inner surface of the chloroplast (e.g. some species of *Chrysochromulina*). In interphase cells, pyrenoids are located approximately centrally on the long axis of the chloroplast appearing as a densely packed aggregation of fine granular material traversed by two-thylakoid bands. There seems to be at least two types of chloroplast division based on the behaviour of the PER.

In *Pleurochrysis haptonemofera* (Hori and Inouye 1981), the first sign of chloroplast division is elongation of the pyrenoid matrix along the long axis of the chloroplast (Fig. 5.1B) and the appearance of the dimple on the inner surface of the chloroplast in the pyrenoid area (Fig. 5.1C). Both the periplastidial endoplasmic reticulum (PER) and the chloroplast envelope form an invagination perpendicular to the long axis of chloroplast at the mid-point of the pyrenoid (Fig. 5.1C–E). Eventually both parent chloroplast and pyrenoid divide to give two daughter chloroplasts

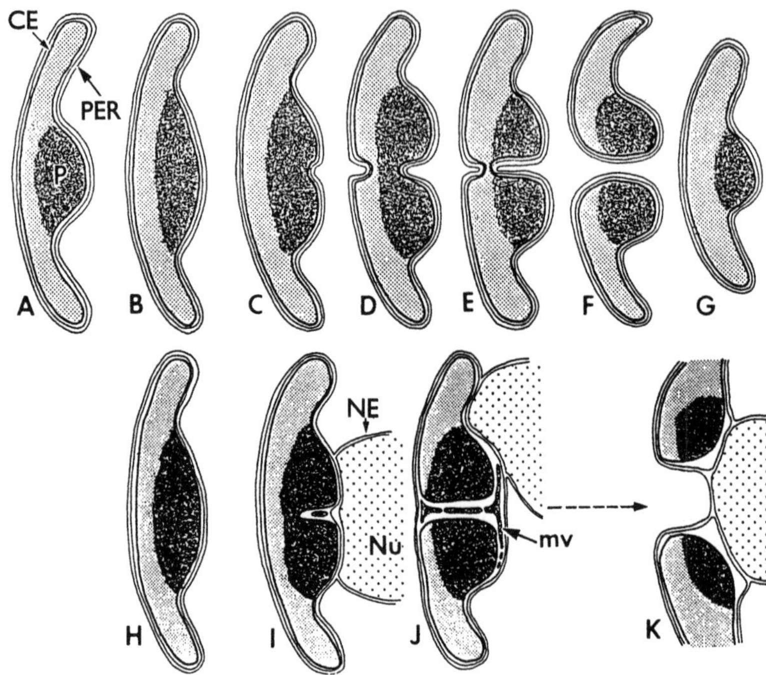

Fig. 5.1 Diagrammatic representation of two possibly different processes of chloroplast and pyrenoid division. The upper series (A–G) represents the simple process observed in *Pleurochrysis haptonemofera* (based on Hori and Inouye 1981) by simultaneous constriction (or furrowing) of both periplastidial-ER (PER) and the chloroplast envelope (CE), resulting in the formation of two daughter chloroplasts each with a pyrenoid (P). In this process, the volume of the pyrenoid matrix increases and expands along the long axis of chloroplast (B) before chloroplast division, and a dimple appears in the central area of the pyrenoid surface (C), followed by deeper invagination and formation of another dimple on the outer surface of the chloroplast (D). Both furrows proceed centripetally (E) and finally fuse, resulting in division of the chloroplast (F). Immediately after division, the daughter pyrenoids become located at the extremities of each chloroplast, each then gradually moves towards the centre of the chloroplast (G, A). The lower series (H–K) represents a more complex method of chloroplast division observed in *Isochrysis galbana* (based on Hori and Green 1985a) and *Phaeocystis globosa* (present study). The PER does not seem to be involved in division of the chloroplast itself until a late stage. As in *P. haptonemofera*, a dimple appears at first on the pyrenoid surface and a membrane-bound vesicle (mv), the origin of which is not known, is formed in the perichloroplast space (I). At the next stage more membrane-bound vesicles appear in the perichloroplast space (J) and chloroplast division is complete. The resulting daughter chloroplasts are encircled by parent PER which retains continuity with the nuclear envelope (NE). The ultrastructural changes in the PER from the stages shown in Fig. 5.1 J–K are not known. For more details, see text. Nu = nucleus. Not to scale.

and pyrenoids (Fig. 5.1F). The nucleus, located at the cell posterior in interphase, separates from the PER before nuclear division and the free nucleus migrates towards the anterior of the cell. The PER and nuclear envelope (NE) are therefore no longer confluent during chloroplast division. However, in *Isochrysis galbana* (Fig. 5.1H–K) and *Phaeocystis globosa* (Hori and Green, unpublished data), only the chloroplast envelope invaginates and divides the chloroplast. Although details of the final stages of the process (Fig. 5.1J, K) are not known, the new daughter PER sacs and the parent NE retain their continuity before nuclear division occurs (Fig. 5.1K).

The pattern of chloroplast division known in other plants and algae (Mita and Kuroiwa 1988; reviews by Possingham and Lawrence 1983 and Kuroiwa 1991) has not been detected in prymnesiophycean algae. Chloroplast of the Prymnesiophyceae and many other chromophyte algae, are enveloped by PER, not found around the chloroplast of higher plants, or in red and green algae, and the maintenance or loss of the PER-NE link clearly affects the way chloroplast and nuclear division take place.

Duplication of flagellar bases and haptonema

Most members of the Prymnesiophyceae have a haptonema and two more or less isokont (anisokont in the Pavlovales) flagella. The first event immediately before cell division is the replication of the basal bodies (Manton 1964; Hori and Inouye 1981). Thus, four flagellar and two haptonematal bases may be seen in cells where the chloroplast has divided, but the nucleus itself may be in prophase or still in interphase. It is not known whether the emergent flagella are excised or resorbed into the cell before duplication. Stacey and Pienaar (1980) suggested that flagella are resorbed in *Pleurochrysis carterae*, but our preliminary observations indicate that the flagella are retained at prophase in *Imantonia* (Hori and Green 1985*c*) and *Pavlova* (Green and Hori 1988). However, in *Imantonia* flagella can be quickly resorbed when the cell is subjected to physical or chemical shock, e.g. centrifugation or fixation (Hori and Green 1985*c*).

The way the haptonema replicates is unclear. In *Isochrysis galbana*, haptonematal traces appear beside the two central flagellar bases to form a row of four flagellar bases and two haptonematal bases (Hori and Green 1985*a*). Probably the two central flagellar bases are the parental basal bodies; the new haptonematal base is, therefore, found near the parental basal bodies. The basal body pairs and the associated haptonematal bases then separate in opposite directions (Stacey and Pienaar 1980; Hori and Inouye 1981; Hori and Green 1985*c*), each set of basal bodies finally being located near a spindle pole, laterally with respect to the long axis of the spindle (Hori and Inouye 1981).

In *Pavlova*, the two sets of basal bodies and haptonematal bases remain close to each other during early stages of nuclear division, separation not occurring until later in the mitotic sequence (Green and Hori 1988).

Division of the mitochondrion

Beech and Wetherbee (1984) demonstrated, by serial reconstruction, the presence of a single reticulate mitochondrion in *Pleurochrysis carterae*. Though the presence of a single mitochondrion in members of the Prymnesiophyceae has not been confirmed in other species, observations on *Imantonia rotunda*, *Emiliania huxleyi*, and *Isochysis galbana* suggest that this is probably the case (Hori and Green 1985*a*, *b*, *c*).

In several species, the mitochondrion extends along the spindle during anaphase-telophase and divides concomitantly with the nucleus (Hori and Green 1985*a*, *c*; Manton 1964; *Phaeocystis*, this chapter). In *Pavlova*, however, it is uncertain when mitochondrial replication takes place because of the complexity of the mitochondrial system (Green and Hori 1988).

Division of the Golgi body

In the Prymnesiophyceae, the interphase cell usually has a single Golgi body (Hori and Green 1985*c*). Division of the Golgi body precedes mitosis in *Chrysochromulina* (Green *et al.* 1989), *Pavlova* (Green and Hori 1988), *Prymnesium parvum* (Manton 1964), and *Pleurochrysis carterae* (Stacey and Pienaar 1980), though the timing of duplication varies. It is usually completed by metaphase in *Emiliania huxleyi* (Hori and Green 1985*b*), but before prophase in *Imantonia rotunda* (Hori and Green 1985*c*).

Nuclear division

Prophase

The behaviour of preprophase and/or prophase nuclei varies from species to species. In some, the formation of extranuclear microtubules (MTs) occurs with the breakdown of the NE and the condensation of heterochromatin, while in others it occurs when the NE is still intact. The MTs are not seen inside the nucleus, but are numerous in one or more of the following regions: (1) in the cytoplasm and ER around the nucleus (Hori and Inouye 1981); (2) in close proximity to the flagellar bases (Hori and Green 1985*a*); (3) in the cytoplasm between the basal bodies and the nucleus (Green and Hori 1988; Green *et al.* 1989); or (4) radiating from centres near the basal bodies and in close proximity to a mitochondrial profile (Hori and Green 1985*c*). Fragmentation of the NE results in large gaps through which extranuclear MTs may pass. The extent of disruption of the NE is variable (see below).

In non-motile coccolith-bearing cells of *Emiliania huxleyi*, a reticulum of heterochromatin expands within the nucleus with a relative decrease of the euchromatin as prophase proceeds. The NE does not seem to break down completely as in the other species of the Prymnesiophyceae, but retains its continuity with the PER (Hori and Green 1985*b*). In *Chrysochromulina chiton*, vesiculation of the NE begins together with condensation of the chromatin and the breakdown of the nucleolus. No contact between the nascent spindle MTs and the flagellar bases has been observed, but it is possible that the MTs have their origin in material associated with the flagellar root system. The MTs radiate from a ribosome-free area near both the nucleus and the flagellar bases (Green *et al.* 1989). Similarly, during prophase in *Isochrysis galbana*, MTs interconnect and radiate from two foci which may represent the future spindle poles (Hori and Green 1985*a*).

In contrast to other members of the Prymnesiophyceae, the spindle MTs in species of *Pavlova* have their origin in the long fibrous root attached to the basal body of the longer flagellum. Penetration of the MTs into the nucleus is accompanied by the breakdown of part of the NE (Green and Hori 1988).

Prometaphase

At prometaphase condensed chromatin moves towards the spindle equator to form the metaphase plate. Some spindle MTs pass between the scattered chromosomes, others extend from the pole either apparently terminating at the chromatin, or still extending into the cytoplasm. There are no membranes around the chromosomes.

Metaphase

The onset of metaphase is indicated by the accumulation of condensed chromatin at the equator, but no individual chromosomes can be resolved. The metaphase plate is traversed by chromatin-free channels through which run bundles of pole-to-pole MTs, the number per channel varying between one and twelve (Hori and Green 1985*b*, *c*; Green *et al.* 1989). There may also be variation in the number and size of the channels. Some spindle MTs terminate at the metaphase plate, but distinct kinetochores have not been detected.

At metaphase, many vesicles without attached ribosomes are scattered among the spindle MTs, but soon move to lie close to the poleward faces of the chromatin where they begin to fuse to form the precursors of the daughter NE, although in many species the metaphase plate remains free of vesicles at its lateral edges (Fig. 5.2A, C, F). The metaphase plate of *Chrysochromulina chiton*, for example, is surrounded by a network of vesicles of rough ER origin and covered with ribosomes, constituting a more or less

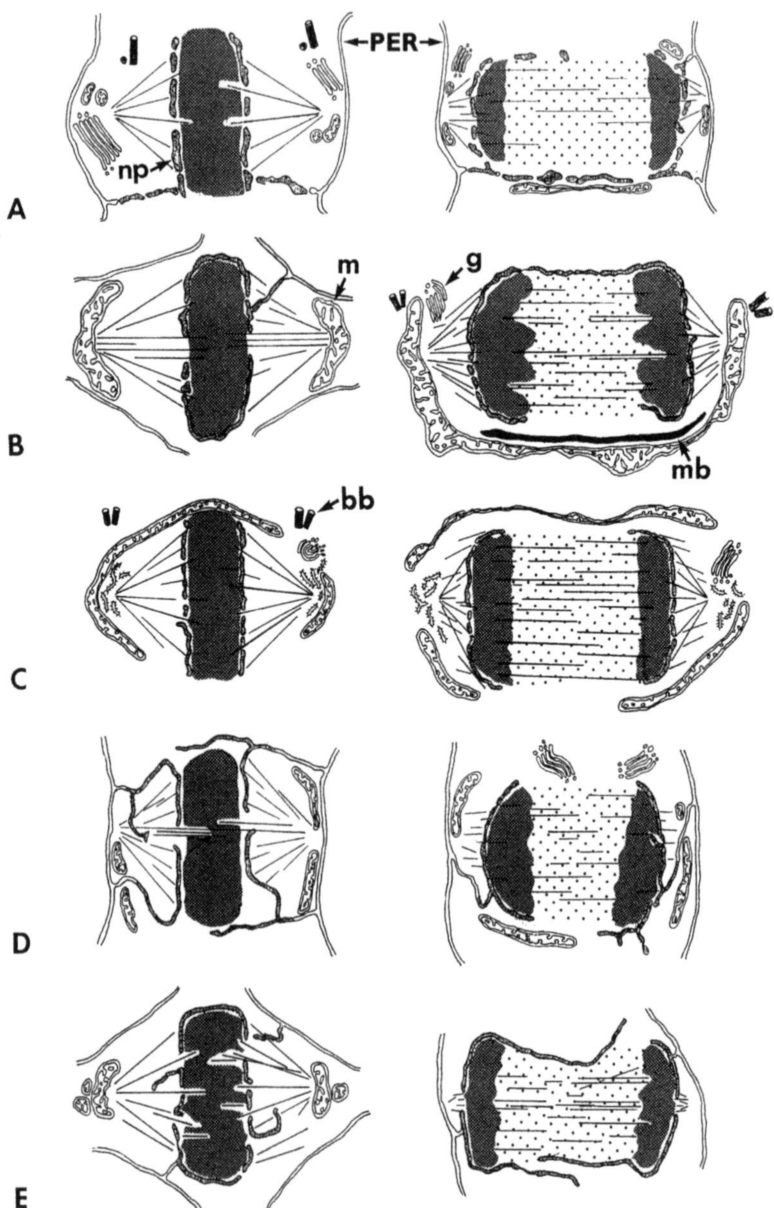
PER
np
A
m
g
B
mb
bb
C
D
E

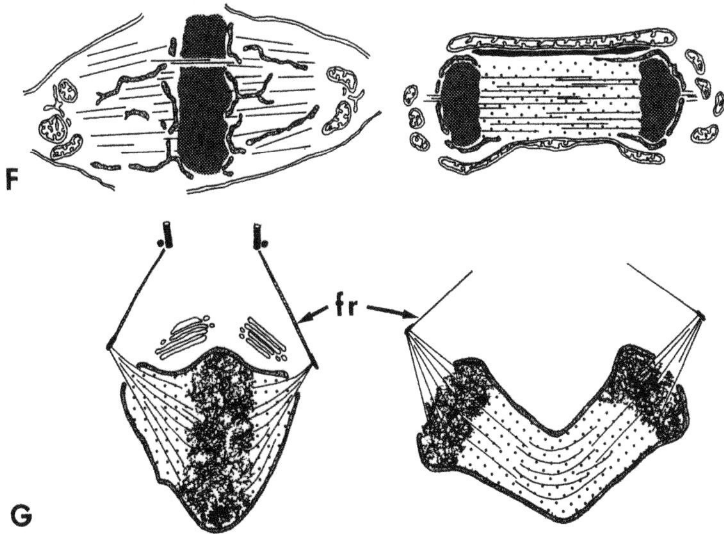

Fig. 5.2 Diagrammatic representation of the ultrastructure of metaphase (left column) and anaphase or early telophase nuclei (right column) of six representatives of the Prymnesiophyceae. (A) *Isochrysis galbana* (based on Hori and Green 1985*a*). (B) *Imantonia rotunda* (based on Hori and Green 1985*c*). (C) *Pleurochrysis* spp. (based on Stacey and Pienaar 1980; Hori and Inouye 1981). (D) *Emiliania huxleyi* (based on Hori and Green 1985*b*). E: *Chrysochromulina chiton* (based on Manton 1966; Green *et al.* 1989). (F) *Prymnesium parvum* (based on Manton 1964). (G) *Pavlova* spp. (based on Green and Hori 1988). bb. = basal bodies; PER = periplastidial-ER; fr = fibrous flagellar root; g = Golgi body; np = precursor of daughter nuclear envelope; m = mitochondrion; mb = microbody. Not to scale.

complete envelope (Fig. 5.2E; Manton 1966; Green *et al.* 1989), but with large gaps through which the spindle MTs pass. Even at this early stage in the formation of the new NE, continuity with the PER is already partially established (Green *et al.* 1989). In contrast, the NE of *Imantonia rotunda* retains much of its integrity, though there are large gaps through which the spindle MTs can pass (Fig. 5.2B; Hori and Green 1985*c*). In *Pleurochrysis haptonemofera* (Fig. 5.2C; Hori and Inouye 1981) and *Emiliania huxleyi* (Fig. 5.2D; Hori and Green 1985*b*), the metaphase spindle is not surrounded by vesicular elements. In *P. haptonemofera* and *E. huxleyi*, however, a long mitochondrial profile extends around the spindle from pole to pole.

In *Prymnesium parvum*, a number of fragments of NE are present both in the spindle and close to the poleward faces of the metaphase plate (Fig. 5.2F). The spindle MTs converge towards ill-defined poles with no resolvable structure. However, the basal bodies, and associated Golgi body and mitochondrial profiles are all situated nearby (Manton 1966).

The NE in *Pavlova salina* (Green and Hori 1988) retains its integrity at metaphase (Fig. 5.2G), although a number of large gaps are present, the largest on either side of the nucleus giving the spindle MTs access to the nucleoplasm. The spindle is bilaterally symmetrical and markedly U- or V-shaped with the chromatin aggregated as a flocculent mass at the equator, and spindle MTs converging towards the two poles on each side. No obvious kinetochores have been observed in this species, but in some specimens, MTs terminating at the chromatin plate were associated with lightly staining amorphous material.

Anaphase

At anaphase, the NE precursors become expanded sheets of membrane as a cap or half-NE covering the poleward surface of the plate, and against which the chromatin is closely appressed. Further coalescence of the vesicles takes place during anaphase, and the resultant envelope extends so that the whole mitotic apparatus becomes surrounded by a more or less convoluted envelope covered with ribosomes.

During anaphase the chromatin separates into two plates and these move towards opposite spindle poles. The separation of the chromatin plates is accompanied by a shortening of the pole-to-chromosome MTs and the development of interzonal MTs. The distance between the spindle poles and the daughter chromatin plates reduces in *Pleurochrysis haptonemofera* (Hori and Inouye 1981), but does not change in *Imantonia rotunda* (Hori and Green 1985*c*). In contrast, the pole-to-pole distance increases during anaphase in many members of the Prymnesiophyceae and, as the daughter plates move apart, vesicles, ribosomes, MTs and, sometimes, coccolithosomes (in coccolithophorids), flow into the region between them. A feature of a number of prymnesiophycean species is the presence of a long mitochondrial profile adjacent to the spindle. (Fig. 5.2B–D, F). In some species, there is also elongation of the cell body during anaphase and, as anaphase progresses, the spindle profile becomes more rectangular. By mid-anaphase interzonal MTs are well-developed between the daughter chromosomes, some MTs passing through the chromatin masses and the gaps in the NE and approaching the spindle poles.

By anaphase, the new NE is usually apparent. In *Imantonia rotunda* the chromatin masses are almost entirely surrounded by NE at anaphase as the NE remains largely conserved during mitosis. However, there are large gaps in the NE on the face towards the large mitochondrion extending between two poles (Fig. 5.2B; Hori and Green 1985*c*). In *Emiliania huxleyi*, separation of the daughter nuclei to opposite poles is associated with the possible shortening of a bridge of ER between the PER and the new NE caps. There is also shortening of the chromosomal MTs (Hori and Green 1985*b*).

The most significant features of anaphase in *Prymnesium* are elongation of the cell-body and the interzonal spindle MTs, and the presence of long mitochondrial profiles and a microbody-like structure referred to as the phragmoplast (Manton 1964). In addition, there are tubular fragments of ER or parent NE along the anaphase spindle (Fig. 5.2F). The polar side of each group of daughter chromosome is covered by re-formed NE, but the spindle side is still open and interzonal spindle MTs persist.

During anaphase in species of *Pavlova*, MTs may be clearly distinguished on the poleward side of the chromatin masses, extending in early anaphase into the chromatin and into the developing interzonal region. Some terminate on the poleward face of the chromatin as in metaphase, some appear to terminate on the inner side of the NE, and others appear to penetrate the nucleoplasm, passing through it and leaving again through pores in the NE. However, by the later stages of anaphase, few MTs are to be seen in the interzonal region.

During anaphase there is a marked elongation of the spindle due principally to the development of the interzonal region, with relatively little increase in the pole-chromatin distance. The U- or V-conformation of the spindle becomes less obvious with the development of the interzonal region. The NE itself may still be traced around the dividing nucleus although by the later stages of anaphase, large gaps may have formed. The connection of the NE with the PER is maintained during anaphase (Green and Hori 1988).

Mitochondrial profiles are conspicuous, a large profile often lying parallel to the long axis of the spindle.

Telophase

The development of interzonal MTs continues during telophase. They extend from pole to pole, passing through the chromatin masses, and sometimes extend through gaps in the NE almost to the periphery of the cell.

The daughter nuclei move apart until they lie peripherally on opposite sides of the cell and, as they do so, the NE is formed anew around the daughter chromosome masses. By the time the chromatin masses reach the spindle poles, there has been a marked shortening of the half-spindles. The final stages of mitosis include the closure of the gaps in the NE and contraction of the PER-NE links to give an intimate association between the nucleus and the chloroplast.

During telophase, the daughter nuclei often rotate so that they lie laterally with respect to the long axis of the spindle. In *Pleurochrysis haptonemofera* (Hori and Inouye 1981) and *Isochrysis galbana* (Hori and Green 1985*a*) this movement leads to abscission of the daughter nuclei from the interzonal spindle, the rotation through 90° causing the MTs of the interzonal spindle to become stretched and eventually disconnected from the nuclei (Hori and Inouye 1981).

In species of *Pavlova*, during telophase, cell elongation also occurs with the separation of the two sets of basal bodies and the Golgi body. In *P. lutheri*, this is sometimes accompanied by rotation of one-half of the cell (i.e. one of the incipient daughter cells) though 180° with respect to the other. However, such reorientation was not observed in *P. salina* (Green and Hori 1988).

Cytokinesis

Separation of the daughter cells at telophase seems to be a very rapid process. Comparisons between coccolith-bearing organisms and scale-bearing species indicate that the former elongate relatively little before cytokinesis, the dividing cell merely becoming constricted at the equator. In the latter species, the dividing cell becomes markedly elliptical as the cell body elongates.

The mechanism of cytokinesis seems to vary. In *Pleurochrysis carterae* (Stacey and Pienaar 1980), *P. haptonemofera* (Hori and Inouye 1981), '*Apistonema*' (Mesquita and Santos 1983), *Emiliania huxleyi* (Hori and Green 1985*b*), and *Imantonia rotunda* (Hori and Green 1985*c*) a large vacuole develops or intrudes into the cytoplasm between the daughter cells. The vacuole then contributes to the septation of the cytoplasm and the separation of the two daughter cells. Similarly, in *Chrysochromulina chiton* cytokinesis is initiated by the formation of large vacuoles in the interzonal region. Some of these may be derived from the Golgi apparatus, but there may also be an intrusion of PER towards the central part of the cell (Green *et al.* 1989). In *Isochrysis galbana*, the separation of the daughter nuclei appears to be effected by the intrusion into the interzonal cytoplasm of scale-free membrane derived from the plasmalemma and the cell vacuole (Hori and Green 1985*a*).

In contrast, cytokinesis in species of *Pavlova* is effected by the formation of a cleavage furrow which pinches the cell in the equatorial plane between the daughter nuclei. Usually cleavage involves the plasmalemma and both membranes of the PER. However, Green and Hori (1988) reported cytokinesis involving only the inner membrane of the PER.

Polar structure and microtubule organizing centres (MTOC)

In many phytoflagellates, the first indication of impending cell division is the proliferation of MTs from the vicinity of the basal bodies (Manton

1964; Slankis and Gibbs 1972; Oakley and Dodge 1976; Oakley and Bisalputra 1977; Heywood 1978; Oakley 1978). These MTs are probably involved in the separation of the spindle poles preceding nuclear division (Manton 1964), and later in the formation of the division spindle. In *Pleurochrysis haptonemofera* (Hori and Inouye 1981) and '*Apistonema*' (Mesquita and Santos 1983), there is a complicated system of flagellar roots, each consisting of a sheet of closely aligned MTs and a crystalline bundle of MTs at right angles to it (Inouye and Pienaar 1985; Beech and Wetherbee 1988). The latter MTs become disorganized during nuclear division and it is possible that their constituent tubulin may be recycled as the spindle MTs.

In all members of the Prymnesiophyceae, other than the Pavlovales, in which the ultrastructure of mitosis has been examined, the spindle poles are ill-defined areas of apparently structureless cytoplasm, near the replicated chloroplasts, surrounded by densely aggregated ribosomes, vesicles, ER, Golgi bodies, and mitochondrial profiles (Hori and Green 1985*a*, *c*; Green *et al.* 1989). The flagellar bases do not act as the poles of the spindle, but each set of basal bodies, with an adjacent Golgi body, lies close by, laterally with respect to the long axis of the spindle (Hori and Inouye 1981; Hori and Green 1985*a*). The basal bodies may also have mitochondrial profiles associated with them (Hori and Green 1985*c*).

In species of *Pavlova*, however, there is a direct link between the spindle MTs and the flagellar apparatus (Green and Hori 1988), the fibrous flagellar roots acting as a spindle organizing centre. Fibrous flagellar roots acting as MT organizing centres (MTOCs) during mitosis are known in a diversity of algal groups. In *Ochromonas danica* (Chrysophyceae) two persistent rhizoplasts, each with a nearby Golgi body, lie at the poles of the spindle (Slankis and Gibbs 1972). There is elaboration of the rhizoplast during mitosis in *Poterioochromonas malhamensis* (Chrysophyceae; Schnepf *et al.* 1977) and *Hydrurus foetidus* (Chrysophyceae; Vesk *et al.* 1984), a persistent rhizoplast in *Nephroselmis olivacea* (= *Heteromastix angulata)* (Prasinophyceae; Mattox and Stewart 1977), and granular material of rhizoplast origin in *Tetraselmis* (=*Platymonas*) *subcordiformis* (Prasinophyceae; Stewart *et al.* 1974). In the diatoms there is a specialized spindle precursor and MTOC, e.g. in *Lithodesmium undulatum* (Manton *et al.* 1969). In contrast, in most prymnesiophycean species except species of *Pavlova*, no obvious polar structure and MTOC has been observed, though the spindle pole area may be defined in other ways.

Recently, however, in *Phaeocystis globosa* (Prymnesiales), we have observed osmiophilic material on the surface of mitochondrial profiles (probably part of a mitochondrial network) from which numerous MTs extend towards the nucleus (Figs 5.3, 5.4). The nucleus appears to be in interphase, but may be at pre-prophase or even prophase. At metaphase

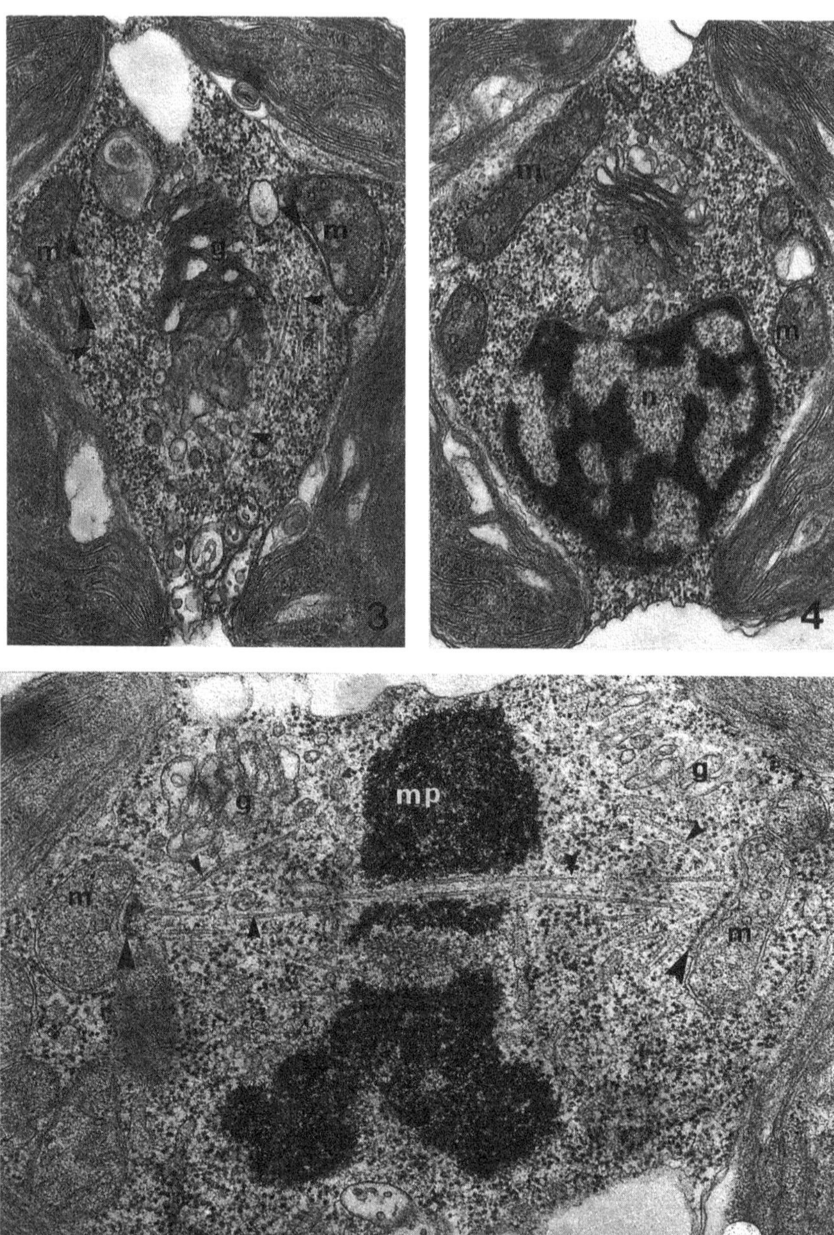
m
g
m
3
m
g
m
n
m
4
mp
g
g
m
m
5

Figs 5.3–5.5 Electron microscopy of mitosis in *Phaeocystis globosa*. (**3–4**) Two sections (no. 1 and 10) from a series through a cell of *Phaeocystis globosa*. These may represent a very early stage of prophase. (**3**) This figure does not include the nucleus but the extranuclear MTs (*small arrow heads*), which seem to be involved in formation of the spindle, extend down from a layer of osmiophilic material (MTOCs; *large arrowheads*) located close to the surface of the mitochondria. (**4**) An early prophase nucleus with chromatin condensed in a reticulate manner, but the NE still intact. g = Golgi body; m = mitochondrion; n = nucleus. × 23 800. (**5**) Metaphase nucleus with many pole-to-pole and chromosomal MTs (*small arrowheads*) extend from MTOCs (*large arrowheads*) on the surface of mitochondria located at the poles; g = Golgi body; m = mitochondrion; mp = metaphase plate. × 31 500.

(Fig. 5.5) chromosomal MTs and pole-to-pole MTs originate from the osmiophilic material present on the mitochondrial surfaces. The osmiophilic material, therefore, represents an MTOC at each spindle pole, the first time such structures have been reported in the Prymnesiophyceae. The possibility that the MTOCs are fixation artefacts cannot be discounted, but the regularity with which they occur suggests that this is not the case.

Behaviour of the NE and PER during nuclear division

The continuity between the NE and the PER is a constant feature in interphase cells of the Prymnesiophyceae. However, at prophase, this continuity is broken, but may be re-established quite early in mitosis. The disorganization and reconstitution of the NE can be summarized as follows: (1) breakdown of the NE by late prophase; (2) reorganization of the new NE by coalescence and fusion of many vesicles, commencing on the poleward surface of the condensed chromatin at or after metaphase; (3) further development of the NE during metaphase and anaphase; (4) re-establishment of the structural continuity of the immature daughter NE with the PER. Thus, throughout mitosis, the chloroplast remains enclosed in the PER.

The degree of disruption of the NE is variable. It is completely broken down into vesicles in *Prymnesium parvum* (Manton 1964), *Pleurochrysis haptonemofera* (Hori and Inouye 1981), *P. carterae* (Stacey and Pienaar 1980), *Isochrysis galbana* (Hori and Green 1985*a*), though fragments of the NE membrane remaining near the chloroplasts may still be confluent with the PER in some species. Even in those species in which disruption of the NE is more or less complete, reorganization may begin at any early stage of mitosis. Manton (1964), Stacey and Pienaar (1980), Hori and Inouye (1981), Mesquita and Santos (1983), and Hori and Green (1985*a*) have all reported vesicles, apparently precursors of the new NEs, situated on the poleward faces of the metaphase plate. For example, in *Pleurochrysis haptonemofera* (Hori and Inouye 1981) vesicles of various sizes aggregate into long cisternae over the spindle surface (but not the lateral edges) of the metaphase plate. However, in *Chrysochromulina chiton*, by metaphase the rough-surfaced ER may extend around the metaphase plate, thus wholly enclosing the chromatin mass. During the later stages of mitosis, therefore, nuclear division takes place in a manner analogous to semi-open systems, with the rough-surfaced ER extending along the spindle (Green *et al.* 1989).

The NE does not always become fully disorganized during prymnesiophycean mitosis. In *Imantonia rotunda* the NE is conserved throughout mitosis, and in both *I. rotunda* (Hori and Green 1985*c*) and in *Emiliania huxleyi* (Hori and Green 1985*b*), NE-PER continuity is maintained during nuclear division (Fig. 5.2B, D). Some NE material is lost during prophase, presumably as the volume of the nucleus decreases with condensation of the chromatin, but the greater part of the mother-cell NE persists around the metaphase plate. A further unusual feature of the NE in *I. rotunda* is its continuity at telophase through the interzonal region between the two daughter nuclei (Fig. 5.2B). In other species, the NE-PER link is re-established at a late stage. At telophase in *Isochrysis galbana*, the daughter nucleus becomes repositioned so that the former polar face with its daughter NE is directed away from the chloroplast and the open face (formerly the interzonal face) lies adjacent to it (Hori and Green 1985*a*). Thus the part of the PER against the open face of the nucleus is probably incorporated as part of the new NE in this species (cf. *Ochromonas danica*; Slankis and Gibbs 1972). A similar mechanism seems to apply in *Imantonia rotunda* (Hori and Green 1985*c*), *Emiliania huxleyi* (Hori and Green 1985*b*), and *Chrysochromulina chiton* (Green *et al.* 1989).

In *Emiliania huxleyi* some larger fragments of the old NE, maintain their continuity with the PER, and, during metaphase, they seem to behave as rough-surfaced ER and approximately delimit the region in which mitosis takes place. During anaphase, the PER grows towards the chromatin mass and re-establishes the PER-NE connection with the nascent daughter NE (Hori and Green 1985*b*). In *Prymnesium parvum* (Manton 1964) and '*Apistonema*' (Mesquita and Santos 1983), however, the connection between the PER and the NE seems to be re-established very late, in the interphase cell after cytokinesis. The renewal of the NE is believed to be completely independent of the PER, fusion of the nuclear and chloroplast membrane systems occurring secondarily (Manton 1964; Mesquita and Santos 1983). In species of *Pavlova*, the metaphase NE remains mostly intact, bounding the nucleoplasm except at the polar fenestrae, and retaining its continuity with the PER. However, the full extent to which the NE is conserved may vary from cell to cell (Green and Hori 1988).

Systematic and phylogenetic considerations

It has recently been argued that the Prymnesiophyceae should be subdivided into two subclasses or even into two classes (Cavalier-Smith 1986, 1993, Chapter 22), and the general similarity of mitotic characteristics in all members of the Prymnesiophyceae other than species of *Pavlova*, supports this. Mitosis in the Pavlovales differs in a number of ways from the general sequence of events of members of the Prymnesiophyceae; for example, it is preceded by the replication of the chloroplasts and the flagellar apparatus, but separation of the replicated systems does not take place until a late stage in cell division. In addition, the spindle is markedly bilaterally symmetrical. Theses differences support the view of the Pavlovales as a coherent natural grouping, and may be of significance with respect to our understanding of the evolution of members of this order in particular, and the Prymnesiophyceae as a whole.

References

Beech, P. L. and Wetherbee, R. (1984). Serial reconstruction of the mitochondrial reticulum in the coccolithophorid, *Pleurochrysis carterae* (Prymnesiophyceae). *Protoplasma*, **123**, 226–9.

Beech, P. L. and Wetherbee, R. (1988). Observations on the flagellar apparatus and peripheral endoplasmic reticulum of the coccolithophorid, *Pleurochrysis carterae* (Prymnesiophyceae). *Phycologia*, **27**, 142–58.

Cavalier-Smith, T. (1986). The Kingdom Chromista: Origin and Systematics. *Progress in Phycological Research*, **4**, 304–47.

Cavalier-Smith, T. (1993). Kingdom Protozoa and its 18 phyla. *Microbiological Reviews*, **57**, 953–94.

Green, J. C. (1980). The fine structure of *Pavlova pinguis* Green and a preliminary survey of the order Pavlovales. *British Phycological Journal*, **15**, 151–91.

Green, J. C. (1989). Relationships between the chromophyte algae: the evidence from studies of mitosis. In *The chromophyte algae: problems and perspectives*, Systematics Association Special Volume No. 38, (ed. J. C. Green, B. S. C. Leadbeater, and W. L. Diver), pp. 189–206. Clarendon Press, Oxford.

Green J. C. and Hori, T. (1988). The fine structure of mitosis in *Pavlova* (Prymnesiophyceae). *Canadian Journal of Botany*, **66**, 1497–509.

Green J. C., Hori, T., and Course, P. A. (1989). An ultrastructural study of mitosis in *Chrysochromulina chiton* (Prymnesipohyceae). *Phycologia*, **28**, 318–30.
Heywood, P. (1978). Ultrastructure of mitosis in the chloromonadophycean alga *Vacuolaria virescens. Journal of Cell Science*, **31**, 37–51.
Hori, T. (1979). Ultrastructure of cell division in the eucaryotic algae exclusive of green algae. *Japanese Journal of Phycology*, **27**, 17–29. (In Japanese with English abstract).
Hori, T. and Green, J. C. (1985*a*). The ultrastructure of mitosis in *Isochrysis galbana* Parke (Prymnesiophyceae). *Protoplasma*, **125**, 140–51.
Hori, T. and Green, J. C. (1985*b*). An ultrastructural study of mitosis in non-motile coccolith-bearing cells of *Emiliania huxleyi* (Lohm.) Hay et Mohler (Prymnesiophyceae). *Protistologica*, **21**, 107–20.
Hori, T. and Green, J. C. (1985*c*). The ultrastructural changes during mitosis in *Imantonia rotunda* Reynolds (Prymnesiophyceae). *Botanica Marina*, **27**, 67–78.
Hori, T. and Inouye, I. (1981). The ultrastructure of mitosis in *Cricosphaera roscoffensis* var. *haptonemofera* (Prymnesiophyceae). *Protoplasma*, **106**, 121–35.
Inouye, I. and Pienaar, R. N. (1985). Ultrastructure of the flagellar apparatus in *Pleurochrysis* (Class Prymnesiophyceae). *Protoplasma*, **125**, 24–35.
Jordan, R. W. and Green, J. C. (1994). A checklist of the extant Haptophyta of the world. *Journal of the Marine Biological Association of the United Kingdom*, **74**, 149–74.
Kuroiwa, T. (1991). The replication, differentiation and inheritance of plastids with emphasis on the concept of organelle nuclei. *International Review of Cytology*, **128**, 1–62.
Manton, I. (1964). Observations with the electron microscope on the division cycle in the flagellate *Prymnesium parvum* Carter. *Journal of Royal Microscopical Society*, **83**, 317–25.
Manton, I. (1966). Further observations on the fine structure of *Chrysochromulina chiton*, with special reference to the pyrenoid. *Journal of Cell Science*, **1**, 187–92.
Manton, I., Kowallik, K., and Stosch, H. A. von (1969). Observations on the fine structure and development of the spindle at mitosis and meiosis in a marine centric diatom (*Lithodesmium undulatum*). I. Preliminary survey of mitosis in spermatogonia. *Journal of Microscopy*, **89**, 292–302.
Mattox, K. R. and Stewart, K. D. (1977). Cell division in the scaly green flagellate *Heteromastix angulata* and its bearing on the origin of the Chlorophyceae. *American Journal of Botany*, **64**, 931–45.
Mattox, K. R. and Stewart, K. D. (1984). Classification of the green algae: a concept based on comparative cytology. In *Systematics of the green algae*, Systematics Association Special Volume No. 27, (ed. D. E. G. Irvine and D. M. John), pp. 29–72. Academic Press, London.
Mesquita, J. F. and Santos, M. F. (1983). Cytological studies in golden algae (Chrysophyceae). III. Fine structure of mitosis and cytokinesis in the 'Apistonema stage' of a Coccolithophoraceae. *Journal of Submicroscopy and Cytology*, **15**, 751–65.

Mita, T. and Kuroiwa, T. (1988). Division of plastids by a plastid-dividing ring in *Cyanidium caldarium*. *Protoplasma*, **Suppl. 1**, 133–52.
Oakley, B. R. (1978). Mitotic spindle formation in *Cryptomonas* and *Chroomonas* (Cryptophyceae). *Protoplasma*, **95**, 333–46.
Oakley, B. R. and Bisalputra, T. (1977). Mitosis and cell division in *Cryptomonas* (Cryptophyceae). *Canadian Journal of Botany*, **55**, 2789–800.
Oakley, B. R. and Dodge, J. D. (1976). The ultrastructure of mitosis in *Chroomonas salina* (Cryptophyceae). *Protoplasma*, **88**, 241–54.
Possingham, J. V. and Lawrence, M. E. (1983). Controls to plastid division. *International Review of Cytology*, **84**, 1–56.
Schnepf, E., Deichgraber, G., Roderer, G., and Herth, W. (1977). The flagellar root apparatus, the microtubular system and associated organelles in the chrysophycean flagellate *Poterioochromonas malhamensis* Peterfi (syn. *Poterioochromonas stipitata* Scherffel and *Ochromonas malhamensis* Pringsheim). *Protoplasma*, **92**, 87–107.
Slankis, T. and Gibbs, S. P. (1972). The fine structure of mitosis and cell division in the chrysophycean alga *Ochromonas danica*. *Journal of Phycology*, **8**, 243–56.
Sluiman, H. J. (1989). The green algal class Ulvophyceae. An ultrastructural survey and classification. *Cryptogamic Botany*, **1**, 83–94.
Stacey, V. J. and Pienaar, R. N. (1980). Cell division in *Hymenomonas carterae* (Braarud et Fagerland) Braarud (Prymnesiophyceae). *British Phycological Journal*, **15**, 365–76.
Stewart, K. D., Mattox, K. R., and Chandler, C. D. (1974). Mitosis and cytokinesis in *Platymonas subcordiformis*, a scaly green monad. *Journal of Phycology*, **10**, 65–79.
Vesk, M., Hoffman, L. R., and Pickett-Heaps, J. D. (1984). Mitosis and cell division in *Hydrurus foetidus* (Chrysophyceae). *Journal of Phycology*, **20**, 461–70.

6. Photosynthetic pigments in the Haptophyta

S.W. JEFFREY
CSIRO Marine Laboratories, Hobart, Tasmania, Australia

and S.W. WRIGHT
Australian Antarctic Division, Kingston, Tasmania, Australia

Abstract

Knowledge of pigment systems is essential to our understanding of taxonomic and biological relationships within algal groups, and is an important parameter for measuring their abundance in aquatic systems. In this paper we review the development of knowledge of haptophyte pigments from the early 1960s to the late 1980s, and then present a new HPLC analysis of 29 species (50 strains) from eight out of 16 haptophyte families. Four pigment types were recognized. All contain chlorophylls *a* and c_1/c_2, diatoxanthin, diadinoxanthin, and β,β-carotene. In addition, Type 1 has fucoxanthin; Type 2, chlorophyll c_3 and fucoxanthin (both types similar to diatoms); Type 3, chlorophyll c_3 and 19′-hexanoyloxyfucoxanthin, and Type 4, chlorophyll c_3 and 19′-butanoyloxfucoxanthin (plus variable amounts of 19′-hexanoyloxyfucoxanthin and fucoxanthin). Seven minor carotenoids, phytylated chlorophyll *c* and β,ε-carotene are also present in some strains. No clear-cut pigment pattern for all haptophytes has emerged, which emphasizes the caution oceanographers must still exercise in identification of this group of planktonic algae from pigment signatures.

Phylogenetic pigment anomalies across haptophyte families are also discussed.

Introduction

Most haptophytes are planktonic unicellular algae, either flagellate or producing flagellate stages in their life history (Hibberd 1980). They are distinguished by organic and/or calcified scales or coccoliths (Moestrup 1979; Vesk *et al.* 1990), and an appendage, the haptonema, which appears to have multiple functions. Species lists have been generated for large ocean areas (e.g. Hasle 1960), as well as localized regions (e.g. Hallegraeff

The Haptophyte Algae (ed. J. C. Green and B. S. C. Leadbeater), Systematics Association Special Volume No. 51, pp. 111–32. Clarendon Press, Oxford, 1994.

1984). It is obvious that these organisms play a key role in the nanoplankton of the world's oceans.

Knowledge of haptophyte pigmentation has not kept pace with taxonomic advances. Algal taxonomy was originally based on colour (pigment content) of cells and tissues (Christensen 1962; Bjørnland and Liaaen-Jensen 1989), and while a small number of haptophyte cultures have been studied for their pigment composition since the early 1960s, the variety of pigments described and their presence in other taxa has not allowed a clear idea of pigmentation of the group to develop. Not only is this important for understanding the taxonomic relationships among organisms grouped within the Haptophyta, their light-harvesting mechanisms, and the way in which they adapt to the optical properties of their aquatic environments, but it is increasingly important to use pigments signatures to quantify algal population types in present-day global ocean studies.

Photosynthetic pigments, chlorophylls, carotenoids, and biliproteins, are extremely useful biomarkers for phytoplankton in aquatic ecosystems, and they can be used to distinguish the presence and abundances of algal groups in mixed phytoplankton populations. Spectrophotometry, thin layer chromatography (TLC), and high performance liquid chromatography (HPLC) have all been used for the analysis of phytoplankton pigments. However, the interpretation of pigment data in field samples requires a thorough knowledge of the pigment composition of each of the likely components of the population (e.g. diatoms; Stauber and Jeffrey 1988).

In spite of the demonstrated significance of the Haptophyta in forming widespread nuisance blooms (Lancelot *et al.* 1987; Moestrup, Chapter 14), causing massive fishkills (Underdal *et al.* 1989), possibly affecting climate change (Malin *et al.* 1992) and playing a key role in the Antarctic marine ecosystem (Davidson and Marchant 1992) the growing requirement for accurate pigment information has not been fulfilled.

Here we review the current knowledge of this field, and we also present a new survey of the pigmentation of 50 strains (29 species) of haptophytes analysed quantitatively by advanced HPLC methodology (Wright *et al.* 1991).

Previous studies of haptophyte pigments

Our knowledge of haptophyte pigmentation depends both on the availability of suitable algal cultures and the sensitivity and resolution of the chromatographic pigment methods used for their study.

Between 1960 and 1968, 15 haptophyte species were examined by various authors using paper, column, and TLC (Table 6.1). The simple pigment pattern of chlorophylls *a* and *c*, carotenes, fucoxanthin,

Table 6.1 Haptophyte species examined for pigments in the period 1960–1968 using simple chromatographic methods

Species examined[1]	Clone	Chromatographic method	Reference
[2]*Isochrysis galbana*	Plym I	Column chromatography	Dales (1960)
	Plym I	TLC	Riley and Wilson (1967)
	Unknown	Paper chromatography	Jeffrey (1961)
	Unknown	TLC	Parsons (1961)
Pseudopedinella sp.	Plym 167	Column chromatography	Dales (1960)
	Plym 214	TLC	Riley and Wilson (1967)
Pavlova gyrans	Plym 93	Column chromatography	Dales (1960)
Phaeocystis pouchetii	Plym 147	Column chromatography	Dales (1960)
	Plym 147	TLC	Riley and Wilson (1967)
Chrysochromulina ericina	Plym 25	Column chromatography	Dales (1960)
Dicrateria inornata	Plym B	Column chromatography	Dales (1960)
	Plym B	TLC	Riley and Wilson (1967)
Hymenomonas sp.	Plym 156	Column chromatography	Dales (1960)

Table 6.1 (*cont.*)

Species examined[1]	Clone	Chromatographic method	Reference
Prymnesium parvum	Plym 65	TLC	Riley and Wilson (1967)
	Unknown	TLC	Allen *et al.* (1960)
Chrysochromulina polylepis	Plym 136	TLC	Riley and Wilson (1967)
Coccolithus (Emiliania) huxleyi	Plym 92	TLC	Riley and Wilson (1967)
	Unknown	Paper chromatography	[3]Jeffrey and Allen (1964)
Cricosphaera sp.	Plym 156	TLC	Riley and Wilson (1967)
Cricosphaera elongata	Plym 62	TLC	Riley and Wilson (1967)
Syracosphaera carterae	Unknown	TLC	Parsons (1961)
[2]*Monochrysis (Pavlova) lutheri*	Unknown	TLC	Jeffrey (1968)
	Unknown	TLC	Parsons (1961)
	Plym 60	TLC	Riley and Wilson (1967)
Sphaleromantis sp.	Unknown	Paper chromatography	Jeffrey (1961)
[4]*Imantonia rotunda*	Plym 133	TLC	[3]Riley and Wilson (1967)

[1] Major pigments found by most authors were chlorophylls *a*, *c*, fucoxanthin, diadinoxanthin, and carotene.

[2] Species in which chlorophyll *c* was subsequently found to contain chlorophyll $c_1 + c_2$ from later work (Jeffrey 1969; 1976). Other species examined which also had chlorophyll c_1 and c_2 were Coccolithophorid sp. 27; *Coccolithus* sp. 21, and *Cricosphaera carterae.*

[3] These authors found chromatographic and spectral evidence for new fucoxanthin derivatives: the first evidence of acylfucoxanthins.

[4] Identification provided by Dr J. C. Green (personal communication).

Table 6.2 Haptophyte species in which full chemical characterization of carotenoids was carried out (1974–1990)

Species examined	Clone	Fucoxanthin derivatives			Reference
		Fuco	19′-hex	19′-but	
[1]*Isochrysis galbana*	WH0I, clone IS0	+	–	–	Berger *et al.* (1977)
[1]*Hymenomonas carterae*	WH0I, clone COCCO	+	–	–	Berger *et al.* (1977)
[1]*Prymnesium parvum*	WH0I, clone PRYM	+	–	–	Berger *et al.* (1977)
[1]*Pavlova lutheri*	WH0I, clone MONO	+	–	–	Berger *et al.* (1977)
[1]*Pavlova* sp.	clone NEP	+	–	–	Berger *et al.* (1977)
[2,3]*Emiliania huxleyi*	WH0I, clone BT6	–	+	–	Nørgad *et al.* (1974)
					Arpin *et al.* (1976)
					Hertzberg *et al.* (1977)
		+	+	+	Wright and Jeffrey (1987)
[1]*Coccolithus pelagicus*	WH0I	+	–	–	Fiksdahl *et al.* (1978)
[1]*Chrysochromulina polylepis*	E. Paasche clone	+	+	–	Bjerkeng *et al.* (1990)
[1]*Chrysochromulina* cf. *herdlensis*	NEPCC 186	+	+	–	Bjerkeng *et al.* (1989)
Corymbellus aureus	field sample		+		Giekes and Kraay (1986)

Code: Fuco = fucoxanthin; 19′-hex = 19′-hexanoyloxyfucoxanthin; 19′-but = 19′-butanoyloxyfucoxanthin

[1]Other carotenoids chemically characterized were β,β-carotene, diatoxanthin, diadinoxanthin, plus various minor carotenoids (includes fucoxanthinol).

[2]Other carotenoids chemically characterized were β,ϵ and β,β-carotene, diatoxanthin, diadinoxanthin, 3′-desacetyl 19′-n-hexanoyloxyfucoxanthin.

[3]Fucoxanthin and 19′-butanoyloxyfucoxanthin were identified in addition to 19′-hexanoyloxyfucoxanthin chromatographically and spectrally, but not chemically, in this clone by Wright and Jeffrey (1987).

diadinoxanthin, and diatoxanthin was found in all species. In *Emiliania huxleyi*, two fucoxanthin derivatives were found by paper chromatography (Jeffrey and Allen 1964), and a new fucoxanthin derivative was found by TLC in an unidentified flagellate, No. 133 (Riley and Wilson 1967), now known to be *Imantonia rotunda* (Dr J. C. Green, personal communication). These 'new' fucoxanthins were probably the first records of the acyl derivatives (discussed below), although they were not identified at the time. In 1969, chlorophyll *c* was found to consist of two components, c_1 and c_2, in a range of algae (Jeffrey 1969), which included five haptophyte species (Jeffrey 1976; Table 6.1).

A significant advance occurred during the period 1974–78 when full chemical characterization of the carotenoids in a small number of haptophytes (7 spp.) was carried out by Professor S. Liaaen-Jensen and co-workers. The simple pattern described above was confirmed in some strains (Berger *et al.* 1977; Table 6.2), but a new acyl fucoxanthin derivative, 19′-hexanoyloxyfucoxanthin, was chemically characterized in *Emiliania huxleyi* (clone BT6) (Norgard *et al.* 1974; Arpin *et al.* 1976; Table 6.2). Two later studies by Bjerkeng *et al.* (1989; 1990) showed 19′-hexanoyloxyfucoxanthin in *Chrysochromulina polylepis* and *Chrysochromulina* cf. *herdlensis*, and Gieskes and Kraay (1986) showed its presence in a field sample dominated by *Corymbellus aureus*. The related 19′-butanoyloxyfucoxanthin was first characterized chromatographically, spectrally (Vesk and Jeffrey 1987) and chemically (Bjørnland *et al.* 1989) in the picoplanktonic marine chrysophyte *Pelagococcus subviridis*, but also in the haptophytes *E. huxleyi* (BT6), and various clones of *Phaeocystis pouchetii* (Wright and Jeffrey 1987). One further phylogenetic complication was the finding of the acyl fucoxanthins in several dinoflagellates, e.g. *Gyrodinium aureoleum* (Tangen and Bjørnland 1981) and *Ptychodiscus brevis* (Bjørnland *et al.* 1984; Bjørnland 1989). There was no sign of haptophyte or other algal endosymbioses from ultrastructural analyses. With this range of phylogenetic distribution, the question remained as to what extent 19′-hexanoyl and 19′-butanoyloxyfucoxanthins could be used as phylogenetic markers for haptophytes, and whether other pigment suites more definitive for the group might be available. It was clear by 1989 that two pigment groups delineated by their carotenoid content existed in the haptophytes: one group had fucoxanthin as the major xanthophyll, and the second group had 19′-hexanoyloxyfucoxanthin (Bjørnland and Liaaen-Jensen 1989).

New literature reports of chlorophyll *c* derivatives

In addition to the well-established distribution patterns of chlorophylls c_1 and c_2 in algae (Jeffrey 1976), a new polar chlorophyll c_3 derivative was found in several haptophytes, diatoms (12% of 71 strains), and one chrysophyte (Jeffrey and Wright 1987; Vesk and Jeffrey 1987; Stauber and Jeffrey 1988; Jeffrey 1989). Chemical characterization (Fookes and Jeffrey 1989) showed that this compound was a carbomethoxy derivative of chlorophyll c_2 (7-dimethyl-7-methoxycarbonyl chlorophyll c_2), which replaced chlorophyll c_1 in *Emiliania huxleyi*, and some, but not all, haptophyte species which were tested (Stauber and Jeffrey 1988). Other reports suggested the presence of new chlorophyll *c* pigments in haptophytes. Nelson and Wakeham (1989) found small quantities of a non-polar chlorophyll *c* fraction in *E. huxleyi* which they suggested might be a phytylated chlorophyll c_2 derivative. Fawley (1989*a, b*) showed that chlorophyll c_1, c_2, and c_3 were present in *Prymnesium parvum* and described a new chlorophyll *c* compound, in addition to c_1 and c_2, in *Pavlova gyrans*. Xiankong *et al.* (1991) presented data on a new red-shifted chlorophyll *c* in *Dicrateria zhanjiangensis*. The latter reports need to be verified and the new compounds chemically characterized.

Pigment functions in haptophytes

Haptophyte pigments fulfil two major functions in the algal cell. The primary function is a light-harvesting and energy-transferring function, as shown by the chlorophyll *a*-*c*-fucoxanthin pigment-protein complexes isolated from *Pavlova lutheri* (Hiller *et al.* 1988), and the demonstration of energy transfer from 19′-hexanoyloxyfucoxanthin to chlorophyll *a* in *Emiliania huxleyi* (Haxo 1985). In addition, certain pigments have a photoprotective function. In haptophytes, an active epoxide cycle converts diadinoxanthin to diatoxanthin in the light, and reforms the epoxide in the dark (Hager and Stransky 1970). This mechanism apparently removes excessive excitation energy from the chloroplast *via* a thermal dissipation process (Demmig-Adams and Adams 1993). A second photoprotective function may occur in *Dicrateria inornata* (CS–254, present work), in which significant quantities of a new unidentified non-polar carotenoid are synthesized in high light. This may be a simple shading effect as found in chlorophytes such as *Dunaliella bardawil*, which synthesize massive quantities of β,β-carotene in high light and low nitrogen conditions (Ben-Amotz and Avron 1983).

Present study: algal cultures

In order to consolidate present knowledge of haptophyte pigments, we now present a new study of 29 species (50 strains) representing the four orders Isochrysidales, Coccosphaerales, Prymnesiales, and Pavlovales (Parke and Green 1976). Representatives of eight out of a possible 16 families were currently available to us in culture. Some of these came from the CSIRO Algal Culture Collection (Jeffrey 1980); others were kindly made available

Table 6.3 List of haptophyte species (including culture code numbers and source) examined in the present work (50 strains; 29 species). Growth temperatures were 15 °C or 17.5 °C for temperate strains; 27 °C for tropical strains. Light intensities were 60 $\mu E\ m^{-2}.sec^{-1}$ on 12 : 12 light : dark cycles

Order, family and species	CSIRO culture no.	Plymouth no.	Source or deposition
Order Isochrysidales			
Family Gephyrocapsaceae			
Dicrateria inornata Parke	CS-254		CCMP355
Dicrateria inornata Parke	CS-267	B	Plymouth, Devon, UK
Emiliania huxleyi (Lohmann) Hay & Mohler	CS-57		Sargasso Sea
Emiliania huxleyi (Lohmann) Hay & Mohler	PCL-06		Pipeclay Lagoon, Tasmania
Emiliania huxleyi (Lohmann) Hay & Mohler	CS-281	5/90/25j	48°20′N 17°03′W
Emiliania huxleyi (Lohmann) Hay & Mohler	CS-283	M186	Durban, South Africa
Emiliania huxleyi (Lohmann) Hay & Mohler	CS-275/2	G1779Ga	60.0°N 020.0°W
Emiliania huxleyi (Lohmann) Hay & Mohler	CS-279	DWN53/74/6	24°27′N 20°24′W
Emiliania huxleyi (Lohmann) Hay & Mohler	CS-282	1A1	32°N 062°W
Emiliania huxleyi (Lohmann) Hay & Mohler	CS-276	B92/11	Espegrend, Norway
Emiliania huxleyi (Lohmann) Hay & Mohler	CS-284	MCH-1	Sargasso Sea
Emiliania huxleyi (Lohmann) Hay & Mohler	CS-280	88E	Gulf of Maine, USA
Emiliania huxleyi (Lohmann) Hay & Mohler	CS-277	B92/18	Espegrend, Norway
Emiliania huxleyi (Lohmann) Hay & Mohler	CS-278	B92/21	Trengereid, Norway
Emiliania huxleyi (Lohmann) Hay & Mohler	CS-274	92d	50°02′N 004°22′W
Gephyrocapsa oceanica Kamptner	JB-02		Jervis Bay, NSW
Imantonia rotunda Kamptner	CS-194		Gulf of Mexico

Table 6.3 (*cont.*)

Order, family and species	CSIRO culture no.	Plymouth no.	Source or deposition
Family Isochrysidaceae			
Isochrysis galbana Parke	CS-22		Halifax, Canada
Isochrysis sp. (T.ISO)	CS-177		Tahiti
Pseudoisochrysis paradoxa Ott	CS-186		California
Chrysotilla lamellosa Anand	CS-272	408	Chelsea Physic, London.
Family Ochrosphaeraceae			
Ochrosphaera neopolitana Schussnig	CS-285	162	Salcombe, Devon, UK
Family Hymenomonadaceae			
Cricosphaera carterae (Braarud & Fagerland) Braarud	CS-40		F. T. Haxo Cr. cart
Pleurochrysis aff. *carterae* (Braarud & Fagerland) Christensen	CS-287	156	Port Erin, Isle of Man, UK
Order Prymnesiales			
Family Prymnesiaceae			
Chrysochromulina camella Leadbeater & Manton	CS-268	297	49°10′N 006°10′W
Chrysochromulina hirta Manton	CS-228		Polar Inst. Japan
Chrysochromulina kappa Parke & Manton	CS-269	K	Port Erin, Isle of Man, UK
Chrysochromulina minor Parke & Manton	CS-270	52	49°19′N 009°26′W
Chrysochromulina strobilus Parke & Manton	CS-271	43a	50°02′N 004°22′W
Chrysochromulina strobilus Parke & Manton	CS-231		Polar Inst., Japan
Chrysochromulina sp.	CS-219		Corner Inlet, Victoria
Prymnesium parvum Carter	CS-288	527	Dorset, UK

Table 6.3 (*cont.*)

Order, family and species	CSIRO culture no.	Plymouth no.	Source or deposition
Family Phaeocystaceae			
Phaeocystis pouchetii (Hariot) Lagerheim	CS-165/4		East Australian Current
Phaeocystis pouchetii (Hariot) Lagerheim	CS-165/6		East Australian Current
Phaeocystis pouchetii (Hariot) Lagerheim	CS-165/7		East Australian Current
Phaeocystis pouchetii (Hariot) Lagerheim	CS-239		Port Esperance, S. E. Tas.
Phaeocystis pouchetii (Hariot) Lagerheim	CS-240	540	47°36.9′N 8°53.0′W
Phaeocystis pouchetii (Hariot) Lagerheim	CS-241		Veldhuis
Order Coccosphaerales			
Family Zygosphaeraceae			
Crystallolithus hyalinus Gaarder & Markali	CS-273	182	50°02′N 004°22′W
Order Pavlovales			
Family Pavlovaceae			
Diacronema vlkianum Prauser	CS-266	244	Ryde, Isle of Wight, UK
Pavlova gyrans Butcher	CS-213	93	Helfard, Cornwall, UK
Pavlova lutheri (Droop) Green	CS-23		Halifax, Canada
Pavlova lutheri (Droop) Green	CS-182		Finland
Pavlova pinguis Green	CS-286	471	32°58.6′N 016°54.8′W
Pavlova salina (N. Carter) Green	CS-49		Sargasso Sea
Pavlova sp.	CS-50		Sargasso Sea
Pavlova sp.	CS-63		Port Phillip Bay, Victoria
Haptophyte (unidentified)	CS-120		Coral Sea
Haptophyte (unidentified)	CS-124		Coral Sea

from the Plymouth Collection by Dr J. C. Green. Table 6.3 gives a full list of species examined, plus the CSIRO Culture Code numbers for each strain, the Plymouth strain number where applicable, and the source of the algal isolate or its deposition. Fourteen strains of *Emiliania huxleyi* and six strains of *Phaeocystis pouchetii* were available to assess variations between strains. Cultures were grown at the CSIRO Algal Culture Laboratory according to procedures detailed in Jeffrey (1980) and Jeffrey and Wright (in preparation).

Pigment analysis

Harvested cells were extracted with buffered methanol according to Wright *et al.* (1991). High performance TLC (HPTLC) was performed on the pigment extracts to screen quickly for the presence of chlorophyll c_3, 19-hexanoyl- and 19′-butanoyloxyfucoxanthins (Vesk and Jeffrey 1987; Stauber and Jeffrey 1988). Quantitative HPLC was performed using the new high resolution method of Wright *et al.* (1991) which separates approximately 50 chlorophylls, carotenoids, and chlorophyll derivatives from mixtures of the most common algal types found in seawater. This method does not separate chlorophyll c_1 and c_2, and the unresolved mixture is hereafter referred to as chlorophyll c_1/c_2. A Spectrophysics Spectra-Focus detector was used to determine the spectrum of each peak as it emerged.

Retention times and spectral absorption characteristics of the pigments may be found in Wright *et al.* (1991). Figure 6.1 shows the absorption

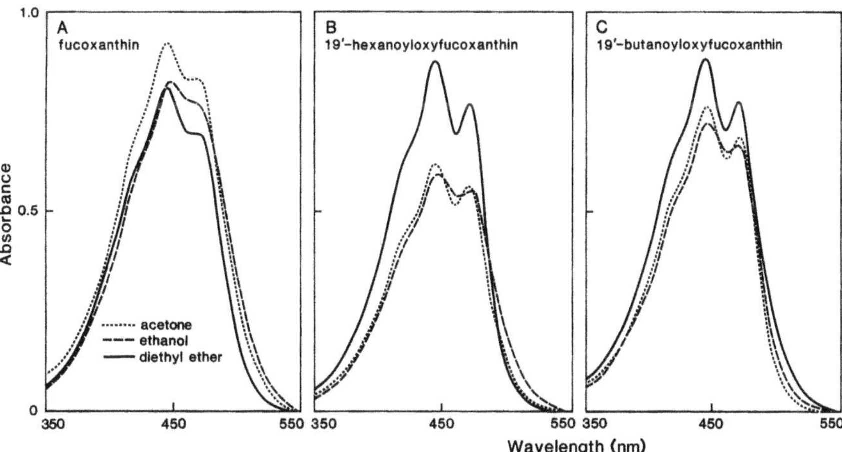

Fig. 6.1 Absorption spectra of (A) fucoxanthin, (B) 19'-hexanoyloxyfucoxanthin, and (C) 19'-butanoyloxyfucoxanthin in three solvents (acetone, ethanol, and diethyl ether). (Adapted from Wright and Jeffrey 1987).

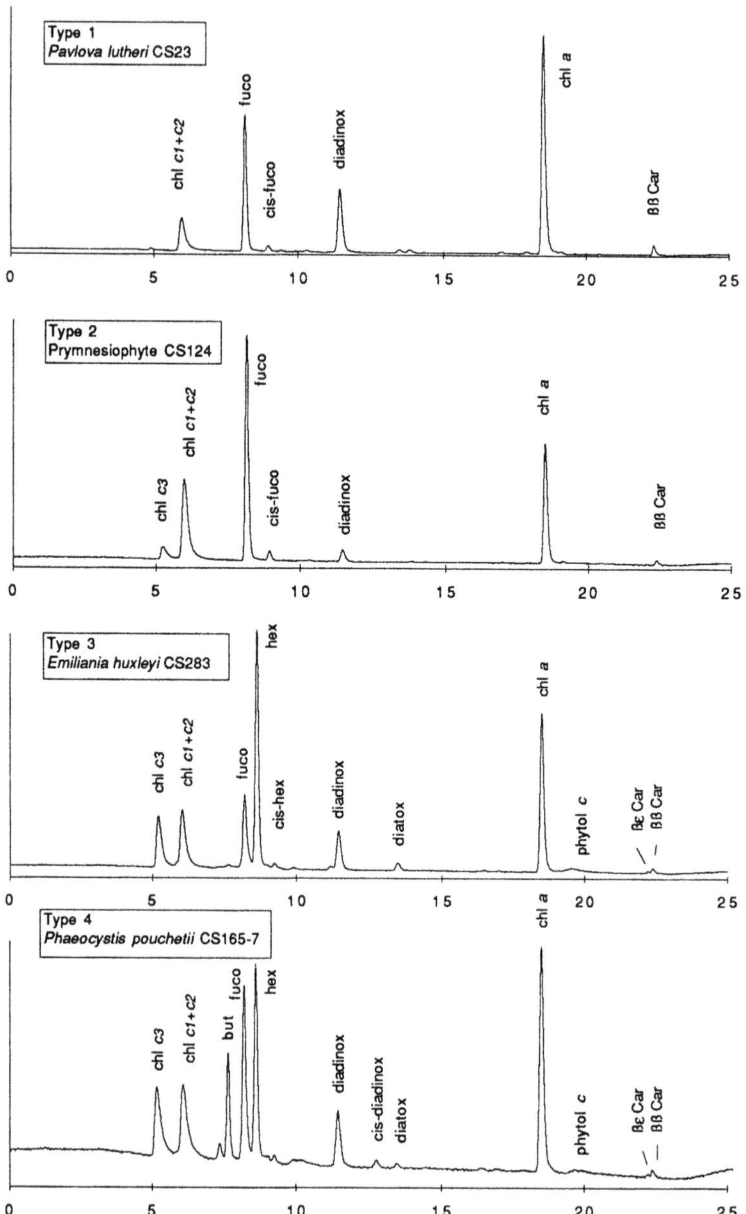

Fig. 6.2 Representative HPLC chromatograms of each pigment group: Type 1, *Pavlova lutheri* (CS 23); Type 2, Haptophyte (CS 124); Type 3, *Emiliania huxleyi* (CS 283); Type 4, *Phaeocystis pouchetii* (CS 165-7). Abbreviations: chl = chlorophyll; but = 19′-butanoyloxyfucoxanthin; fuco = fucoxanthin; hex = 19′-hexanoyloxyfucoxanthin; diadinox = diadinoxanthin; diatox = diatoxanthin; phytol c = phytolated chlorophyll *c*; β, ϵ car = β, ϵ-carotene; β, β car = β, β-carotene.

spectra of the three important fucoxanthin derivatives: fucoxanthin, 19′-hexanoyloxyfucoxanthin, and 19′-butanoyloxyfucoxanthin in standard solvents. The lack of structure in the fucoxanthin spectra compared to those of 19′-hexanoyl- and 19′-butanoyloxyfucoxanthin should be noted.

Pigment distribution within haptophytes: present work

The HPLTC screen of 50 strains showed that approximately half had chlorophyll c_3 and 19′-hexanoyloxyfucoxanthin, and provided confirmation of the HPLC analyses. Figure 6.2 shows representative HPLC chromatograms of the four pigment types encountered. All haptophytes had the core pigments chlorophylls *a*, c_1/c_2, β,β-carotene, and diadinoxanthin. Variations arose in the presence or absence of chlorophyll c_3, phytylated chlorophyll *c*, β,ε-carotene, and the acylfucoxanthins (19′-hexanoyl- and 19′-butanoyloxyfucoxanthin). Seven minor carotenoids were also detected, but were not identified. In addition to the five core pigments, pigment Type 1 haptophytes had fucoxanthin; Type 2 had chlorophyll c_3 plus fucoxanthin; Type 3 had chlorophyll c_3, 19′-hexanoyloxyfucoxanthin, fucoxanthin, and occasionally traces of the 19′-butanoyl derivative; and Type 4 had chlorophyll c_3, 19′-butanoyloxyfucoxanthin, plus variable amounts of 19′-hexanoyloxyfucoxanthin and fucoxanthin. Type 3 and Type 4 often had phytylated chlorophyll *c* present in trace quantities (<5% total chlorophyll *a*), as well as both β,ε- and β,β-carotene.

Table 6.4 gives quantitative analyses of the pigments in 16 representative haptophytes, as ng/ml culture. The last two columns give the content of chlorophyll *a* per 10^6 cells and our designation of each species to a pigment type. These are summarized for all 50 strains across the eight families in Table 6.5. All members of the Pavlovaceae (order Pavlovales), Isochrysidaceae, and Hymenomonadaceae (order Isochrysidales) belonged to the simplest haptophyte pigment Type 1. One member of each of the Ochrosphaeraceae (order Prymnesiales), Prymnesiaceae, and Phaeocystaceae (order Prymnesiales), and one unidentified species, belonged to Type 2. The majority of the Gephyrocapsaceae (16 out of 18 strains, order Isochrysidales), Prymnesiaceae (seven out of eight strains, order Prymnesiales), and one member of the Zygosphaeraceae (order Coccosphaerales) belonged in Type 3, where 19′-hexanoyloxyfucoxanthin was dominant. Most members of the Phaeocystaceae (five out of six, order Prymnesiales), and one member of the Gephyrocapsaceae (order Isochrysidales) belonged in Type 4, where 19′-butanoyloxyfucoxanthin was a major component. It is obvious that the range of pigment suites existing in the Haptophyta is not always strictly delimited by the present taxonomic assignments. Table 6.6. shows the most frequently found

Table 6.4 Concentration of pigments[a] (ng/ml culture) determined by HPLC in 16 species of haptophytes

Species	CSIRO culture no.	chl c_3	chl c_1+c_2	19′-but	Pigment fuco[b]
Chrysochromulina strobilus	CS-271	76	60	7	220
Crystallolithus hyalinus	CS-273	45	69		384
Diacronema vlkianum	CS-266		43		195
Dicrateria inornata[f]	CS-254		103		358
Dicrateria inornata	CS-267	139	123	170	94
Emiliania huxleyi	CS-57	67	113	7	19
Emiliania huxleyi	CS-275/2	48	79	6	39
Imantonia rotunda	CS-194	137	157	169	166
Isochrysis sp. (T.ISO)[g]	CS-177		129		399
Ochrosphaera neopolitana	CS-285	84	414		632
Pavlova gyrans	CS-213		86		392
Phaeocystis pouchetii	CS-165/6	22	52	90	22
Phaeocystis pouchetii	CS-239	90	112		709
Pleurochrysis aff. *carterae*	CS-287		1333		4858
Haptophyte (unidentified)	CS-124	26	173		503
Prymnesium parvum	CS-288	74	246		1090

[a] Pigment name codes: chl = chlorophyll; 19′-but = 19′-butanoyloxyfucoxanthin; fuco = fucoxanthin; 19′-hex = 19′-hexanoyloxyfucoxanthin; diadino = diadinoxanthin; diato = diatoxanthin; phytol *c* = phytolated chlorophyll c.

[b] including *cis*-fuco (average 3.1% of total fuco)

[c] including *cis*-19′-hex (average 3.7% of total 19′-hex)

[d] including *cis*-diadino (average 3.7% of total diadino)

[e] including chlorophyllide *a*, allomer, epimer (average 1.6% of total chl *a*)

[f] *Dicrateria inornata* (CS-254) contained in addition 77.0 and 169.1 $\mu g\ l^{-1}$ of an unknown carotenoid and a carotenoid resembling echinenone, respectively.

[g] *Isochrysis* sp. (CS-177) also contained 13.5 $\mu g\ l^{-1}$ of the echinenone-like carotenoid.

pigments in the 50 strains. All strains (100%) contained fucoxanthin, 68% contained chlorophyll c_3, 56% contained 19′-hexanoyloxyfucoxanthin (mostly Gephyrocapsaceae and Prymnesiaceae), and only 14% contained 19′-butanoyloxyfucoxanthin (mostly Phaeocystaceae). While certain pigments or pigment suites are characteristic of individual species (e.g. *Emiliania* or *Phaeocystis*), no one pigment pattern is characteristic of the class.

The concentration of chlorophyll *a* found in the 50 strains of haptophytes is of interest, with a mean of 0.41 μg chlorophyll *a* and a range of 0.049 to 4.5 μg per 10^6 cells. These reflect pigment concentrations under

Table 6.4 (*cont.*)

19′-hex[c]	diadino[d]	diato	chl *a*[e]	phytol *c*	carotenes	Chl *a* µg per 10^6 cells	Pigment type
350	51	11	449	4.0	12		3
21	38	4.8	369		8.7	0.785	3
	113	4.3	486		74	0.124	1
	154	8.7	803		22	0.184	1
916	104		858	9.2	15	0.049	4
640	41	1.1	497	12	13	0.151	3
399	35	9.8	320	7.9	7.2	0.087	3
837	176	26	1002	13	16		4
	78	4.4	422		15	0.154	1
	75	8.7	1359		36	0.432	2
	190	11	769		60	0.281	1
111	30	1.0	172	2.6	3.7	0.149	4
	67	0.7	636	9.1	16	0.208	2
	1132	42	9492		359	4.551	1
	25		457		7.5		2
0.1	138	20	1369		45	0.387	2

a light regime of 60 µE m^{-2} sec^{-1} (12:12h light:dark cycles) and will obviously increase and decrease with lower or higher light regimes, respectively (Vesk and Jeffrey 1977; Brown *et al.* 1993), and sometimes with light quality (Jeffrey and Vesk 1981). The values may be compared with those of 50 diatoms, cultured under similar conditions (Stauber and Jeffrey 1988), in which the range for chlorophyll *a* plus *c* was 0.02 µg in the smallest diatom, *Extubocellulus spinifer*, to 174.39 µg per 10^6 cells in the largest diatom, *Coscinodiscus* sp. A similar study with 22 cultured dinoflagellates (Jeffrey *et al.* 1975) gave a range for chlorophylls *a* plus *c* of 1.03 in the smallest dinoflagellate, *Gymnodinium punctatum*, to 141 µg per 10^6 cells in the largest dinoflagellate, *Gonyaulax polyedra*. The haptophytes are obviously in the lower range of pigment concentration per cell, corresponding to their smaller size (2–20 µm).

Strain variations in pigment composition

Variations in pigment composition between strains of a single species were assessed from ratios of fucoxanthin and derivatives to chlorophyll *a* in 14 strains of *Emiliania huxleyi* and six strains of *Phaeocystis pouchetii*, grown

Table 6.5 Pigment types and haptophyte families

	No. of strains in pigment types*			
	1	2	3	4
Order Isochrysidales				
Family Gephyrocapsaceae	1	–	16	1
Isochrysidaceae	4	–	–	–
Ochrosphaeraceae	–	1	–	–
Hymenomonadaceae	2	–	–	–
Order Prymnesiales				
Family Prymnesiaceae	–	1	7	–
Phaeocystaceae	–	1	–	5
Order Coccosphaerales				
Family Zygosphaeraceae	–	–	1	–
Order Pavlovales				
Family Pavlovaceae	8	–	–	–
Unidentified	1	1	–	–

* All groups have chlorophyll *a*, c_1/c_2, ββ-carotene, diadinoxanthin and diatoxanthin. In addition: Type 1 has fucoxanthin; Type 2 has chlorophylls c_3 and fucoxanthin; Type 3 has chlorophyll c_3 and 19′-hexanoyloxyfucoxanthin, fucoxanthin ± <5% 19′-butanoyloxyfucoxanthin; and Type 4 has chlorophylls c_3 and 19′-butanoyloxyfucoxanthin, plus variable quantities of fucoxanthin and 19′-hexanoyloxyfucoxanthin.

under identical conditions. In nine strains of *Emiliania huxleyi*, 19′-hexanoyloxyfucoxanthin was the major pigment compared to fucoxanthin, in the other five *Emiliania* strains, fucoxanthin dominated, while in all strains 19′-butanoyloxyfucoxanthin was either absent or present only in trace quantities. Wide variations were also seen in the ratios of *Phaeocystis pouchetii* pigments, with 19′-hexanoyloxyfucoxanthin dominant in three strains, present in minor quantities in two strains, and absent in one strain. In one strain (CS–239) fucoxanthin was the major pigment with 19′-butanoyl and 19′-hexanoyloxyfucoxanthins both undetectable. Whether all these *Phaeocystis* strains are indeed of the species *P. pouchetii* needs to be addressed, to attempt to explain the divergent pigment contents.

A further anomaly was seen in the pigment composition of the two *Dicrateria inornata* strains where CS–257 had Type 1 pigmentation and CS–267 had that of Type 4. We presume that these strains had been correctly identified.

Table 6.6 Number of haptophyte strains (total = 50) with definitive pigments

Pigment	No. of strains
chlorophyll c_3	34
fucoxanthin	50
19′-hexanoyloxyfucoxanthin (>5% of total fucoxanthins)	28
19′-hexanoyloxyfucoxanthin (minor pigment, <5% of total fucoxanthins)	2
19′-butanoyloxyfucoxanthin (>5% of total fucoxanthins)	7
19′-butanoyloxyfucoxanthin (minor or trace pigment, <5% of total fucoxanthins)	14
phytolated chlorophyll *c* (<5% total chlorophyll *a*)	28

Minor carotenoids and chlorophyll c_1

Some unidentified carotenoids were encountered in the 50 strains whose retention times, spectral characteristics, and abundances will be listed elsewhere (Jeffrey and Wright, in preparation). Unknown carotenoids 1 and 3, which were never abundant, resembled fucoxanthin spectrally, but eluted before it. Unknown 2, which was found mainly in strains of *Phaeocystis pouchetii*, resembled, but ran ahead of, 19′-butanoyloxyfucoxanthin. Unknown 4 was found only in strains of *Emiliania huxleyi*, whereas unknown 5, which followed diatoxanthin, was found only in the Pavlovaceae and one species of *Chrysochromulina*. Unknown 6 was found only in one strain of *Dicrateria inornata*, but was a significant peak in that sample. Unknown 7, which resembled echinenone, was also a major peak in *Dicrateria inornata*, and was found in conjunction with unknown 6. It was a minor component in a few other species.

The HPLC method used here does not separate chlorophylls c_1 and c_2. On the basis of previous work (reviewed by Jeffrey 1989), it is expected that where chlorophyll c_3 is absent, the chlorophyll *c* fraction will contain both c_1 and c_2, and where chlorophyll c_3 is present it replaces c_1. The precise distribution of the chlorophyll *c* fractions of the 50 strains will be examined and described in Jeffrey and Wright (in preparation).

Summary of the present status of haptophyte pigmentation

The present study has added a further step to our understanding of the photosynthetic pigments of the haptophytes. Professor S. Liaaen-Jensen and co-workers had designated two haptophyte pigment groups, those which contained fucoxanthin or 19′-hexanoyloxyfucoxanthin (Bjerkeng *et al.* 1989). We have extended this to four types, by including both the distribution of the chlorophyll *c* pigments, and 19′-butanoyloxyfucoxanthin. While certain families now fit neatly into only one pigment type (Table 6.5), others do not, and not all strains possess the same suites of pigments that could be considered characteristic of the group as a whole (Table 6.6). Haptophyte pigment Types 1 and 2 are identical to those of diatoms (Stauber and Jeffrey 1988), and Types 3 and 4, with chlorophyll c_3 and the acyl fucoxanthins, are also characteristic of a marine chrysophyte and three bloom-forming dinoflagellates (Vesk and Jeffrey 1987; Bjørnland *et al.*, 1989; Bjørnland and Liaaen-Jensen 1989). The minor pigments, phytolated chlorophyll *c* and β,ε-carotene, are linked with Types 3 and 4, and may provide additional chemotaxonomic clues. It is obvious that our search for a definitive pigment signature for microalgae classed as haptophytes has not succeeded, and possibly taxonomic re-assignments may be necessary. From the point of view of oceanographic applications, it appears that the taxonomic identification of haptophyte populations will still be necessary to validate HPLC-derived pigment signatures.

References

Allen, M. B., Goodwin, T. W., and Phagpolngarm, S. (1960). Carotenoid distribution in certain naturally occurring algae and in some artificially induced mutants of *Chlorella pyrenoidosa. Journal of General Microbiology*, **23**, 93–103.

Arpin, N., Svec, W. A., and Liaaen-Jensen, S. (1976). New fucoxanthin-related carotenoids from *Coccolithus huxleyi. Phytochemistry*, **15**, 529–32.

Ben-Amotz, A. and Avron, M. (1983). On the factors which determine massive β-carotene accumulation in the halotolerant alga *Dunaliella bardawil. Plant Physiology*, **72**, 593–7.

Berger, R., Liaaen-Jensen, S., McAlister, V., and Guillard, R. R. L. (1977). Carotenoids of Prymnesiophyceae (Haptophyceae). *Biochemical Systematics and Ecology*, **5**, 71–5.

Bjerkeng, B., Haugan, J. A., and Liaaen-Jensen, S. (1989). Microalgal carotenoids and chemosystematics: updating the Prymnesiophyceae. *Biological Oceanography*, **6**, 271–8.

Bjerkeng, B., Vernet, M., Nielsen, M. V., and Liaaen-Jensen, S. (1990) Carotenoids of *Chrysochromulina polylepis* (Prymnesiophyceae). *Biochemical Systematics and Ecology*, **18**, 303–6.

Bjørnland, T. (1989). Carotenoid structures and lower plant phylogeny. In *Carotenoids: chemistry and biology*, (ed. N. I. Krinsky, M. M. Mathews-Roth, and R. F. Taylor), pp. 21–37. Plenum, New York

Bjørnland, T. and Liaaen-Jensen, S. (1989). Distribution patterns of carotenoids in relation to chromophyte phylogeny and systematics. In *The chromophyte algae: problems and perspectives*, (ed. J. C. Green, B. S. C. Leadbeater, and W. L. Diver), pp. 37–60. Clarendon Press, Oxford.

Bjørnland, T., Pennington, F. C., Haxo, F. T., and Liaaen-Jensen, S. (1984). Carotenoids of Chrysophyceae and Dinophyceae 'Coc. Min Haltenbanken' and *Gymnodinium breve* (Florida red tide). In *7th International IUPAC Symposium on Carotenoids*, Munich, p. 21.

Bjørnland, T., Liaaen-Jensen, S., and Throndsen, J. (1989). Carotenoids of the marine chrysophyte *Pelagococcus subviridis*. *Phytochemistry*, **28**, 3347–53.

Brown, M. R., Dunstan, G. A., Jeffrey, S. W., Volkman, J. K., Barrett, S. M., and LeRoi, J. M. (1993). The influence of irradiance on the biochemical composition of *Isochrysis* sp. clone T.ISO. *Journal of Phycology*, **29**, 601–12.

Christensen, T. (1962). Alger. In *Botanik*, Bd. 2, *Systematisk Botanik*, Nr. 2, (ed. T. W. Böcher, M. Lange, and T. Sørensen), pp. 1–178. Munksgaard, Copenhagen.

Dales, R. P. (1960). On the pigments of the Chrysophyceae. *Journal of the Marine Biological Association of the United Kingdom*, **39**, 693–9.

Davidson, A. T. and Marchant, H. J. (1992). The biology and ecology of *Phaeocystis* (Prymnesiophyceae). *Progress in Phycological Research*, **8**, 1–45.

Demmig-Adams, B. and Adams, W. W. (1993). The xanthophyll cycle. In *Carotenoids in photosynthesis*, (ed. A. Young and G. Britton), pp. 206–51. Chapman & Hall, Bury St. Edmunds.

Fawley, M. W. (1989*a*). A new form of chlorophyll *c* involved in light-harvesting. *Plant Physiology*, **91**, 727–32.

Fawley, M. W. (1989*b*). Detection of chlorophyll c_1, c_2, c_3 in pigment extracts of *Prymnesium parvum* (Prymnesiophyceae). *Journal of Phycology*, **25**, 601–4.

Fiksdahl, A., Liaaen-Jensen, S., and Siegelman, H. W. (1978). Carotenoids of *Coccolithus pelagicus*. *Biochemical Systematics and Ecology*, **7**, 47–8.

Fookes, C. J. R. and Jeffrey, S. W. (1989). The structure of chlorophyll c_3, a novel marine photosynthetic pigment. *Journal of the Chemical Society, Chemical Communications*, **23**, 1827–8.

Gieskes, W. W. C. and Kraay, G. W. (1986). Analysis of phytoplankton pigments by HPLC before, during and after mass occurrence of the microflagellate *Corymbellus aureus* during the spring bloom in the open North Sea in 1983. *Marine Biology* (Berlin), **92**, 45–52.

Hager, A. and Stransky, H. (1970). The carotenoid pattern and the occurrence of the light-induced xanthophyll cycle in various classes of algae. *Archiv für Mikrobiologie*, **73**, 77–89.

Hallegraeff, G. M. (1984). Coccolithophorids (calcareous nanoplankton) from Australian waters. *Botanica Marina*, **27**, 229–47.

Hasle, G. R. (1960). Plankton coccolithophorids from the Subantarctic and Equatorial Pacific. *Nytt Magasin for Botanikk*, **8**, 77–91.

Haxo, F. T. (1985). Photosynthetic action spectrum of the coccolithophorid *Emiliania huxleyi* (Haptophyceae): 19′-hexanoyloxyfucoxanthin as antenna pigment. *Journal of Phycology*, **21**, 282–7.

Hertzberg, S., Mortensen, T., Borch, G., Siegelman, H. W., and Liaaen-Jensen, S. (1977). On the absolute configuration of 19'-hexanoyloxyfucoxanthin. *Phytochemistry*, **16**, 587–90.

Hibberd, D. J. (1980). Prymnesiophytes (= Haptophytes). In *Developments in marine biology*, Vol. 2, *Phytoflagellates*, (ed. E. R. Cox), pp. 273–317. Elsevier North Holland, New York.

Hiller, R. G., Larkum, A. W. D., and Wrench, P. M. (1988). Chlorophyll proteins of the prymnesiophyte *Pavlova lutheri* (Droop) comb. nov.: identification of the major light-harvesting complex. *Biochimica Biophysica Acta*, **932**, 223–31.

Jeffrey, S. W. (1961). Paper-chromatographic separation of chlorophylls and carotenoids from marine algae. *Biochemical Journal*, **80**, 336–42.

Jeffrey, S. W. (1968). Quantitative thin-layer chromatography of chlorophylls and carotenoids from marine algae. *Biochimica Biophysica Acta*, **162**, 271–85.

Jeffrey, S. W. (1969). Properties of two spectrally different components in chlorophyll *c* preparations. *Biochimica Biophysica Acta*, **177**, 456–67.

Jeffrey, S. W. (1976). The occurrence of chlorophyll c_1 and c_2 in algae. *Journal of Phycology*, **12**, 349–54.

Jeffrey, S. W. (1980). Cultivating uni-cellular marine plants. *CSIRO Division of Fisheries and Oceanography Annual Report*, 1977–79, pp. 22–43.

Jeffrey, S. W. (1989). Chlorophyll *c* pigments and their distribution in the chromophyte algae. In *The chromophyte algae: problems and perspectives*, (ed. J. C. Green, B. S. C. Leadbeater and W. L. Diver), pp. 13–36. Clarendon Press, Oxford.

Jeffrey, S. W. and Allen, M. B. (1964). Pigments, growth and photosynthesis in cultures of two Chrysomonads, *Coccolithus huxleyi* and a *Hymenomonas* sp. *Journal of General Microbiology*, **36**, 277–88.

Jeffrey, S. W., Sielicki, M. and Haxo, F. T. (1975). Chloroplast pigment patterns in dinoflagellates. *Journal of Phycology*, **11**, 374–84.

Jeffrey, S. W. and Vesk, M. (1977). The effect of blue-green light on photosynthetic pigments and chloroplast structure in the marine diatom *Stephanopyxis turris*. *Journal of Phycology*, **13**, 271–9.

Jeffrey, S. W. and Vesk, M. (1981). Blue-green light effects in microalgae: enhanced thylakoid and chlorophyll synthesis. *Photosynthesis VI*, (ed. G. Akoyunoglou), pp. 435–42. Balaban International Science Services, Philadelphia.

Jeffrey S. W. and Wright, S. W. (1987). A new spectrally distinct component in preparations of chlorophyll *c* from the microalga *Emiliania huxleyi* (Prymnesiophyceae). *Biochimica Biophysica Acta*, **187**, 180–8.

Lancelot, C., Billen, G., Sournia, A., Weisse, T., Colijn, F., Veldhuis, M. J. W., *et al.* (1987). *Phaeocystis* blooms and nutrient enrichment in the continental coastal zones of the North Sea. *Ambio* **16**, 38–46.

Malin, G., Turner, S. M., and Liss, P. S. (1992). Sulphur: the plankton/climate connection. *Journal of Phycology*, **28**, 590–7.

Moestrup, Ø. (1979). Identification by electron microscopy of marine nanoplankton from New Zealand, including the description of four new species. *New Zealand Journal of Botany*, **17**, 61–95.

Nelson, J. R. and Wakeham, S. G. (1989). A phytol-substituted chlorophyll *c* from *Emiliana huxleyi* (Prymnesiophyceae). *Journal of Phycology*, **25**, 761–6.

Norgard, S., Svec, W. A., and Liaaen-Jensen, S. (1974). Algal carotenoids and chemotaxonomy. *Biochemical Systematics and Ecology*, **2**, 7–9.

Parke, M. and Green, J. C. (1976). Haptophyta. In *Checklist of British marine algae — third revision*, (ed. M. Parke and P. S. Dixon). *Journal of the Marine Biological Association of the United Kingdom*, **56**, 551–5.

Parsons, T. R. (1961). On the pigment composition of eleven species of marine phytoplankters. *Journal of the Fisheries Research Board of Canada*, **18**, 1017–25.

Riley, J. P. and Wilson, T.R.S. (1967). The pigments of some marine phytoplankton species. *Journal of the Marine Biological Association of the United Kingdom*, **47**, 351–62.

Stauber, J. L. and Jeffrey, S.W. (1988). Photosynthetic pigments in fifty-one species of marine diatoms. *Journal of Phycology*, **24**, 158–72.

Tangen, K. and Bjørnland, T., (1981). Observations on pigments and morphology of *Gyrodinium aureolum* Hulbert, a marine dinoflagellate containing 19′-hexanoyloxyfucoxanthin as main carotenoid. *Journal of Plankton Research*, **3**, 389–401.

Underdal, B., Skulberg, O. M., Dahl, E., and Aune, T. (1989). Disastrous bloom of *Chrysochromulina polylepis* (Prymnesiophyceae) in Norwegian coastal waters 1988 — mortality in marine biota. *Ambio*, **18**, 265–70.

Vesk, M., Jeffrey, S. W., and Hallegraeff, G. M. (1990). Golden-brown algae: Prymnesiophyta and Chrysophyta. In *Biology of marine plants*, (ed. M. N. Clayton and R. J. King), pp. 96–114. Longman-Cheshire, Melbourne.

Vesk, M. and Jeffrey, S. W. (1977). Effect of blue-green light on photosynthetic pigments and chloroplast structure in uni-cellular algae from six classes. *Journal of Phycology*, **13**, 280–8.

Vesk, M. and Jeffrey, S. W. (1987). Ultrastructure and pigments of two strains of the picoplanktonic alga *Pelagococcus subviridis* (Chrysophyceae). *Journal of Phycology*, **23**, 322–36.

Wright, S. W. and Jeffrey, S. W. (1987). Fucoxanthin pigment markers of marine phytoplankton analysed by HPLC and HPTLC. *Marine Ecology Progress Series,* **38**, 259–66.

Wright, S. W., Jeffrey, S. W., Mantoura, R. F. C., Llewellyn, C. A., Bjørnland, T., Repeta, D. *et al.* (1991). Improved HPLC method for the analysis of chlorophylls and carotenoids from marine phytoplankton. *Marine Ecology Progress Series,* **77**, 183–96.

Xiankong, Z., Junmin, P., and Qifang, L. (1991). Some properties of a new chlorophyll *c* component from *Dicrateria zhanjiangensis. Abstracts* of the 4th *International Phycological Congress, Duke University, USA,* p. 80.

7. Cellular regulation during calcification in *Emiliania huxleyi*

C. BROWNLEE
Marine Biological Association, The Laboratory, Plymouth, UK

N. NIMER, L. F. DONG, and M. J. MERRETT
School of Biological Sciences, University College of Swansea, UK

Abstract

The evidence for the coordination of calcification and cellular metabolism in the coccolithophorid, *Emiliania huxleyi*, is reviewed. Experimental evidence strongly suggests that the carbon source for photosynthesis is CO_2, while calcification uses HCO_3^- as an external substrate. Calcification is dependent on photosynthesis both as a supply of energy and reducing power, and as a sink for CO_2 and protons produced during the net calcification reaction. Calcification, on the other hand, can provide a source of CO_2 for photosynthesis in the absence of any CO_2 concentrating mechanism. Spatial separation of calcite formation and CO_2 production occurs, with calcite formation occurring in the Golgi/coccolith vesicle. Measurements of intracellular pH and intracellular dissolved inorganic carbon indicate that significant CO_2 production can occur in the cytosol. Calcification and photosynthesis appear to be under both thermodynamic and kinetic control. The opposing effects of calcification and photosynthesis on cytosolic pH invoke a likely role for pH in coordination of these processes. Tight control of calcification and metabolism can also occur via control of membrane transport of HCO_3^- and Ca^{2+} ions. Calcium ions probably play a dual role as a substrate for calcification and as a regulator of cytosolic processes.

Introduction

Calcification is widespread among the algae (see Borowitzka 1982, 1987; Simkiss and Wilbur 1989; Crick 1989 for reviews). In most species, calcification occurs extracellularly and is the combined result of spatially separated fluxes of CO_2 and ions (particularly H^+, OH^-, and HCO_3^-) across

The Haptophyte Algae (ed. J. C. Green and B. S. C. Leadbeater), Systematics Association Special Volume No. 51, pp. 133–48. Clarendon Press, Oxford, 1994.

the plasma membranes of specialized cells, such as occurs in *Halimeda* (Borowitzka and Larkum 1976) and *Potamogeton* (Prins *et al.* 1980, 1982), or in different regions of the same cell as in *Chara*, (e.g. Lucas 1983, 1985). The unicellular coccolithophorids represent a special case, producing $CaCO_3$ deposits as complete coccoliths in intracellular compartments (Golgi/coccolith vesicles) which are subsequently transported to the outer cell surface. This requires the close coordination of calcification with the rest of cellular metabolism, including photosynthesis, and places special requirements and constraints on both metabolism and calcification.

Calcification and photosynthesis can be subject to both thermodynamic and kinetic control. With respect to thermodynamic control, availability of substrates and energy supply for their transport (or removal of products) determines the rates at which reactions can proceed. Kinetic control involves direct regulation of enzyme or transporter activity by substrates, products, second messengers, or other regulatory molecules in the cell. Examples of both types of regulation appear to operate in the co-ordination of calcification and metabolism in *E. huxleyi*, and these will be discussed below.

Carbon sources and products in calcifying cells

The major products of inorganic carbon uptake by calcifying cells of *Emiliania huxleyi* are organic carbon (fixed primarily by photosynthesis) and calcite (calcification). The major enzyme responsible for organic carbon fixation is the photosynthetic enzyme, ribulose bisphosphate carboxylase oxygenase (Rubisco; see Edwards and Walker 1983). In marine algae, as in all photosynthetic plants, Rubisco uses CO_2 exclusively as its carbon source (Raven and Johnston 1991), producing 3-phosphoglycerate as the primary product which then enters the Calvin cycle. Several lines of evidence indicate that the external substrate for calcification in *E. huxleyi* is HCO_3^- while CO_2 is the substrate for photosynthesis. (see Borowitzka 1987; Sikes and Fabry 1993 for recent reviews). Sikes *et al.* (1980) exploited the relatively slow kinetics of the hydration and dehydration of CO_2 (K_1 in equation (7.1): $t_{1/2}$=20 s at 20 °C; Sikes *et al.* 1980; Lehman 1978) to discriminate between the carbon sources for photosynthesis and calcification. Thus, in short term experiments, ^{14}C label supplied as $^{14}CO_2$ will accumulate relatively slowly (over several seconds) in HCO_3^- according to the reaction

$$CO_2 + H_2O \overset{K_1}{\leftrightarrow} H_2CO_3 \overset{K_2}{\leftrightarrow} HCO_3^- + H^+ \overset{K_3}{\leftrightarrow} CO_3^{2-} + 2H^+ \qquad (7.1)$$

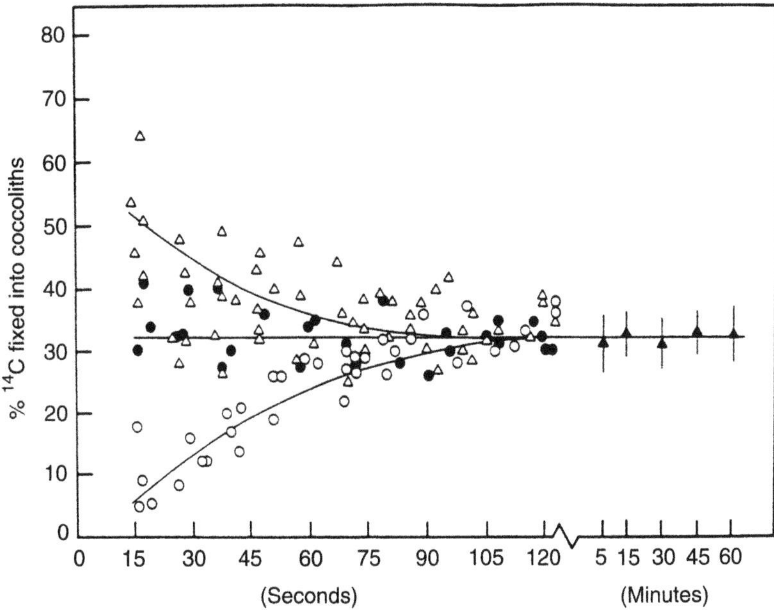

Fig. 7.1 Short-term fixation of ^{14}C into coccoliths of *E. huxleyi* (isotopic disequilibrium) when label was supplied as $H^{14}CO_3^-$(Δ), $^{14}CO_2$(○) or at equilibrium concentrations of HCO_3^-, CO_2 and CO_3^{2-}(●). (▲ = mean for all treatments.) (Reproduced, with permission, from Sikes *et al.* 1980).

and *vice versa* for label supplied as HCO_3^-, assuming no external carbonic anhydrase activity. In experiments with calcifying cells where ^{14}C was supplied in the form of $H^{14}CO_3^-$, an initial enhancement of carbon fixation into coccoliths occurred, concomitant with a delay in appearance of label into photosynthate (Fig. 7.1). Likewise, when supplied as $^{14}CO_2$, label was initially preferentially fixed into photosynthate with a lag in the appearance of label in calcite. When ^{14}C was added at equilibrium levels of CO_2, HCO_3^- and CO_3^{2-}, no enhancement or lag in the appearance of label into photosynthate or calcite occurred. The conclusion from these experiments was that CO_2 was the external substrate for photosynthesis while HCO_3^- was the source of carbon for calcification. This conclusion was supported in the same study by the demonstration that calcifying cells reduced the alkalinity of the medium by two moles per mole of calcite produced. Non-calcifying cells did not change the alkalinity of the medium.

Further support for HCO_3^- as the external carbon source for calcification in *E. huxleyi* came from measurements of the $\delta^{13}C$ of calcifying and non-calcifying cells (Sikes and Wilbur 1982). The $\delta^{13}C$ value is a measure of the discrimination of the process in favour of the stable carbon isotopes, ^{12}C and ^{13}C. Processes which discriminate in favour of ^{12}C have low $\delta^{13}C$ values and those showing no discrimination or discrimination in favour of ^{13}C, have higher $\delta^{13}C$ values. Both the dissolution of CO_2 and its fixation by Rubisco discriminate in favour of ^{12}C, whereas conversion of CO_2 to HCO_3^- discriminates in favour of ^{13}C (Deuser and Degens 1967; Farquahar *et al.* 1989). Sikes and Wilbur (1982) showed that coccoliths from calcifying cells had higher $\delta^{13}C$ values than cells with coccoliths removed.

Calcification-photosynthesis relations

Dependence of calcification on photosynthesis

Since the early work of Paasche (1962, 1963, 1964, 1965), a large body of evidence has accumulated demonstrating the dependence of calcification on photosynthesis in *Emiliania huxleyi.* This evidence has been summarized and discussed in several recent reviews (Borowitzka 1987; Simkiss and Wilbur 1989; Sikes and Fabry 1993). Calcification in *E. huxleyi* is strongly light dependent (Paasche 1962, 1968; Nimer and Merrett 1993), both with respect to photon flux density (Paasche 1962; Nimer and Merrett 1993) and light-dark regime (Linschooten *et al.* 1991), though calcification was shown by Paasche (1964) to reach saturation at lower photon flux density than photosynthesis. The photosynthesis inhibitor CMU was shown by Paasche (1965) to inhibit coccolith formation to a lesser extent than photosynthesis, though more recent experiments (Nimer and Merrett, unpublished data) suggest a much greater inhibition of calcification by DCMU in a high-calcifying strain, in excess of the inhibition of photosynthetic carbon fixation.

There are two obvious explanations for the overall dependence of calcification on photosynthesis in *E. huxleyi.* As a sink for CO_2, photosynthesis will act to drive the net reaction of calcification

$$2HCO_3^- + Ca^{2+} \rightarrow CaCO_3 + CO_2 + H_2O \qquad (7.2)$$

by maintaining an intracellular CO_2 gradient and a flux of CO_2 away from the site of calcification. However, the small cell size and relatively rapid rate of diffusion of CO_2 imply that calcification should not be completely inhibited in the absence of photosynthesis, and other factors are likely to be involved in the dependence of calcification on photosynthesis. Photosynthesis will maintain elevated ATP/ADP ratios in the cytosol (Santarius and Heber 1965), though this has not been demonstrated directly in *E. huxleyi*, and a supply of reducing power in the form of NADH and NADPH (Heber and Heldt 1981). These may be essential for the provision of energy for transport processes at the plasma membrane and the Golgi vesicle membrane. Additional control of calcification and photosynthesis may be exerted via cytosolic pH and $[Ca^{2+}]$ (see below).

Dependence of photosynthesis on calcification

Carbon dioxide produced during calcification (reaction 7.2) can potentially be used by photosynthesis to supplement CO_2 diffusion from the external medium if other carbon accumulation mechanisms are absent. This has led to the hypothesis that the primary function of calcification is to supply CO_2 for photosynthesis where the concentration of available CO_2 may normally be limiting (Sikes *et al.* 1980), though calcification alone does not bypass any need for carbonic anhydrase (see below). Raven and Johnston (1991) summarized evidence for active carbon accumulation (indicating the presence of a dissolved inorganic carbon (DIC) pump) in several marine phytoplankton species.

Criteria for the presence of a DIC pump include saturation of photosynthesis at sea water concentrations of HCO_3^- and a $K_{1/2}(CO_2)$ (i.e. the concentration of dissolved CO_2 at which half the maximum rate of photosynthesis, measured as photosynthetic O_2 evolution or ^{14}C uptake in long-term experiments, occurs) significantly lower than the K_m for Rubisco. Two species, however, (*Stichococcus minor* and *E. huxleyi*) had $K_{1/2}(CO_2)$ values similar to the K_m for Rubisco, i.e. low affinity, suggesting the absence of a CO_2-concentrating mechanism. It was not clear, however whether the *E. huxleyi* cells referred to in this study (Raven and Johnston 1991) were calcifying. Earlier work by Steemann-Nielsen (1966) showed that photosynthesis in *E. huxleyi* at saturating light levels was not saturated at sea water $[CO_2]$, suggesting the absence of a CO_2-concentrating mechanism. More recently, Nimer and Merrett (1993), measuring photosynthesis using ^{14}C interpreted a similar result with calcifying cells of *E. huxleyi* at external pH 8.3, in terms of uptake of HCO_3^-, and calcification. At high irradiance, and high external [DIC] ($[DIC]_{ext}$), CO_2 may not be produced via calcification at a rate sufficient to saturate photosynthesis, allowing diffusive entry of CO_2 to supply photosynthetic requirements.

There is no absolute dependence of photosynthesis on calcification. Paasche (1964) for example, demonstrated a marked inhibition of calcification in the presence of low $[Ca^{2+}]_{ext}$. Photosynthesis, on the other hand, could be maintained at maximal rate at $[Ca^{2+}]_{ext}$ which markedly inhibited calcification. Lower $[Ca^{2+}]_{ext}$ (< 1.0 mM) caused a sharp inhibition of photosynthesis. However, previously decalcified cells, while showing similar inhibition of coccolith formation in low $[Ca^{2+}]_{ext}$, did not show reduced photosynthesis at low $[Ca^{2+}]_{ext}$, suggesting that the inhibition of photosynthesis at low $[Ca^{2+}]_{ext}$ in calcified cells was *via* some effect unrelated to calcification. However, in these experiments, at external pH 8.0 and $[DIC]_{ext}$ 3.4 mM, $[CO_2]_{ext}$ would be around 30 μM (Paasche 1964). This is higher than the generalized K_m for algal Rubisco (~12 μM; Raven and Johnston 1991) and it is likely that diffusive CO_2 entry could account for the photosynthetic rates observed. Paasche (1964), comparing photosynthesis at varying $[CO_2]_{ext}$, also deduced that both external CO_2 and HCO_3^- could be used, when available, as external photosynthetic substrates in calcifying cells. Recently, we have observed a close correlation between calcification and photosynthesis at different $[Ca^{2+}]_{ext}$ at higher pH_{ext} (8.3) (Fig. 7.2) where diffusive CO_2 entry may have a more limiting effect on photosynthesis ($[CO_2]_{ext} < 10$ μM) and where CO_2 released by calcification may supply a larger proportion of the substrate for Rubisco.

Comparison of high- and low-calcifying strains of *E. huxleyi* has provided additional insights into the role of calcification as a possible source of CO_2 for photosynthesis. Nimer and Merrett (1992) showed that, at high photon flux density and external pH 8.3, photosynthesis by high-calcifying cells increased with external pH while that of low-calcifying cells was severely inhibited above external pH 7.0. Low-calcifying cells had a higher affinity for external CO_2 than high-calcifying cells. Photosynthetic O_2

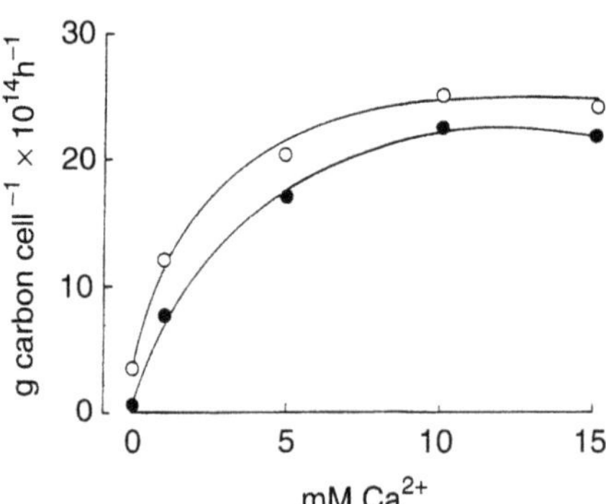

Fig. 7.2 Rates of calcification (●) and photosynthetic CO_2 fixation (○) in a calcifying strain of *E. huxleyi* (strain 88E) in response to different $[Ca^{2+}]$ext at pH 8.3 (2.0 mM DIC, 20 μM NO_3^-, photon flux density 50 μmolm^{-2}s^{-1}, 15 °C). Photosynthesis and calcification were monitored as described in Nimer and Merrett (1992, 1993).

evolution was saturated at $[DIC]_{ext}$ below 1.0 mM in high-calcifying cells at pH 8.3 and high irradiance (Nimer and Merrett 1992). Using the silicone oil centrifugation technique, Nimer and Merrett (1992) showed that at external pH 8.3, high- and low-calcifying cells had similar $[DIC]_i$ but differed greatly in photosynthetic ability. It was concluded that lack of calcification in low-calcifying cells was not due to their inability to take up HCO_3^-. While high-calcifying cells had higher $[DIC]_i$ than low-calcifying cells, they were unable to accumulate DIC significantly above $[DIC]_{ext}$, suggesting the absence of an active HCO_3^- accumulation mechanism.

In air-equilibrated medium, both high- and low-calcifying strains were able to grow at comparable rates (Dong *et al.* 1993). However, while the low-calcifying strain did not affect the $[DIC]_{ext}$ or pH, the high-calcifying strain produced a significant decline in $[DIC]_{ext}$ and $[CO_2]$. In cultures bubbled with CO_2-free air, low-calcifying cells were unable to grow while high-calcifying cells were able to sustain growth rates comparable to those of air-equilibrated cells. This strongly supports the hypothesis that high-calcifying cells can utilize external HCO_3^- for production of CO_2 in the absence of a HCO_3^- concentrating mechanism or any measurable carbonic anhydrase (Sikes and Wilbur 1982) to facilitate the production of CO_2 from HCO_3^- in the cytosol. In a high-calcifying strain, when HCO_3^- provides the bulk of carbon for photosynthesis, the ratio of carbon fixed in photosynthesis to that in calcite was close to 1:1 (Nimer and Merrett 1993).

The overall interpretation of these results is summarized in Fig 7.3, which illustrates the likely fluxes of substrates and products of organic and inorganic carbon fixation. The data summarized in the preceding sections are best interpreted in a model which describes the partial reactions of calcification:

$$HCO_3^- \leftrightarrow CO_3^{2-} + H^+ \tag{7.3}$$

$$Ca^{2+} + CO_3^{2-} \leftrightarrow CaCO_3 \tag{7.4}$$

and the hydration and dehydration of CO_2 (reaction 7.1) occurring in separate compartments (calcite formation in the Golgi/coccolith vesicle and CO_2 production in the cytosol and/or the chloroplast). This is necessary in order to explain the isotopic disequilibrium experiments (Sikes *et al.* 1980), since if the partial reactions of calcification (reactions 7.1, 7.3 and 7.4) occurred in a homogeneous compartment, the overall rate of the net reaction (reaction 7.2) and hence the rate of CO_3 formation (reaction 7.3) would be limited by the rate of removal of H^+ in the slow hydration/dehydration equilibrium, eliminating the possibility of rapid incorporation of ^{14}C into calcite when label was supplied as $H^{14}CO_3^-$ (Sikes *et al.* 1980).

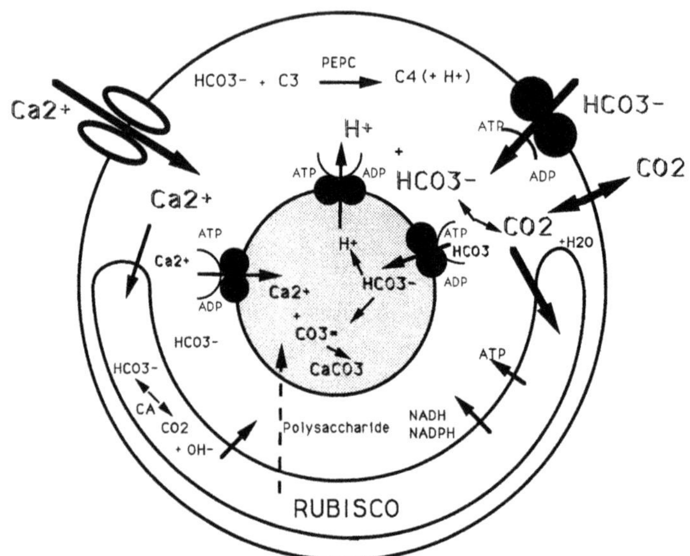

Fig. 7.3 Representation of the likely fluxes of carbon and Ca^{2+}, together with their reactions and compartmentalization during calcification and photosynthesis in *E. huxleyi.* Stippled region = Golgi/coccolith vesicle; RUBISCO = ribulose bisphosphate carboxylase oxygenase in the chloroplast; CA = carbonic anhydrase; PEPC = phosphoenol pyruvate carboxylase.

Intracellular CO_2 production

The balance between H^+ produced by calcification and the net removal of cytosolic H^+ by CO_2 production, and removal by photosynthesis is likely to be important in cytosolic pH regulation. Maintaining cytosolic pH at near neutral values may allow significant CO_2 production from HCO_3^- in the cytosol. Knowledge of cytosolic pH and $[DIC]_i$ allows estimates of the maximum potential uncatalysed rate of CO_2 production in the cytosol:

$$\frac{d[CO_2]}{dt} = k_f \frac{K_1}{K}[H^+] \frac{[DIC]}{1 + \frac{[H^+]}{K} + \frac{K_3}{[H^+]}} \tag{7.5}$$

Where k_f = the forward rate constant for the hydration of CO_2 (24 s^{-1}; Watt and Paasche 1963), $K_1 = 2.59 \times 10^{-3}$, $K_3 = 4.69 \times 10^{-11}$, as defined in (reaction 7.1), $K = K_2K_1 = 4.45 \times 10^{-7}$ (Lucas 1975). For intracellular pH = 7.0 (Dixon *et al.* 1989), cell diameter 5 μm, and $[DIC]_i$ = 0.25–0.5 mM (Nimer and Merrett 1992), the maximum uncatalysed rate of CO_2

production = 10^{-6} pmol/cell/s. This is comparable to the measured rate of photosynthetic carbon fixation by a calcifying cell (8×10^{-6} pmol/cell/s; Nimer and Merrett 1992, 1993), and suggests that the uncatalysed production of CO_2 in the cytosol can provide some CO_2 for photosynthesis at measured intracellular pH values. However, this calculation probably overestimates the rate of CO_2 production since it only gives a maximum instantaneous rate, assuming that $[CO_2]$ tends to zero. Sikes and Wheeler (1982) found no measurable carbonic anhydrase activity in *E. huxleyi* indicating that CO_2 production is not catalysed. This also agrees with the isotopic disequilibrium experiments (Sikes *et al.* 1980, see above) which strongly suggested that the rate of hydration and dehydration of CO_2 was not increased by carbonic anhydrase. Recently, however, carbonic anhydrase has been detected in intact chloroplasts, isolated by isopycnic density gradient centrifugation from crude cell homogenates of high-calcifying *E. huxleyi* cultures (Nimer, Guan and Merrett, unpublished). The precise role of this carbonic anhydrase activity is at present unclear.

A role for phosphoenol pyruvate (PEP) carboxylase in carbon fixation?

PEP carboxylase is the principal enzyme in C_4 metabolism and fixes carbon initially into 4-carbon organic acids, using HCO_3^- as a substrate (O'Leary 1982). The presence of PEP carboxylase has not been demonstrated directly in *Emiliania huxleyi*, but has been shown to be present in *Coccolithus pelagicus* (Glover and Morris 1979). β-carboxylation by PEP carboxykinase has, however, been demonstrated in *E. huxleyi* (Descolas-Gross and Oriol 1992). Indirect evidence for the occurrence of C_4-type metabolism in *E. huxleyi* was obtained by Sikes and Wilbur (1982), comparing $\delta^{13}C$ values of carbon fixed into coccoliths, organic carbon, and non-calcified cells. Non-calcifying cells had $\delta^{13}C$ values higher than expected for carbon fixation by Rubisco, possibly indicating fixation by PEP carboxylase, though not by PEP carboxykinase (Descolas-Gros and Oriol 1992) since this enzyme has similar carbon isotope discrimination to Rubisco (Arnelle and O'Leary 1992). A more likely explanation may be that CO_2 supply to Rubisco in non-calcifying cells was limiting, reducing the discrimination in favour of ^{12}C. Irrespective of whether or not β-carboxylation may be responsible for a significant fraction of carbon fixation, the implications for control of cytosolic pH, and the corresponding inter-relations between calcification and photosynthesis are likely to be highly significant. Fixation of carbon into organic acids will have a direct acidifying effect on the cytosol. Indeed PEP carboxylase has been postulated to have a major pH-stat role in plant cells (Davies 1973, 1979).

Membrane transport and calcification

Carbonate and Ca^{2+} will need to be maintained in the coccolith vesicle at concentrations which exceed the solubility product of $CaCO_3$. H^+ will also need to be removed in order to prevent acidification. Central to this model is the presence of a net outwardly-directed transport of H^+ at the Golgi/coccolith vesicle membrane. This could be either via an outwardly directed H^+ pump or some other transporter maintaining a sufficiently positive potential across the Golgi/coccolith vesicle membrane (Raven 1980). The problem here is that most plant cells maintain acidic Golgi vesicles *via* the action of an *inwardly*-directed H^+ pump (Chanson and Taiz 1985). The solubility product of calcite in the Golgi/coccolith vesicles will also depend greatly on the presence of other factors, such as the acidic polysaccharide which has been postulated to regulate calcite precipitation in a complex manner (Westbroek *et al.* 1984).

DIC transport

The mechanism of HCO_3^- uptake across the *Emiliania huxleyi* plasma membrane is not known. The electrochemical gradient for HCO_3^- across the plasma membrane can be determined from $[HCO_3^-]_i$ and membrane potential. At $[HCO_3^-]_i$ of 0.25 mM, compatible with measurements of $[DIC]_i$ using the silicone oil centrifugation technique (Nimer and Merrett 1992; Dong *et al.* 1993), the equilibrium potential for $HCO_3^-(E_{HCO_3^-})$ is –63 mV. Measurement of the membrane potential using ^{14}C TPP^+ gave an estimate of –60 mV for the plasma membrane potential (V_m) (Nimer and Merrett 1992), albeit in a non-calcifying strain. This suggests equilibrium between intracellular and extracellular HCO_3^- with no requirement for active uptake of HCO_3^-. Sikes and Wilbur (1982), however, measured V_m at –81 mV, using the dye dis-C3-5, and –145 mV, based on intracellular and extracellular $[K^+]$. If this were the case, then active HCO_3^- uptake would be required to maintain measured intracellular concentrations. However, this work must be considered together with evidence that the use of both dyes and TPP^+ to measure V_m may give inaccurate estimates of V_m (Johnstone *et al.* 1982; Ritchie 1984). If active HCO_3^- uptake across the plasma membrane did occur, then this would ultimately require ATP, which could be provided by photosynthesis.

Bicarbonate availability has been shown to have significant and rapid effects on intracellular pH which can be related to HCO_3^- transport across the plasma membrane. Cytoplasmic pH has been measured at between 7.0 and 7.3 in the presence of 2 mM external HCO_3^- (Dixon *et al.* 1989). Removal of external DIC results in acidification of the cytosol by as much as 0.3 pH units (Dixon *et al.* 1989; Nimer, Brownlee and Merrett, unpublished data). This fall in intracellular pH indicates a role for HCO_3^- in buffering cytosolic pH. It also suggests that $[DIC]_i$ falls to very low levels in the absence of external HCO_3^-, since reduction in $[HCO_3^-]_i$ would only be expected to reduce pH_i directly if levels fell to sub-micromolar levels. Re-addition of HCO_3^- produces rapid (within seconds) re-alkalinisation of the cytosol. This occurs independently of illumination or the presence of the photosynthesis inhibitor DCMU (Nimer, Brownlee and Merrett, unpublished), suggesting that the increase in intracellular pH is primarily a consequence of HCO_3^- influx and re-establishment of HCO_{3i}^- levels.

Ca^{2+} transport

Calcium, as a major substrate for calcification needs to be transported into the Golgi/coccolith vesicle. This will require the diffusion of Ca^{2+} into the cell, most likely through Ca^{2+}-selective channels in the plasma membrane (Brownlee and Sanders 1992), and through the cytosol.

The cytosolic free Ca^{2+} concentration in *E. huxleyi* is very low (0.1 μM, Davies and Brownlee, unpublished). It can be assumed that Ca^{2+} needs to be pumped into the Golgi/coccolith vesicle against a concentration gradient, though the extent of this will depend on the Golgi $[Ca^{2+}]$ required for $CaCO_3$ precipitation which will, in turn, depend on the Golgi $[HCO_3^-]$. The concentrations of HCO_3^- or Ca^{2+} in the Golgi/coccolith vesicle are unknown. High total $[Ca^{2+}]$ has been shown in the Golgi/coccolith vesicle system (van der Wal *et al.* 1985) and transfer of Ca^{2+} into the Golgi/coccolith vesicle may occur via fusion with other Ca^{2+}-containing vesicles. Ca^{2+} accumulation into plant vacuoles occurs via H^+/Ca^{2+} exchange driven by the H^+ gradient established by the inwardly-directed vacuolar H^+ pump (Blackford *et al.* 1990). H^+/Ca^{2+} exchange has been postulated to drive Ca^{2+} accumulation in *E. huxleyi* Golgi vesicles, Ca^{2+} being taken up in exchange for H^+ released during calcification (Wal *et al.* 1985). However, the $\Delta\mu H^+$ across the *E. huxleyi* Golgi vesicle membrane may well be in the wrong direction to fuel Ca^{2+} influx, since Golgi pH needs to be kept high and H^+ may need to be actively extruded. Information on the pH, HCO_3^- and Ca^{2+} gradients, and Golgi membrane potential are needed to address this problem further.

A high calcifying cell can produce one coccolith per hour (Paasche 1962), containing between 3.5–5.5 $\times$ 10^{-13} g Ca^{2+} (Paasche, 1962). For a cell, diameter 5 μm, this requires a *net* influx of 3–5 pmol Ca^{2+} $cm^{-2}s^{-1}$ (and presumably a unidirectional influx considerably higher than this). This is significantly higher than most unidirectional measurements of Ca^{2+} influx measured across the plasma membrane of other plant cells. Reid and Tester (1992), for example, measured unidirectional Ca^{2+} influx between 0.04 and 0.2 pmol $cm^{-2}s^{-1}$ across the plasma membrane of *Chara corallina* using $^{45}Ca^{2+}$. Direct measurements of net Ca^{2+} influx, using a Ca^{2+}-specific vibrating electrode, gave values ranging from 0.2 pmol

$cm^{-2}s^{-1}$ at the rhizoid tip of *Pelvetia fastigiata* to 4 pmol $cm^{-2}s^{-1}$ at the growing tip of a tobacco pollen tube (Kuhtreiber and Jaffe 1990). High calcifying cells thus need to maintain a high net flux of Ca^{2+} through the cytosol while maintaining low cytosolic [Ca^{2+}]. Raven (1980) postulated on the basis of reasonable assumptions as to free Ca^{2+} activities in the cytosol, diffusion distances, and fluxes, that Ca^{2+} diffusion (requiring a source-sink free Ca^{2+} gradient of around 1 μM) to the Golgi could be maintained by the presence of mobile Ca^{2+} chelators exchanging rapidly with free Ca^{2+}. The gradient of Ca^{2+} can be maintained by a gradient of Ca^{2+}-bound and Ca^{2+}-free chelators. If, however, the distance required for Ca^{2+} diffusion across the cytosol is small, which could be the case if Ca^{2+}-sequestering vesicles were located near the periphery of the cell and were able to move to and fuse with the central Golgi, then the problem of Ca^{2+} flux across a virtual Ca^{2+} vacuum would not be so severe.

Calcium is a ubiquitous regulator of many cellular processes and a role for cytoplasmic Ca^{2+} in the regulation of photosynthesis and calcification must be considered. Photosynthesis has been shown to reduce cytosolic [Ca^{2+}] in *Nitellopsis* (Miller and Sanders 1987), and this in turn may serve to gear cytosolic metabolism to photosynthesis (Miller and Sanders 1987; Brauer *et al.* 1990). Plasma membrane Ca^{2+} transport also appears to be necessary for HCO_3^- uptake in *E. huxleyi.* Our own unpublished observations show that re-alkalinization of cytosolic pH upon addition of HCO_3^- to carbon-starved cells can be prevented by the presence of the Ca^{2+} channel blocker Gd^{3+}. Gd^{3+} also prevented ^{14}C uptake by calcifying cells. Further measurements of cytosolic Ca^{2+} and HCO_3^- uptake will provide more information on this potentially important link between calcification and photosynthesis.

In summary, calcification, photosynthesis, membrane transport, and metabolism appear to interact in a complex manner. In addition to mutual limitations and enhancements imposed by energetic and flux constraints, both coarse and fine kinetic regulation is also likely *via* cytosolic Ca^{2+} and other regulators. We are only beginning to characterize and quantify the components of this complex system.

Acknowledgements

This work was supported by The Natural Environment Research Council, The European Commission (Contract No. MAS2–CT92–0038) and the Marine Biological Association, UK. We are grateful to Drs M. Whitfield and M. Davies (M.B.A.) for helpful discussions and comments.

References

Arnelle, J. and O'Leary, M. H. (1992). Binding of carbon dioxide to phosphoenolpyruvate carboxykinase deduced from carbon kinetic isotope effects. *Biochemistry*, **31**, 4363–8.

Blackford, S., Rea, P., and Sanders, D. (1990). Voltage sensitivity of H^+/Ca^{2+} antiport in higher plant tonoplast suggests a role in vacuolar Ca^{2+} accumulation. *Journal of Biological Chemistry*, **265**, 9617–20.

Borowitzka, M. A. (1982). Mechanisms in algal calcification. *Progress in Phycological Research*, **1**, 137–77.

Borowitzka, M. A. (1987). Carbonate calcification in algae: initiation and control. In *Biomineralization: chemical and biochemical perspectives*, (ed. S. Mann, J. Webb, and R. J. P. Williams), pp. 63–94. VCH, Weinheim.

Borowitzka, M. A. and Larkum, A. W. D. (1976). Calcification in the green alga *Halimeda*. II. The exchange of Ca^{2+} and the occurrence of age gradients in calcification and photosynthesis. *Journal of Experimental Botany*, **27**, 864–78.

Brauer, M., Sanders, D., and Stitt, M. (1990). Regulation of photosynthetic sucrose synthesis: a role for calcium? *Planta*, **182**, 236–43.

Brownlee, C. and Sanders, D. (ed.) (1992). *Calcium channels in plants. Philosophical Transactions of the Royal Society of London*, Series B, **338**.

Chanson, A. and Taiz, L. (1985). Evidence for an ATP-dependent proton pump on the Golgi of corn coleoptiles. *Plant Physiology*, **78**, 232–40.

Crick, R. E. (ed.) (1989). *Origin, evolution and modern aspects of biomineralization in plants and animals*. Plenum, New York.

Davies, D. D. (1973). Control of and by pH. *Symposium of the Society of Experimental Biology*, **27**, 513–30.

Davies, D. D. (1979). The central role of phosphoenolpyruvate in plant metabolism. *Annual Review of Plant Physiology*, **30**, 131–58.

Descolas-Gros, C. and Oriol, L. (1992). Variations in carboxylase activity in marine phytoplankton cultures. β-carboxylation in carbon flux studies. *Marine Ecology Progress Series*, **85**, 163–9.

Deuser, W. G. and Degens, E. T. (1967). Carbon isotope fractionation in the system CO_2 (gas) – CO_2 (aqueous) -HCO_3^- (aqueous). *Nature*, **215**, 1033–5.

Dixon, G. K., Brownlee, C., and Merrett, M. J. (1989). Measurement of internal pH in the coccolithophore *Emiliania huxleyi* using 2′, 7′-bis-(2-carboxyethyl)-5 and-6) carboxyfluorescein acetoxymethyl ester (BCECF) and digital imaging microscopy. *Planta*, **178**, 443–9

Dong, L. F., Nimer, N. A., Okus, E., and Merrett, M. J. (1993). Dissolved inorganic carbon utilization in relation to calcite production in *Emiliania huxleyi*. *New Phytologist*, **123**, 679–84.

Edwards. G. E. and Walker, D. A. (1983). *C3, C4 mechanisms, and cellular and environmental regulation of photosynthesis*. Blackwell, Oxford.

Farquahar, G. D., Ehrleringer, J. R., and Hubick, K. T. (1989). Carbon isotope discrimination and photosynthesis. *Annual Reviews of Plant Physiology and Plant Molecular Biology*, **40**, 503–37.

Freedman, R. A., Weiser, M. M., and Isselbacher, K. J. (1977). Calcium translocation by Golgi and lateral-basal membrane vesicles from rat intestine: decrease in vitamin D-deficient rats. *Proceedings of the National Academy of Science of the United States of America*, **74**, 3612–6.

Glover, H. E. and Morris, I. (1979). Photosynthetic carboxylating enzymes in marine phytoplankton. *Limnology and Oceanography*, **24**, 510–19.

Heber, U. and Heldt, H. W. (1981). The chloroplast envelope: structure, function and role in leaf metabolism. *Annual Reviews of Plant Physiology*, **32**, 139–68.

Johnstone, R. M., Laris, P. C., and Eddy, A. A. (1982). The use of fluorescent dyes to measure membrane potentials: a critique. *Journal of Cell Physiology*, **112**, 298–301.

Kuhtreiber, W. M. and Jaffe, L. F. (1990). Detection of extracellular calcium gradients with a calcium-specific vibrating electrode. *Journal of Cell Biology*, **110**, 1565–73.

Lehman, J. T. (1978). Enhanced transport of inorganic carbon into algal cells and its implication for the biological fixation of carbon. *Journal of Phycology*, **14**, 33–42.

Linschooten, C., van Bleijswijk, J. D. L., van Emburg, P. R., de Vrind, J. P. M., Kempers, E. S., Westbroek, P. *et al.* (1991). Role of the light-dark cycle and medium composition on the production of coccoliths by *Emiliania huxleyi* (Haptophyceae). *Journal of Phycology*, **27**, 82–6.

Lucas, W. J. (1983). Photosynthetic assimilation of exogenous carbon by aquatic plants. *Annual Review of Plant Physiology*, **34**, 71–104.

Lucas, W. J. (1975). Photosynthetic fixation of 14 carbon by internodal cells of *Chara corallina*. *Journal of Experimental Botany*, **26**, 331–46.

Lucas, W. J. (1985). Bicarbonate utilization by *Chara*: a re-analysis. In *Inorganic carbon uptake by aquatic photosynthetic organisms*, (ed. W. J. Lucas and J. A. Berry), pp. 229–54. American Society of Plant Physiologists, Rockville, Maryland.

Miller, A. J. and Sanders, D. (1987). Depletion of cytosolic free calcium induced by photosynthesis. *Nature*, **326**, 397–400.

Nimer, N. and Merrett, M. J. (1992) Calcification and utilization of inorganic carbon by the coccolithophorid *Emiliania huxleyi* (Lohmann). *New Phytologist*, **121**, 173–7.

Nimer, N. and Merrett, M. J. (1993). Calcification rate in *Emiliania huxleyi* Lohmann in response to light, nitrate and inorganic carbon availability. *New Phytologist*, **123**, 673–7.

O'Leary, M. H. (1982). Phosphoenolpyruvate carboxylase: an enzymologist's view. *Annual Review of Plant Physiology*, **33**, 297–315.

Paasche, E. (1962). Coccolith formation. *Nature*, **193**, 1094–5.

Paasche, E. (1963). The adaptation of the carbon-14 method for the measurement coccolith production in *Coccolithus huxleyi. Physiologia Plantarum*, **16**, 186–200.

Paasche E. (1964). A tracer study of the inorganic carbon uptake during coccolith formation and photosynthesis in the coccolithophorid *Coccolithus huxleyi. Physiologia Plantarum*, **Supplement III**, 5–82.

Paasche, E. (1965). The effect of 3-(p-chlorophenyl)-1, 1-dimethylurea (CMU) on photosynthesis and light-dependent coccolith formation in *Coccolithus huxleyi. Physiologia Plantarum*, **18**, 138–45.

Paasche, E. (1968). The effect of temperature, light intensity and photoperiod on coccolith formation. *Limnology and Oceanography*, **13**, 178–81.

Prins, H. B. A., Snel, J. F. H., Helder, R. J., and Zanstra, P. E. (1980). Photosynthetic HCO_3^- utilization and OH^- excretion in aquatic angiosperms. Light-induced pH changes at the leaf surface. *Plant Physiology*, **66**, 818–22.

Prins, H. B. A., Snel, J. F. H., Zanstra, P. E., and Helder, R. J. (1982). The mechanism of bicarbonate assimilation by the polar leaves of *Potamogeton* and *Elodea*. CO_2 concentration at the leaf surface. *Plant Cell Environment*, **5**, 207–14.

Raven, J. A. and Johnston, A. M. (1991). Mechanisms of inorganic carbon acquisition in marine phytoplankton and their implications for the use of other resources. *Limnology and Oceanography*, **36**, 1701–4.

Raven, J. A. (1980). Nutrient transport in microalgae. *Advances in Microbial Physiology*, **21**, 47–226

Reid, R. J. and Tester, M. (1992). Measurements of Ca^{2+} fluxes in intact plant cells. *Philosophical Transaction of the Royal Society of London*, Series B, **338**, 83–9.

Ritchie, R. R. (1984). A critical assessment of the role of lipophilic cations as membrane potential probes. *Progress in Biophysical and Molecular Biology*, **43**, 1–32.

Santarius, K. A. and Heber, U. (1965). Changes in the intracellular levels of ATP, ADP, AMP and P_i and regulatory function of the adenylate system in leaf cells during photosynthesis. *Biochimica et Biophysica Acta*, **102**, 39–54.

Sikes, C. S., and Fabry, V. J. (1993). Photosynthesis, $CaCO_3$ deposition, coccolithophorids and the global carbon cycle. In *Photosynthetic carbon metabolism and regulation of atmospheric CO_2 and O_2*, (ed. N. E. Tolbert and J. Priess). Oxford University Press (In press).

Sikes C. S. and Wheeler, A. P., (1982). Carbonic anhydrase and carbon fixation in coccolithophorids. *Journal of Phycology*, **18**, 423–6.

Sikes, C. S. and Wilbur, K. M. (1982). Functions of coccolith formation. *Limnology and Oceanography*, **27**, 18–26.

Sikes, C. S., Roer, R. D., and Wilbur, K. M. (1980). Photosynthesis and coccolith formation: inorganic carbon sources and net inorganic reaction of deposition. *Limnology and Oceanography*, **25**, 248–61.

Simkiss, K. and Wilbur, K. M. (1989). *Biomineralization: cell biology and mineral deposition*. Academic Press, San Diego.

Steeman-Nielsen, E. (1966). The uptake of free CO_2 and HCO_3^- during photosynthesis of planktonic algae with special reference to the coccolithophorid *Coccolithus huxleyi*. *Physiologia Plantarum*, **19**, 232–40.

Wal, P. van der, de Bruin, W. C., and Westbroek, P. (1985). Cytochemical and X-ray microanalysis studies of intracellular calcium pools in scale-bearing cells of the coccolithophorid *Emiliania huxleyi*. *Protoplasma*, **124**, 1–9.

Watt, W. D. and Paasche, E. (1963). An investigation of the conditions for distinguishing between CO_2 and bicarbonate utilization by algae according to the methods of Hood and Park. *Physiologia Plantarum*, **16**, 674–81.

Westbroek, P., de Jong, E. W., van der Wal, P., Borman, A. H., de Vrind, J. P. M., Kok, D., *et al.* (1984). Mechanisms of calcification in the marine alga *Emiliania huxleyi*. *Philosophical Transactions of the Royal Society of London*, Series B, **304**, 435–44.

8. Mechanisms of calcification: *Emiliania huxleyi* as a model system

E. W. DE VRIND-DE JONG, P. R. VAN EMBURG, and J. P. M. DE VRIND
Department of Biochemistry, Gorlaeus Laboratory, Leiden University, The Netherlands

Abstract

Coccolithophorids are haptophytes which form calcified scales called coccoliths. Heterococcoliths consist of unit elements of an often complex shape and are formed intracellularly. Holococcoliths contain rhombohedral calcite crystals and are apparently formed extracellularly. We assume that similar mechanisms play a role in the nucleation and growth control of both heterococcoliths and holococcoliths. The formation of heterococcoliths in the species *Emiliania huxleyi* is described as a model system. Coccoliths were observed with transmission electron microscopy. They were subjected to morphometric and crystallographic analysis and a model approach for the morphogenesis of coccoliths is briefly described. It is concluded that the majority of faces of the unit elements of coccoliths represent biologically inhibited faces instead of crystal faces of calcite. An acidic polysaccharide isolated from coccoliths is thought to play a role in crystal growth control. A similar function is attributed to the organic skins associated with the mineral elements of coccoliths of other coccolithophorid species.

Introduction

Members of the Prymnesiophyceae which form calcified scales are known as coccolithophorids. The coccolithophorids form a substantial component of the oceanic phytoplankton and some of the species such as *Emiliania huxleyi*, sometimes accompanied by *Coccolithus pelagicus*, can occur in extensive blooms covering thousands of square kilometres (Ackleson *et al.* 1988; Holligan 1986). The observation of the enormous amounts of carbon dioxide fixed by these blooms as organic carbon and calcium carbonate has resulted in an integrated interdisciplinary research programme being carried out into the interaction of the coccolithophorid

The Haptophyte Algae (ed. J. C. Green and B. S. C. Leadbeater), Systematics Association Special Volume No. 51, pp. 149–66. Clarendon Press, Oxford, 1994.

blooms with global climate (Westbroek *et al.*, Chapter 17). Calcium carbonate is laid down in often delicately sculptured structures, the coccoliths (see, for instance Green *et al.* 1990). Two main types of coccoliths exist. In one type regular rhombohedral calcite elements are deposited in varying configurations on an organic base-scale (see Leadbeater, Chapter 2). These are called holococcoliths. In the so-called heterococcoliths the component unit elements can have a complex shape with rounded crystal faces. In general they are also deposited on an organic base-scale (Green *et al.* 1990; Leadbeater, Chapter 2). In some species holococcoliths as well as heterococcoliths can be formed depending on the life stage of the organism (Billard, Chapter 9). The alternating holo- and heterococcolith formation in *Coccolithus pelagicus* is a clear example of this phenomenon (Rowson *et al.* 1986). So far, electron microscopical observations indicate that holococcoliths are formed extracellularly, viz. on the periplasmic side of the plasma membrane, and that heterococcoliths are formed intracellularly. Intuitively one would suspect a higher degree of cellular control in the formation of heterococcoliths than in that of holococcoliths. In both cases the sites of crystal nucleation on the organic base plate are not random and the ultimate size of the unit crystal elements of the coccoliths is strictly determined. As far as crystal nucleation and size limitation is concerned, similar principles will probably apply to the formation of both types of coccoliths. This chapter will primarily describe the observations and proposed mechanisms of heterococcolith formation in one of the most ubiquitous and extensively studied coccolithophorid, *Emiliania huxleyi*. Where appropriate, observations on coccolith formation in other species of the Prymnesiophyceae will be discussed in view of the current ideas on the mechanism of calcification in *Emiliania huxleyi*.

Characteristics of the cell of *Emiliania huxleyi*

Site of coccolith formation

A schematic representation of the most conspicuous features of the calcifying cell of *Emiliania huxleyi* is given in Fig. 8.1a. The nucleus and chloroplast are usually situated closely together. The Golgi apparatus can be clearly distinguished, and vesicles budding off from this organelle probably fuse to form the compartment where the coccoliths are produced. This coccolith production compartment is tightly apposed to the nucleus during most stages of coccolith production (see also below). It consists of the coccolith vesicle (called coccolith room in Fig. 8.1a) in which calcification takes place, and a reticular body of anastomosing tubes. The latter is believed to supply the precursors of the growing

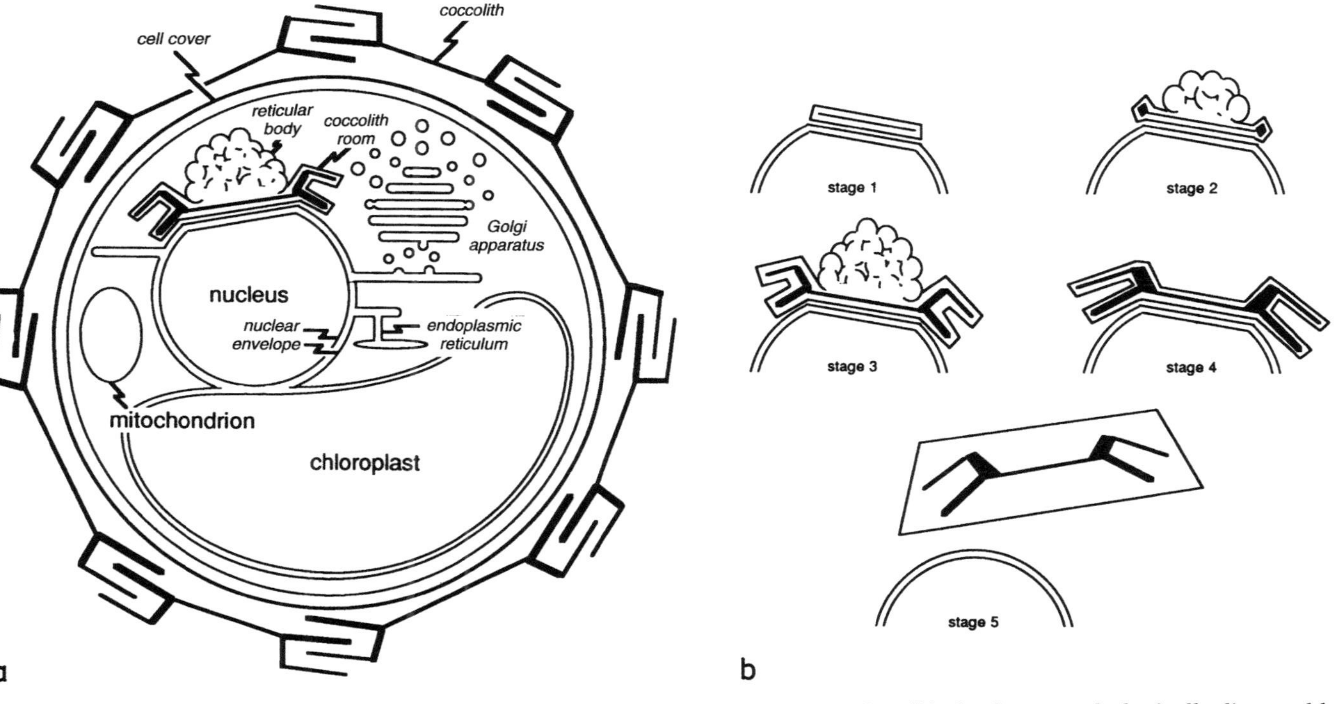

Fig. 8.1 (a) Schematic representation of a section through a cell of *Emiliania huxleyi*; (b) the five morphologically discernable stages of the coccolith production compartment.

coccolith and its enveloping membrane. The cell is surrounded by the plasmamembrane and the so-called cell cover (Klaveness 1972; van der Wal *et al.* 1983*a*). The latter seems to consist of a proximal lipid bilayer and a distal biomembrane (van der Wal *et al.* 1985). Completed coccoliths are gathered in a coherent spherical shell outside the cell cover. Non-calcified organic scales, flagella, and a haptonema are never present in calcifying cells of *Emiliania huxleyi.* These are constituents of most other prymnesiophycean species.

In the formation of a coccolith, five stages can be identified on the basis of the morphology of the coccolith production compartment and its luminal contents (Fig. 8.1b; cf. van der Wal *et al.* 1983*a*). In stage 1 a flat pancake-like vesicle is attached to the nuclear envelope; a thin organic base plate is often observed within its lumen. In stage 2 a reticular body is added to the vesicle which can now be recognized as the coccolith vesicle by initiation of calcification on the rim of the organic base plate. Stage 3 is defined by proximal and distal outgrowth of the coccolith vesicle and continuing calcification; the coccolith vesicle encloses the mineral as a loose glove and is still attached to a reticular body. In stage 4 calcification is completed and a reticular body is absent. In the final stage 5 the coccolith vesicle loses its close fit around the mineral and is detached from the nucleus. Soon after the coccolith will be extruded and incorporated into the coccosphere. The organic base-plate on which the coccolith mineral is nucleated seems to bear no structural resemblance to the organic scales of other prymnesiophyte coccoliths.

Formation of coccoliths

A model approach based on morphometrical and crystallographical analysis

1. Description of the model Coccoliths are composed of a radial array of unit elements. One unit element is a single crystal of calcite (Watabe 1967; Mann and Sparks 1988). In a completed coccolith these crystals have rounded faces and edges. Figure 8.2a illustrates the position of three such unit elements in a coccolith. Different parts can be distinguished in a unit element (Fig. 8.2b): the inner area element (2), the block element (3), the proximal shield element (4), and the hammer-like distal shield element (5). All originate from the original nucleation point (1) which will be called the *root* in the following description (cf. van Emburg 1989). It is assumed that the roots are situated on the rim of the organic base-plate mentioned above and that the initial nuclei are calcite rhombohedra (Klaveness 1976; cf. Fig. 8.3a inset). Recent observations suggest that initially nuclei are alternated with c-axes oriented perpendicular to each other (Young *et al.* 1992 and, Chapter 20). These are called V and R units.

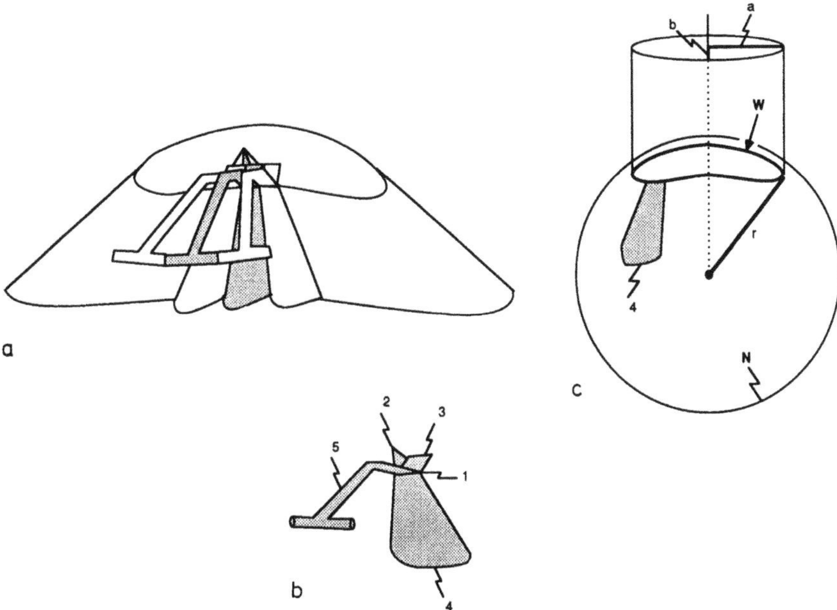

Fig. 8.2 Morphology of a coccolith of *Emiliania huxleyi*: (a) position of three unit elements in a coccolith; (b) morphology of a unit element, for description of component parts, see text; (c) initial nuclei are supposed to be located on a curve (wreath, W) resulting from the intersection of an elliptical cylinder (semiaxes of transverse section: a and b) with a sphere (nucleus, N, with radius r).

This type of nucleation is attributed to special properties (possibly plication) of the underlying base plate and seems to be preserved in geological time (Young *et al.* 1992). In coccoliths of *Emiliania huxleyi* the final unit elements appear to stem from R-units solely, the V-units being totally overgrown and not discernible in completed coccoliths. The stages of growth of coccoliths can be monitored by transmission electron microscopy. When external coccoliths are removed from cells by acidification the formation of a new coccosphere will be initiated (Linschooten *et al.* 1991). By applying a procedure for the isolation of coccoliths to such a cell preparation, a considerable part of the isolated coccoliths are found to be *in statu nascendi* (van Emburg 1989). Successive growth stages are shown in Fig. 8.3a–h. Figure 8.3a was obtained from a whole cell preparation after extensive fixation of decalcified cells; this permitted a view of a protococcolith ring in its native form stabilized by its attached support, the nucleus. In the successive growth stages straight contours can be distinguished at the growth fronts of the different coccolith parts (Fig. 8.3b–h).

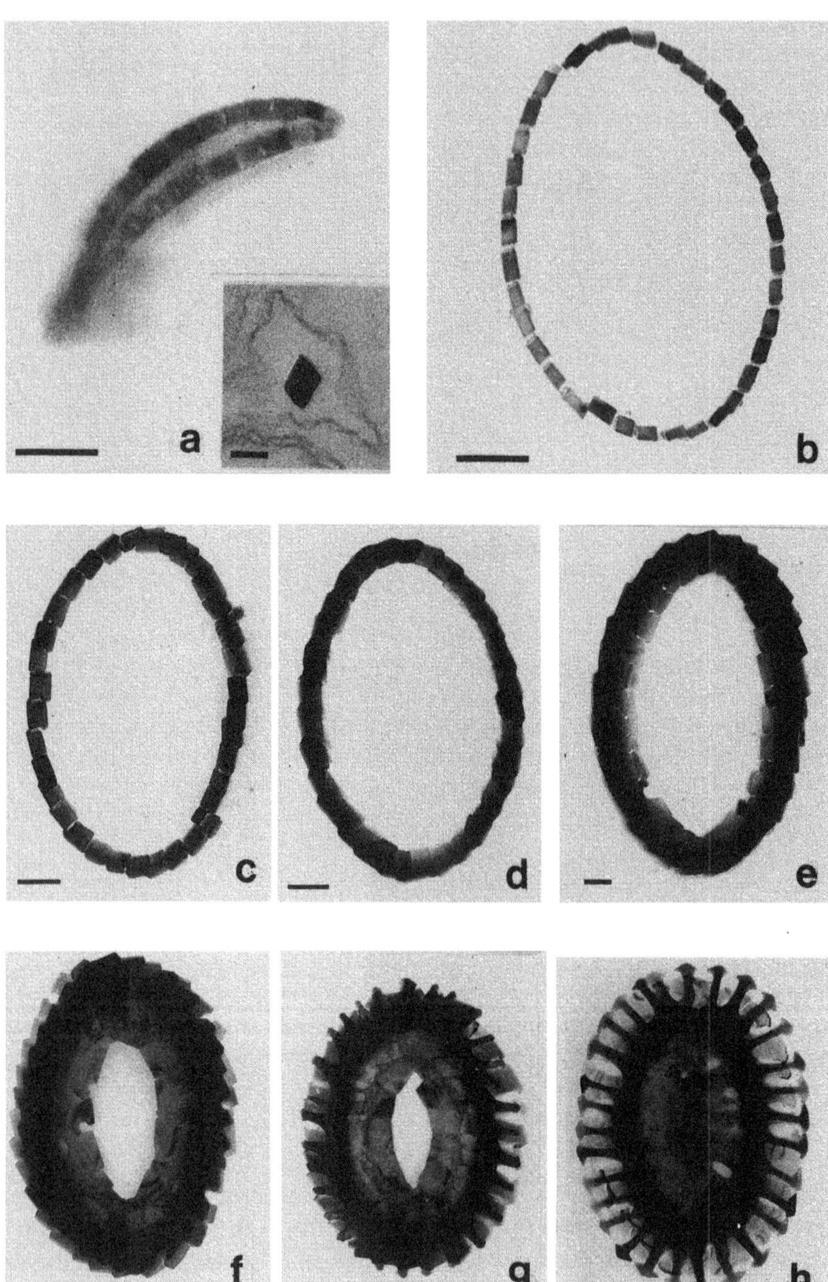
a
b
c
d
e
f
g
h

Fig. 8.3 Transmission electron micrographs of coccoliths *in statu nascendi.* (a) Whole cell preparation showing a projection of the wreath of calcite crystals with small protrusions, in its native three-dimensional shape (bar = 0.2 μm). Inset: transverse section through initial calcite rhombohedron in an ultrathin section of a resin-embedded cell (bar = 0.5 μm). (b) Mineral growth stage considered to be the earliest. The oblong crystals have no protrusions (bar = 0.2 μm, micrograph by courtesy of J. R. Young). (c) Small protrusions develop, and (d) crystals interlock (bars = 0.2 μm). (e–g) Development of inner shield area and distal shield elements, straight contours mark growth fronts (bars in e and f = 0.1 μm: bar in g = 0.2 μm). (h) Proximal and distal shield elements lose straight contours (bar = 0.2 μm).

The completed coccolith has a complex shape. It has chirality and details of its morphology can be strain-specific (van Emburg 1989; van Bleijswijk *et al.* 1991; cf. Fig. 8.4a and Fig. 8.4b). By considering the chirality of the coccolith in a transmission electron micrograph one can determine the orientation of a coccolith sedimented on a grid film (Fig. 8.4c, d). A coccolith sedimented with its proximal shield elements on the grid (Fig. 8.4d) is said to have the standard orientation in the following description. The distal shield composed of the hammer elements is far more fragile than the proximal shield. The former is easily removed by gentle sonication during sample preparation, or by slight etching of the calcite particles. Prolonged etching can result in the collapse of the proximal shield in the plane of the grid (Fig. 8.4d). Van Emburg (1989) has proposed a model approach to explain the morphogenesis of this complex biomineral. It concentrates on the growth of the proximal shield elements (4 in Fig. 8.2b). Here we wish to explain the basis of this approach and to discuss some of the predictions the model makes about geometrical and crystallographical features of the proximal shield element. For details the reader is referred to van Emburg (1989).

Because the coccolith has an overall oval shape and is always formed closely apposed to the nucleus, it is supposed to originate from a series of roots which lie on the curve resulting from the intersection of an elliptical cylinder and a sphere (Fig. 8.2c). This curve will be called the wreath (W in Fig. 8.2c; the sphere is the representation of the nucleus N with radius r). The elliptical cylinder is defined by the semiaxes a and b of the elliptic transverse section. This model is corroborated by the image of the native protococcolith ring in Fig. 8.3a. The transmission electron micrographs of a coccolith can be analyzed as a two-dimensional projection of the three-dimensional model. Figure 8.5a shows the frame of reference chosen for the analysis of the proximal shield elements based on the symmetry of the coccolith (van Emburg 1989). An orthogonal triaxial system was defined. The X_c-axis and Y_c-axis were chosen to be colinear

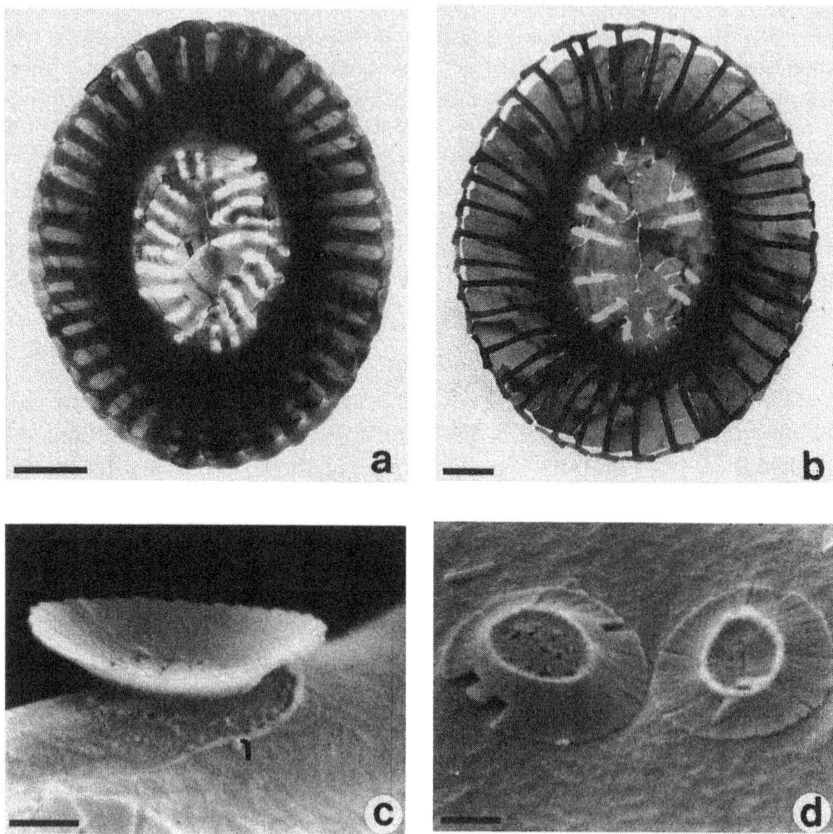

Fig. 8.4 Transmission and scanning electron micrographs of coccoliths of *Emiliania huxleyi.* (a) Transmission electron micrograph of a coccolith of strain L (bar = 0.5 μm); (b) Idem of a coccolith of strain D (bar = 0.5 μm); (c) Scanning electron micrograph of a proximal shield sedimented in non-standard orientation (bar = 1 μm); (d) Idem of proximal shield elements sedimented in standard orientation, original shape (*left*) and collapsed (*right*) (bar = 1 μm); (c) and (d) by courtesy of P. van der Wal.

with the semiaxes W_a and W_b of the wreath W respectively; the Z_c-axis to be perpendicular to both (Fig. 8.5a). The Z_c-axis represents the diad rotation axis of the coccolith. All unit elements are supposed to grow from series of roots arranged on the wreath W. Each proximal shield element is assumed to lie in the plane tangent to the nucleus N in W (Fig. 8.5a). Thus, in a full-grown coccolith, the proximal shield elements fill a quasi-conical surface as indicated in Fig. 8.5a. Each proximal shield element is characterized by a number of morphometric parameters. Two of them are

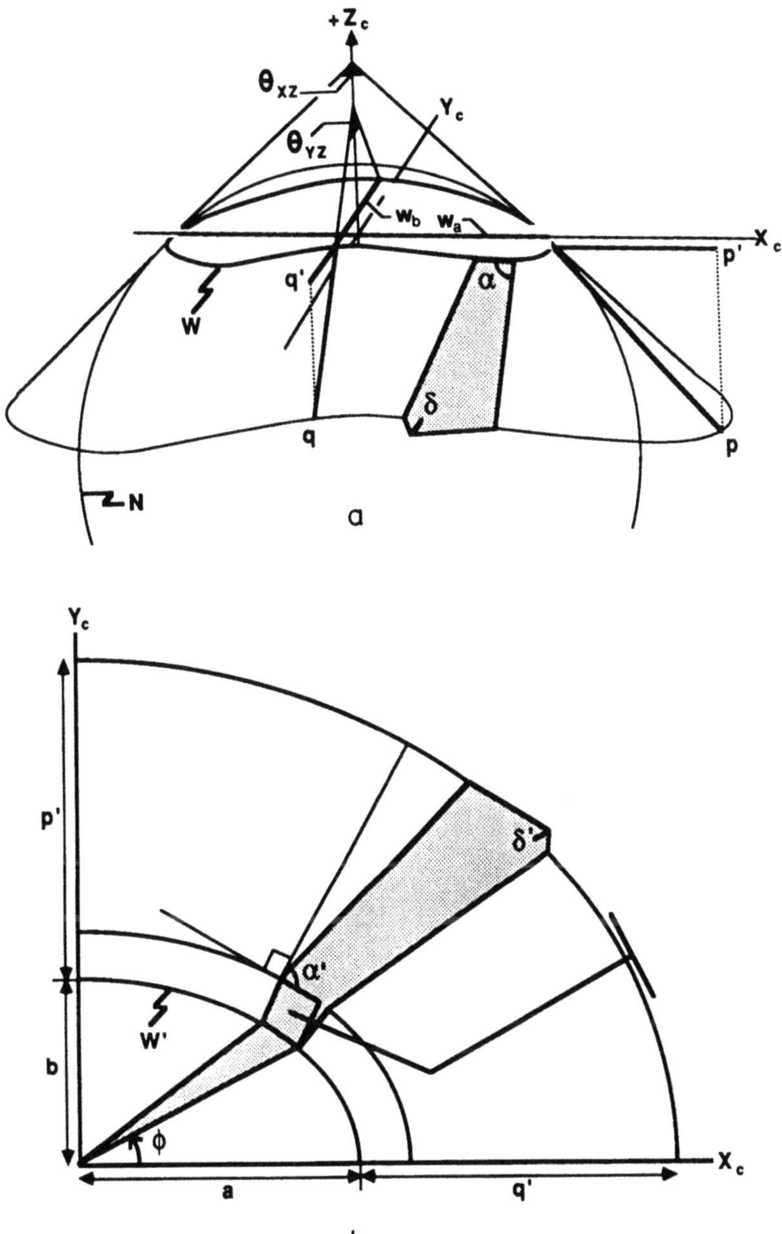

Fig. 8.5 Frame of reference for morphometric analysis of the proximal shield element of a coccolith: (a) orthogonal triaxial system, X_c, Y_c, Z_c, and relative position of a proximal shield element; (b) projection of proximal shield element in X_cY_c plane. For detailed description, see text.

indicated in Fig. 8.5a. The coccolith is viewed in its standard orientation. First we define the distance by which a proximal shield element separates its two neighbours in the wreath as the root chord. The angle α is defined as the angle between the border of a proximal shield element and the first root chord going in a clockwise direction along the wreath. The angle δ is defined as the angle between the straight contours at the tip of a proximal shield element (cf. Fig. 8.3e–g). In a transmission electron micrograph the angles α' and δ' represent α and δ in the projection of the wreath and the proximal shield element in the X_c–Y_c-plane (Fig. 8.5b). As a consequence of its form, the positions on the wreath are not equivalent. This is best illustrated by considering the orientation of the proximal shield elements at two extreme positions, i.e. at the intersection of the wreath with the semiaxes W_a and W_b. An element at the position of the W_a-axis will have higher inclination with respect to the X_c–Y_c-plane than one at the position of the W_b-axis; in other words the cone angle Θ_{xz} is smaller than the cone angle Θ_{yz} in Fig. 8.5a. The cone angles will depend on the values of the radius r of the nucleus and those of the semiaxes W_a and W_b and were estimated from scanning electron microscope images to be about 120° and 150° respectively. The variation in the cone angle will influence the apparent morphometric parameters in the projection in the X_c–Y_c plane. In order to determine the morphometric parameters as a function of the position of a proximal shield element on the wreath, the position coordinate ϕ was defined in the projection in the X_c–Y_c-plane (Fig. 8.5b). The cone angle Θ is dependant on ϕ with Θ at a maximum at ϕ = 90° (Θ_{yz}) and Θ at a minimum at ϕ = 0° (Θ_{xz}). The model makes certain predictions on the values of α' and δ' as a function of ϕ.

2. *The angles α and α'* Suppose that the angle α has the same value in all the proximal shield elements of one coccolith. If the angle α' is measured in transmission electron micrographs the arithmetic mean of α' can be calculated and the deviation from the arithmetic mean ($\Delta\alpha'$) can be plotted as a function of ϕ. The model predicts that α' will be maximal at ϕ = 0° ($\Delta\alpha'$ = positive) and minimal at ϕ = 90° ($\Delta\alpha'$ = negative). The range of $\Delta\alpha'_{\phi=0} - \Delta\alpha'_{\phi=90}$ is predicted to be about 2% of the mean value of α'. The values of $\Delta\alpha'$ were measured in the proximal shield elements of several coccoliths as a function of ϕ. A regression line was calculated and the predicted trend was indeed observed, but the measured range appeared to be significantly greater (7%) than the predicted one (van Emburg 1989). This result suggests that α is not constant within one coccolith but depends on ϕ, i.e. on the position of the proximal shield element on the wreath. This feature of the angle α is included in a model for the growth of the proximal shield elements (see 5 below).

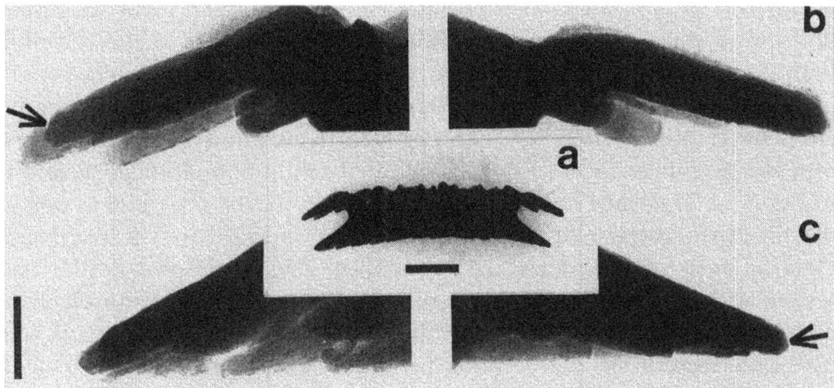

Fig. 8.6 Transmission electron micrograph of a coccolith *in statu nascendi* sedimented with its X_cY_c plane perpendicular to the plane of the grid (a, bar = 0.5 μm). Arrows mark the traces of a third facet at the growth fronts of distal (b) and proximal (c) shield elements (bar = 0.1 μm).

3. The angles δ and δ' The values of δ' were measured and the arithmetic mean was calculated. Analogous as described for the angle α', the deviation from the arithmetic mean of δ' ($\Delta\delta'$) was plotted as a function of ϕ. The model predicts that δ' will be maximal at $\phi = 0°$ and minimal at $\phi = 90°$. If the angle δ is independent of ϕ the predicted range of $\Delta\delta'$ will be about 2% of the mean value. The predicted trend was observed. The range of $\Delta\delta'$ appeared to be about 3% of the mean value, indicating that the angle δ is independent of the position of the proximal shield element on the wreath. This was confirmed by measurement of δ' in the elements of collapsed proximal shield cones (cf. Fig. 8.6d). In a collapsed cone $\delta' = \delta$, and no trend of the deviation of δ' from the mean value will be found. This appeared to be the case (van Emburg 1989).

The dispersion in the values of δ', in a standardly oriented coccolith as well as in a collapsed proximal shield cone, appeared to be large, varying from 96° to 124°. This dispersion appeared to be too large to be accounted for by measurement errors. A possible explanation for the dispersion in the values of δ' and thus δ will be given below.

4. Analysis of the growth front of the proximal shield elements The angle δ is the angle between the straight contours observed at the tip of a growing proximal shield element. This tip represents the growth front of the element. The straight contours are supposed to represent the traces of two crystal faces. Experiments in which slightly etched coccoliths were overgrown *in vitro* with calcite crystals of known crystal form suggested that the crystal faces of the coccoliths' growth fronts are of the rhombohedral form $\{10\bar{1}1\}$ of calcite (van Emburg 1989). The c-axis was concluded to be approximately parallel to the plane of the proximal shield element. It can be shown that the growth front will display three facets representing three faces of the $\{10\bar{1}1\}$ calcite rhombohedron (f_1, f_2 and f_3 in Fig. 8.7). The trace of a third facet at the tips of the proximal as well as the distal shield element was thought to be observed in a transmission electron microscope image of a coccolith *in statu nascendi* which had settled with its X_c–Y_c plane perpendicular to the plane of the electron microscope grid (Fig. 8.6 *arrows*). Such a position of a coccolith is very rare and cannot be deliberately arranged in a transmission electron microscope preparation.

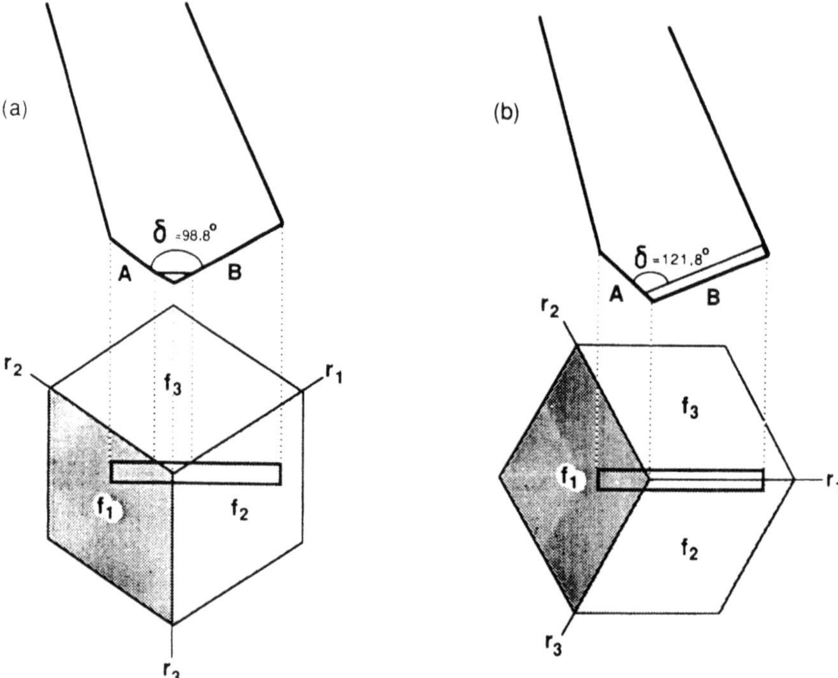

Fig. 8.7 Variation of plane of proximal shield element with respect to a calcite $\{10\bar{1}1\}$ rhombohedron with faces f_1, f_2, f_3 and edges r_1, r_2, r_3. The plane of the element is supposed to contain the short body diagonal (c-axis) of the rhombohedron. Then the orientation of the rhombohedron can vary between to extreme positions (a) and (b) resulting in two extreme values for the angle δ between the straight contours A and B of the proximal shield element.

The configuration at the growth front may be interpreted as follows. Because the proximal shield element is very thin it may be considered as a plane. The straight contours at the front (A and B in Fig. 8.7) then represent the traces of intersection of this plane with the calcite $(10\bar{1}1)$ rhombohedron, i.e. the traces of two rhombohedral faces. Because the plane of the proximal shield element is parallel to the c-axis of calcite, it must contain the short body diagonal of the rhombohedron. The vertex of the two straight contours A and B represents in fact the vertex of three rhombohedron faces. The short body diagonal of the rhombohedron contains this vertex. If the only prerequisite for the orientation of planes intersecting the rhombohedron is that they contain the short body diagonal of the latter, then the traces of two rhombohedral faces which intersect in a vertex lying on the short body diagonal will make angles with each other of which the values are confined to a limited interval. The extremes of this interval with the corresponding values of the angle δ are shown in Fig. 8.7. In a transmission electron micrograph δ' is measured of which the value depends on the position of the proximal shield element on the wreath and thus on the cone angle (see section 3 above). If the extremes of the cone angle are assumed to be $\Theta_{xz} = 120°$ and $\Theta_{yz} = 150°$ the values of δ' can theoretically range from 99.1° and 124.1°. These values match the measured range of δ' of 96° to 124°. These results suggest that the plane of the proximal shield element can have a varying orientation with respect to the crystal lattice of calcite. This would imply that the surface of an element only rarely represents a crystal face of calcite (cf. Mann and Sparks 1988). These surfaces probably represent biologically inhibited faces. Possibly an organic constituent of the coccoliths (see below) plays a role in crystal growth inhibition.

5. Mathematical reconstruction of the electron microscope image of a proximal shield cone After the identification of the crystal faces at the growth front a simple mathematical model was formulated for the dependancy of the angle α on the cone angle Θ (van Emburg 1989). Relative growth velocities of the traces A and B were deduced ($V_A = 2{,}1V_B$), and it was assumed that $\Theta = 180°$ for all ϕ (yielding a collapsed proximal shield cone corresponding to its projection in the X_c–Y_c plane). The number of proximal shield elements was taken to be 36 (cf. Fig. 8.4a and Fig. 8.4b) and the ratio W_b/W_a (Fig. 8.5a) was taken to be 0.69, which were the mean values in the coccoliths of the two strains used here. These data were used to reconstruct the projection of a proximal shield cone in the X_c–Y_c plane (Fig. 8.8). This reconstruction yielded a fair representation of a proximal shield *in statu nascendi* as observed with the electron microscope (cf. Fig. 8.3a–h).

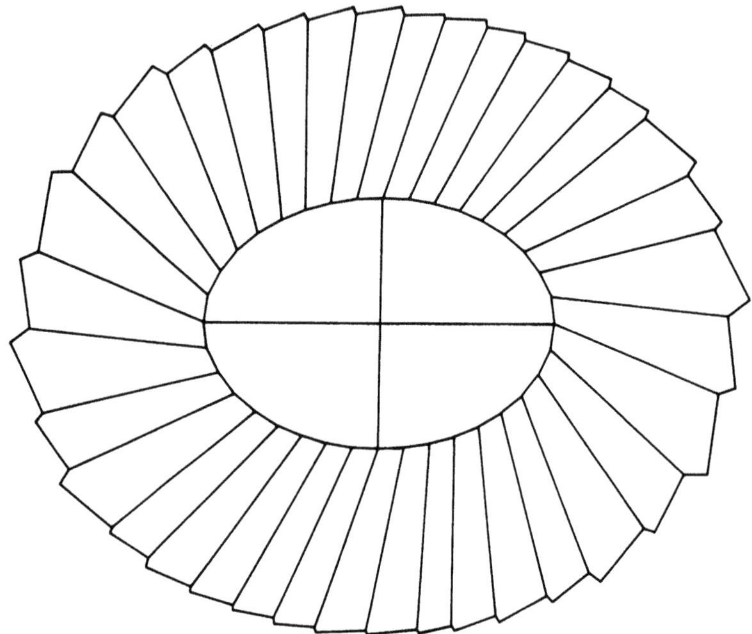

Fig. 8.8 Reconstruction drawing of a projection of the proximal shield of a coccolith in the X_cY_c plane. For a more detailed description, see text.

The model briefly described here, represents a novel approach to understand the morphogenesis of a complex biomineral. The study of morphogenesis in biology has been mainly descriptive. It calls for mathematical methods which can be supported by computer analysis. When experimental observations on the development of a biological structure can be translated into parameters and fitted into a mathematical model, it will be possible to predict the response of the morphogenetic process to variations in those parameters. The validity of the model will appear from subsequent experimental observations.

Organic constituents of coccoliths

So far, two organic constituents of coccoliths have been identified. One is the organic base-plate on which crystals are nucleated. The second is a polysaccharide which can be obtained from isolated coccoliths by dissolving the $CaCO_3$ (de Vrind-de Jong *et al.* 1976). The polysaccharide has been characterized in some detail (Fichtinger-Schepman *et al.* 1981). It contains at least 13 different monosaccharides, among which methylated, sulfated, and carboxylated sugars. It is able to bind Ca^{2+} ions and it inhibits the crystallization of $CaCO_3$ *in vitro* (Borman *et al.* 1982). The carboxyl groups are responsible for the inhibition of crystallization. Antibodies raised against the isolated polysaccharide demonstrate its close association with the mineral of the coccoliths (van Emburg *et al.* 1986). It was also localized in the lumina of the Golgi apparatus, the reticular body, and the coccolith vesicle (van Emburg *et al.* 1986). This localization suggests that the polysaccharide is synthesized and processed in the Golgi apparatus, and delivered to the coccolith production compartment by fusion of the latter with Golgi vesicles. It is thought to be involved in coccolith morphogenesis by inhibition of the growth of certain crystal faces. The polysaccharide was also observed at the cell cover (van Emburg *et al.* 1986) and surrounding the coccoliths (van Bleijswijk, pers. comm.). It probably also functions as an organic matrix for maintaining a coherent coccosphere.

Less is known about the composition of the organic base-plate. Special staining procedures indicate that it contains polysaccharide (van der Wal *et al.* 1983*a*). Whether this polysaccharide is identical to the acidic methylated polysaccharide isolated from coccoliths is uncertain. When coccoliths are dissolved and the residue is simultaneously fixed by an acidic solution of glutaraldehyde, a coherent oval disc surrounded by scattered flocculent material remains (van Emburg 1989). The oval disc approximately occupies the area of deposition of the inner elements (2 in Fig. 8.5b) and may represent the base-plate on which coccolith formation is initiated. It is only slightly labelled with antibodies raised against the coccolith polysaccharide. Thus the latter does probably not form a covalent part of the organic base-plate. Preliminary tests indicate that the organic base plate contains protein. The total protein content of isolated coccoliths amounts to 0.1% (w/w). As mentioned before, the organic base plates of coccoliths of *Emiliania huxleyi* do not show structural features such as the fibrillar material characteristic of the organic scales of other haptophytes (see Leadbeater, Chapter 2).

Calcification in other haptophytes

Pleurochrysis carterae is another heterococcolith-forming species which has been studied in detail. Its coccoliths consist of mineralized organic scales which are produced in the Golgi apparatus (Pienaar 1969; Leadbeater, Chapter 2). Calcium carbonate elements are deposited in alternating orientation on the rim of the organic scale (Pienaar 1969; van der Wal *et al.* 1983*b*). This deposition is reminiscent of the alternating orientation of crystal nuclei on the organic support as proposed by Young *et al.* (1992). The organic scale is formed prior to calcification. In a later stage, so-called coccolithosomes fuse with the Golgi vesicle containing the organic scale and calcification is initiated. The coccolithosomes were shown to contain calcium and polysaccharide (van der Wal *et al.* 1983*b*). The $CaCO_3$ elements of the coccoliths are associated with an organic skin also containing polysaccharide (van der Wal *et al.* 1983*b*). The coccolithosomes are therefore thought to be the precursors of the $CaCO_3$ elements. The polysaccharide skin may function to determine the eventual size of the mineral elements by inhibition of the growth of certain crystal faces, in analogy with the supposed function of the acidic polysaccharide from coccoliths of *Emiliania huxleyi.* If so, it can be expected to contain acidic groups. Acidic polysaccharides have been isolated from coccoliths of *Pleurochrysis carterae*, among which is a novel one containing glyoxylate and tartrate residues (Marsh *et al.* 1992). Acidic polysaccharides are constituents of the organic scale supporting the $CaCO_3$ elements (Romanovicz 1981). Whether one of the isolated acidic polysaccharides represents the organic skin associated with the mineral has to be answered by localization experiments using specific antibodies.

Coccolithus pelagicus has a non-motile heterococcolith-bearing form and a motile holococcolith-bearing form. The holococcoliths consist of an organic scale on which rhombohedral calcite elements are deposited in a more or less regular configuration (Rowson *et al.* 1986). The calcite elements are confined to a certain size and they were shown to be associated with an organic skin (Rowson *et al.* 1986). The organic scales are formed inside the Golgi apparatus, but calcification proceeds outside the plasmalemma in a space covered by an organic envelope. After dissolution of the holococcoliths of *Coccolithus pelagicus*, the first mineralized scales during remineralization were observed at a specific site of the cell near the flagellar bases. Apparently the decalcified scales already present extracellularly were not remineralized. This implies that the presence of an organic scale is not sufficient to induce coccolith formation. It is tempting to assume that the formation of holococcoliths can only proceed by a coordinated excretion of the organic support and an organic substance regulating the growth of the crystals. In spite of the extracellular site of calcification the cell apparently exerts a high degree of control over the mineralization process, analogous to the prymnesiophycean cells producing heterococcoliths.

References

Ackleson, S., Balch, W. M., and Holligan, P. M. (1988). White waters of the Gulf of Maine. *Oceanography*, **1**, 18–22.

Bleijswijk, J. van, van der Wal, P., Kempers, R., Veldhuis, M., Young, J. R., Muyzer, G., *et al.* (1991). Distribution of two types of *Emiliania huxleyi* (Prymnesiophyceae) in the northern Atlantic region as determined by immunofluorescence and coccolith morphology. *Journal of Phycology*, 27, 566–70.

Borman, A. H., de Vrind-de Jong, E. W., Huizinga, M., Kok, D. J., Westbroek, P., and Bosch, L. (1982). The role in $CaCO_3$ crystallization of an acid Ca^{2+}-binding polysaccharide associated with coccoliths of *Emiliania huxleyi. European Journal of Biochemistry*, **192**, 179–83.

Emburg, P. R. van (1989). Coccolith formation in *Emiliania huxleyi*. Ph. D. thesis. University of Leiden, The Netherlands, 145 pp.

Emburg, P. R. van, de Vrind-de Jong, E. W., and Daems, W. Th. (1986). Immunochemical localization of a polysaccharide from biomineral structures (coccoliths) of *Emiliania huxleyi. Journal of Ultrastructure and Molecular Structure Research*, **94**, 246–59.

Fichtinger-Schepman, A. M. J., Kamerling, J. P., Versluis, C., and Vliegenthart, J. F. G. (1981). Structural studies of the methylated acidic polysaccharide associated with coccoliths of *Emiliania huxleyi* (Lohmann) Kamptner. *Carbohydrate Research*, **93**, 105–23.

Green, J. C., Perch-Nielsen, K., and Westbroek, P. (1990). Phylum Prymnesiophyta. In *Handbook of Protoctista*, (ed. L. Margulis, J. O. Corliss, M. Melkonian, and D. J. Chapman), pp 293–317. Jones & Bartlett, Boston.

Holligan, P. M. (1986). Phytoplankton distributions along the shelf break. *Proceedings of the Royal Society of Edinburgh*, **88B**, 239–63.

Jong E. W. de, Bosch, L., and Westbroek, P. (1976). Isolation and characterization of a Ca^{2+}-binding polysaccharide associated with coccoliths of *Emiliania huxleyi* (Lohmann) Kamptner. *European Journal of Biochemistry*, **70**, 611–21.

Klaveness, D. (1972). *Coccolithus huxleyi* (Lohmann) Kamptner. I. Morphological investigations on the vegetative cell and the process of coccolith formation. *Protistologica*, **8**, 335–46.

Klaveness, D. (1976). *Emiliania huxleyi* (Lohmann) Kamptner. III. Mineral deposition and the origin of the matrix during coccolith formation. *Protistologica*, **12**, 217–24.

Linschooten, C., van Bleijswijk, D. L., van Emburg, P. R., de Vrind, J. P. M., Kempers, E. S., Westbroek, P., *et al.* (1991). Role of the light-dark cycle and medium composition on the production of coccoliths by *Emiliania huxleyi* (Haptophyceae). *Journal of Phycology*, **27**, 82–6.

Mann, S. and Sparks, N. H. C. (1988). Single crystalline nature of coccolith elements of the marine alga *Emiliania huxleyi* as determined by electron diffraction and high resolution microscopy. *Proceedings of the Royal Society*, Series B, **234**, 441–53.

Marsh, M. E., Chang, D.-K., and King, G. C. (1992). Isolation and characterization of a novel acidic polysaccharide containing tartrate and glyoxylate residues from the mineralized scales of a unicellular coccolithophorid alga *Pleurochrysis carterae. Journal of Biological Chemistry*, **267**, 20507–12.

Pienaar, R. N. (1969). The fine structure of *Cricosphaera carterae*. I. External morphology. *Journal of Cell Science*, **4**, 561–7.

Romanovicz, D. K. (1981). Scale formation in flagellates. In *Cytomorphogenesis in plants*, (ed. O. Kiermayer), pp 27–62. Springer-Verlag, Vienna.

Rowson, J. D., Leadbeater, B. S. C., and Green, J. C. (1986). Calcium carbonate deposition in the motile (Crystallolithus) phase of *Coccolithus pelagicus* (Prymnesiophyceae). *British Phylological Journal*, **21**, 359–70.

Wal, P. van der, de Jong, E. W., and Westbroek, P. (1983*a*). Ultrastructural polysaccharide localization in calcifying and naked cells of the coccolithophorid *Emiliania huxleyi*. *Protistologica*, **118**, 157–68.

Wal, P. van der, de Jong, E. W., Westbroek, P., de Bruijn, W. C., and Mulder-Stapel, A. A. (1983*b*). Polysaccharide localization, coccolith formation and Golgi dynamics in the coccolithophorid *Hymenomonas carterae*. *Journal of Ultrastructure Research*, **85**, 139–58.

Wal, P. van der, Leunisse-Bijveld, A. J., and Verkley, A. J. (1985). Ultrastructure of the membranous layers enveloping the cell of the coccolithophorid *Emiliania huxleyi*. *Journal of Ultrastructure Research*, **91**, 24–9.

Watabe, N. (1967). Crystallographical analysis of the coccolith in *Coccolithus huxleyi*. *Calcified Tissue Research*, **1**, 114–21.

Young, J. R., Didymus, J. M., Bown, P. R., Prins, B., and Mann, S. (1992). Crystal assembly and phylogenetic evolution in heterococcoliths. *Nature*, **356**, 516–18.

9. Life cycles

C. BILLARD

Laboratoire de Biologie et Biotechnologies Marines, Université de Caen, France

Abstract

Alternations of morphologically distinct generations are frequent in the Haptophyta, but the existence and place of sexuality, if applicable, often remains unknown. In most Pavlovales a nonmotile stage exists as an alternative to flagellate cells and is assumed to respond to environmental conditions. Similarly palmelloid and motile stages are recorded in the Isochrysidales where the dominant stage is planktonic (*Isochrysis*) or benthic (*Chrysotila*). In the Prymnesiales a complex cycle is suspected for *Phaeocystis* involving colonial and motile stages, although observations on the nuclear ploidy levels of the different phases are lacking. Except in *Platychrysis* where amoeboid cells produce swarmers, members of the Prymnesiaceae are flagellates in culture, with resting cysts recorded in *Prymnesium*. Two motile cell types, with different scales, recently described in *Chrysochromulina polylepis* cultures could possibly indicate sexuality in *Chrysochromulina*. Many Coccolithophorales have heteromorphic life histories. In coastal genera (*Pleurochrysis, Hymenomonas, Ochrosphaera*), diploid coccolith-bearing cells and haploid scale-covered forms alternate, with different patternings on the organic scales of both phases. Life histories combining hetero- and holococcolithophorids are noteworthy: increasing numbers of oceanic holococcolithophorids, previously considered as autonomous species, are now shown to be alternate phases in the life cycle of heterococcolithophorids, but, as often in the Prymnesiophyceae, information on the nuclear cytology is needed.

Introduction

Although our knowledge of the Haptophyta (with one class, the Prymnesiophyceae) has expanded in the last decade, and interest in this group has increased with recent focus on its toxic members, information is still needed on the basic biology of these organisms. Life cycles have received less attention than ultrastructural investigations as a whole, except maybe for the coccolithophorids in which heteromorphic life histories were

The Haptophyte Algae (ed. J. C. Green and B. S. C. Leadbeater), Systematics Association Special Volume No. 51, pp. 167–86. Clarendon Press, Oxford, 1994.

reported very early (Parke and Adams 1960; Stosch 1967). Indeed as it will be shown below, very convincing examples of alternations of generations are found in the coccolithophorids, but with growing evidence for haplo-diploid life cycles in some members of the Prymnesiales also. In contrast in the remaining (more primitive ?) Prymnesiophyceae, life histories appear more simple and there is as yet no indication that alternating phases with changes in ploidy levels are a general rule in the class.

In the subsequent account the main groups traditionally recognized in the class will be followed, except for the coccolithophorids which are gathered into a single order to accommodate all species which produce coccoliths at some stage of their life cycle (see Chrétiennot-Dinet 1990).

Pavlovales

Members of this distinctive order differ notably from other haptophytes in various cytological features such as lack of flat body scales on the cell surface (see Green 1980 for a review; Leadbeater, Chapter 2), and are considered a natural assemblage. Recent biochemical evidence, based on fatty acid and sterol compositions (Volkman *et al.* 1991), confirms this view.

At present, three genera are included by Green (1980) in the order, namely *Pavlova, Diacronema,* and *Exanthemachrysis.* The last two genera are monospecific: *E. gayraliae* is a estuarine palmelloid organism; while *D. vlkianum,* a markedly euryhaline species, is strictly flagellate (Green and Hibberd 1977). Another euryhaline flagellate, *Boekelovia hooglandii,* inhabiting temporary saline ponds, probably also belongs to the Pavlovales (see Barclay *et al.* 1991). Members of the type genus *Pavlova* have been recorded in various habitats, from freshwater lakes to marine environments, although most thrive in brackish waters. *Pavlova* species are predominantly nonmotile or motile, with only two, *P. helicata* and *P. lutheri,* recorded so far as strictly flagellate (see Green 1980). The general organization of the non-motile stages may vary, from cells surrounded with homogenous mucilage, as in *P. virescens* (Billard 1976), to cells ensheathed in stratified mucilage and forming stalks, as in *P. noctivaga* (Veer and Leewis 1977).

Appendages are present in the nonmotile stages of the Pavlovales, but usually in a more or less abbreviated form; thus, flagellar bases only are recorded in *Exanthemachrysis* (Gayral and Fresnel 1979). All species in the order are known to produce motile cells in culture, but without any indication of a sexual process. Anyone familiar with coastal species of *Pavlova* has witnessed the most remarkable, and in some instances, very rapid transition from nonmotile cells grown on agar plates, to motile cells after transferring to liquid medium. So far it has been assumed that the swimming cells are swarmers released through asexual processes, and that the planktonic and benthic stages recorded in most Pavlovales are merely responses triggered by environmental conditions. Cytological studies to confirm or refute this assumption would be welcome.

Isochrysidales

With the removal of coccolith-bearing genera, this order appears less heterogenous and conveniently groups organisms with motile cells featuring a very reduced haptonema which might be altogether absent. All genera except *Dicrateria* have a covering of one or several layers of simple organic body scales. Members of the Isochrysidales are typically marine haptophytes except for one species, *Chrysotila lamellosa*, which is also found in terrestrial habitats with high saline concentrations.

Four genera remain in the order circumscribed as above, *Imantonia*, *Chrysotila*, *Isochrysis*, and *Dicrateria*. The only species recorded in the first genus, *Im. rotunda* (Reynolds 1974), is an oceanic nanoflagellate with no alternative stage reported in culture. In contrast *Chrysotila* is a distinctive genus of benthic epilithic organisms, with two species distinguished by the dendroid habit of the vegetative cells. A motile stage was described for both species in culture (Billard and Gayral 1972; Green and Parke 1975). The motile cells are *Isochrysis*-like and are produced in numbers (8–16 or more) inside thick-walled zoosporangia; after a short period of active swimming the swarmers settle, round off and resume the vegetative condition. There is no indication of sexuality in the genus and the motile cells released appear as an effective mean of asexual propagation.

The three species described by Parke (1949), *D. inornata*, *D. gilva*, and *Isochrysis galbana* are flagellates, but nonmotile stages are recorded for all three in culture. Asexual reproduction by longitudinal division occurs in both stages. Parke (1949) described a form of reproduction in older cultures of the three species above, where the product of fusion of two nonmotile (?) individuals was a large cell which developed a gelatinous envelope. The latter was considered to be a zygote dividing to produce four motile cells, after possible reduction division. Nevertheless the figures by Parke illustrating this last sequence in *I. galbana* (Figs 41–43 in Parke 1949) do not show the distinctive tetrads indicative of meiosis, and the earlier fusion stages have not been observed with certainty. Clearly further studies are required to confirm this presumptive sexual cycle in *Isochrysis* or *Dicrateria*. Furthermore, in the other species of *Isochrysis*, *I. litoralis*, there is no indication of a sexual process. In *I. litoralis* the dominant stage is palmelloid with older cells irregular in size surrounded by thick mucilage; swarmer production is stimulated by transferring the culture to fresh medium and swarmers may persist for a while (Billard and Gayral 1972). Thin sections have shown that the palmelloid cells are covered by one or more proximal layers of body scales underlying a thick coat of mucilage; ornamentation of the body scales are identical to those of the scales covering the motile cells (Billard, unpublished observations).

Earlier reports of chrysophycean-like cysts in *Isochrysis* and *Dicrateria* (Parke 1949) were shown to be erroneous and are to be discounted (Hibberd 1976). To this date there are no firm indications of sexuality in the Isochrysidales. The motile cells observed in species of *Chrysotila* and in *I. litoralis* are clearly asexual swarmers, while the nonmotile cells observed in cultures of *I. galbana* and *Dicrateria* spp. are probably responses to adverse conditions. In the three species of truly coastal Isochrysidales, *I. litoralis* and both species of *Chrysotila*, the benthic and alternative motile stages can be considered as an adaptation to their ecological niche.

Prymnesiales

Motile cells of this order are characterized by a conspicuous haptonema, which may be much longer than the flagella as in *Chrysochromulina* spp. (Inouye and Kawachi, Chapter 4), and a covering of at least two types of organic scales. As a rule, the scales show a pattern of radiating ridges on the proximal face and concentric or spiral fibres on the distal face (Leadbeater, Chapter 2). When different layers of scales are present, the distal scales are generally characteristic of the species (see Green *et al.* 1989). The great majority of the Prymnesiales are marine or brackish except for three species of *Chrysochromulina* recorded in freshwater lakes. The Prymnesiales are the only haptophytes known to have toxic members; these are found in two closely related genera, *Prymnesium* and *Chrysochromulina* (see Moestrup and Larsen 1992; Moestrup, Chapter 14).

Authors generally recognize two families in the order, the Phaeocystaceae with a single genus, *Phaeocystis*, and the Prymnesiaceae with the remaining four genera, *Prymnesium*, *Platychrysis*, *Chrysochromulina*, and *Corymbellus*.

Phaeocystaceae

Phaeocystis is one of the few colonial planktonic haptophytes known. The distinctive bloom-forming colonies, where the cells embedded in mucilage are distributed in a single layer at the periphery, may reach several millimeters in size. The morphology of the colonies (globular or lobular) combined with arrangements of the cells at the periphery, could be valid taxonomic criteria in distinguishing some species of *Phaeocystis* in the colonial stage, according to Jahnke and Baumann (1987). Nevertheless, the question of the number species in the genus, especially in the North Sea, is still a matter of debate (Moestrup and Larsen 1992). The colonial cells of *P. pouchetii* lack appendages and scales (Chang 1984); this absence of scales on the colonial stage has further been confirmed in other strains of *Phaeocystis* in culture (Chrétiennot-Dinet, personal communication).

Although in natural situations the colonial stage is the most commonly encountered form, alternative unicellular flagellate stages have been known in the life history of *Phaeocystis* since the work of Kornmann (1955). Motile cells of *P.* aff. *pouchetii* have typical haptophycean features. Besides two flagella and a short haptonema, they are covered by two types of scale (Parke *et al.* 1971), but some clones contain unusual trichocyst-like structures which after discharge display characteristic pentagonal stars. Recent observations on motile cells from the southern hemisphere (Moestrup 1979; Pienaar 1991) indicate that both patterning of the trichocyst discharge and scale morphology are most certainly important taxonomic characteristics for discriminating species of *Phaeocystis* in the flagellate stage. *Phaeocystis scrobiculata* (Moestrup 1979) and the unnamed species studied by Pienaar (1991) are recorded so far only as motile cells, and the presumptive colonial stages are yet to be found.

Despite the fact that in certain *Phaeocystis* species colonial and flagellate stages alternate, the life cycle has not been fully elucidated. Field observations show that colonies of *P. globosa* develop from single cells which have settled on suitable substrates such as *Chaetoceros* setae. Mature colonies may divide by fragmentation or release solitary cells, some of which are motile cells that initiate new colonies. Some flagellate cells are also capable of self-replication (Lancelot *et al.* 1991). Kornmann (1955) described different types of motile cells in culture, including smaller ones which he suspected to be presumptive gametes. However, he himself and subsequent workers were unable to confirm this experimentally (Veldhuis 1987). The existence of two distinct cell types in the life history of *Phaeocystis*, i.e. flagellate cells producing scales (with or without trichocysts), and colonial cells lacking scales, could indeed be linked with changes of ploidy levels. Studies using modern quantitative techniques, such as flow cytometry enabling the estimation of DNA contents of cells from clonal cultures of *Phaeocystis* are currently under way. Preliminary results (Vaulot, personal communication) indicate that the colonial cells are diploid, whereas motile cells may either be diploid or haploid, depending on their size. Evidence is therefore accumulating, indicating a sexual heteromorphic cycle in the genus *Phaeocystis*. More work will be necessary to confirm the existence of two generations with different ploidy levels and to determine the place of sexuality. Furthermore the question of whether this heteromorphic cycle is the rule in all *Phaeocystis* species will need to be answered.

Prymnesiaceae

Corymbellus, the other colonial planktonic genus known in the Prymnesiales, differs from *Phaeocystis* in the form and composition of its motile colonies (Green 1976). In *C. aureus*, the only species described so far, colonies result from the aggregation of flagellate cells. The cells have a short haptonema and are covered by dimorphic body scales. Following the disruption of mature colonies, individual cells or small groups of cells are liberated which in turn develop new colonies. No alternative stage has been reported, although the ability to form colonies is progressively lost in culture (Green 1976).

Prymnesium is a genus of typically coastal or brackish flagellates. The cells feature a short, non-coiling haptonema, two heterodynamic flagella, and a covering of two types of scales. At the light microscopic level there is little morphological variation between species, and these can only be reliably distinguished by examination of the ornamentation of their distal scales (Green *et al.* 1982; Billard 1983). Two or possibly three toxic species are recorded (Green *et al.* 1982; Chang 1985). *Prymnesium* species are easily grown in culture (Billard 1987) and propagation takes place by division of the motile cells. The only alternative stage known so far is a type of resting cyst reported in certain, but not all species. *Prymnesium* cyst walls are composed of layers of scales, with electron-dense siliceous material deposited on the outermost scales (Pienaar 1981). Salinity of the medium seems to influence the production of cysts (Green *et al.* 1982), and there is no evidence that these might be the result of a sexual phenomenon.

In contrast to *Prymnesium*, two stages are known in the life history of the closely related, but rarely reported genus *Platychrysis* which shares the same ecological niche as *Prymnesium*. The dominant stage is neustonic with nonmotile amoeboid cells, flattened anterio-posteriorly. In amoeboid cells of *P. pigra* the appendages are visible at the light microscope level, with both flagella coiled around the haptonema, and transition to swimming cells can be very rapid (Chrétiennot 1973). The three other described species of *Platychrysis* share the common feature of lacking obvious flagella in the amoeboid stage (Norris 1967; Gayral and Fresnel 1983*a*), but following transfer to fresh medium, motile cells are usually produced. Motile cells of *Platychrysis* have two homodynamic flagella and a short, non-coiling haptonema. Both motile and amoeboid cells are covered with organic scales. As in *Prymnesium* the scales are dimorphic, and the outermost scales bear species specific ornamentation. Scales of both stages are identical (Chrétiennot 1973; Gayral and Fresnel 1983*a*), and there is no indication of alternation of generations or sexuality in the genus. Cysts have never been reported.

The last member of the family Prymnesiaceae is the large genus *Chrysochromulina* with approximately fifty species, the great majority of which are oceanic nanoflagellates (Estep *et al.* 1984). A long, coiling haptonema and homodynamic flagella are well-known characteristics of members of the genus, along with the production of elaborate scale types. Following the outbreak of the toxic bloom of *C. polylepis* on the Scandinavian coasts in May and June 1988, attention has focussed on the genus as a whole, and other species of *Chrysochromulina* are now also suspected to be toxic (Moestrup and Larsen 1992; see also Moestrup, Chapter 14). Some species were reported earlier to form amoeboid stages in culture (Parke *et al.* 1955), but such stages are never observed in natural environments. In contrast, recent studies on cultured material of *C. polylepis* indicate that two morphologically distinct flagellate stages alternate in the life cycle of this species (Edvardsen and Paasche 1992). Both stages, known as α and β, differ in cell size, light and temperature tolerances, and toxicity. Furthermore, α and β cells may be distinguished by their scaly covering; α cells, considered 'authentic' *C. polylepis* clones, produce four different scale types, including those scales with the lateral 'fish-tail' extension characteristic of the species in the wild; β cells, the deviant cell-type, carry three kinds of scales, including spiked scales, all of which are different from the four kinds seen on the authentic cells (Edvardsen and Paasche 1992). Some authors (see Moestrup and Larsen 1992) consider the β cells, which have only been observed in laboratory cultures, as aberrant cell-types and the possibility exists that both α and β cells are morphotypes representing the same stage in the life history of *C. polylepis.* Nevertheless, the preliminary results of Edvardsen and Paasche (1992), following cytometric flow analyses, indicate that ploidy levels in α and β stages could be different. Moreover, it must be emphasized that in coccolithophorids where diploid and haploid phases alternate, the patterning of the organic body scales is different in both stages (Gayral and Fresnel 1983*b*; Fresnel 1989). Since both flagellate stages of *C. polylepis* may be distinguished by their scaly covering, and a presumption exists that the two stages have different ploidy levels, it is conceivable that a sexual haplo-diploid life cycle exists in this species. In any case the existence of two distinct flagellate cell-types in the life history of *C. polylepis* has been demonstrated. Such a life cycle is highly unusual in haptophytes, where alternation between nonmotile and motile stages is the general rule. More studies will be necessary to confirm the presumed alternation of nuclear phases in *C. polylepis* and other *Chrysochromulina* species. Direct evidence for syngamy or meiosis will certainly be difficult if not impossible to obtain, considering the small size of most *Chrysochromulina* species and their overall fragility. Flow cytometers, with sorting facilities, are certainly the best analytical tools in such cases. Clonal cultures are a prerequisite in such investigations, and single cell capillary isolations are to be preferred to isolations by the serial dilution method.

In view of the existence of a presumptive haplo-diploid life cycle in some Prymnesiales (*Phaeocystis*, *Chrysochromulina*), the possibility that alternation of generations may also exist in other genera of the order must not be neglected. Studies on clonal strains of these genera, such as the research currently undertaken with *C. polylepis*, are needed.

Coccolithophorales

Haptophytes of this almost exclusively marine order produce coccoliths at some stage of their life cycle. A large number of families are distinguished, generally based on coccolith morphology. Two fundamental types exists; holococcoliths, made of a single type of calcified elements, and heterococcoliths, formed of elements of different sizes and shapes. Heterococcoliths are formed intracellularly, whereas the calcification step in holococcoliths seems to take place extracellularly (see Green *et al.* 1989 for a review). Although this traditional distinction between hetero- and holococcolithophorids might be convenient for descriptive purposes, investigations have shown that both types may be linked in the life histories of certain species (e.g. Thomsen *et al.* 1991). This distinction may, therefore, not have any value in coccolithophorid phylogenetic taxonomy.

In most cases the coccolith is associated with a base-plate scale and has an organic matrix. Generally, with very few exceptions in the species studied so far, body scales are present beneath the coccoliths, overlying the plasmalemma. Coccolithophorids are generally solitary organisms, their cells either nonmotile or motile. The latter are biflagellate and the haptonema is either well-developed and of the coiling type (e.g. *Syracosphaera*), vestigial (e.g. *Cruciplacolithus*), or altogether missing (motile scale-bearing cells of *Emiliania*).

A limited number of coccolithophorids can be grown in the laboratory, most of them coastal species. It is not surprising, therefore, that the few studies devoted to life cycles have been done on marine littoral species (see Fresnel 1989).

Life cycles involving coccolith- and non-coccolith-bearing stages (Fig. 9.1)

For practical purposes, the life cycle of the Pleurochrysidaceae will be reviewed first since valid information exists for a number of species of this family of heterococcolithophorids. Early observations by Rayns (1962) showed the coccolith-bearing stage was diploid, whereas the non-coccolith-bearing form was haploid, in the species now known as *Pleurochrysis carterae*. Details of its life history were added by von Stosch (1967) and the fine

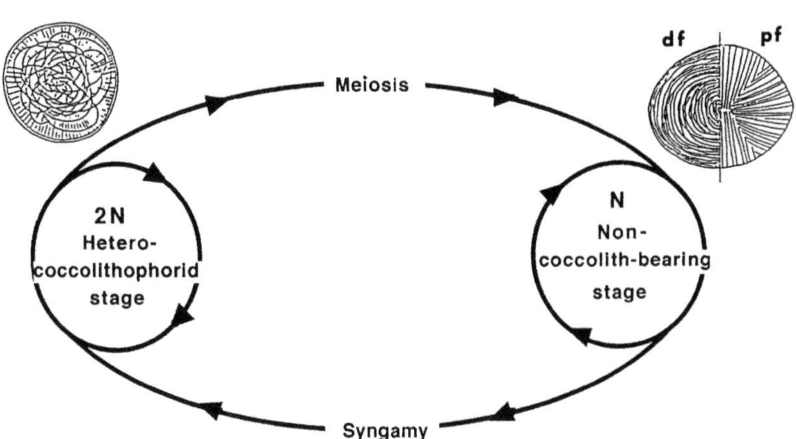

Fig. 9.1 Life cycle of Coccolithophorales involving coccolith- and non-coccolith-bearing stages. Schematic drawings illustrating respective body scale morphologies are provided. df = distal face; pf = proximal face.

structure of both stages was investigated by Leadbeater (1970). Furthermore, syngamy and meiosis were carefully studied in *P. pseudoroscoffensis* (Gayral and Fresnel 1983*b*), and other species of *Pleurochrysis* were also shown to have a heteromorphic life history (Fresnel and Billard 1991). The family was created for coccolithophorids with diploid cricolith-bearing cells, generally alternating with haploid benthic pseudofilaments lacking coccoliths (Fresnel and Billard 1991). In the genus *Pleurochrysis* the life cycle is completed, and both generations are present. Besides presence or absence of coccoliths, each generation can be characterized by a distinctive type of unmineralized body scale which is best observed in shadowcast material; those of the diploid phase are circular with identical ornamentation on both sides and a distinct rim; haploid phase scales are usually more elliptic, rimless, with a proximal pattern of radiating ridges arranged in four quadrants, and a pattern of concentric fibers on the distal face (Leadbeater 1970; Gayral and Fresnel 1983*b*; Fresnel and Billard 1991). Haptonematal scales of haploid swarmers are smaller. Both scales types, diploid and haploid, are illustrated in Fig. 9.4. In Pleurochrysidaceae one of the generations may be missing. The genus *Cricosphaera* is currently maintained for species where only the cricolith-bearing (diploid) form is recorded (e.g. *C. elongata*). Conversely benthic pseudofilamentous forms are known which are unable to produce coccoliths: these *Apistonema*-like stages are covered with haploid type body scales. The example of *P. roscoffensis* (see Fresnel and Billard 1991) illustrates these possible deviations. In France only the diploid stage is recorded (Gayral and Fresnel 1976), but, in Japan, the species completes its life cycle (Inouye and Chihara 1979).

Recently, another type of haplo-diploid life cycle was described, involving heterococcolithophorids with tremaliths and cells lacking coccoliths, but covered by two distinct types of organic scale (Fresnel 1989). This cycle exists in a number of species belonging to the genera *Hymenomonas* and *Ochrosphaera*. As confirmed by nuclear staining, the tremalith-bearing cells are diploid and, in addition to coccoliths, are covered by body scales identical to those of the diploid stage of *Pleurochrysis*. The alternate cells, which are not pseudofilamentous, lack coccoliths and are covered with dimorphic scales. There are rimless body scales with proximal and distal patternings identical to those of the haploid stage of *Pleurochrysis*, together with supplementary external scales with species specific ornamentation and rim morphology. Although syngamy was not directly observed by Fresnel (1989) in *O. neapolitana*, her chromosome counts clearly show that the tremalith-covered cells are diploid, whereas the cells lacking coccoliths and featuring dimorphic scales are haploid. The life cycle proposed earlier by Schwarz (1932) was not confirmed and must, therefore, be discounted. At this point it is important to stress that in the life cycles described above, for species of either *Pleurochrysis* or *Hymenomonas*/*Ochrosphaera*, only one type of coccolith is involved, either cricoliths or tremaliths. The so-called cycle reported by Lefort (1975), linking forms with *Ochrosphaera*-type coccoliths to forms either *Apistonema*-like or with *Pleurochrysis*-type coccoliths, has never been confirmed in our laboratory or by any other worker. Lefort's observations, most probably based on non-unialgal (or contaminated) cultures, are probably, therefore, invalid.

Fresnel (1989) was unable to demonstrate an alternation of generations in the strain of *H. globosa* she investigated, only the coccolith-bearing (diploid) phase being recorded so far for this species (Gayral and Fresnel 1976). Furthermore, of the different strains of *O. neapolitana* she studied, only one showed an alternation of generations; the other were blocked in the diploid, coccolith-bearing, stage. These functional irregularities, already mentioned for some Pleurochrysidaceae, are probably common in the life cycles of coccolithophorids. Either one generation is lost or could be produced under certain environmental conditions. Similar examples are reported for various macroalgae (Rhodophyceae or Phaeophyceae) where the life cycle is completed in restricted geographical ranges only.

Figure 9.1 summarizes this first type of life cycle recorded in the Coccolithophorales.

Life cycles involving hetero- and holococcolith-bearing stages (Fig. 9.2)

The first example of this life cycle was reported in 1960 by Parke and Adams who demonstrated that two species previously considered as autonomous, the heterococcolithophorid *Coccolithus pelagicus* and the holococcolithophorid *Crystallolithus hyalinus*, were linked in the same life cycle. The nonmotile cells, *C. pelagicus*, produce placoliths whereas the flagellate cells synthesize crystalloliths. Although the nuclear cytology of both phases has still not been worked out, useful information can be obtained by a careful examination of their organic body scales. The scales of both phases are in fact illustrated from shadowcast material by Manton and Leedale (1969); the body scales of *C. pelagicus*, as indicated in Figures 12–14 of Manton and Leedale (1969), are circular and rimmed, and their ornamentation, with a sparse pattern of roughly concentric threads, are identical to those of the diploid cricolith-bearing phase of *Pleurochrysis*; in contrast, the elliptical rimless body scales with their radiating ridges arranged in four quadrants, illustrated in Figure 15 of Manton and Leedale (1969), were considered by the authors to be exceptional for *C. pelagicus*. These unusual scales are here interpreted as body scales (viewed from their proximal face) of the *Crystallolithus* stage; the two smaller scales visible in Manton and Leedale's figure 15 are in fact haptonematal scales. The body scales of the *Crystallolithus* stage had formerly been seen and correctly interpreted in a previous paper by Manton and Leedale (1963). These scales are identical to the body scales of haploid, non-coccolith-bearing cells of *Pleurochrysis*. From the evidence presented above, based on scale morphology, it is

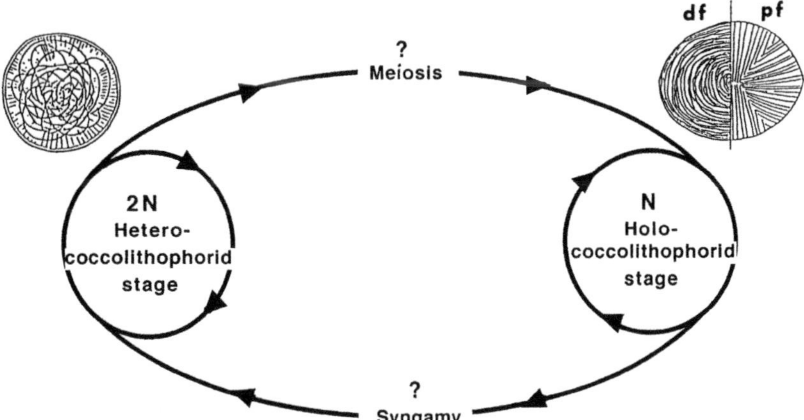

Fig. 9.2 Hypothetical life cycle of Coccolithophorales linking hetero- and holo-coccolith-bearing stages. Schematic drawings illustrate respective body scale morphologies. df = distal face; pf = proximal face.

postulated that *C. pelagicus* has a haplo-diploid life cycle. Kleijne (1991) and Janin (1992) recently linked another species with crystalloliths, *Crystallolithus rigidus*, to *Calcidiscus leptoporus*, a heterococcolithophorid with placoliths. Further examples are given by Kleijne (1991) linking heterococcolithophorid and holococcolithophorid genera, e.g. *Syracosphaera/Zygosphaera*. No evidence based on scale morphology is available since Kleijne's observations are based on natural specimens examined by SEM. However, illustrations of body scales of the type species of *Syracosphaera*, *S. pulchra*, exist (Leadbeater and Morton 1973; Inouye and Pienaar 1988). As Leadbeater and Morton (1973) point out, they are similar in appearance to body scales of coccolith-bearing cells of *Coccolithus pelagicus*, *Pleurochrysis carterae*, and *Ochrosphaera neapolitana*, and, in my opinion, they are indicative of a diploid stage. In situations where only one stage is known, examination of the body scales can, therefore, be informative. In the holococcolithophorid genus *Calyptrosphaera*, the ornamentation of the body scales (see Klaveness 1973; Leadbeater and Morton 1973) is similar to those of *Cr. hyalinus* indicating that *Calyptrosphaera* species could be haploid stages of heterococcolithophorids. As in the case of *Pleurochrysis* or *Hymenomonas/Ochrosphaera* life cycles, it is conceivable that in some geographical areas one generation dominates, or that within the hetero/holococcolithophorid assemblages, some 'matching couples' remain to be discovered.

Other striking examples of heteromorphic life histories have recently been described by Thomsen *et al.* (1991) for flagellate, arctic coccolithophorids. Holococcolithophorid species belonging to the genera *Turrisphaera*, *Trigonaspis*, and probably also *Calciarcus*, and previously thought to be autonomous, were shown to be part of the life cycle of certain heterococcolithophorid genera, namely *Papposphaera*, *Pappomonas*, and *Wigwamma*. Cells showing both distinctive holo- and heterococcoliths were observed in all the above genera by means of shadowcast preparations of natural samples. Although conclusive evidence is at present lacking with regard to the patterning of the respective body scales, these alternative changes are most probably linked to changes in ploidy levels. By reference to the *Coccolithus/Crystallolithus* life cycle, discussed above, it might be hypothesized that the holococcolith-bearing stages represent the haploid generation. Again some species of the above genera are known so far in one life history stage, but clearly further linkages between holo- and heterococcolithophorids will be discovered in the future. Thomsen *et al.* (1991) go as far as to predict that few if any arctic/antarctic nanoflagellate coccolithophorids are autonomous species, while Kleijne (1991) suggests the holococcolithophorid family Calyptrosphaeraceae will eventually have to be eliminated. The hypothetical life cycle proposed involving hetero- and holococcolithophorids is illustrated Fig. 9.2.

The Emiliania *life cycle* (Fig.9.3)

The heterococcolithophorid genus *Emiliania* is considered to occupy an isolated position within the Prymnesiophyceae (Green *et al.* 1989). Klaveness (1972) clearly demonstrated two stages in the life history of the ubiquitous *E. huxleyi*, i.e. nonmotile coccolith-bearing cells (C-cells) and alternate scaly motile cells (S-cells) without coccoliths. His preliminary studies on the nuclear cytology of both phases indicated that the C-cells could be diploid whereas the S-cells could be haploid. The life cycle of *Emiliania* differs, nevertheless, from apparently similar cycles involving coccolith-bearing and non-coccolith-bearing cells (e.g. *Pleurochrysis*) for two main reasons : (1) absence of organic body scales on the C-cells; and (2) the body scales of the S-cells are unusual, reminiscent of *Isochrysis* body scales, and their patterning is seemingly unlike the one found in non-coccolith-bearing (haploid) stages examined so far. *Emiliania* body scales are difficult to observe, being 'glued' to the cell surface (Klaveness 1972), and more shadowcast preparations are needed to see if patternings on the distal and proximal faces are different. In any case, absence of body scales on the C-cells do not allow further comparisons. Nevertheless, the existence of two distinct cell types in the life history of *E. huxleyi*, one of which acquires the ability to form organic scales is, in my opinion, indicative of a haplo-diploid cycle. Conclusive information on the ploidy levels of both stages is urgently needed, but meanwhile a tentative life cycle is proposed Fig. 9.3.

Umbilicosphaera sibogae var. *foliosa* is the only other coccolithophorid at present known to lack organic body scales, as Inouye and Pienaar (1984) demonstrated using sectioned material. Although motile cells were never observed in their cultures, the flagellar apparatus was persistently present

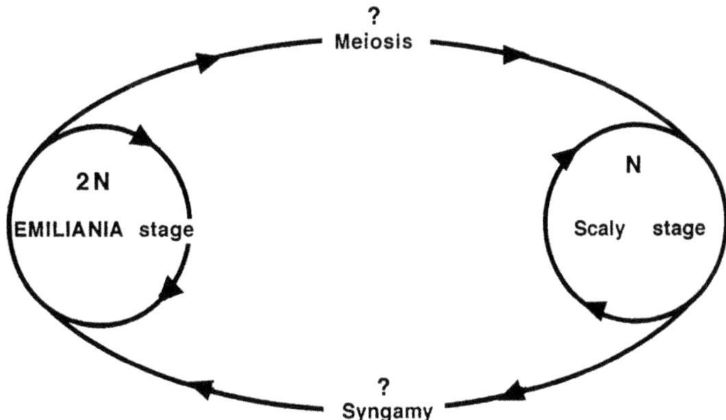

Fig. 9.3 Hypothetical life cycle of *Emiliania huxleyi*.

indicating that an alternate motile stage might exist for this heterococcolithophorid. This does not imply, however, that both taxa, *E. huxleyi* and *U. sibogae* var. *foliosa*, could have the same type of life history.

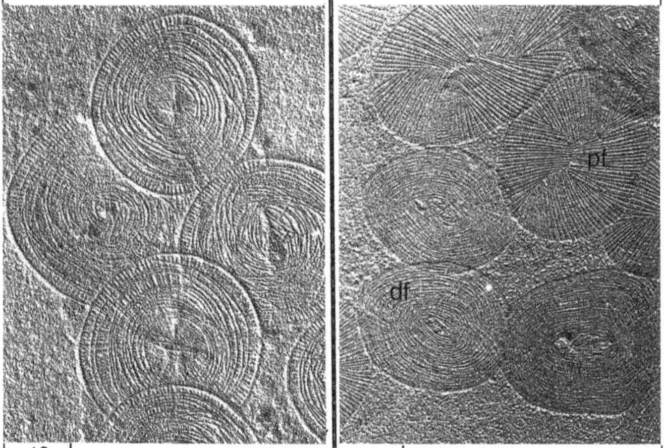

	DIPLOID (2N) GENERATION		HAPLOID (N) GENERATION
	df		pf
HETEROCOCCOLITHOPHORIDS	*Pleurochrysis Cricosphaera *Hymenomonas *Ochrosphaera	No coccoliths	*Pleurochrysis *Hymenomonas *Ochrosphaera
	Coccolithus Calcidiscus Syracosphaera ?	Holococcoliths	Crystallolithus Crystallolithus Zygosphaera ? Calyptrosphaera
	'Umbilicosphaera' hulburtiana Jomonlithus	?	? ?

Fig. 9.4 Body scale morphologies in coccolithophorid genera where a haplo-diploid life cycle is recorded (*) or suspected. df = distal face; pf = proximal face.

At present it is not known whether a life history with alternation of generations is the general rule in all coccolithophorids. Some are known to reproduce in culture via binary fission and/or by production of asexual swarmers. Examination of their body scales, whenever possible might be informative. In *Jomonlithus littoralis* nonmotile coccospheres and flagellate cells bear identical heterococcoliths and body scales (Inouye and Chihara 1983); the ornamentation of their scales, identical on both sides, is typical of 'diploid' body scales. Scales of *Umbilicosphaera hulburtiana* are also of the 'diploid' type (see Figure 8c in Gaarder 1970) and, according to Inouye and Pienaar (1984), this species should be removed from the genus *Umbilicosphaera* since the type species, *U. sibogae,* has no organic body scales. In the heterococcolithophorid *Cruciplacolithus neohelis,* coccospheres and motile cells also bear identical coccoliths and body scales, according to Fresnel (1986). The body scales, with distinct proximal and distal faces, are unusual for placolith-bearing cells. This species should be further investigated to check that no alternative generation is present.

From the evidence gathered so far, derived mainly from our knowledge of the *Pleurochrysis* haplo-diploid cycle, Fig. 9.4 is proposed, in which the morphology of unmineralized body scales is emphasized as it seems a useful indicator of ploidy levels in certain coccolithophorids. Only genera for which information on the body scales is available, are mentioned. These represent only a small fraction of the extant coccolithophorid flora. As frequently noted, data on nuclear cytology are lacking in most cases. The scheme is, therefore, tentative, but might be useful as a working hypothesis. *Emiliania* does not appear in this scheme since its life cycle, although probably also haplo-diploid, does not involve alternating scale types. *Cruciplacolithus* is not mentioned for the reasons given above.

Conclusions and perspectives

Our knowledge of the biology of the Haptophyta is, in many ways, still incomplete. When a life cycle is suspected, the position of syngamy and meiosis is difficult to observe. Little information on the nuclear cytology of these algae is available as studies using nuclear stains and chromosome counts are generally not feasible due to the small size of the cells. It is hoped that new analytical techniques, such as flow cytometry, will help solve some of these problems. Moreover, information relative to the different orders in the class is unbalanced. In the Pavlovales and Isochrysidales (excluding coccolithophorid genera) the question of sexuality has rarely been addressed and it is intuitively thought that these haptophytes propagate asexually, although their ploidy levels have never been investigated. Except for *Phaeocystis,* sexuality in other genera in the Prymnesiales has never been considered seriously. Current research on

Chrysochromulina and *Phaeocystis* indicates that there are reasons to believe a haplo-diploid life cycle exists in both genera, and this should stimulate further research in related genera in the Prymnesiales. Probably most Coccolithophorales have a heteromorphic life history and a haplo-diploid cycle has clearly been demonstrated for some genera, e.g. *Pleurochrysis, Hymenomonas,* and *Ochrosphaera.* In such cases, body scale characteristics, examined after shadowcasting, are useful phenotypic indicators of ploidy levels. It appears that this reasoning can be extended to other Coccolithophorales, including those for which only one stage is recorded. As a result, it is suggested that in life cycles involving coccolith polymorphy, the holococcolithophorid stage could be haploid. Major changes in coccolithophorid taxonomy are thus to be expected.

If haplo-diploid life cycles are indeed more common in the Haptophyta than previously thought, further investigation will be needed to determine the external factors that govern the change from one phase to the other. Generally a haplo-diploid cycle is considered as an adaptation to an environment that is seasonally variable or that contains two different niches (see the review by Valero *et al.* 1992). Examples exist in the Haptophyta that agree well with this idea: in *Pleurochrysis,* the haploid filaments are clearly adapted to the epilithic habitats where they are found, while the diploid cells ensure dispersion. In contrast to heterococcolithophorids, holococcolithophorids are recorded with higher frequencies in oligotrophic waters (Kleijne 1991). The two flagellate stages reported in the life cycle of *C. polylepis* have different physiological requirements and could be adapted responses to changing conditions in the water column. In view of the apparent benefits of a haplo-diploid life cycle, the question might be asked why is it seemingly absent in certain orders? Better understanding of haptophyte life cycles and their evolution may therefore help to provide functional accounts of haploidy and diploidy in planktonic algae.

References

Barclay, W. R., Johansen, J. R., Terry, K. L., and Toon, S. P. (1991). Influence of ionic parameters on the growth and the distribution of *Boekelovia hooglandii* (Chromophyta). *Phycologia,* **30**, 355–64.

Billard, C. (1976). Sur une nouvelle espèce de *Pavlova, P. virescens* nov. sp. (Haptophycées). *Bulletin de la Société Phycologique de France,* **21**, 18–27.

Billard, C. (1983). *Prymnesium zebrinum* sp. nov. et *P. annuliferum* sp. nov., deux nouvelles espèces apparentées à *P. parvum* Carter (Prymnesiophyceae), *Phycologia,* **22**, 141–51.

Billard, C. (1987). L'algothèque du Laboratoire d'Algologie fondamentale et appliquée de l'Université de Caen. *Cryptogamie, Algologie,* **8**, 79–90.

Billard, C. and Gayral, P. (1972). Two new species of *Isochrysis* with remarks on the genus *Ruttnera*. *British Phycological Journal*, **7**, 289–97.

Chang, F. H. (1984). The ultrastructure of *Phaeocystis pouchetii* (Prymnesiophyceae) vegetative colonies with special reference to the production of new mucilaginous envelopes. *New Zealand Journal of Marine and Freshwater Research*, **18**, 303–8.

Chang, F. H. (1985). Preliminary toxicity test of *Prymnesium calathiferum* n. sp. isolated from New Zealand. In *Toxic dinoflagellates*, (ed. D. M. Anderson, A. W. White, and D. G. Baden), pp. 109–12. Elsevier, New York.

Chrétiennot, M.-J. (1973). The fine structure and taxonomy of *Platychrysis pigra* Geitler (Haptophyceae). *Journal of the Marine Biological Association of the United Kingdom*, **53**, 905–14.

Chrétiennot-Dinet, M.-J. (1990). *Atlas du phytoplancton marin*, Vol. 3. Editions du CNRS, Paris.

Edvardsen, B. and Paasche, E. (1992). Two motile stages of *Chrysochromulina polylepis* (Prymnesiophyceae): morphology, growth, and toxicity. *Journal of Phycology*, **28**, 104–14.

Estep, K. W., Davis, P. G., Hargraves, P. E., and Sieburth, J. McN. (1984). Chloroplast containing microflagellates in natural populations of North Atlantic nanoplankton, their identification and distribution; including a description of five new species of *Chrysochromulina* (Prymnesiophyceae). *Protistologica*, **20**, 613–34.

Fresnel, J. (1986). Nouvelles observations sur une Coccolithacée rare: *Cruciplacolithus neohelis* (McIntyre and Bé) Reinhardt (Prymnesiophyceae). *Protistologica*, **22**, 193–204.

Fresnel, J. (1989). Les coccolithophorides (Prymnesiophyceae) du littoral. Genres: *Cricosphaera, Pleurochrysis, Cruciplacolithus, Hymenomonas* et *Ochrosphaera*. Ultrastructure, cycle biologique, systématique. Ph. D. thesis, Université de Caen.

Fresnel, J. and Billard, C. (1991). *Pleurochrysis placolithoides* sp. nov. (Prymnesiophyceae), a new marine coccolithophorid with remarks on the status of cricolith-bearing species. *British Phycological Journal*, **26**, 67–80.

Gaarder, K. R. (1970). Three new taxa of Coccolithinae. *Nytt Magasin for Botanikk*, **17**, 113–26.

Gayral, P. and Fresnel, J. (1976). Nouvelles observations sur deux Coccolithophoracées marines : *Cricosphaera roscoffensis* (Dangeard) comb. nov. et *Hymenomonas globosa* (F. Magne) comb. nov. *Phycologia*, **15**, 339–55.

Gayral, P. and Fresnel, F. (1979). *Exanthemachrysis gayraliae* Lepailleur (Prymnesiophyceae, Pavlovales). Ultrastructure et discussion taxinomique. *Protistologica*, **15**, 271–82.

Gayral, P. and Fresnel, J. (1983*a*). *Platychrysis pienaarii* sp. nov. et *P. simplex* sp. nov. (Prymnesiophyceae): description et ultrastructure. *Phycologia*, **22**, 29–45.

Gayral, P. and Fresnel, J. (1983*b*). Description, sexualité et cycle de développement d'une nouvelle Coccolithophoracée (Prymnesiophyceae) : *Pleurochrysis pseudoroscoffensis* sp. nov. *Protistologica*, 19, 245–61.

Green, J. C. (1976). *Corymbellus aureus* gen. et sp. nov., a new colonial member of the Haptophyceae. *Journal of the Marine Biological Association of the United Kingdom*, **56**, 31–8.

Green, J. C. (1980). The fine structure of *Pavlova pinguis* Green and a preliminary survey of the order Pavlovales (Prymnesiophyceae). *British Phycological Journal*, **15**, 151–91.

Green, J. C. and Hibberd, D. J. (1977). The ultrastructure and taxonomy of *Diacronema vlkianum* (Prymnesiophyceae) with special reference to the haptonema and flagellar apparatus. *Journal of the Marine Biological Association of the United Kingdom*, **57**, 1125–36.

Green, J. C. and Parke, M. (1975). New observations upon members of the genus *Chrysotila* Anand, with remarks upon their relationships within the Haptophyceae. *Journal of the Marine Biological Association of the United Kingdom*, **55**, 109–21.

Green, J. C., Hibberd, D. J., and Pienaar, R. N. (1982). The taxonomy of *Prymnesium* (Prymnesiophyceae) including a description of a new cosmopolitan species, *P. patellifera* sp. nov., and further observations on *P. parvum* N. Carter. *British Phycological Journal*, **17**, 363–82.

Green, J. C., Perch-Nielsen, K., and Westbroek, P. (1989). Phylum Prymnesiophyta. In *Handbook of the Protoctista*, (ed. L. Margulis, J. Corliss, M. Melkonian, and D. Chapman), pp. 293–317. Jones and Bartlett, Boston.

Hibberd, D. J. (1976). The ultrastructure and taxonomy of the Chrysophyceae and Prymnesiophyceae (Haptophyceae) : a survey with some observations on the ultrastructure of the Chrysophyceae. *Botanical Journal of the Linnean Society*, **72**, 55–80.

Inouye, I. and Chihara, M. (1979). Life-history and taxonomy of *Cricosphaera roscoffensis* var. *haptonemofera*, var. nov. (Class Prymnesiophyceae) from the Pacific. *The Botanical Magazine*, Tokyo, **92**, 75–87.

Inouye, I. and Chihara, M. (1983). Ultrastructure and taxonomy of *Jomonlithus littoralis* gen. et sp. nov. (Class Prymnesiophyceae), a coccolithophorid from the northwest Pacific. *The Botanical Magazine*, Tokyo, **96**, 365–76.

Inouye, I. and Pienaar, R. N. (1984). New observations on the coccolithophorid *Umbilicosphaera sibogae* var. *foliosa* (Prymnesiophyceae) with reference to cell covering, cell structure and flagellar apparatus. *British Phycological Journal*, **19**, 357–69.

Inouye, I. and Pienaar, R. N. (1988). Light and electron microscope observations on the type species of *Syracosphaera*, *S. pulchra* (Prymnesiophyceae). *British Phycological Journal*, **23**, 205–17.

Jahnke, J. and Baumann, M. (1987). Differentiation between *Phaeocystis pouchetii* (Har.) Lagerheim and *Phaeocystis globosa* Scherffel. *Hydrobiological Bulletin*, **21**, 141–7.

Janin, M.-C. (1992). Miocene variability of *Calcidiscus* gr. *leptoporus* and possible evolutionary relationship with another Coccolithaceae: *Umbilicosphaera* gr. *sibogae*. *BioSystems*, **28**, 169–78.

Klaveness, D. (1972). *Coccolithus huxleyi* (Lohm.) Kamptn. II. The flagellate cell, aberrant cell types, vegetative propagation and life cycles. *British Phycological Journal*, **7**, 309–18.

Klaveness, D. (1973). The microanatomy of *Calyptrosphaera sphaeroidea*, with some supplementary observations on the motile stage of *Coccolithus pelagicus*. *Nytt Magasin for Botanikk*, **20**, 151–62.

Kleijne, A. (1991). Holococcolithophorids from the Indian Ocean, Red Sea, Mediterranean Sea and North Atlantic Ocean. *Marine Micropaleontology*, **17**, 1–76.

Kornmann, P. (1955). Beobachtungen an *Phaeocystis*-Kulturen. *Helgoländer wissenschaftliche Meeresuntersuchungen*, **5**, 218–33.

Lancelot, C., Billen, G., and Barth, H. (1991). The dynamics of *Phaeocystis* blooms in nutrient enriched coastal zones. Water Pollution Report, No. 23. Commission of the European Communities.

Leadbeater, B. S. C. (1970). Preliminary observations on differences of scale morphology at various stages in the life cycle of '*Apistonema-Syracosphaera*' *sensu* von Stosch. *British Phycological Journal*, **5**, 57–69.

Leadbeater, B. S. C. and Morton, C (1973). Ultrastructural observations on the external morphology of some members of the Haptophyceae from the coast of Jugoslavia. *Nova Hedwigia*, **24**, 207–33.

Lefort, F. (1975). Etude de quelques Coccolithophoracées marines rapportées aux genres *Hymenomonas* et *Ochrosphaera*. *Cahiers de Biologie Marine*, **16**, 213–29.

Manton, I. and Leedale, G. F. (1963). Observations on the microanatomy of *Crystallolithus hyalinus* Gaarder and Markali. *Archiv für Mikrobiologie*, **47**, 115–36.

Manton, I. and Leedale, G. F. (1969). Observations on the microanatomy of *Coccolithus pelagicus* and *Cricosphaera carterae* with special reference to the origin and nature of coccoliths and scales. *Journal of the Marine Biological Association of the United Kingdom*, **49**, 1–16.

Moestrup, Ø. (1979). Identification by electron microscopy of marine nanoplankton from New Zealand, including the description of four new species. *New Zealand Journal of Botany*, **17**, 61–95.

Moestrup, Ø. and Larsen, J. (1992). Potentially toxic phytoplankton. 1. Haptophyceae (Prymnesiophyceae). ICES Identification Leaflets, No. 179, (ed. J. A. Lindley). International Council for the Exploration of the Sea, Copenhagen.

Norris, R. E. (1967). Microalgae in enrichment cultures from Puerto Peñasco, Sonora, Mexico. *Bulletin of the Southern California Academy of Sciences*, **66**, 233–50.

Parke, M. (1949). Studies on marine flagellates. *Journal of the Marine Biological Association of the United Kingdom*, **28**, 255–86.

Parke, M. and Adams, I. (1960). The motile (*Crystallolithus hyalinus* Gaarder and Markali) and non-motile phases in the life-history of *Coccolithus pelagicus* (Wallich) Schiller. *Journal of the Marine Biological Association of the United Kingdom*, **39**, 263–74.

Parke, M., Green, J. C., and Manton, I. (1971). Observations on the fine structure of zoids of the genus *Phaeocystis* (Haptophyceae). *Journal of the Marine Biological Association of the United Kingdom*, **51**, 927–41.

Parke, M., Manton, I., and Clarke, B. (1955). Studies on marine flagellates. II. Three new species of *Chrysochromulina*. *Journal of the Marine Biological Association of the United Kingdom*, **34**, 579–609.

Pienaar, R. N. (1981). Ultrastructural studies on the cysts of *Prymnesium* (Prymnesiophyceae). *Phycologia*, **20**, 112.

Pienaar, R. N. (1991). Thread formation in the motile cells of *Phaeocystis*. *Proceedings of the Electron Microscopy Society of Southern Africa*, **21**, 135–6.

Rayns, D. G. (1962). Alternation of generations in a coccolithophorid, *Cricosphaera carterae* (Braarud and Fagerl.) Braarud. *Journal of the Marine Biological Association of the United Kingdom*, **42**, 481–4.

Reynolds, N. (1974). *Imantonia rotunda* gen. et sp. nov., a new member of the Haptophyceae. *British Phycological Journal*, **9**, 429–34.

Schwarz, E. (1932). Beiträge zur Entwicklungsgeschichte der Protophyten. IX. Der Formwechsel von *Ochrosphaera neapolitana*. *Archiv für Protistenkunde*, **77**, 434–62.

Stosch, H. A. von (1967). Haptophyceae. In *Vegetative Fortpflanzung, Parthenogenese and Apogamie bei Algen. Encyclopedia of plant physiology*, Vol. 18, (ed. W. Ruhland), pp. 646–56. Springer-Verlag, Berlin.

Thomsen, H. A., Østergaard, J. B., and Hansen, L. E. (1991). Heteromorphic life histories in arctic coccolithophorids (Prymnesiophyceae). *Journal of Phycology*, **27**, 634–42.

Valero, M., Richerd, S., Perrot, V., and Destombe, C. (1992). Evolution of alternation of haploid and diploid phases in life cycles. *Trends in Ecology and Evolution*, **7**, 25–9.

Veer, J. van der and Leewis, R. J. (1977). *Pavlova ennorea* sp. nov. a haptophycean alga with a dominant palmelloid phase from England. *Acta Botanica Neerlandica*, **26**, 159–76.

Veldhuis, M. J. W. (1987). The eco-physiology of the colonial alga. *Phaeocystis pouchetii*. D. Phil. Thesis, Rijksuniversiteit, Groningen.

Volkman, J. K., Dunstan, G. A., Jeffrey, S. W., and Kearney, P. S. (1991). Fatty acids from microalgae of the genus *Pavlova*. *Phytochemistry*, **30**, 1855–9.

10. Haptophytes as components of marine phytoplankton

HELGE A. THOMSEN
Department of Phycology, Botanical Institute, University of Copenhagen, Denmark

KURT R. BUCK and FRANCISCO P. CHAVEZ
Monterey Bay Aquarium Research Institute, Pacific Grove, California, USA

Abstract

The Haptophyta comprises 11 unmineralized genera with approximately 80 species, and more than 200 coccolithophorid species allocated to about 70 different genera. Results from Pacific Ocean surveys indicate, based on epifluorescence microscopical counts, that the biomass and relative contribution of haptophytes to autotrophic biomass is relatively low, 13 ± 9%, and constant over a wide variety of environmental settings. Also the biomass contribution of haptophytess to the nanoplankton, 37 ± 20%, is remarkably consistent worldwide.

Introduction

Marine haptophytes occupy an important position within the food web of the upper water column and their metabolic products may have an impact on global climate. In oceanic waters they comprise a large proportion of nanoplankton biomass (Chavez *et al.* 1990; Hoepffner and Haas 1990; Sieracki *et al.* 1993), and therefore are major agents in the cycling of organic matter within the oceanic microbial food web. With the proven capacity of some haptophytes to graze upon picoplankton (Kawachi *et al.* 1991; Sanders 1991), they occupy the ecological niches of both primary producers and grazers. The capacity of certain haptophytes (coccolithophorids) to produce refractive and fossilizable $CaCO_3$ body parts makes them important contributors to the vertical flux of carbon in the ocean (Broecker and Peng 1982). The production of $CaCO_3$ modifies alkalinity and, in short time intervals, can increase the partial pressure of CO_2 in the upper water column (Balch *et al.* 1992). In addition, they are one of the major phytoplankton groups implicated in

The Haptophyte Algae (ed. J. C. Green and B. S. C. Leadbeater), Systematics Association Special Volume No. 51, pp. 187–208. Clarendon Press, Oxford, 1994.

DMSP (Keller *et al.* 1989) and thence DMS production. The haptophytes, therefore, with their cosmopolitan distribution, have the potential to affect factors involved in global climate change (Malin *et al.* 1992). The importance of coccoliths in interpreting palaeoclimatological trends is inarguable. Evaluating the importance of haptophytes in these processes relies on the ability of oceanographers to assess accurately both the assemblage composition and abundance/biomass of haptophytes.

For the purpose of this review, three major groups of haptophytes are defined: (a) unmineralized forms with only organic scales (i.e. lacking $CaCO_3$); (b) weakly calcified forms (mostly holococcolithophorids and, e.g. 'polar' heterococcolithophorids); and (c) coccolithophorids (organisms with heavy $CaCO_3$ deposits, i.e. mostly heterococcolithophorids).

Scanning electron microscopy (SEM) and transmitted light microscopy (LM), utilizing the inverted microscope and settling chambers (Reid 1980) on samples from the water column and sediment traps, have increased understanding of the diversity, abundance, and biogeography of coccolithophorids in major parts of the world ocean (Hasle 1959; Okada and Honjo 1973; Reid 1980; Boysen 1991).

Identification of the species of weakly calcified and the unmineralized forms generally requires transmission electron microscopy (TEM) (e.g. Estep *et al.* 1984; Thomsen *et al.* 1988). Most surveys of the flagellated nanoplankton have been based upon near-shore samples, and present knowledge of the biogeographical ranges of weakly calcified and unmineralized forms is, therefore, incomplete.

Complete enumeration of haptophytes relies upon the epifluorescence microscopy (EFM) of samples filtered on to polycarbonate filters (Booth 1987). The unmistakable haptophyte signature of cells viewed with EFM (chloroplasts reminiscent of the wings of butterflies with two equal length flagella and a haptonema) has facilitated the acquisition of data on the spatial and temporal distribution of the group. While EFM is a major tool for the enumeration of haptophytes, it does not allow for species-specific identifications, although it appears that the coccolithophorids can be distinguished from the weakly calcified and unmineralized forms. The present lack of techniques to identify and quantify the weakly calcified and unmineralized forms simultaneously means that knowledge of their ecology is sparse. A major part of this paper summarizes the existing data on the group with particular emphasis on the Pacific Ocean.

Assemblages

Haptophyte taxonomy is currently in a state of flux. New species are being discovered continuously and the validity of generic circumscriptions is, in many cases, questionable because of limited information on type specimens and the recent examples of cells, simultaneously carrying coccoliths or scales, previously assigned to different species or genera. Only a small percentage (< 5%) of the heterococcolithophorid species (and mostly coastal forms) have been established in culture, a step which may be necessary to assign organisms unequivocally to taxa.

At present the Prymnesiophyceae comprises 11 unmineralized genera with approximately 80 species, and more than 200 coccolithophorid species allocated to about 70 different genera. This chapter will focus on certain unmineralized genera, and aims to present new information on the abundance of haptophytes in time and space.

Coccolithophorids and weakly calcified forms

The coccolithophorids constitute a conspicuous element of the plankton in oceanic regions. Species diversity is high in tropical regions but considerably lower in temperate regions. Only a few coccolithophorids (e.g. *Coccolithus pelagicus* and *Emiliania huxleyi*) have been reported from subpolar regions.

Extensive surveys of coccolithophorids using SEM techniques have contributed significantly to a thorough understanding of both morphological and distributional characteristics of individual species (e.g. Okada and Honjo 1975; Gaarder and Heimdal 1977; Okada and McIntyre 1977, 1979; Conley 1979; Heimdal and Gaarder 1980, 1981; Reid 1980; Hallegraeff 1984; Kleijne 1991, 1992; Samtleben and Schröder 1992). Distinct assemblages have been identified for both the Atlantic Ocean (tropical, subtropical, transitional, subarctic, and subantarctic; McIntyre and Bé 1967; Okada and McIntyre 1979) and the Pacific Ocean (subarctic, transitional, central north, equatorial north, equatorial south, central south; Okada and Honjo 1973). In addition several studies from the Pacific have identified 'shade' assemblages, those groups of organisms found exclusively or predominantly at greater depths than 100 m (Okada and Honjo 1973; Fryxell *et al.* 1979; Reid 1980; Boysen 1991).

Due to sampling being biased towards coastal and polar regions, the lightly calcified genera (e.g. *Papposphaera, Pappomonas, Wigwamma*) are thought to be found mainly in regions of the world oceans that border the main biogeographical provinces of coccolithophorid distribution, i.e. the edges of the ocean basins. However, recent surveys of nanoflagellates from temperate, subtropical, and tropical regions have shown that these lightly calcified genera are also found in regions where typical coccolithophorids are abundant (Thomsen, unpublished results).

Figures 10.1–10.9 show coccolithophorid diversity and illustrate some of the species encountered during the September 1989, California Undercurrent Cruise (see also Fig. 10.18 and below). The lightly calcified

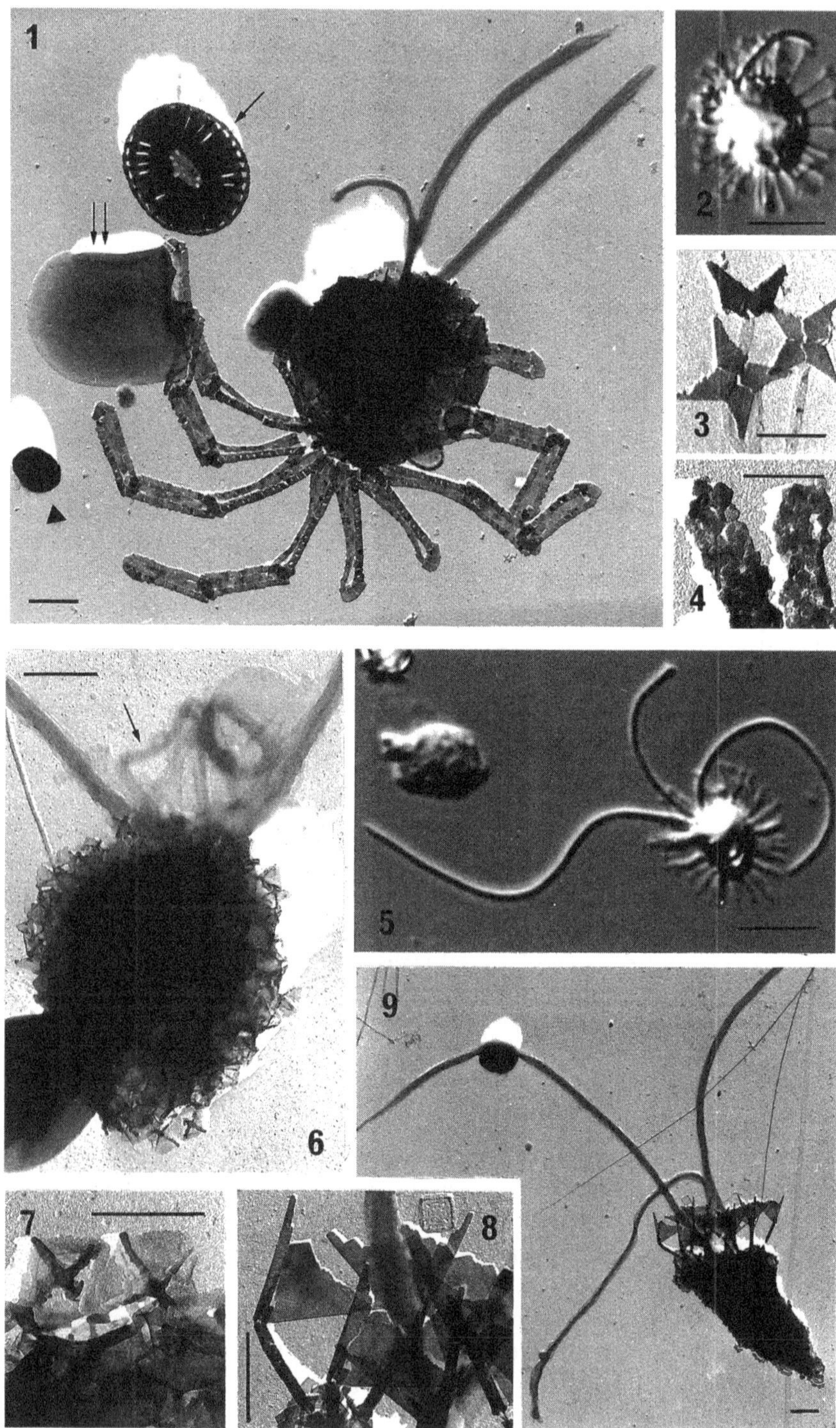
1
2
3
4
5
6
7
8
9

Figs 10.1–10.9 A selection of coccolithophorids encountered along a Californian transect (see also Fig. 10.18). TEM shadowcast whole mounts (**1, 3, 4, 6–9**); light micrographs (**2, 5**). (**1**) Complete cell of *Ophiaster hydroides* showing flagella and short haptonema. Single coccoliths from *Emiliania huxleyi* (arrow) and *Florisphaera* (*double arrow*) are pointed out. A likely *Synechococcus* cell (*arrowhead*) serves as a natural scale bar (1 μm). (**2**) *Papposphaera* sp.; dried cell. (**3**) Detail of tips of coccolith appendages (see also **2**). (**4, 5**) *Turrisphaera* sp.; notice (**4**) hexagonal crystallites. (**6, 7**) *Polycrater galapagensis*; the *arrow* (**6**) points to the haptonema. (**8, 9**) *Pappomonas flabellifera* var. *flabellifera*. Scale bar = 0.5 μm (**3, 4, 7**); 1 μm (**1, 6, 9**); 5 μm (**2, 5**).

'polar' genera include *Papposphaera* (Figs 10.2, 10.3), *Turrisphaera* (Figs 10.4, 10.5), and *Pappomonas* (Figs 10.8, 10.9). Other organisms illustrated are *Ophiaster hydroides, Emilliania huxleyi, Florisphaera* sp. (Fig. 10.1), which represent types of coccolithophorids commonly encountered in subtropical samples. *Polycrater galapagensis* (Figs 10.6, 10.7) is previously recorded only from the Galapagos Islands (Manton and Oates 1980) and the Atlantic Ocean (Chrétiennot-Dinet 1990). It is unusual in having a microcrystalline substructure based on calcium carbonate as aragonite instead of the more usual calcite (Manton and Oates 1980). The material illustrated here (Fig. 10.6) documents for the first time the presence of a haptonema in this species.

Unmineralized forms

The diversity among unmineralized haptophytes is much less pronounced than among the coccolithophorids. Only some genera, e.g. *Chrysochromulina* Lackey, *Phaeocystis* Lagerheim (Figs 10.14, 10.15; Marchant and Thomsen, Chapter 11; Lancelot and Rousseau, Chapter 12), *Corymbellus* Green and *Imantonia* Reynolds appear to be abundant in the open ocean (Estep *et al.* 1984; Hoepffner and Haas 1990; Table 10.1).

Table 10.1 Unmineralized haptophytes encountered in oceanic regions

	1	2	3	4
Chrysochromulina				
C. aff. *acantha* Leadbeater & Manton			*	
C. apheles Moestrup & Thomsen			*	
C. bergenensis Leadbeater			*	
C. chiton Parke & Manton	*			
C. cyathophora Thomsen			*	
C. discophora Manton	*		*	*

Table 10.1 (*cont.*)

	1	2	3	4
Chrysochromulina				
C. elegans Estep *et al.*	*			
C. ephippium Parke & Manton	*			*
C. fragilis Leadbeater			*	
C. herdlensis Leadbeater	*		*	
C. hirta Manton			*	
C. latilepis Manton				*
C. leadbeateri Estep *et al.*	*	*	*	
C. mantoniae Leadbeater			*	*
C. pachycylindra Manton & Oates	*	*	*	*
C. parkeae Green & Leadbeater	*			
C. pelagica Estep *et al.*	*	*		
C. pringsheimii Parke & Manton	*		*	
C. pyramidosa Thomsen			*	
C. simplex Estep *et al.*	*	*	*	
Chrysochromulina 'Plymouth 384' (Moestrup 1979)			*	
C. tenuispina Manton	*			
C. tenuisquama Estep *et al.*	*			
C. vexillifera Manton & Oates	*	*	*	*
Chrysochromulina spp.	*(3)		*(>10)	
Corymbellus				
C. aureus Green	*	*	*	
Phaeocystis				
P. pouchetii (Hariot) Lagerheim	*	*	*	
P. scrobiculata Moestrup	*	*	*	
Platychrysis				
P. pienaarii Gayral & Fresnel	*			
Prymnesium				
P. patelliferum Green *et al.*	*			

(1) Trans-Atlantic stations (Estep *et al.* 1984); (2) North Pacific central gyre (Hoepffner and Haas 1990); (3) Stations off California (Thomsen and Buck, unpublished results; see also Figs 10–16); (4) Galapagos (Manton 1982, 1983; Manton and Oates 1983).

Chrysochromulina *Lackey, 1939* The genus *Chrysochromulina* is a ubiquitous component of marine nanoplankton. The history of research in this genus largely reflects the introduction of electron microscopical techniques to marine biology. Early species descriptions were based on light- and electron microscopical studies of culture strains (Parke *et al.* 1955, 1956). However, the astonishing diversity within this group of organisms was emphasized when freshly collected samples were prepared for the first time directly for TEM work (Leadbeater 1972; Manton and Leadbeater 1974). The number of species currently described is close to 50. However, the number of unidentifiable *Chrysochromulina*-like groups of scales that appear in almost any whole mount preparation from a marine environment (Thomsen, unpublished results), make it likely that the actual number of species may exceed 100.

Species of the genus *Chrysochromulina* are highly variable in body shape and dimensions. Some species (e.g. *C. apheles*, *C. minor*, *C. elegans*, *C. pyramidosa*) approach the cell-size category of picoplankton with a diameter only slightly larger than 2 μm, while other species may be up to 20–30 μm in length (e.g. *C. parkeae*, *C. pringsheimii* and the enigmatic *C. birgeri*). The flagella, 6–40 μm long, and in particular the haptonema, 4–180 μm long, display an astonishing size variability among the species described. In addition to cell size and shape, the length of the haptonema relative to either flagellar length or cell body diameter are useful criteria for LM identification of particular species. The haptonema is, in most species of *Chrysochromulina*, capable of coiling (see Inouye and Kawachi, Chapter 4).

Differences in morphology and number of types of organic scales covering the cell body form the prime criteria for distinguishing species on a routine basis from TEM whole mounts (See Leadbeater, Chapter 2). In some species, e.g. *C. elegans*, the scale case consists of only two types of scales (Fig. 10.10), whereas in other species, e.g. *C. parkeae*, a total of five different types may be distinguished. The range in scale morphology within species of *Chrysochromulina* is almost endless. The main types are plates (Fig. 10.10), cups, cylinders (Fig. 10.13), and spines (Fig. 10.12). Some scales are large enough to be visible clearly in the LM when live cells are studied either by phase contrast or Nomarski optics. This, in some cases, enables a precise identification of the species, e.g. *C. parkeae* and *C. hirta*, and sometimes it narrows down the possible identification to between a few species e.g. *C. ericina* or *C. spinifera*.

Most species of *Chrysochromulina* are at present known from only a handful of localities due to the fact that a TEM examination of whole mounts of cells is necessary to confirm their identity. Much of the material studied so far originates from coastal localities. The number of major

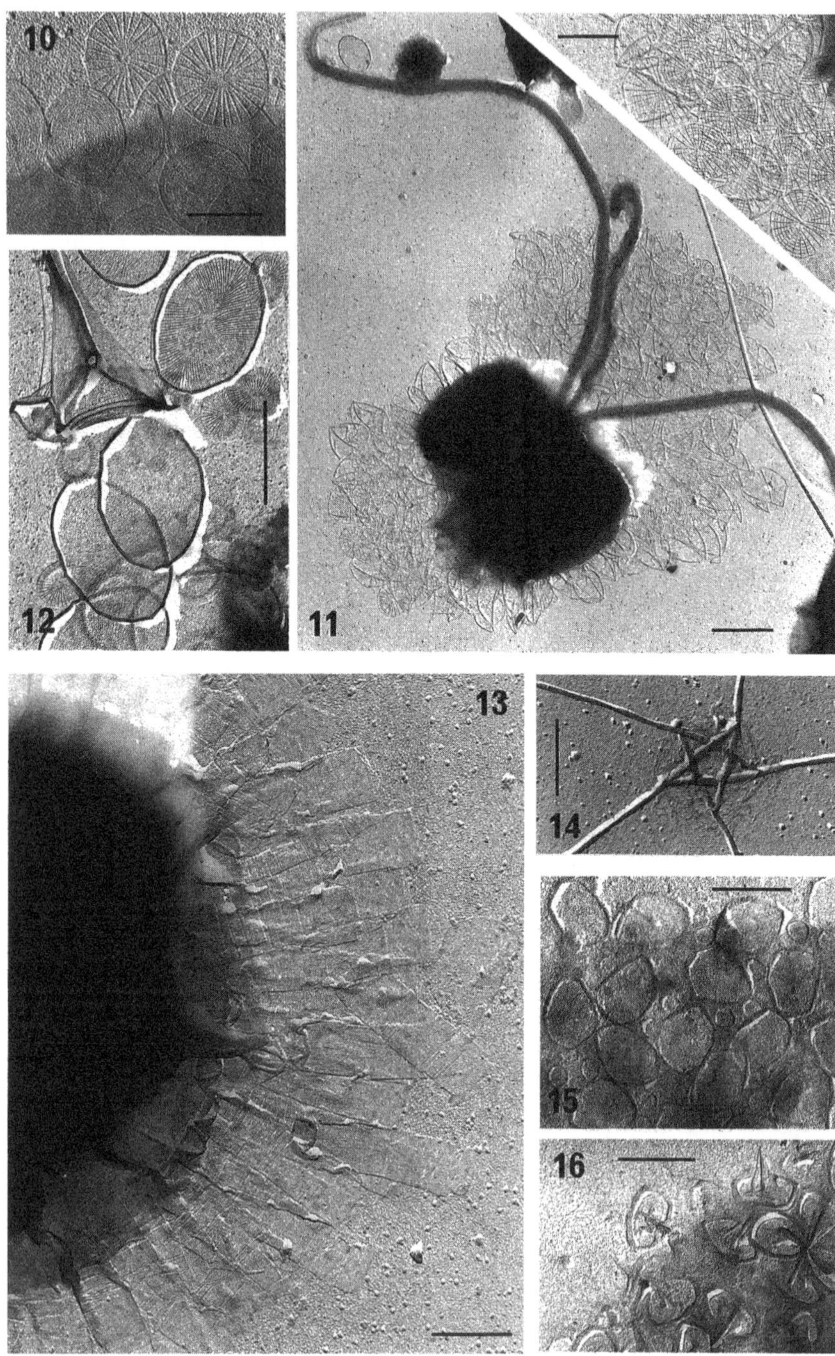
10
11
12
13
14
15
16

Figs 10.10–10.16 A selection of unmineralized prymnesiophytes encountered along a Californian transect (see also Fig. **10.18**). TEM shadowcast whole mounts. (**10**) Detail of *Chrysochromulina elegans* periplast showing plate scales with concentric and radiating surface markings. (**11**) Complete cell of *Chrysochromulina pyramidosa*; the inset shows plate scales and pyramidal scales at a higher magnification. (**12**) *Chrysochromulina mantoniae* plate scales of two sizes and basal part of spine scale. (**13**) *Chrysochromulina cyathophora* cell surrounded by numerous flattened cylinder scales. (**14**) Ejectile structure from *Phaeocystis pouchetii*. (**15**) Detail of *Phaeocystis scrobiculata* periplast. (**16**) *Corymbellus aureus* spine scales. Scale bars = 0.5 μm (**10, 11** (inset), **13, 15, 16**); 1 μm (**11, 12, 14**).

surveys of nanoflagellates from oceanic regions, including routine identification of cells from TEM whole mounts, is limited (Estep *et al.* 1984; Hoeppfner and Haas 1992). Table 10.1 lists findings of species of *Chrysochromulina* from oceanic regions. However, the general picture emerging when analyzing *Chrysochromulina* distributional data is that most species are apparently ubiquitous in the world's oceans (see also Marchant and Thomsen, Chapter 11).

Corymbellus *Green, 1976* This monotypic genus is fairly easy to recognize in the LM because of the tendency to form large, spherical or annuloid, motile colonies (up to 200 μm). The single cell is about 8 × 10 μm with a short, non-coiling haptonema. The scale covering consists of two types of slightly differently sized minute scales. One type of scale has a central four-strutted spine bridging a central pore (Fig. 10.16).

The species was first described from the English Channel (Green 1976) and has since been reported from New Zealand (Moestrup 1979), Australia (Hallegraeff 1983), the oceanic parts of the Atlantic Ocean (Estep *et al.* 1984), North Sea (Gieskes and Kraay 1986), Beagle Channel (Thomsen, unpublished results), and California (Thomsen and Buck, unpublished results). Evidently *Corymbellus aureus* is a widespread organism, with the potential of forming local blooms (see Moestrup, Chapter 14).

Imantonia *Reynolds, 1974* *Imantonia rotunda* is a minute flagellate (2–4 μm in diameter) with two chloroplasts, two flagella, and a much reduced haptonema. A small protuberance containing endoplasmic reticulum, and no microtubules, is sometimes present between the flagella (Green and Pienaar 1977). Scales of one or two types ('bicycle-wheels'), distinguished by their size and the presence or absence of an upturned rim, cover the cell body. Aspects of the ultrastructure of this organism have been described by Hori and Green (1985) and Green and Hori (1986).

Originally *Imantonia rotunda* was described from the arctic Atlantic, Spitsbergen (Reynolds 1974), but has since been reported from a wide range of localities indicating a cosmopolitan distribution. These include Friday Harbor, USA (Green and Pienaar 1977), Australia (Hallegraeff 1983), the English Channel and Cape Town (Chrétiennot-Dinet 1990), Denmark (Thomsen *et al.* 1992), and the Weddell Sea (Thomsen, unpublished results).

A recent extensive TEM survey of phytoplankton from Danish brackish waters (Thomsen *et al.* 1992), showed this organism to be present in almost all samples analyzed. The small size of the cell and the fact that TEM of whole mounts of cells is needed for proper identification has prevented the frequent recognition of *I. rotunda* as a significant component of haptophyte plankton.

Abundance and distribution

Estimates derived from microscopy

Quantitative estimation of haptophyte abundance has been obtained using two principal techniques. The first focuses on coccolithophorids, while the second attempts to provide abundance and biomass estimates of all haptophytes. The former relies on transmitted light microscopy of sedimented water samples (Reid 1983) to identify and quantify material (Fryxell *et al.* 1979; Reid 1980), although SEM has also been used for this purpose (e.g. Boysen 1991). Such techniques are not suitable for estimates of total haptophytes due to inadequate preservation of unmineralized and lightly mineralized forms. Total haptophyte abundance and biomass estimates are usually undertaken by EFM after filtering water samples through polycarbonate filters (Booth 1987, 1988; Chavez *et al.* 1990, 1991). Scanning electronmicroscopy has also been used as a quantitative technique for studying total haptophyte numbers (e.g. Booth *et al.* 1982).

As a consequence of this methodological dichotomy, it is difficult to make general statements regarding, for example, the relative importance of coccolithophorids. Furthermore, the inability to identify individual species by EFM prevents the acquisition of detailed quantitative information on communities of unmineralized haptophytes as has been possible for the coccolithophorids.

Coccolithophorids Blooms ($> 10^6$ cells 1^{-1}) of coccolithophorids are persistent features in certain regions of the Atlantic. Satellite remote sensing in the 550 nm band by the Coastal Zone Color Scanner indicates extensive blooms ($> 50\,000$ km^2). *Emiliania huxleyi* may be dominant with population abundances exceeding 2×10^6 cells 1^{-1} recorded in the Gulf of

Maine and off the south of Iceland (Balch *et al.* 1991, 1992). The contribution to the total carbon fixation in the Gulf of Maine by the prymnesiophyte bloom was calculated to be 0.5% of the annual planktonic production. However, the contribution to the vertical flux of carbon from the surface of the ocean to depths was calculated to be 25% of the annual total, presumably via the contribution of the $CaCO_3$ in coccoliths (Balch *et al.* 1992). However, blooms of coccolithophorids are not confined only to regions of the open ocean. In Norwegian coastal waters, the North Sea, and the Skagerrak conspicuous blooms of *Emiliania huxleyi* appear to be an almost annual phenomenon with cell numbers up to 115×10^6 cells l^{-1} (Berge 1962; Holligan *et al.* 1989).

Large scale spatial and temporal studies carried out during the 1970s in the Pacific Ocean point to a contrasting picture. Quantitative data obtained using inverted transmitted light microscopy by Hasle (1959), Okada and Honjo (1973), Fryxell *et al.* (1979), and Reid (1980) document heterococcolithophorid abundances in the range of 10^4 cells l^{-1} (Table 10.2). Higher abundances have been reported from the Pacific, but only north of 45°N (Okada and Honjo 1973). The method used in these studies is recognized now as inappropriate for assessment of the total autotrophic biomass due to undersampling of the picoplankton and an inability to differentiate autotrophs from non-autotrophs. Therefore, it is impossible to ascertain the relative contribution of coccolithophorids to total phytoplankton biomass. However, counts made with EFM on equatorial Pacific (10°N–10°S) samples indicate a maximum coccolithophorid contribution of 4% of total haptophyte numbers (Table 10.2). These estimates of coccolithophorid population densities made using EFM are similar to those made by inverted microscopy of settled samples, and are several orders of magnitude less than the numbers of the unmineralized (and weakly mineralized) forms of haptophytes.

Weakly calcified and unmineralized forms A total of 180 surface samples have been analyzed by EFM from three areas in the Pacific Ocean (Table 10.2). Haptophytes have been enumerated as a taxonomic group and their cell-sizes measured to enable conversion to biovolume and subsequently carbon content (see Chavez *et al.* 1991 for methodology). The confidence limits obtained after counting three replicate transects on three replicate filters, counting approximately 100 cells for each, was 7% of the mean, using EFM analysis of samples taken off the coast of California in March 1993. Coccolithophorids were not separated into a distinct group in counts made of water samples from the temperate continental shelf stations, but they were assumed to be relatively rare in these areas. The results of these surveys show that the relative contribution of haptophytes to autotrophic biomass and abundance is relatively low, 13 ± 9%, and

Table 10.2 Abundance (cells l^{-1}) of unmineralized haptophytes, weakly calcified prymnesiophytes and coccolithophorids from several studies conducted in the Pacific Ocean

	Unmineralized and weakly calcified haptophytes		Fully mineralized coccolithophorids	
	Range	Mean	Range	Mean
Pacific Stratispheric Investigation	1.4×10^5– 7.5×10^6	1.7×10^6	NC	NC
California Undercurrent	2.3×10^5– 3.7×10^6	1.4×10^6	NC	NC
EquaPac spring	4.3×10^5– 1.8×10^6	9.3×10^5	7.6×10^2– 4.6×10^4	5.55×10^3
EquaPac autumn	3.8×10^5– 1.9×10^6	1.0×10^6	1.2×10^3– 6.6×10^4	8.40×10^3
Hasle (1959)			$1.1–2.0 \times 10^4$	
Reid (1980)			3.1×10^3– 1.1×10^4	5.08×10^3
Fryxell *et al.* (1979)			9.4×10^2– 1.8×10^4	6.12×10^3

Notes: The Pacific Stratispheric Investigation was conducted off the coast of Washington in April of 1989–1991. The California Undercurrent Investigation took place off Monterey Bay from August 1989–January 1991, and the EquaPac cruises were part of NOAA's Global Climate Change Cruises to the Equatorial Pacific 110°W–140°W, 10°S–10°N in 1992. These estimates were performed using EFM. The coccolithophorid estimates are similar to those made using the inverted microscope and settled volumes, and are several orders of magnitude less than the weekly calcified and unmineralized forms.
NC = not counted as a separate group.

constant over a wide variety of environmental conditions (Table 10.2, 10.3; Fig. 10.17). Also the contribution of haptophytes to the nanoplankton is remarkably consistent over both a wide range of environmental conditions and assemblages (37 ± 20%; Table 10.3). Table 10.3 additionally shows the average contributions from other groups of phytoplankton to the total autotrophic biomass for the three study sites.

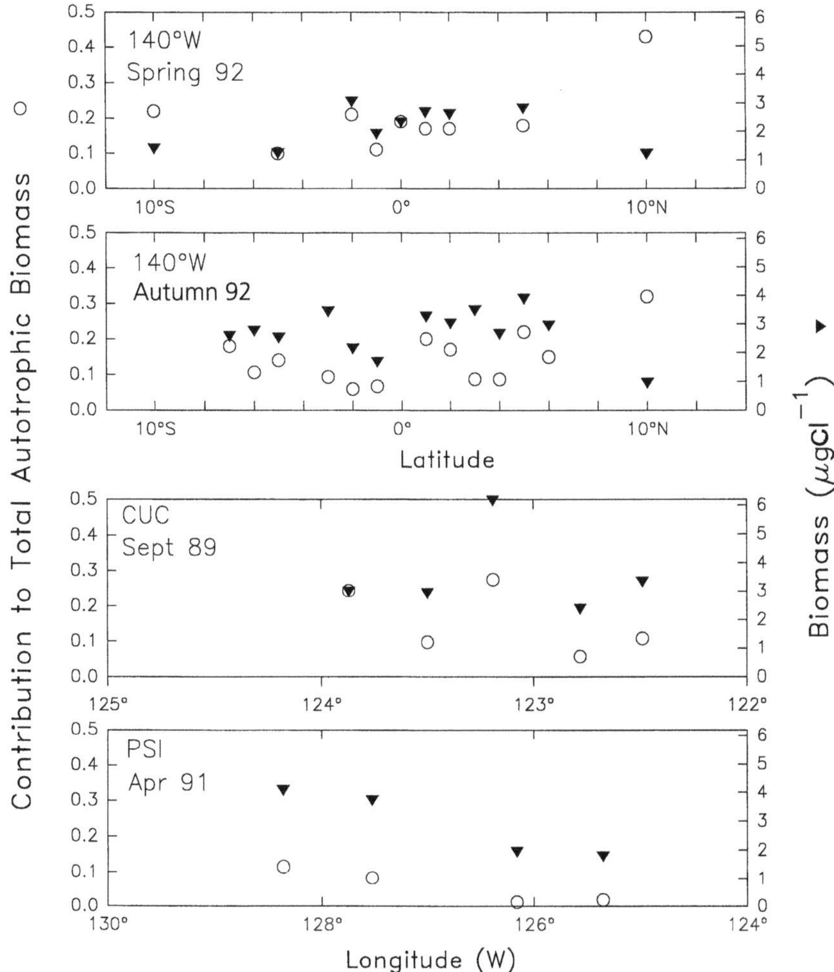

Fig. 10.17 Haptophyte biomass (▼) and contribution to the total autotrophic biomass (○) for the EquaPac spring and autumn 1992, California Undercurrent (CUC), and Pacific Stratispheric Investigation (PSI) cruises. Autotrophic biomass does not take into consideration prochlorophytes which can account for up to 50% of the chlorophyll in equatorial regions (Chavez *et al.* 1991).

The inability to resolve even generic differences by EFM, and the dearth of TEM studies of the weakly calcified and unmineralized forms from the open ocean means that fundamental questions about the species and generic composition of haptophyte communities remain unresolved. Of primary importance is the relative abundance of *Phaeocystis* motile zooids. Motile zooids of both *P. pouchetii* and *P. scrobiculata* are cosmopolitan in

Table 10.3 Average contribution (± S.D.) of the haptophytes to the surface autotrophic nanoplankton and total autotrophic biomass (μgC l^{-1}) from four Pacific Ocean Investigations. Cruises are the same as for Table 10.2. Autotrophic nanoplankton inlcudes haptophytes, cryptophytes, prasinophytes, dinophytes, and *Phaeocystis* (colonial)

	Nanoplankton	Total autotrophic biomass	n
Pacific Stratispheric Investigation	0.38 ± 0.22	0.07 ± 0.05	44
California Undercurrent cruises	0.38 ± 0.20	0.13 ± 0.11	80
EquaPac spring	0.33 ± 0.11	0.17 ± 0.09	34
EquaPac autumn	0.26 ± 0.08	0.15 ± 0.07	22
Grand mean	0.37 ± 0.20	0.13 ± 0.09	180

Average contribution from taxonomic groups of organisms to the total autotrophic biomass for the three study sites.

	1	2	3	4	5	6
PSI	0.42	0.07	0.03	0.14	0.24	0.08
CUC	0.44	0.13	0.13	0.08	0.13	0.08
Equapac	0.47	0.13	0.31	0.07	0.03	0.00

1 = picoplankton, 2 = haptophytes, 3 = dinoflagellates, 4 = pennate diatoms, 5 = centric diatoms, 6 = cryptomonads.

distribution. Given the potential noxious bloom threat and production of DMS associated with *P. pouchetii* colonial forms, information on the quantitative distribution of *P. pouchetii* motile zooids is imperative to understanding the autecology of this organism. The cosmopolitan distribution of the motile zooid implies that blooms of the colonial form are possible on short time-scales once conditions that favour this form are realized. Estep *et al.* (1984) indicated that both species of *Phaeocystis* are occasional or even frequent components of mid-Atlantic phytoplankton assemblages. Based solely on SEM counts Booth *et al.* (1982) found that *Phaeocystis* periodically dominated the unmineralized component of the phytoplankton in the subarctic Pacific. Quantitative estimations from the

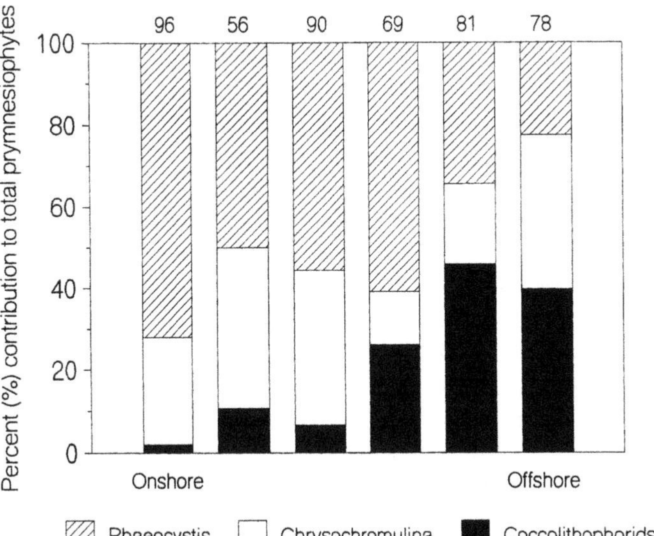

Fig. 10.18 Relative abundances of the two principal components of unmineralized haptophytes (=prymnesiophytes) (*Phaeocystis* and *Chrysochromulina*) and coccolithophorids on a 200 km transect conducted during a California Undercurrent cruise in September 1989 (see Fig. 10.17 for accompanying biomass data). The total number of cells examined for each station are located along the top axis.

California Undercurrent cruise in September 1989 (Fig. 10.18), show a decrease in the relative contribution of *Phaeocystis* motile zooids to the unmineralized forms with distance from shore to open water accompanied by a concommitant increase in coccolithophorids. Unfortunately, in this study there was no separation between coccolithophorids and the weakly calcified forms. The hypothesis of the relative unimportance of coccolithophorids would imply that most haptophytes enumerated are weakly calcified or unmineralized forms. However, it does raise the possibility that the TEM studies are overlooking a significant fraction of the unmineralized forms. Other high resolution surveys are necessary before these questions can be fully resolved.

Estimates derived from the serial dilution culture technique

One of the few ways of obtaining species specific abundance data for nonmineralized members of the Haptophyta is through the technique of serial dilution culture (Throndsen 1978*a*). In studies of nanoflagellate abundance in Norwegian coastal waters by Throndsen (1976, 1978*b*) he quotes abundance data for *Chrysochromulina* spp. of 10^3–10^5 cells l^{-1} and

Dicrateria inornata, which according to Throndsen (1976) may also have been *Imantonia rotunda*, of 10^3–10^5 cells l^{-1}. In dilution cultures, the cosmopolitan coccolithophorid *Emiliania huxleyi* was found to outgrow most flagellates reaching population densities exceeding 10^6–10^7 cells l^{-1} (Throndsen 1976).

A season of haptophyte abundance data based on the serial dilution culture technique is available from the Kiel Bight, Western Baltic (Jochem 1990). *Dicrateria* and/or *Imantonia* were present from April until October at temperatures ranging from 2–20 °C, with maximum numbers of 10^5 cells l^{-1} recorded during July. Species of *Chrysochromulina* were observed throughout the period studied, but at a lower abundance of 10^2–10^4 cells l^{-1}.

In a recent survey of protist plankton from the Kattegat (Thomsen 1992) the quantitative technique of serial dilution was used in conjunction with TEM. Three species of *Chrysochromulina* (*C. brevifilum*, *C. hirta*, and *C. spinifera*) were observed. The maximum cell numbers noted were 10^4–10^5 cells l^{-1} during the late summer. The coccolithophorid *Emiliania huxleyi* simultaneously reached a maximum of 10^5–10^6 cells l^{-1}. *Imantonia rotunda* occurred throughout the year with between 10^2–10^4 cells l^{-1}.

Unfortunately this quantitative culture approach has not yet been used in oceanic situations.

Estimates derived from algal pigments

Investigations of marine phytoplankton distribution using pigments as diagnostic markers for specific groups of phytoplankton have been published recently (e.g. Siegel *et al.* 1990; Ondrusek *et al.* 1991). The marker specific for haptophytes is 19′-hexanoyloxyfucoxanthin. Ondrusek *et al.* (1991) found this particular pigment to be the most cosmopolitan of the acetone-extractable carotenoids in the open North Pacific Ocean. The highest concentrations of this pigment, together with chlorophyll *b*, were observed near the base of the euphotic zone. A somewhat similar trend was evident in the Sargasso Sea (Siegel *et al.* 1990). Here the distribution of the haptophyte pigment marker was similarly parallel to that of chlorophyll *b*, but the subsurface maximum was at a depth of 5% incident PAR, somewhat nearer the surface than in the North Pacific Ocean.

Mixotrophy

The capacity of autotrophs to ingest particles (mixotrophy) considerably increases the complexity of the microbial food web (see Jones *et al.*, Chapter 13). The magnitude and occurrence of mixotrophy in the euphotic zone of the open ocean needs to be assessed accurately before reasonable modelling efforts may proceed. There are indications that a significant percentage of autotrophs may be capable of mixotrophy. Caron (cited in Sanders 1991) observed that up to 53% of the autotrophic flagellates from the Sargasso Sea contained ingested food particles. No estimates were given for the composition of these assemblages, but it is likely that haptophytes were a significant component.

Unmineralized forms from the EquaPac cruises have been observed frequently containing ingested *Synechoccocus*. Up to a maximum of 10% of the unmineralized forms of haptophytes contain such autotrophic picoplankton (unpublished data). This is a conservative estimate for total mixotrophy, since heterotrophic bacteria are potential prey and are much more abundant than the minute picoalgae. Green (1991) reports 24 species of *Chrysochromulina* ingesting particulate material.

Conclusions

Information on the diversity and importance of the Prymnesiophyceae has increased in phase with the advances in technology available for their study and observation. Light, transmission electron, scanning electron, and epifluorescence microscopy, together with satellite remote sensing have all contributed to major advances in the understanding of this group of organisms. The Haptophyta are accepted as a ubiquitous and important component of autotrophic nanoplankton in the open ocean. However, despite recent technological advances, several important and unanswered questions remain with respect to open water haptophytes. The qualitative composition of haptophyte assemblages in the open ocean remains virtually unaddressed mainly due to the inaccessibility of samples from the deep ocean basins. With the increase in large multinational, multidisciplinary integrated research programmes and an appreciation of the importance of this information, the lack of taxonomic knowledge of oceanic phytoflagellates is being corrected slowly. It is possible only to infer the relative abundances of the various groups that comprise the haptophytes with the quantitative techniques currently in use, e.g. EFM analysis of organisms concentrated on to filters. While it is currently the most commonly used method of enumeration, it does not enable differentiation between the lightly calcified, unmineralized, and the heavily calcified forms. For instance we have difficulty reconciling the EFM analyses, which indicate low coccolithophorid abundances, e.g. from the California Undercurrent cruises, with relative abundance counts from the TEM which indicate 40% of the haptophytes to be coccolithophorids. A single method combining EFM with light microscopy (with polarization) of FTF (filter-transfer-freeze; Hewes and Holm-Hansen 1983) samples, or the use of aluminium filters (McKenzie *et al.* 1992) may be appropriate for this purpose (W. Balch, personal communication). The use of molecular techniques (Campbell *et al.* 1989) may also soon increase the understanding we have of the distribution of members of this group.

References

Balch, W. M., Holligan, P. M., Ackleson, S. G., and Voss, K. J. (1991). Biological and optical properties of mesoscale coccolithophore blooms in the Gulf of Maine. *Limnology and Oceanography*, **36**, 629–43.

Balch, W. M., Holligan, P. M., and Kilpatrick, K. A. (1992). Calcification, photosynthesis and growth of the bloom-forming coccolithophore, *Emiliania huxleyi*. *Continental Shelf Research*, **12**, 1353–74.

Berge, G. (1962). Discoloration of the sea due to *Coccolithus huxleyi* 'bloom'. *Sarsia*, **6**, 27–40.

Booth, B. C. (1987). The use of autofluorescence for analysing oceanic phytoplankton communities. *Botanica Marina*, **30**, 101–8.

Booth, B. C. (1988). Size classes and major taxonomic groups of phytoplankton at two locations in the subarctic Pacific Ocean in May and August, 1984. *Marine Biology*, **97**, 275–86.

Booth, B. C., Lewin, J., and Norris, R. E. (1982). Nanoplankton species dominant in the subarctic Pacific in May and June 1978. *Deep-Sea Research*, **29**, 185–200.

Boysen, M. P. (1991) Differential export from a two-layered euphotic zone. Ph.D. dissertation, University of California.

Broecker, W. S. and Peng, T.-H. (1982). *Tracers in the sea.* Lamont-Doherty Geological Observatory, Palisades, NY.

Campbell, L., Shapiro, L. P., Haugen, E. M., and Morris, L. (1989). Immunochemical approaches to the identification of the ultraplankton: assets and limitations. In *Novel phytoplankton blooms*, (ed. E. M. Cosper, V. M. Bricelj, and E. J. Carpenter), pp. 39–56. Springer-Verlag, Berlin.

Chavez, F. P., Buck, K. R., and Barber, R. T. (1990). Phytoplankton taxa in relation to primary production in the equatorial Pacific. *Deep-Sea Research*, **11**, 1733–52.

Chavez, F. P., Buck, K. R., Coale, K. H., Martin, J. H., Ditullio, G. R., and Welschmeyer, N. A. (1991). Growth rates, grazing, sinking and iron limitation of equatorial Pacific phytoplankton. *Limnology and Oceanography*, **36**, 1816–33.

Chrétiennot-Dinet, M.-J. (1990). *Atlas du phytoplancton marin.* Vol. 3. Éditions du CNRS, Paris.

Conley, S. M. (1979). Recent coccolithophores from the Great Barrier Reef-Coral Sea region. *Micropaleontology*, **25**, 20–43.

Estep, K. W., Davis, P. G., Hargraves, P. E., and Sieburth, J. McN. (1984). Chloroplast containing microflagellates in natural populations of North Atlantic nanoplankton, their identification and distribution; including a description of five new species of *Chrysochromulina* (Prymnesiophyceae). *Protistologica*, **20**, 613–34.

Fryxell, G. A., Taguchi, S., and El-Sayed, S. Z. (1979). Vertical distribution of diverse phytoplankton communities in the central Pacific. In *Marine geology and oceanography of the Pacific manganese nodule province,* (ed. J. L. Bischoff and D. Z. Piper), pp. 203–39. Plenum, New York.

Gaarder, K. R. and Heimdal, B. R. (1977). A revision of the genus *Syracosphaera* Lohmann (Coccolithineae). *'Meteor' Forschungs-Ergebnisse,* Reihe D, **24**, 54–71.

Gieskes, W. W. and Kraay, G. W. (1986). Analysis of phytoplankton pigments by HPLC before, during and after mass occurrence of the microflagellate *Corymbellus aureus* during the spring bloom in the open northern North Sea in 1983. *Marine Biology,* **92**, 45–52.

Green, J. C. (1976). *Corymbellus aureus* gen. et sp. nov., a new colonial member of the Haptophyceae. *Journal of the Marine Biological Association of the United Kingdom,* **56**, 31–8.

Green, J. C. (1991). Phagotrophy in prymnesiophyte flagellates. In *The biology of free-living heterotrophic flagellates,* (ed. D. J. Patterson and J. Larsen), pp. 401–14. Clarendon Press, Oxford.

Green, J. C. and Hori, T. (1986). The ultrastructure of the flagellar root system of *Imantonia rotunda* (Prymnesiophyceae). *British Phycological Journal,* **21**, 5–18.

Green, J. and Pienaar, R. N. (1977). The taxonomy of the order Isochrysidales (Prymnesiophyceae) with special reference to the genera *Isochrysis* Parke, *Dicrateria* Parke and *Imantonia* Reynolds. *Journal of the Marine Biological Association of the United Kingdom,* **57**, 7–17.

Hallegraeff, G. M. (1983). Scale-bearing and loricate nanoplankton from the East Australian Current. *Botanica Marina,* **26**, 493–515.

Hallegraeff, G. M. (1984). Coccolithophorids (calcareous nanoplankton) from Australian waters. *Botanica Marina,* **27**, 229–47.

Hasle, G. R. (1959). A quantitative study of phytoplankton from the equatorial Pacific. *Deep-Sea Research,* **6**, 38–59.

Heimdal, B. R. and Gaarder, K. R. (1980). Coccolithophorids from the northern part of the eastern central Atlantic. I. Holococcolithophorids. *'Meteor' Forschungs-Ergebnisse,* Reihe D, **32**, 1–14.

Heimdal, B. R. and Gaarder, K. R. (1981). Coccolithophorids from the northern part of the eastern central Atlantic. II. Heterococcolithophorids. *'Meteor' Forschungs-Ergebnisse,* Reihe D, **33**, 37–69.

Hewes, C. D. and Holm-Hansen, O. (1983). A method for recovering nanoplankton from filters for identification with the microscope: the filter-transfer-freeze (FTF) technique. *Limnology and Oceanography,* **28**, 389–94.

Hoepffner, N. and Haas, L. W. (1990). Electron microscopy of nanoplankton from the North Pacific central gyre. *Journal of Phycology,* **26**, 421–39.

Holligan, P. M., Aarup, T., and Groom, S. B. (1989). The North Sea: satellite colour atlas. *Continental Shelf Research,* **9**, 667–765.

Hori, T. and Green, J. (1985). The ultrastructural changes during mitosis in *Imantonia rotunda* Reynolds (Prymnesiophyceae). *Botanica Marina,* **27**, 67–78.

Jochem, F. J. (1990). On the seasonal occurrence of autotrophic naked nanoflagellates in Kiel Bight, Western Baltic. *Estuarine, Coastal and Shelf Science*, **31**, 189–202.

Kawachi, M., Inouye, I., Maeda, O., and Chihara, M. (1991). The haptonema as a food-capturing device: observations on *Chrysochromulina hirta* (Prymnesiophyceae). *Phycologia*, **30**, 563–93.

Keller, M. D., Bellows, W. K., and Guillard, R. R. L. (1989). Dimethyl sulfide production in marine phytoplankton. In *Biogenic sulfur in the environment*, (ed. E. S. Saltzman, and W. J. Cooper), pp. 183–200. American Chemical Society, Washington.

Kleijne, A. (1991). Holococcolithophorids from the Indian Ocean, Red Sea, Mediterranean Sea and North Atlantic Ocean. *Marine Micropaleontology*, **17**, 1–76.

Kleijne, A. (1992). Extant Rhabdosphaeraceae (Coccolithophorids, class Prymnesiophyceae) from the Indian Ocean, Red Sea, Mediterranean Sea and North Atlantic Ocean. *Scripta Geologica*, **100**, 1–63.

Lackey, J. B. (1939). Notes on plankton flagellates from the Scioto River. *Lloydia*, **2**, 128–43.

Leadbeater, B. S. C. (1972). Fine structural observations on six new species of *Chrysochromulina* (Haptophyceae) from Norway with preliminary observations on scale production in *C. microcylindra* sp. nov. *Sarsia*, **49**, 65–80.

Malin, G., Turner, S. M., and Liss, P. S. (1992). Sulphur: the plankton/climate connection. *Journal of Phycology*, **28**, 590–7.

Manton, I. (1982). *Chrysochromulina latilepis* sp. nov. (Prymnesiophyceae = Haptophyceae) from the Galapagos Islands, with preliminary comparisons with relevant taxa from South Africa. *Botanica Marina*, **25**, 163–9.

Manton, I. (1983). Nanoplankton from the Galapagos Islands: *Chrysochromulina discophora* sp. nov. (Haptophyceae = Prymnesiophyceae), another species with exceptionally large scales. *Botanica Marina*, **26**, 15–22.

Manton, I. and Leadbeater, B. S. C. (1974). Fine-structural observations on six species of *Chrysochromulina* from wild Danish marine nanoplankton, including a description of *C. campanulifera* sp. nov. and a preliminary summary of the nanoplankton as a whole. *Det Kongelige Danske Videnskabernes Selskabs Biologiske Skrifter*, **20** (5), 1–26.

Manton, I. and Oates, K. (1980). *Polycrater galapagensis* gen. et sp. nov., a putative coccolithophorid from the Galapagos Islands with an unusual aragonite periplast. *British Phycological Journal*, **15**, 95–103.

Manton, I. and Oates, K. (1983). Nanoplankton from the Galapagos Islands: *Chrysochromulina vexillifera* sp. nov. (Haptophyceae = Prymnesiophyceae), a species with semivestigal body spines. *Botanica Marina*, **26**, 517–25.

McIntyre, A. and Bé, A. W. H. (1967). Modern Coccolithophoridae of the Atlantic Ocean. 1. Placoliths and cyrtoliths. *Deep-Sea Research*, **14**, 561–97.

McKenzie, C. H., Helleur, R., and Deibel, D. (1992). Use of inorganic membrane filters (Anopore) for epifluorescence and scanning electron microscopy of nanoplankton and picoplankton. *Applied and Environmental Microbiology*, **58**, 773–6.

Moestrup, Ø. (1979). Identification by electron microscopy of marine nanoplankton from New Zealand, including the description of four new species. *New Zealand Journal of Botany*, **17**, 61–95.

Okada, H. and Honjo, S. (1973). The distribution of oceanic coccolithophorids in the Pacific. *Deep-Sea Research*, **20**, 355–74.

Okada, H. and Honjo, S. (1975). Distribution of coccolithophores in marginal seas along the western Pacific Ocean and in the Red Sea. *Marine Biology*, **31**, 271–85.

Okada, H. and McIntyre, A. (1977). Modern coccolithophores of the Pacific and North Atlantic Oceans. *Micropaleontology*, **23**, 1–55.

Okada, H. and McIntyre, A. (1979). Seasonal distribution of modern cocolithophores in the western North Atlantic Ocean. *Marine Biology*, **54**, 319–28.

Ondrusek, M. E., Bidigare, R. R., Sweft, S. T., Defreitas, D. A., and Brooks, J. M. (1991). Distribution of phytoplankton pigments in the North Pacific Ocean in relation to physical and optical variability. *Deep-Sea Research*, **38**, 243–66.

Parke, M., Manton, I., and Clarke, B. (1955). Studies on marine flagellates. II. Three new species of *Chrysochromulina*. *Journal of the Marine Biological Association of the United Kingdom*, **34**, 579–609.

Parke, M., Manton, I., and Clarke, B. (1956). Studies on marine flagellates. III. Three further species of *Chrysochromulina*. *Journal of the Marine Biological Association of the United Kingdom*, **35**, 387–414.

Reid, F. M. H. (1980). Coccolithophorids of the North Pacific Central Gyre with notes on their vertical and seasonal distribution. *Micropaleontology*, **26**, 151–76.

Reid, F. M. H. (1983). Biomass estimation of components of the marine nanoplankton and picoplankton by the Uthermöhl method. *Journal of Plankton Research*, **5**, 235–51.

Reynolds, N. (1974). *Imantonia rotunda* gen. et sp. nov., a new member of the Haptophyceae. *British Phycological Journal*, **9**, 429–34.

Samtleben, C. and Schröder, A. (1992). Living coccolithophore communities in the Norwegian-Greenland Sea and their record in sediments. *Marine Micropaleontology*, **19**, 333–54.

Sanders, R. W. (1991). Trophic strategies among heterotrophic flagellates. In *The biology of free-living heterotrophic flagellates*, (ed. D. J. Patterson and J. Larsen), pp. 21–38. Clarendon Press, Oxford.

Siegel, D. A., Iturriaga, R., Bidigare, R. B., Smith, R. C., Pak, H., Dickey, T. D., *et al.* (1990). Meridional variations of the springtime phytoplankton community in the Sargasso Sea. *Journal of Marine Research*, **48**, 379–412.

Sieracki, M. E., Verity, P. G., and Stoecker, D. K. (1993). Plankton community response to sequential silicate and nitrate depletion during the 1989 North Atlantic spring bloom. *Deep-Sea Research*, **40**, 213–26.

Thomsen, H. A. (ed.) (1992). *Plankton i de indre danske farvande.* Miljøministeriet Miljøstyrelsen, Copenhagen.

Thomsen, H. A., Buck, K. R., Coale, S. L., Garrison, D. L., and Gowing, M. M. (1988). Nanoplanktonic coccolithophorids (Prymnesiophyceae, Haptophyceae) from the Weddell Sea, Antarctica. *Nordic Journal of Botany*, **8**, 419–36.

Thomsen, H. A., Hansen, G., Larsen, J., Moestrup, Ø., and Vørs, N. (1992). Fytoplankton og heterotroft nanoplankton. In *Planktondynamik og stofomsætning i Kattegat*, (ed. T. Fenchel), pp. 31–59. Miljøministeriet Miljøstyrelsen, Copenhagen.

Throndsen, J. (1976). Occurrence and productivity of small marine flagellates. *Norwegian Journal of Botany*, **23**, 269–93.

Throndsen, J. (1978*a*). The dilution-culture method. In *Phytoplankton manual*, (ed. A. Sournia), pp. 218–24. UNESCO, Paris.

Throndsen, J. (1978*b*). Productivity and abundance of ultra- and nanoplankton in Oslofjorden. *Sarsia*, **63**, 273–84.

11. Haptophytes in polar waters

HARVEY J. MARCHANT
Australian Antarctic Division, Kingston, Tasmania, Australia

and HELGE A. THOMSEN
Department of Phycology, Botanical Institute, University of Copenhagen, Denmark

Abstract

Haptophytes are abundant and important constituents of the plankton of polar waters. The colonial stage of *Phaeocystis* commonly dominates the phytoplankton of the marginal ice zone where it is a major source of particulate and dissolved organic carbon. The extent to which this species contributes to carbon flux to the deep sea is equivocal, but recent evidence indicates that, despite heavy sedimentation, much is utilized by microbial activity and grazing in the upper aphotic zone. *Phaeocystis* is also a major source of the precursor of dimethyl sulfide, which in air is rapidly oxidized to a number of products including sulfate. It has been proposed that sulfate particles act as cloud condensation nuclei thereby influencing global albedo and therefore climate. Lightly calcified coccolithophorids are a characteristic, but not endemic, element of polar nanoplankton. Originally considered to be autotrophic, it has recently been found that they are heterotrophic. The mixotrophic genus *Chrysochromulina* occurs associated with the sea ice and in the water column. Finding scales of these organisms in gut contents of Antarctic krill, *Euphausia superba*, collected from under the ice during winter, has prompted speculation on the importance of mixotrophy in light limited polar waters.

Introduction

For the purpose of this paper we are considering polar waters to be those parts of the Southern Ocean south of the Antarctic Convergence and in the northern hemisphere, the Arctic Ocean and its surrounding ice-covered seas including the Bering, Greenland, Kara, Barents, East Siberian, Chuckchi, and Beaufort Seas. Because of the amount of work carried out in the Gulf of Alaska, some aspects of this region are also considered.

Historically, the study of phytoplankton in polar waters has concentrated on the large robust diatoms, the organisms most likely to be retained by plankton nets and survive storage in preservatives. As a

The Haptophyte Algae (ed. J. C. Green and B. S. C. Leadbeater), Systematics Association Special Volume No. 51, pp. 209–28. Clarendon Press, Oxford, 1994.

consequence there has been a long standing view that the primary productivity and standing crop of phytoplankton in polar waters is dominated by large diatoms. Although *Phaeocystis* was recognized by investigators early this century, Hasle (1969) was among the first to stress the abundance of nanoplankton and 'monads and flagellates' in Antarctic waters. It is now clear that small organisms, representing a great diversity of taxa, play a role in polar waters and frequently are dominant (Weber and El-Sayed 1987; Booth 1988). Diatoms, especially nanoplanktonic forms, constitute the greatest biomass of phytoplankton in polar waters and account for most of the species diversity. However, the diatoms are rivalled in abundance by blooms of chlorophyll *c*-containing organisms, e.g. haptophytes and cryptophytes (Booth *et al.* 1982; Buma *et al.* 1992).

Haptophytes are a major component of the phytoplankton throughout the world's oceans (Thomsen *et al.*, Chapter 10), and in polar waters *Phaeocystis pouchetii* is highly conspicuous and important. Other haptophytes are generally less abundant but are of considerable interest and importance. These organisms include coccolithophorids, including the lightly calcified forms that have recently been found to be heterotrophic, and *Chrysochromulina*. Here we report on the variety of haptophytes from both Arctic and Antarctic waters and discuss their role in these cold water environments.

Characteristics of polar waters

Polar marine oceans are characterized by low and relatively invariant temperatures (–2 to +5 °C), and extreme variation in irradiance and day-length. The seasonal ice cover of these waters also contributes to the prolonged periods of low light conditions, especially if it is snow covered. Sea ice has other profound effects on polar waters by inhibiting wind driven vertical mixing, and by contributing to the stability of the water column in springtime by the production of a surface layer of less saline water from the melting ice. The marginal ice edge zone that accompanies the springtime retreat of the sea ice is a region of elevated biological activity at all trophic levels (Ainley *et al.* 1986; Smith and Nelson 1986). This region is the site of around 60% of the annual pelagic primary productivity of the Southern Ocean (Smith *et al.* 1988). These authors have also suggested that there is very high interannual variability (around ± 25%) in the primary productivity of this region. Concentrations of nitrate, phosphate, and silicate are high in polar waters. Maximum nutrient concentrations in Arctic waters are typically lower than the minimum concentrations of Antarctic waters (Sakshaug and Holm Hansen 1984). While nutrient concentration apparently only rarely limits phytoplankton growth in Antarctic waters (Lizotte and Sullivan

1992), the low phytoplankton biomass following the ice edge bloom is a result of nutrient limitation in Arctic waters (Sakshaug and Skjoldal 1989). In Antarctic waters, phytoplankton biomass in open water is low, resulting from light limitation as a consequence of deep vertical mixing. The factors controlling growth and distribution of phytoplankton and ice algae have been reviewed by Smith and Sakshaug (1990). The recent reviews by Sakshaug and Slagstadt (1991), Harrison and Cota (1991), and Sakshaug *et al.* (1991) deal with the role of light, nutrients, and other factors on phytoplankton productivity in polar waters.

Organisms

Phaeocystis

Although diatoms are usually the principal constituents of the microbial community of Antarctic pack ice, a variety of autotrophic nanoflagellates are regularly present. The most abundant of these organisms are both the flagellate and colonial stages in the life cycle of *Phaeocystis* (Garrison and Buck 1989; Lancelot and Rousseau, Chapter 12; Medlin, Chapter 21). That *Phaeocystis* sp. is a major component of polar phytoplankton assemblages has been well established (Fryxell 1989; Verity *et al.* 1991; Davidson and Marchant 1992*a*). Garrison *et al.* (1987) found that *Phaeocystis* and the diatom *Nitzschia cylindrus* dominate the algal populations in the pack ice community and the water column of the marginal ice edge zone. Despite the recent plethora of studies on *Phaeocystis*, its taxonomy, life cycle, and ultrastructure remain poorly understood (Davidson and Marchant 1992*b*; Verity *et al.* 1991).

In Antarctic waters *Phaeocystis* is usually the first alga to bloom in the marginal ice edge zone. We have found that, although the abundance of diatoms and *Phaeocystis* begin to increase at the same time, the growth of diatom populations slows during the peak of maximum growth of *Phaeocystis* increasing again after the *Phaeocystis* peak (Davidson and Marchant 1992*a*). Factors accounting for this dominance of *Phaeocystis* are yet to be fully resolved. In Arctic waters *Phaeocystis* is abundant, comprising more than 95% of the plankton cells, from the onset of the spring bloom, or follows the diatom peak when silicate becomes limiting (Vernet 1991). In the Gulf of Alaska, Booth *et al.* (1982) reported high concentrations of nanoplanktonic organisms which were dominated by motile cells of *Phaeocystis pouchetii*. This is in contrast to other regions where blooms of this alga principally comprise the colonial stage of the life cycle. Preceding the bloom, other haptophytes, especially the coccolithophorid *Emiliania huxleyi*, were present.

The brown ice photosynthetic community is dominated by diatoms, dinoflagellates, and occasionally haptophyte colonies (Figs 11.1– 11.5) routinely identified as *Phaeocystis* sp. The release of swarmers from such

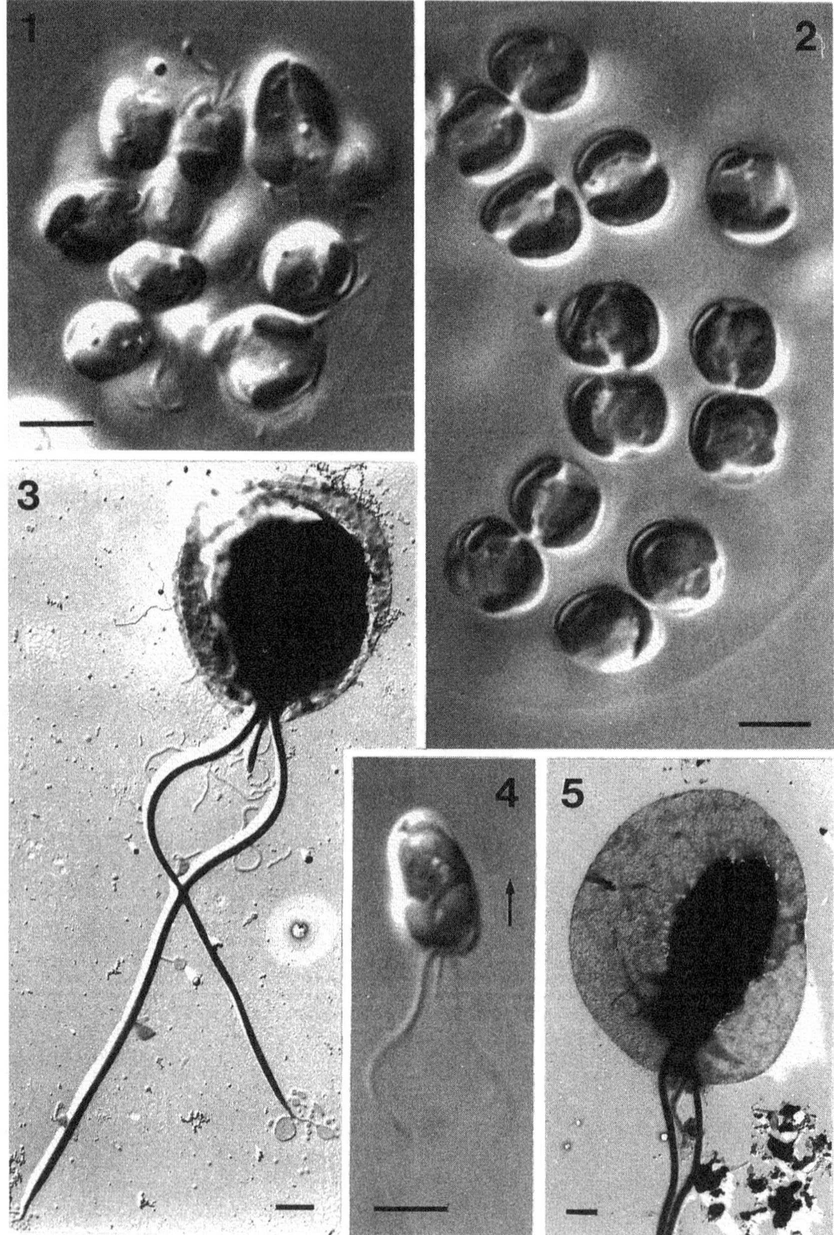

Figs 11.1–11.5 Colonial and flagellate stages of haptophyte species from the Antarctic ice biota. (**1**) Formation of swarmers inside colony. (**2**) Part of vegetative colony with spherical cells embedded in mucilage; the boundary of the colony is particular evident in the lower part of the micrograph. (**3**) Shadowcast whole

colonies has been studied on three occasions (Garrison and Thomsen 1993; Thomsen, unpublished results). Whereas the colonies are basically indistinguishable, the flagellates emerging from these, following the slow melting of ice-cores, were found to be different at the generic level, and in all cases unlike typical *Phaeocystis* swarmers. Figures 11.1–11.5 illustrate one such type of flagellate emanating from an ice biota colony. These particular swarmers were subsequently (5–6 hours after their formation) found to attach themselves to the spines and processes of large diatoms also present in the sample. Realizing that large diatoms are very likely trapped during frazil ice formation and incorporated into sea ice, this strategy suggests that this particular type of organism is primarily adapted to a life within sea ice.

Pigment analysis has proved to be most useful in ascertaining the distribution of phytoplankton which contain fucoxanthin and its derivatives, 19′-hexanoyloxyfucoxanthin and 19′-butanoyloxyfucoxanthin. Fucoxanthin is found in diatoms, haptophytes, raphidophytes, and chrysophytes (Jeffrey and Vesk 1981; Jeffrey and Wright, Chapter 6), and 19′-hexanoyloxyfucoxanthin is a major carotenoid of coccolithophorids and some other haptophytes, including *Phaeocystis* and *Corymbellus aureus*. A 19′-butanoyloxyfucoxanthin-like pigment has been found in Antarctic strains of *Phaeocystis* (Wright and Jeffrey 1987). Both the 19′-hexanoyloxyfucoxanthin and the 19′-butanoyloxyfucoxanthin-like pigments were found to be concentrated in the < 5 μm fraction, and fucoxanthin in the > 5 μm size fraction in water samples taken in the Prydz Bay region of Antarctica, indicating the likely distributions of haptophytes and diatoms (Wright and Jeffrey 1987). Gieskes and Elbrächter (1986), on the basis of the relative abundance of 19′-hexanoyloxyfucoxanthin among the carotenoids, concluded that most of the nanoplankton from the Antarctic Peninsular region consisted principally of haptophytes. *Phaeocystis* is likely to be largely responsible for the high concentration of this pigment in Antarctic waters, whilst north of the Antarctic Convergence both coccolithophorids and *Phaeocystis* contribute.

Coccolithophorids

Coccolithophorids are an abundant component of the phytoplankton in temperate and tropical waters forming dense blooms of $1–2 \times 10^6$ cells l^{-1}. Detached coccoliths, the periplast calcite plates produced by coccoli-

mount (TEM) of swarmer showing flagella and short haptonema. (**4**) Light micrograph of swarmer (interference contrast optics); the *arrow* indicates direction of swimming. (**5**) Detail of cell (TEM shadowcast whole mount) showing the cell periplast which is devoid of scales. Scale bar = 1 μm (**3, 5**); 5 μm (**1, 2, 4**).

thophorids, reach concentrations of 4×10^8 l^{-1} which is sufficiently dense to be easily seen from aircraft and sensed by satellites (Balch *et al.* 1991). They are likely to be the largest single carbonate sink in marine biogeochemical cycles contributing to the massive accumulation of carbonate in sediments. More than 90% of the total carbon being accumulated in marine sediments is in the form of calcite (Aiken *et al.* 1992). Detailed studies on the latitudinal and depth distribution of coccolithophorids in the Southern Ocean have revealed that the abundance of these organisms decreased with increasing latitude and south of 60 °S they were essentially absent (Nishida 1986). This distribution of coccolithophorids has been independently confirmed by Sikes and Volkman (1993), who reported the concentration of alkenones synthesized by *Emiliania huxleyi* decreasing with increasing latitude across the Southern Ocean south of Australia to about 60°S from where these long-chain unsaturated ketones were essentially undetectable. Jacques and Panouse (1991) also report *Emiliania huxleyi* at only the northern end of series of north-south transects in the Weddell–Scotia Confluence area.

The presence of coccoliths in sediments has been regarded as indicating sedimentation from outside polar waters. As indicated by Honjo (1990), while this notion is applicable to Antarctic waters, year-round sedimentation of coccoliths and coccolithophorids, dominated by *Coccolithus pelagicus*, occurs in northern waters, including the Greenland Basin and the Fram Strait. The most abundant coccolithophorid found in the Gulf of Alaska was *Emiliania huxleyi* at a concentration of 3×10^5 cells l^{-1} (Booth *et al.* 1982), which is the most abundant coccolithophorid in the subarctic Pacific. *Coccolithus pelagicus* is apparently the most stenothermal species of coccolithophorid occurring over a temperature range of 0 to 15 °C (Okada and McIntyre 1979). It has been found abundant in the water column as far north as 86° (Honjo 1990). Finding a high flux of *Coccolithus pelagicus* during winter and early spring has prompted speculation that this organism is heterotrophic at this time (Paasche 1968; Okada and Honjo 1973) rather than being advected under, or released from sea ice. However, as Honjo (1990) points out, if heterotrophy is the mechanism used in boreal waters why does this also not occur in Antarctic waters where insolation, temperature, and ice conditions are similar?

Figs 11.6–11.9 Two lightly calcified Antarctic coccolithophorids (shadowcast whole mounts/TEM). (**6**) *Wigwamma arctica*; complete cell with two flagella and haptonema. (**7**) High magnification of *Wigwamma arctica* heterococcoliths shaped as 'wigwams'. Note organic base plate patterning (*arrow*). (**8**) Detail of *Trigonaspis melvillea* flagellar pole holococcoliths. (**9**) *Trigonaspis melvillea*; complete cell with flagella and coiled up haptonema. Notice threads from *Phaeocystis pouchetii*. Scale bar 1 μm.

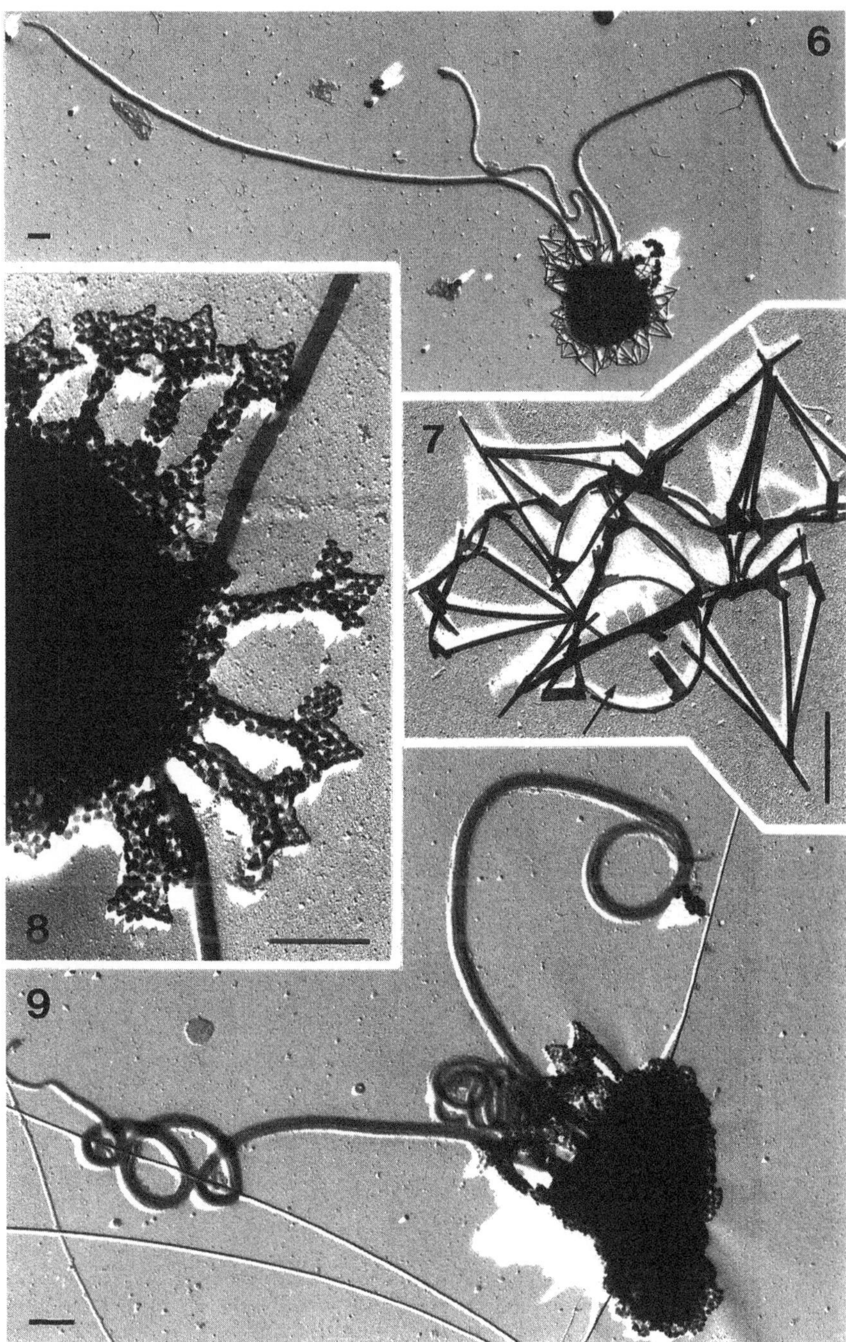
6
7
8
9

Table 11.1 Occurrence of lightly calcified coccolithophorids in polar regions

	Greenland	Resolute Bay	Homer Alaska	Antarctica
Balaniger				
B. balticus	*			
Calciarcus				
C. alaskensis	*		*	
C. aff. *alaskensis*	*			*
Pappomonas				
P. flabellifera var. *flabellifera*	*			
P. flabellifera var. *borealis*	*		*	
P. virgulosa	*		*	
P. weddellensis				*
Pappomonas sp.				*
Papposphaera				
P. obpyramidalis				*
P. sagittifera	*		*	*
P. sarion	*			
P. simplicissima				*
Papposphaera spp.	*			*
Quaternariella				
Q. obscura	*			(*)
Trigonaspis				
T. diskoensis	*			
T. melvillea				*
T. minutissima	*			(*)
Turrisphaera				
T. arctica	*	*	*	(*)
T. borealis	*	*	*	(*)
T. polybotrys	*			(*)
T. spp.				*
Wigwamma				
W. annulifera	*		*	*
W. antarctica				*
W. arctica	*	*		*
W. scenozonion	*			
W. triradiata				*
W. spp.				*
Coccolithoph. sp1 (Thomsen *et al.* 1988)				*
Coccolithoph. sp2 (Thomsen *et al.* 1988)				*

(*) indicates that the species identification is uncertain. For simplicity the listing does not reflect taxonomic changes recently made to acknowledge life history events (Thomsen *et al.* 1991). Based on data from: Manton and Oates 1975; Manton and Sutherland 1975; Manton *et al.* 1976*a*, *b*; Manton *et al.* 1977; Thomsen 1980*a*, *b*, *c*; Thomsen 1981; Thomsen *et al.* 1988; Thomsen, unpublished results; Hansen *et al.* 1989; Østergaard 1993.

The preparation of freshly collected samples for TEM has resulted in the finding of a contingent of lightly calcified coccolithophorids which appear to have their main distribution in polar regions (Figs 11.6–11.9). These organisms were first encountered in subarctic and arctic samples, and were later found to form a characteristic nanoplanktonic element also in Antarctic waters (see Thomsen *et al.* 1988 for review). Table 11.1 summarizes the findings from polar regions. It should be emphasized that these lightly calcified coccolithophorids are not endemic to polar regions, but are occasionally found in temperate and subtropical regions (Norway: Espeland and Throndsen (1986); Baltic Sea: Thomsen (1979), Thomsen, unpublished data; North Pacific Central Gyre: Hoepffner and Haas (1990); Mexico: Thomsen, Kosman, and Buck, unpublished data; South Africa: Manton and Oates (1975), Manton *et al.* (1977)).

Several taxa of Arctic coccolithophorids previously considered to be autonomous species are, in fact, part of life histories combining hetero- and holococcolithophorid forms (Thomsen *et al.* 1991). Additional examples of combination cells, with holococcoliths and heterococcoliths forming part of the same periplast, have recently been encountered in Arctic (Østergaard, personal communication) and Antarctic samples (Thomsen, unpublished results).

All species of Prymnesiophyceae, with the exception of the enigmatic coccolithophorid *Balaniger balticus* (Thomsen and Oates 1978; Thomsen 1986), have been considered photosynthetic. However it has become evident that all Antarctic coccolithophorids listed in Table 11.1 are in fact heterotrophic organisms (Garrison and Thomsen 1993; Thomsen, unpublished results). It remains to be verified whether these taxa are genuine heterotrophic organisms, or forms that have secondarily lost the photosynthetic apparatus. The heterotrophic nature of these organisms that thrive on the outskirts of the coccolithophorid pastures is likely to be linked to the stressed environment of the polar regions (low light, low temperature, low salinity). The polar lightly calcified coccolithophorids are most abundant in water column samples (1×10^3–5×10^3 cells l^{-1}), and only rarely observed in preparations from recently formed sea ice. However, healthy looking cells have been encountered in Antarctic grease ice, platelet ice, and thin pancake ice (Thomsen, unpublished results), but never in multi-year brown ice samples.

Chrysochromulina

Chrysochromulina is a genus consisting of some 50 species that are apparently ubiquitous in the world's oceans. Estep *et al.* (1984) found the Prymnesiophyceae, of which the majority were species of *Chrysochromulina*, to be the dominant nanoplanktonic organisms on a transect along 24° 30′ N latitude in the Atlantic Ocean. Although species of *Chrysochromulina* have been reported to be widely distributed, there are few data on their abundance; which is not surprising considering their size, fragility, and the difficulty of recognizing many species without electron microscopy. Although not reported as occurring in high numbers, *Chrysochromulina* is a component of polar waters of both hemispheres. Thomsen (1982) reported nine described species of *Chrysochromulina* from Disko Bay, West Greenland, and at least another 11 undescribed species. *Chrysochromulina hirta* has been reported from Alaskan waters and the North West Passage at water temperatures of 6 and –1 °C respectively (Manton 1978). Although species of this genus have been seen in Antarctic waters, there have been no detailed investigations reported. In the Southern Ocean there are at least 10 species including undescribed forms (Marchant, unpublished data). It has been reported from pack ice microbial assemblages (Garrison and Buck 1989), and Jacques and Panouse (1991) list its possible occurrence in the Weddell-Scotia Confluence area.

Other haptophytes

On the Pro-Mare cruises in the Barents Sea, Throndsen and Kristiansen (1991) reported finding *Dicrateria inornata* and *Imantonia rotunda.* These organisms, which are 3–5.5 μm and 2–4 μm in size respectively, were difficult to separate by light microscopy. Together with the prasinophyte *Mantoniella squamata* and the chrysophyte *Pseudopedinella tricostata,* they reached maximum abundance in open pack ice and in open water rather than close pack ice, but were never found to dominate the plankton. The colonial flagellate *Corymbellus aureus,* which has been reported to dominate the spring bloom in the North Sea (Gieskes and Kraay 1986), does not appear to have been reported as a component of polar waters.

Role of haptophytes in polar waters

Studies in lakes have shown that algal phagotrophy is a significant component of their food webs. Bird and Kalff (1986, 1987) reported that the chrysophycean alga, *Dinobryon,* depended more on grazed bacteria than on photosynthesis for nutrition, removing more bacteria from the water column than the crustaceans, rotifers, and ciliates combined. *Ochromonas* and *Chromulina,* also chrysophytes, which accounted for about 25% of the planktonic biomass in a small Finnish lake, ingested 75–203% of their body carbon per day from bacteria (Salonen and Jokinen 1988). *Uroglena americana,* another chrysophyte, was reported to be photoheterotrophic, requiring bacteria for growth (Kimura and Ishida 1985). However, the ecological importance of mixotrophs in the sea is not clear (Porter *et al.*

1985; Estep *et al.* 1986, Sanders 1991). Mixotrophy would be expected to increase the efficiency of the transfer of energy and nutrients to higher trophic levels.

Many species of *Chrysochromulina* have been found to be photoheterotrophic (able to utilize organic carbon in the light), both as dissolved organic carbon (Pintner and Provasoli 1968) and by phagocytosis of particulate material (Green 1991, Jones *et al.* Chapter 13). Of the some 20 species of this alga on which observations of phagocytosis have been reported, only in *C. mantoniae* and *C. fragilis* has ingestion of particles not been unequivocally shown (Manton and Leadbeater 1974). In polar waters which are ice and snow covered for much of the year, light limits photosynthesis. The ability of autotrophs to feed phagotrophically would be expected to permit growth of these organisms when there is insufficient light for photosynthesis. In the Southern Ocean, organisms such as bacteria have been found in the vacuoles of *Chrysochromulina.* Periplast scales of this alga have been found in the gut contents and faeces of different developmental stages of Antarctic krill (Marchant, unpublished). As krill and other crustacea feed under the ice in Antarctic waters (Daly 1990), mixotrophs, including *Chrysochromulina,* may provide an important food source during winter.

One of the most important phytoplanktonic organisms in polar waters is *Phaeocystis.* This alga may influence the growth of other autotrophs by mediating the availability of manganese (Davidson and Marchant 1987), although this is disputed by Lubbers *et al.* (1990). These authors demonstrate that colonial *Phaeocystis* can remove most of the manganese from the medium in which it is growing. Lubbers *et al.* (1990) suggest that this alga is likely to play an important role in the manganese cycle and probably the cycles of other trace metals such as zinc, iron and copper. At least in Antarctic waters, this alga provides substrates for heterotrophs by the secretion of a large proportion of its photoassimilated carbon as particulate and dissolved organic matter (Davidson and Marchant 1992*a*). The role of *Phaeocystis* in polar waters in the global carbon cycle is unclear. As well as being a major component of the phytoplankton bloom in the Antarctic marginal ice edge zone (Bodungen *et al.* 1986; Fryxell and Kendrick 1988; Davidson and Marchant 1992*a*) this alga has been reported to form massive blooms in the Barents, Greenland, and Icelandic Seas (Stefánsson 1990; Wassmann *et al.* 1990; Smith *et al.* 1991). From the removal of nitrate, Smith *et al.* (1991) calculated that the new production was some 40 g C m^{-2} over the 35 days duration of the bloom. The extent of the contribution of organic carbon from this bloom to deep water is equivocal. Hebbeln and Wefer (1991) reported a vertical flux of 33 mg C m^{-2} d^{-1} in the Greenland Sea following the *Phaeocystis* bloom, and Estep *et al.* (1990) found that copepods in this region graze *Phaeocystis.* In the

Barents Sea where macrozooplankton also graze *Phaeocystis* colonies, extremely high sedimentation rates have been reported (Wassmann *et al.* 1990). Together with diatoms, *Phaeocystis* is likely to contribute substantially to marine snow and vertical flux from the photic zone. Wassmann *et al.* (1990) consider that, despite the heavy sedimentation of this alga, it is likely to play a relatively minor role in carbon flux to the deep sea and the sequestering of CO_2 because much of the particulate and dissolved material derived from the photoassimilated carbon of this alga is utilized by microbial activity and zooplankton grazing in the upper aphotic zone. The extent of the loss of material from the photic zone of Antarctic waters is presently under investigation. Short term shallow sediment trap deployments indicate low vertical loss rates of around 1% of suspended particle load per day (Smetacek *et al.* 1990). Higher rates generally were found only when euphausiid faeces were the principal content of the sediment trap. Although *Phaeocystis* is a principal component of the spring phytoplankton bloom in some regions within the Bransfield Strait, sediment trap data did not indicate that this alga contributes directly to vertical flux (Bodungen *et al.* 1986). In this region sedimentation is dominated largely by grazers, especially euphausiids. In addition, the importance of protozoan faecal pellets in the vertical flux of carbon and other elements in polar waters has recently been recognized (Nöthig and Bodungen 1990). *Phaeocystis* has been found in faecal pellets of Antarctic sea ice dinoflagellates (Buck *et al.* 1990).

The extent to which haptophytes are a constituent in the diet of metazoan grazers in polar waters has yet to be ascertained. It is hardly surprising that this topic has received little attention because of the technical difficulties involved. Marchant and Nash (1986) demonstrated the presence of *Phaeocystis* in gut contents and faecal material of Antarctic krill (*Euphausia superba*). *Phaeocystis* has also been found in faecal material in the Scotia and Weddell Seas by Gonzáles (1992), who also reported finding coccoliths of *Emiliania huxleyi* in salp faeces from this area. Periplast scales of *Chrysochromulina* have also been found in krill faeces (Marchant, unpublished data). Although *Phaeocystis* is grazed by herbivores including *Euphausia superba* (Marchant and Nash 1986), the effect of grazing on this alga and its food value are equivocal (Verity and Smayda 1989). In an investigation on the impact of copepod grazing on a phytoplankton bloom in which *Phaeocystis* comprised about 97% of the biomass and the remainder was mainly diatoms, the diatoms accounted for some 74% of the copepod diet (Claustre *et al.* 1990). Only 1.5% of the biomass of *Phaeocystis* was grazed by the copepods, the remainder apparently being lost to the pelagic food web. In addition, Claustre *et al.* (1990) reported that the low nutritional value of *Phaeocystis* was due to its fatty acid to chlorophyll *a* ratio being much lower than was found in diatoms. This was also the case for amino acids and vitamin C. *Phaeocystis* from Antarctic sea ice has been found to have significantly lower concentrations of neutral lipids than diatom assemblages dominated by *Nitzschia* and *Navicula* (Priscu *et al.* 1990). In an investigation of the fatty acids of krill and Antarctic *Phaeocystis*, Virtue *et al.* (1993) concluded that *Phaeocystis* was deficient in a number of the essential fatty acids. However, this alga may be an adequate food source for krill because of the crustacean's apparent ability to convert exogenous short chain fatty acids to long chain polyunsaturated fatty acids. Antarctic euphausiids reportedly have a dietary preference for diatoms (Miller and Hampton 1989). At an Antarctic inshore site very little of the carbon attributable to *Phaeocystis* is apparently utilized by metazoa (Davidson and Marchant 1992*a*) and, as was found by Claustre *et al.* (1990), most of the carbon was not used *in situ*.

Emission of dimethyl sulfide (DMS) from the oceans account for the largest biogenic source of reduced sulfur compounds to the atmosphere (Andreae and Raemdonck 1983). DMS is produced, together with acrylic acid, by the enzymatic cleavage of dimethyl sulfoniopropionate (DMSP), which is likely to have osmoregulatory and bacteriocidal activity (Sieburth 1960; Barnard *et al.* 1984). DMS is rapidly oxidized in the atmosphere to a number of products including sulfur dioxide, methanesulfonate, and sulfate (Anderson *et al.* 1992). It is proposed that sulfate particles constitute a major source of cloud condensation nuclei (CCN), and that the abundance of CCN determines global albedo thereby establishing a mechanism for the regulation of climate by marine biological activity (Charlson *et al.* 1987). Some members of the Prymnesiophyceae, such as *Hymenomonas carterae* and *Phaeocystis* produce three orders of magnitude more DMS per cell than most other groups of phytoplankton (Barnard *et al.* 1984). Admittedly based on few studies, the abundance of *Phaeocystis* at higher latitudes correlates with elevated concentration of DMS in the water column (Andreae and Raemdonck 1983; Barnard *et al.* 1984). To date, the highest recorded concentration of DMS in seawater is 290 nM measured at an Antarctic coastal site, coinciding with a bloom of colonial *Phaeocystis* (Gibson *et al.* 1990). These authors have estimated that Antarctic *Phaeocystis* may contribute as much as 10% of the total global flux of DMS to the atmosphere.

The marked depletion in stratospheric ozone over Antarctica during spring produces an extended period when incident UV radiation is as high or higher than at the summer solstice (Frederick and Snell 1988). This finding has prompted considerable activity to ascertain the impact of UV on the productivity and survival of Antarctic marine phytoplankton. It has become apparent from various investigations that there is substantial interspecific variability to UV exposure (Karentz 1991; Smith *et al.* 1992). The colonial stage in the life cycle of Antarctic strains of *Phaeocystis pouchetii* have five to ten times the concentration of UV-B absorbing compounds than a number of strains of *P. pouchetii* from other parts of the world's ocean (Marchant *et al.* 1991). It was also found that the motile cells of Antarctic *Phaeocystis* lacked UV-B absorbing compounds. When exposed to a gradient in intensity of UV irradiation the percent of motile cells surviving was very much lower than cells of the colonial phase (Marchant *et al.* 1991). Thus, as well as there being a high level of interspecific variation among diatoms in the impact of UV-B, the two principal stages in the life cycle of *Phaeocystis* also demonstrate substantial variation in their survival. The consequence of such a vulnerable stage to UV exposure has yet to be explored. It appears that colonial *Phaeocystis* is less tolerant of UV exposure than all of the species of Antarctic diatoms that have been investigated to date (Smith *et al.* 1992). The extent to which differences in the tolerance of components of phytoplankton communities could lead to changes in community structure cannot presently be assessed. Further, we are not aware of investigations on the UV-B tolerance of other taxa of haptophytes.

Conclusions

Haptophytes, especially *Phaeocystis*, are frequently the dominant autotrophs in polar waters and have been identified as having key roles in a variety of microbial processes, some of which are of global significance. Despite this, however, many aspects of their biology are only superficially known. Clearly, further investigations are required before we have an understanding of these environmentally and economically significant organisms.

References

Aiken, J., Moore, G. F., and Holligan P. M. (1992). Remote sensing of oceanic biology in relation to global climate change. *Journal of Phycology*, **28**, 579–90.

Ainley, D. G., Fraser, W. R., Sullivan, C. W., Torres, J. J., Hopkins, T. L., and Smith, W. O. Jr. (1986). Antarctic mesopelagic micronekton: evidence from seabirds that pack ice affects community structure. *Science*, **232**, 847–9.

Anderson, T. L., Wolfe, G. V., and Warren, S. G. (1992). Biological sulfur, clouds and climate. In *Encyclopedia of earth system science*, Vol. 1, pp. 363–76. Academic Press, London.

Andreae, M. O. and Raemdonck, H. (1983). Dimethyl sulfide in the surface ocean and the marine atmosphere: a global view. *Science*, **221**, 744–7.

Balch, W. M., Holligan, P. M., Ackleson, S. G., and Voss, K. J. (1991). Biological and optical properties of mesoscale coccolithophore blooms in the Gulf of Maine. *Limnology and Oceanography*, **36**, 629–43.

Barnard, W. R., Andreae, M. O., and Iverson, R. L. (1984). Dimethylsulfide and *Phaeocystis pouchetii* in the southeastern Bering Sea. *Continental Shelf Research*, **3**, 103–13.

Bird, D. F. and Kalff, J. (1986). Bacterial grazing by planktonic lake algae. *Science*, **231**, 493–5.

Bird, D. F. and Kalff, J. (1987). Algal phagotrophy: Regulating factors and importance relative to photosynthesis in *Dinobryon* (Chrysophyceae). *Limnology and Oceanography*, **32**, 277–84.

Bodungen, B. von., Smetacek, V. S., Tilzer, M. M., and Zeitzschel, B. (1986). Primary production and sedimentation during spring in the Antarctic Peninsula region. *Deep-Sea Research*, **33**, 177–94.

Booth, B. C. (1988). Size classes and major taxonomic groups of phytoplankton at two locations in the subarctic Pacific ocean in May and August, 1984. *Marine Biology*, **97**, 275–86.

Booth, B. C., Lewin, J., and Norris, R. E. (1982). Nanoplankton species predominant in the subarctic Pacific in May and June 1978. *Deep-Sea Research*, **29**, 185–200.

Buck, K. R., Bolt, P. A., and Garrison, D. L. (1990). Phagotrophy and fecal pellet production by an athecate dinoflagellate in Antarctic sea ice. *Marine Ecology Progress Series*, **60**, 75–84.

Buma, A. G. J., Gieskes, W. W. C., and Thomsen, H. A. (1992). Abundance of Cryptophyceae and chlorophyll *b*-containing organisms in the Weddell–Scotia Confluence area in the spring of 1988. *Polar Biology*, **12**, 43–52.

Charlson, R. J., Lovelock, J. E., Andreae, M. O., and Warren, S. G. (1987). Oceanic phytoplankton, atmospheric sulphur, cloud albedo and climate. *Nature*, **326**, 655–61.

Claustre, H., Poulet, S. A., Williams, R., Marty, J-C., Coombs, S., Ben Mlih, F., *et al.* (1990). A biochemical investigation of a *Phaeocystis* sp. bloom in the Irish Sea. *Journal of the Marine Biological Association of the United Kingdom*, **70**, 197–207.

Daly, K. L. (1990). Overwintering development, growth, and feeding of larval *Euphausia superba* in the Antarctic marginal ice zone. *Limnology and Oceanography*, **35**, 1564–76

Davidson, A. T. and Marchant, H. J. (1987). Binding of manganese by Antarctic *Phaeocystis pouchetii* and the role of bacteria in its release. *Marine Biology*, **95**, 481–7.

Davidson, A. T. and Marchant, H. J. (1992a) Protist interactions and carbon dynamics of a *Phaeocystis*-dominated bloom at an Antarctic coastal site. *Polar Biology*, **12**, 387–95.

Davidson, A. T. and Marchant, H. J. (1992b) The biology and ecology of *Phaeocystis* (Prymnesiophyceae). In *Progress in phycological research*, Vol. 8, (ed. F. E. Round and D. J. Chapman), pp. 1–45. Biopress, Bristol.

Espeland, G. and Throndsen, J. (1986). Flagellates from Kilsfjorden, southern Norway, with description of two new species of Choanoflagellida. *Sarsia*, **71**, 209–26.

Estep, K. W., Davis, P. G., Hargraves, P. E., and Sieburth, J. McN. (1984). Chloroplast containing microflagellates in natural populations of North Atlantic nanoplankton, their identification and distribution; including a description of five new species of *Chrysochromulina* (Prymnesiophyceae). *Protistologica*, **20**, 613–34.

Estep, K. W., Davis, P. G., Hargraves, P. E., and Sieburth, J. McN. (1986). How important are algal nanoflagellates in bactivory? *Limnology and Oceanography*, **31**, 646–50.

Estep, K. W., Nejstgaard, J. C., Skjoldal, H. R., and Rey, F. (1990). Predation of copepods upon natural populations of *Phaeocystis pouchetii* as a function of the physiological state of the prey. *Marine Ecology Progress Series*, **67**, 235–49.

Frederick, J. E. and Snell, H. E. (1988). Ultraviolet radiation levels during the antarctic spring. *Science*, **241**, 438–40.

Fryxell, G. A. (1989). Marine phytoplankton at the Weddell Sea ice edge: seasonal changes at the specific level. *Polar Biology*, **10**, 1–18.

Fryxell, G. A. and Kendrick, G. A. (1988). Austral spring microalgae across the Weddell Sea ice edge: spatial relationships found along a northward transect during AMERIEZ 83. *Deep-Sea Research*, **35**, 1–20.

Garrison, D. L. and Buck, K. R. (1989). The biota of Antarctic pack ice in the Weddell Sea and Antarctic Peninsula regions. *Polar Biology*, **10**, 211–19.

Garrison, D. L. and Thomsen, H. A. (1993). Ecology and biology of ice biota. *Berichte zur Polarforschung*, **121**, 68–74.

Garrison, D. L., Buck, K. R., and Fryxell, G. A. (1987). Algal assemblages in Antarctic pack ice and in the ice-edge plankton. *Journal of Phycology*, **23**, 564–72.

Gibson, J. A. E., Garrick, R. C., Burton, H. R., and McTaggart, A. R. (1990). Dimethylsulfide and the alga *Phaeocystis pouchetii* in Antarctic coastal waters. *Marine Biology*, **104**, 339–46.

Gieskes, W. W. C. and Elbrächter, M. (1986). Abundance of nanoplankton-size chlorophyll-containing particles caused by diatom disruption in surface waters of the Southern Ocean (Antarctic Peninsula region). *Netherlands Journal of Sea Research*, **20**, 291–303.

Gieskes, W. W. C. and Kraay, G. W. (1986). Analysis of phytoplankton pigments by HPLC before, during and after mass occurrence of the microflagellate *Corymbellus aureus* during the spring bloom in the open northern North Sea in 1983. *Marine Biology*, **92**, 45–52.

Gonzáles, H. E. (1992). The distribution and abundance of krill faecal material and oval pellets in the Scotia and Weddell Seas (Antarctica) and their role in particle flux. *Polar Biology*, **12**, 81–91.

Green, J. C. (1991). Phagotrophy in prymnesiophyte flagellates. In *The biology of free-living heterotrophic flagellates*, (ed. D. J. Patterson and J. Larsen), pp. 401–14. Clarendon Press, Oxford.

Hansen, L. E., Nielsen, D., Skovgaard, K., and Østergaard, J. B. (1989). Taxonomiske og kvantitative undersøgelser af de frie vandmasser ved Disko, Grønland. In *Feltkursus arktisk biologi*, (ed. M. Jørgensen), pp. 51–142. Zoologisk Museum, Københavns Universitet, 1988.

Harrison, W. G. and Cota, G. F. (1991). Primary production in polar waters: relation to nutrient availability. *Polar Research*, **10**, 87–104.

Hasle, G. R. (1969). An analysis of the phytoplankton of the Pacific Southern Ocean: abundance, composition and distribution during the Brategg Expedition 1947–48. *Hvalrådets Skrifter*, **52**, 1–168.

Hebbeln, D. and Wefer, G. (1991). Effects of ice coverage and ice-rafted material on sedimentation in the Fram Strait. *Nature*, **350**, 409–11.

Hoepffner, N. and Haas, L. W. (1990). Electron microscopy of nanoplankton from the North Pacific central gyre. *Journal of Phycology*, **26**, 421–439.

Honjo, S. (1990). Particle fluxes and modern sedimentation in the polar oceans. In *Polar oceanography*, (ed. W. O. Smith Jr.), pp. 687–739. Academic Press, San Diego.

Jacques, G., and Panouse, M. (1991). Biomass and composition of size fractionated phytoplankton in the Weddell-Scotia confluence area. *Polar Biology*, **11**, 315–28.

Jeffrey, S. W. and Vesk, M. (1981). The phytoplankton — systematics, morphology and ultrastructure. In *Marine botany — an Australasian perspective*, (ed. M. N. Clayton and R. J. King), pp. 138–79. Longman-Cheshire, Melbourne.

Karentz, D. (1991). Ecological considerations of Antarctic ozone depletion. *Antarctic Science*, **3**, 3–11.

Kimura, B. and Ishida, Y. (1985). Photophagotrophy in *Uroglena americana*, Chrysophyceae. *Japanese Journal of Limnology*, **46**, 315–18.

Lizotte, M. P. and Sullivan, C. W. (1992). Biochemical composition and photosynthate distribution in sea ice microalgae of McMurdo Sound, Antarctica: evidence for nutrient stress during the spring bloom. *Antarctic Science*, **4**, 23–30.

Lubbers, G. W., Gieskes, W. W. C., del Castilho, P., Salomons, W., and Bril, J. (1990). Manganese accumulation in the high pH microenvironment of *Phaeocystis* sp. (Haptophyceae) colonies from the North Sea. *Marine Ecology Progress Series*, **59**, 285–93.

Manton, I. (1978). *Chrysochromulina hirta* sp. nov., a widely distributed species with unusual spines. *British Journal of Phycology*, **13**, 3–14.

Manton, I. and Leadbeater, B. S. C. (1974). Fine structural observations on 6 species of *Chrysochromulina* from wild Danish marine nanoplankton including a description of *Chrysochromulina campanulifera* sp. nov. and a preliminary summary of the nanoplankton as a whole. *Det Kongelige Danske Videnskabernes Selskab Biologiske Skrifter*, **20**, 1–26.

Manton, I., and Oates, K. (1975). Fine-structural observations on *Papposphaera* Tangen from the southern hemisphere and on *Pappomonas* gen. nov. from South Africa and Greenland. *British Phycological Journal*, **10**, 93–109.

Manton, I. and Sutherland, J. (1975). Further observations on the genus *Pappomonas* Manton et Oates with special reference to *P. virgulosa* sp. nov. from West Greenland. *British Phycological Journal*, **10**, 377–85.

Manton, I., Sutherland, J., and McCully, M. (1976*a*). Fine structural observations on coccolithophorids from South Alaska in the genera *Papposhaera* Tangen and *Pappomonas* Manton and Oates. *British Phycological Journal*, **11**, 225–38.

Manton, I., Sutherland, J., and Oates, K. (1976*b*). Arctic coccolithophorids: two species of *Turrisphaera* gen. nov. from West Greenland, Alaska and the North West Passage. *Proceedings of the Royal Society of London,* Series B, **194**, 179–94.

Manton, I., Sutherland, J., and Oates, K. (1977). Arctic coccolithophorids: *Wigwamma arctica* gen. et ap. nov. from Greenland and arctic Canada, *W. annulifera* sp. nov. from South Africa and S. Alaska and *Calciarcus alaskensis* gen. et sp. nov. from S. Alaska. *Proceedings of the Royal Society, London,* Series B, **197**, 145–68.

Marchant, H. J. and Nash, G. V. (1986). Electron microscopy of gut contents and faeces of *Euphausia superba* Dana. *Memoirs of the National Institute for Polar Research, Special. Issue,* **40**, 167 -77.

Marchant, H. J., Davidson, A. T., and Kelly, G. (1991). UV-B protecting pigments in the alga *Phaeocystis pouchetii* from Antarctica. *Marine Biology,* **109**, 391–5.

Miller, D. G. M. and Hampton, I. 1989. Biology and ecology of the Antarctic krill (*Euphausia superba* Dana): a review. BIOMASS Scientific Series No. 9. SCAR and SCOR, Scott Polar Research Institute, Cambridge.

Nishida, S. (1986). Nannoplankton flora in the Southern Ocean, with special reference to siliceous varieties. *Memoirs of the National Institute for Polar Research, Special Issue,* **40**, 56–68.

Nöthig, E.-M. and Bodungen, B. von. (1989). Occurrence and vertical flux of faecal pellets of probably protozoan origin in the Southeastern Weddell Sea (Antarctica). *Marine Ecology Progress Series,* **56**, 281–9.

Okada, H. and Honjo, S. (1973). The distribution of oceanic coccolithophorids in the Pacific. *Deep-Sea Research,* **20**, 355–74.

Okada, H. and McIntyre, A. (1979). Seasonal distribution of modern coccolithophores in the western North Atlantic ocean. *Marine Biology,* **54**, 319–28.

Østergaard, J. B. (1993). Nanoplanktoniske coccolithophorer (Prymnesiophyceae) og nanoplanktoniske, loricabærende choanoflagellater (Acanthoecidae) i farvandet omkring Godhavn, Grønland. MS Thesis, University of Copenhagen.

Paasche, E. (1968). Biology and physiology of coccolithophorids. *Annual Review of Microbiology,* **22**, 71–86.

Pintner, I. J. and Provasoli, L. (1968). Heterotrophy in subdued light of 3 *Chrysochromulina* species. *Bulletin of Misaki Marine Biological Institute of Kyoto University,* **12**, 25–31.

Porter, K. G., Sherr, E. B., Sherr, B. F., Pace, M., and Saunders, R. W. (1985). Protozoa in planktonic food webs. *Journal of Protozoology,* **32**, 409–15.

Priscu, J. C., Priscu, L. R., Palmisano, A. C., and Sullivan C. W. (1990). Estimation of neutral lipid levels in Antarctic sea ice microalgae by Nile Red fluorescence. *Antarctic Science,* **2**, 149–55.

Sakshaug, E. and Holm-Hansen, O. (1984). Factors governing pelagic production in polar oceans. In *Marine phytoplankton and productivity,* (ed. O. Holm-Hansen, L. Bolis, and R. Gilles), pp. 1–18. Springer-Verlag, Berlin.

Sakshaug, E. and Skjoldal, H. R. (1989). Life at the ice edge. *Ambio,* **18**, 60–7.

Sakshaug, E. and Slagstad, D. (1991). Light and productivity of phytoplankton in polar marine ecosystems: a physiological view. *Polar Research*, **10**, 69–85.

Sakshaug, E., Slagstad, D., and Holm-Hansen, O. (1991). Factors controlling the development of phytoplankton blooms in the Antarctic Ocean — a mathematical model. *Marine Chemistry*, **35**, 259–71.

Salonen K. and Jokinen, S. (1988). Flagellate grazing on bacteria in a small dystrophic lake. *Hydrobiologia*, **161**, 203–9.

Sanders, R. W. (1991) Mixotrophic protists in marine and freshwater ecosystems. *Journal of Protozoology*, **38**, 76–81.

Sieburth, J. McN. (1960) Acrylic acid, an 'antibiotic' principal in *Phaeocystis* blooms in Antarctic waters. *Science* **132**, 676–7.

Sikes, E. L. and Volkman, J. K. (1993). Calibration of alkone unsaturation ratios ($U_{37}^{k'}$) for paleotemperature estimation in cold polar waters. *Geochimica et Cosmochimica Acta*, **57**, 1883–9.

Smetacek, V., Scharek, R., and Nöthig, E.-M. (1990). Seasonal and regional variation in the pelagial and its relationship to the life history cycle of krill. In *Antarctic ecosystems. Ecological change and conservation*, (ed. K. R. Kerry and G. Hempel), pp. 103–14. Springer-Verlag, Berlin.

Smith, R. C., Prezelin, B. B., Baker, K. S., Bidigare, R. R., Boucher, N. P., Coley, T., *et al.* (1992). Ozone depletion: ultraviolet radiation and phytoplankton biology in Antarctic waters, *Science*, **255**, 952–9.

Smith, W. O. Jr. and Nelson, D. M. (1986). Importance of ice edge phytoplankton production in the Southern Ocean. *BioScience*, **36**, 251–7.

Smith, W. O. Jr. and Sakshaug, E. (1990). Polar phytoplankton. In *Polar oceanography*, (ed. W. O. Smith Jr.), pp. 477–525. Academic Press, San Diego.

Smith, W. O. Jr., Keene, N. K., and Comiso, J. C. (1988). Interannual variability in estimated primary productivity of the Antarctic marginal ice zone. In *Antarctic Ocean and resources variability*, (ed. D. Sahrhage), pp. 131–39. Springer-Verlag, Berlin.

Smith, W. O. Jr., Codispoti, L. A., Nelson, D. M., Manley, T., Buskey, E. J., Niebauer, H. J., *et al.* (1991). Importance of *Phaeocystis* blooms in the high-latitude ocean carbon cycle. *Nature*, **352**, 514–16.

Stefánsson, U. (1990). Anomalous silicate-nitrate relationships associated with *Phaeocystis pouchetii* blooms. *Eos*, **71**, 77.

Thomsen, H. A. (1979). Electron microscopical observations on brackish-water nannoplankton from the Tvärminne area, SW coast of Finland. *Acta Botanica Fennica*, **110**, 11–37.

Thomsen, H. A. (1980*a*). *Turrisphaera polybotrys* sp. nov. (Prymnesiophyceae) from West Greenland. *Journal of the Marine Biological Association of the United Kingdom*, **60**, 529–37.

Thomsen, H. A. (1980*b*). Two species of *Trigonaspis* gen. nov. (Prymnesiophyceae) from West Greenland. *Phycologia*, **19**, 218–29.

Thomsen, H. A. (1980c). *Wigwamma scenozonion* sp. nov. (Prymnesiophyceae) from West Greenland. *British Journal of Phycology*, **15**, 335–42.

Thomsen, H. A., (1981). *Quaternariella obscura* gen. et sp. nov. (Prymnesiophyceae) from West Greenland. *Phycologia*, **19**, 260–5.

Thomsen, H. A. (1982). Planktonic choanoflagellates from Disko Bugt, West Greenland, with a survey of the marine nanoplankton of the area. *Meddelelser om Grønland, Bioscience*, **8**, 1–35.

Thomsen, H. A. (1986). A survey of the smallest eukaryotic organisms of the marine phytoplankton. In *Photosynthetic picoplankton*, (ed. T. Platt and W. K. Li). *Canadian Bulletin of Fisheries and Aquatic Sciences*, **214**, 121–58.

Thomsen, H. A. and Oates, K. (1978). *Balaniger balticus* gen. et sp. nov. (Prymnesiophyceae) from Danish coastal waters. *Journal of the Marine Biological Association of the United Kingdom*, **58**, 773–9.

Thomsen, H. A., Buck, K. R., Coale, S. L., Garrison, D. L., and Growing, M. M. (1988). Nanoplanktonic coccolithophorids (Prymnesiophyceae Haptophyceae) from the Weddell Sea, Antarctica. *Nordic Journal of Botany*, **8**, 419–36.

Thomsen, H. A., Østergaard, J. B., and Hansen, L. E. (1991). Heteromorphic life histories in Arctic coccolithophorids (Prymnesiophyceae). *Journal of Phycology*, **27**, 634–42.

Throndsen, J. and Kristiansen, S. (1991). *Micromonas pusilla* (Prasinophyceae) as part of the pico- and nanoplankton communities of the Barents Sea. *Polar Research*, **10**, 201–7.

Verity, P. G. and Smayda, T. J. (1989). Nutritional value of *Phaeocystis pouchetii* (Prymnesiophyceae) and other phytoplankton for *Acartia* spp. (Copepoda): ingestion, egg production and growth of nauplii. *Marine Biology*, **100**, 161–71.

Verity, P. G., Smayda, T. J., and Sakshaug, E. (1991). Photosynthesis, excretion, and growth rates of *Phaeocystis* colonies and solitary cells. *Polar Research*, **10**, 117–28.

Vernet, M. (1991). Phytoplankton dynamics in the Barents Sea estimated from chlorophyll budget models. *Polar Research*, **10**, 129–45.

Virtue, P., Nichols, P. D., McMinn, A., and Sikes, E. L. (1993). The lipid composition of *Euphausia superba* Dana in relation to the nutritional value of *Phaeocystis pouchetii* (Hariot) Lagerheim. *Antarctic Science*, **5**, 169–77.

Wassmann, P., Vernet, M., Mitchell, B. G., and Rey, F. (1990). Mass sedimentation of *Phaeocystis pouchetii* in the Barents Sea. *Marine Ecology Progress Series*, **66**, 183–95.

Weber, L. H. and El-Sayed, S. Z. (1987). Contributions of the net, nano- and picoplankton to the phytoplankton standing crop and primary productivity in the Southern Ocean. *Journal of Plankton Research*, **9**, 973–94.

Wright, S. W. and Jeffrey, S. W. (1987). Fucoxanthin pigment markers of marine phytoplankton analysed by HPLC and HPTLC. *Marine Ecology Progress Series*, **38**, 259–66.

12. Ecology of *Phaeocystis*: the key role of colony forms

CHRISTIANE LANCELOT and VÉRONIQUE ROUSSEAU
Université Libre de Bruxelles,
Groupe de Microbiologie des Milieux Aquatiques, Belgium

Abstract

Species of *Phaeocystis* exhibit phase alternation between individual cells and gelatinous colonies. They regularly form dense, nearly specific blooms, in very contrasting nutrient-rich areas of the world's oceans. The uniqueness of this genus of marine phytoplankters rests not only in its ubiquity but mostly in its peculiar physiology and ecology. No other marine phytoplankter has ever been shown to dominate an entire ecosystem; no other marine species distinguishes itself by a complex polymorphic life cycle that induces dramatic changes in the structure and functioning of planktonic and benthic food-webs as well as in the biogeochemistry of trace elements. The main features of the ecology of *Phaeocystis*-dominated ecosystems are analysed with regard to the *Phaeocystis* life cycle, and to recent data on the biochemistry and nutrient (major and trace element) metabolism of the different morphological forms that succeed each other during *Phaeocystis* bloom development, in relationship to the behaviour of bacteria and micro-, meso-, and meta-zooplankton and the physical structure of the marine habitat. Particular emphasis is given to the biological functioning of *Phaeocystis* colonies that constitute by far the most important morphological forms in natural environments, as determined from the analysis of the structure and function of the mucilaginous matrix embedding the cells. Evidence is presented that the most remarkable ecological and biogeochemical properties of *Phaeocystis*-dominated ecosystems are attributable to the capacity of *Phaeocystis* colonial cells to synthesize, in nutrient-deprived conditions, exopolysaccharides capable of gelation.

The Haptophyte Algae (ed. J. C. Green and B. S. C. Leadbeater), Systematics Association Special Volume No. 51, pp. 229–45. Clarendon Press, Oxford, 1994.

Phaeocystis: a very widespread genus but still poorly understood

Phaeocystis is one of the most widespread marine genera and also one of the most intriguing. *Phaeocystis* species are among the few phytoplankters exhibiting phase alternation between free-living solitary cells (3–9 μm in diameter) and gelatinous colonies (palmelloid stage, reaching several millimetres in diameter) (Kornmann 1955; Verity *et al.* 1988, 1992; Rousseau *et al.* 1994). Species of *Phaeocystis* are euryhaline and eurythermal, and, in the spring, they regularly forms dense, near monospecific blooms in very contrasting nutrient-rich areas of the world ocean (Davidson and Marchant 1992). In most of these areas, the colonial forms largely dominate and are sustained by new sources of nutrient which are either of natural (Buck and Garrison 1983; Smith *et al.* 1991) or anthropogenic origin (Al Hassan 1990; Cadée and Hegeman 1990, 1991; Lancelot *et al.* 1992).

The uniqueness of *Phaeocystis* lies not only in its massive blooms, but also in its exceptional physiology and ecology. No other marine phytoplankters have been ever shown to dominate an entire ecosystem, or to distinguish themselves by a complex polymorphic life cycle (Kornmann 1955; Rousseau *et al.* 1994) that induces dramatic changes in the structure and functioning of the planktonic (e.g. Lancelot *et al.* 1987; Davidson and Marchant 1992; Weisse *et al.* 1994) and benthic food-webs (e.g. Pieters *et al.* 1980; Rogers and Lockwood 1990), as well as in the biogeochemistry of trace elements (Davidson and Marchant 1987; Lubbers *et al.* 1990) and sulfur (Liss *et al.* 1994).

The biology of *Phaeocystis* and the ecology of *Phaeocystis*-dominated systems have recently been reviewed by Davidson and Marchant (1992). Other recent publications have dealt with more specific aspects of *Phaeocystis* ecology such as species diversity and associated biochemistry (Baumann *et al.* 1994); the life cycle (Rousseau *et al.* 1994), the fate of *Phaeocystis* colonies (Weisse *et al.* 1994; Thingstad and Billen 1994; Wassmann 1994), and DMS production (Liss *et al.* 1994). All these authors concluded that the main features of *Phaeocystis*-dominated ecosystems are driven by the physiology, biochemistry, and peculiar life cycle of this marine phytoplankter. Surprisingly, our basic knowledge in this field is still limited and fragmentary due to uncertainties surrounding species identity, and the incomplete morphological and biochemical descriptions of *Phaeocystis* life forms found during bloom development.

This paper constitutes an attempt to establish the cause and effect relationship between the peculiar physiology of *Phaeocystis*, and the structure and functioning of *Phaeocystis*-dominated ecosystems. This will be done on the basis of recent investigations on the *Phaeocystis* life cycle, and on the carbon and nutrient (major and trace element) metabolism of the different successional morphotypes occurring during a *Phaeocystis* bloom.

Particular emphasis is given to the biological functioning of *Phaeocystis* colonies since they constitute by far the most important morphological form in natural environments. In order to avoid confusion due to inter-population variability, and so leave aside the unresolved species diversity problem, all the reported data refer to microscopic descriptions, chemical analyses, and process-oriented studies carried out on *Phaeocystis* populations originating from one area, the southern North Sea. On the basis of this analysis, current knowledge on *Phaeocystis* in the world ocean is briefly re-appraised and some general features of the factors controlling the structure and functioning of *Phaeocystis*-dominated ecosystems are put forward.

The sequence of *Phaeocystis* life forms in the southern North Sea and its ecological implications

The complex sequence of morphological forms exhibited in a pure culture of *Phaeocystis* colonies (German Bight strain) during their growth (Kornmann 1955) has been shown to correspond to those during a *Phaeocystis* bloom development in the southern North Sea (Rousseau *et al.* 1994; Fig. 12.1). Under natural conditions, however, the successive phases of a *Phaeocystis* bloom are accompanied by the development of a large variety of heterotrophic organisms feeding selectively on some *Phaeocystis* morphotypes, so producing complex and dynamic food webs (Fig. 12.2). The main features of the *Phaeocystis* bloom in the southern North Sea can be summarized as follows.

Phaeocystis succeeds an early spring diatom bloom and dominates the phytoplankton community at more than 90% of cell number, nearly wholly of the colonial form (Lancelot and Mathot 1987). Colonies originate from the transformation of free-living cells and multiply by budding or division (Fig. 12.1; Kornmann 1955). A low density of free-living cells is always present, being controlled by colonial lysis and grazing by microzooplankton (Fig. 12.2; Martens 1981; Admiraal and Venekamp 1986; Weisse and Scheffel-Möser 1990).

Colony forms exhibit a marked temporal evolution, from small (20–50 μm in diameter) spherical colonies, often localized on *Chaetoceros* setae, to large (mm) healthy colonies of various sphere-derived forms devoid of attached heterotrophic microorganisms during the exponential phase of the bloom development (Fig. 12.1), to senescent irregular colonies (Fig. 12.1) progressively invaded by protozoa actively grazing on colonial cells (Fig. 12.3) and, finally, at the end of the bloom, to the formation of sticky aggregates (Fig. 12.1) colonized by various heterotrophic organisms developing complex microbial networks (Fig. 12.3). This progressive transformation of homogeneous biological entities to heterogeneous

Some events of *Phaeocystis* sp. life cycle in natural environments

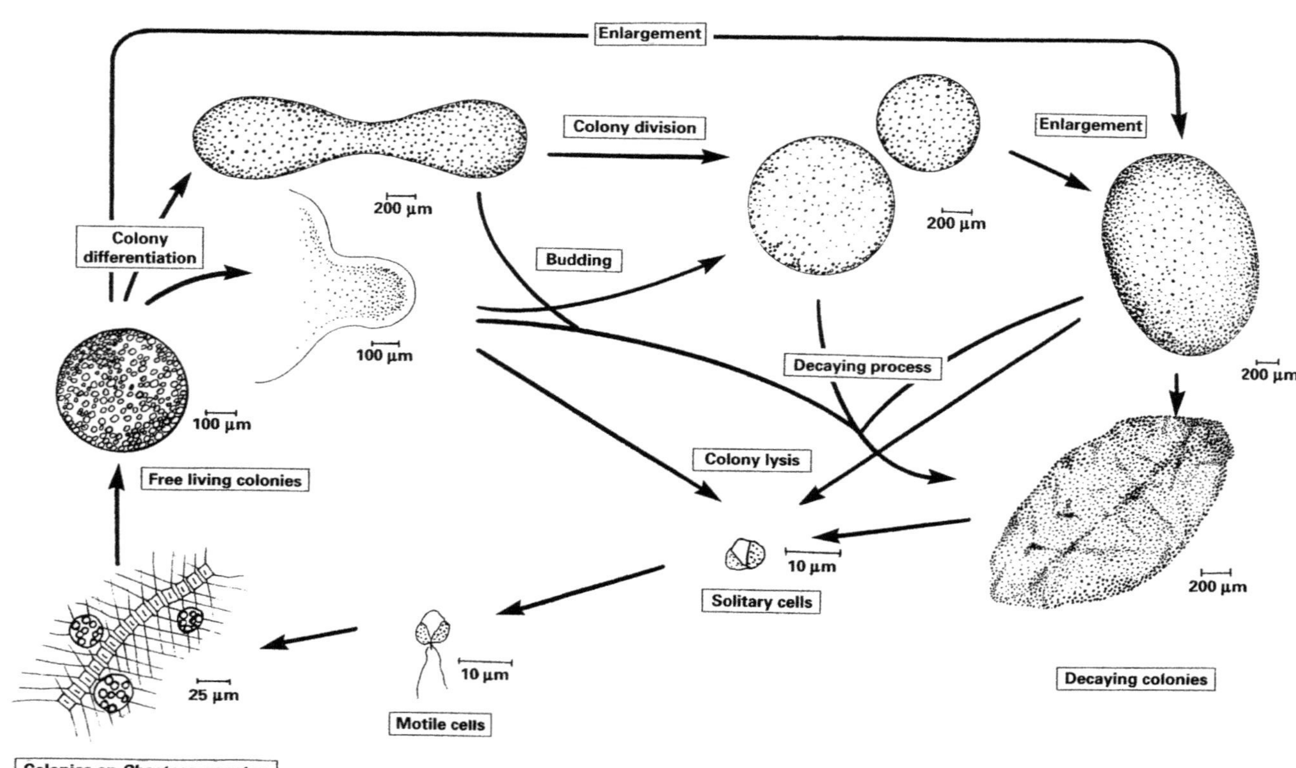

Fig. 12.1 The sequence of *Phaeocystis* morphotypes during spring bloom development in Belgian coastal waters (southern North Sea). Redrawn from Rousseou *et al.* (1994).

microbial aggregates appears to be driven by the maturation of the colony itself. This is made possible by the unpalatibility of healthy colonies for co-occurring mesozooplankton (Hansen and Boekel 1992; Fransz *et al.* 1992). Knowledge of environmental factors controlling these transformations is very scarce.

In the southern North Sea, as in most *Phaeocystis*-dominated environments, the termination of the bloom is characterized by the sudden complete disappearance of senescent colonies and their derived aggregates due to either accelerated sedimentation (possibly resulting from increasing density due to colonization), microbial disintegration in the water column, consumption by mesozooplankton, or advective export. Little is known at present about the relative importance of these mechanisms which depend on the physical characteristics of the marine habitat, the food quality and density of the aggregates, the feeding behaviour of mesozooplankton, and the biodegradability of the organic matter from decaying colonies. Thus, the biochemical composition of the primary colonies may be important. Evidence for low biodegradability of the *Phaeocystis*-derived polymeric material is given by the large accumulation of sea foam observed at bloom decline in the open sea (Rogers and Lockwood 1990), and on the beaches (Bätje and Michaelis 1986; Lancelot *et al.* 1987) of the shallow turbulent southern North Sea.

This simple visual description of the *Phaeocystis* event in the southern North Sea highlights the key role of *Phaeocystis* colonies in determining ecosystem structure and functioning. The appraisal of its ecological function is now approached through the determination of the biological functioning of *Phaeocystis* colonies.

The biological functioning of *Phaeocystis* colonies

A *Phaeocystis* colony originates from the transformation of one free-living cell (Kornmann 1955; Rousseau *et al.* 1994). Once formed, each colony constitutes an entity inside which the non-motile cells grow and divide (Lancelot and Mathot 1985). How far this aggregation makes *Phaeocystis* colonies particularly well adapted to growth in nutrient-rich conditions and to outcompete other phytoplankters is examined on the basis of the biological functioning of *Phaeocystis* colonies (Fig. 12.4). This in turn is determined from an analysis of the structure and function of the mucilaginous matrix.

Planktonic food-web of *Phaeocystis*-dominated ecosystem

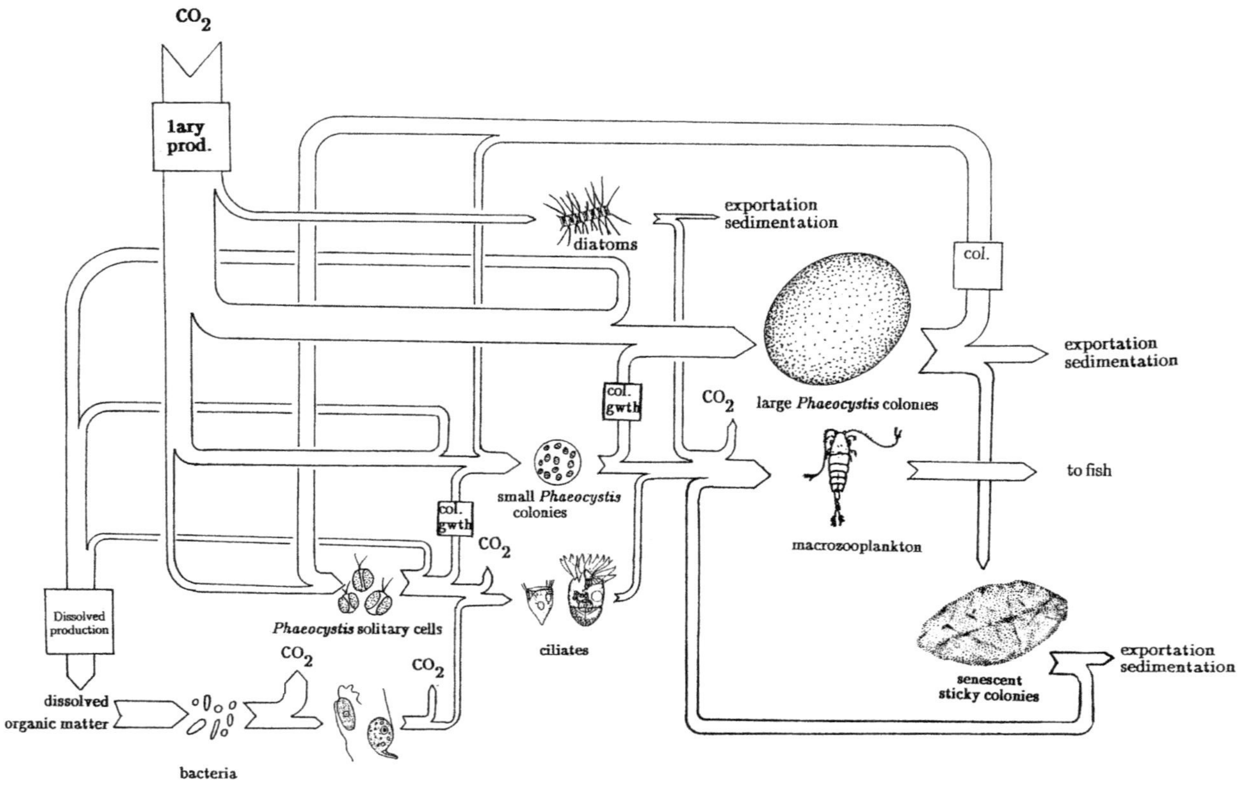

CO_2

1ary prod.

diatoms

exportation sedimentation

col.

large *Phaeocystis* colonies

exportation sedimentation

col. gwth

CO_2

small *Phaeocystis* colonies

to fish

col. gwth

CO_2

macrozooplankton

ciliates

Dissolved production

Phaeocystis solitary cells

CO_2

CO_2

senescent sticky colonies

exportation sedimentation

dissolved organic matter

bacteria

Fig. 12.2 Schematic representation of the structure of the planktonic food-web of the *Phaeocystis*-dominated ecosystem of the southern North Sea.

The mucilaginous matrix: chemical characterization and biosynthesis

Various microscopic and chemical methods have been used to determine the biochemical composition of *Phaeocystis* cells and colony matrix. Current knowledge, although preliminary, indicates that the mucilaginous matrix is formed through gelation of carboxylated and sulphated polysaccharide chains promoted by salt (calcium and magnesium) bridges (Boekel 1992). These polysaccharides are actively secreted by the colony cells, under the control of light and inorganic nutrients (Lancelot and Billen 1985; Lancelot *et al.* 1986). The exopolymeric synthesis is, however, not triggered by nutrient depletion (Lancelot 1983). At the height of the bloom, when nutrients are depleted, more than 80% of the photo-assimilated carbon is devoted to the synthesis of exopolymeric substances, compared with about 50% when nutrients are not limiting. Thus, the contribution of the mucilaginous matrix to the *Phaeocystis* colony biomass increases dramatically from about 50% to 90% during bloom development (Rousseau *et al.* 1990). Also, the seawater content of the gel increases with the size of the colony, suggesting that the gel compactness decreases with colony growth (Fig. 12.1). To what extent this apparent modification of gel consistency is accompanied by changes in gel properties (e.g. gel strength, swelling ability) has not been determined, but might be a clue to understanding the progressive transformation of healthy colonies to aggregates as observed under natural conditions (Fig. 12.1). Determination of gel firmness requires, however, a complete analysis of the composition and structure of the polysaccharide chains, i.e. the type and sequence of sugars in the polymeric units that make up the mucilaginous matrix. Basic knowledge in this field is, however, still limited and would benefit from further investigations.

The physiological function of the mucilaginous matrix.

The energy storage function of the mucilaginous matrix is now well agreed. Numerous process studies have demonstrated that the polysaccharides composing the matrix constitute an energy-storing substrate, which is catabolized by the colony cells during the light-limited period to meet their biosynthetic requirements (Fig. 12.4; Lancelot and Mathot 1985; Veldhuis and Admiraal 1985; Lancelot *et al.* 1986; Veldhuis *et al.* 1991). This reservoir thus gives to colonial cells a selective advantage over free-living cells to benefit from high nutrient concentrations in low-light environments by increasing the energy storage capacity of each cell. The

Food-web in *Phaeocystis*-derived aggregates

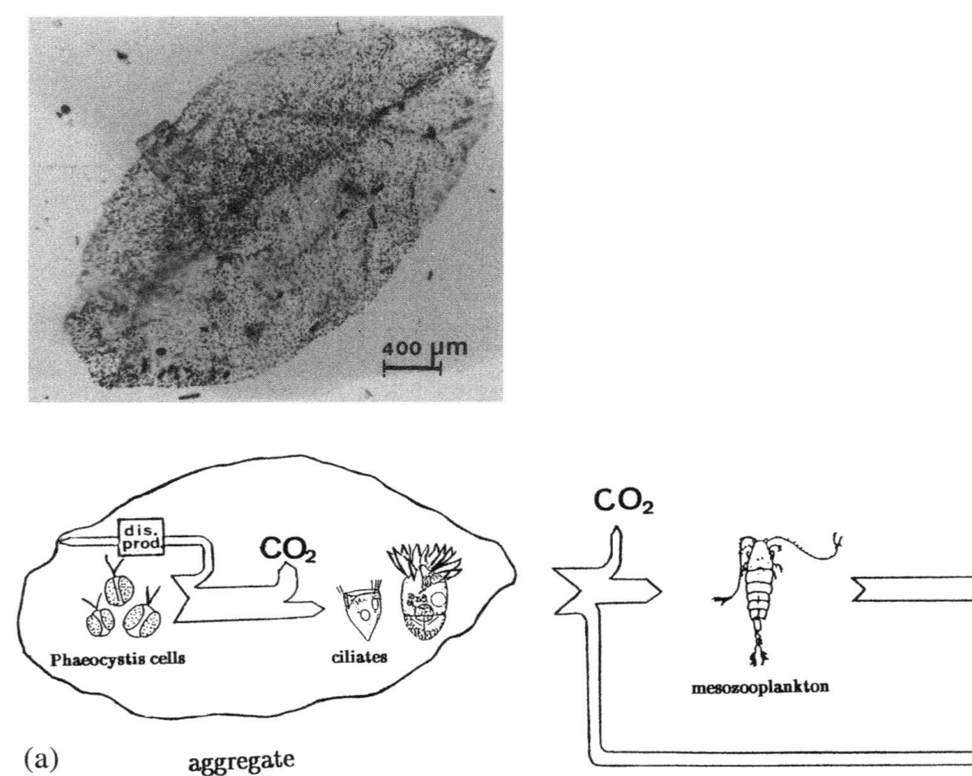

Food-webs in *Phaeocystis*-derived aggregates

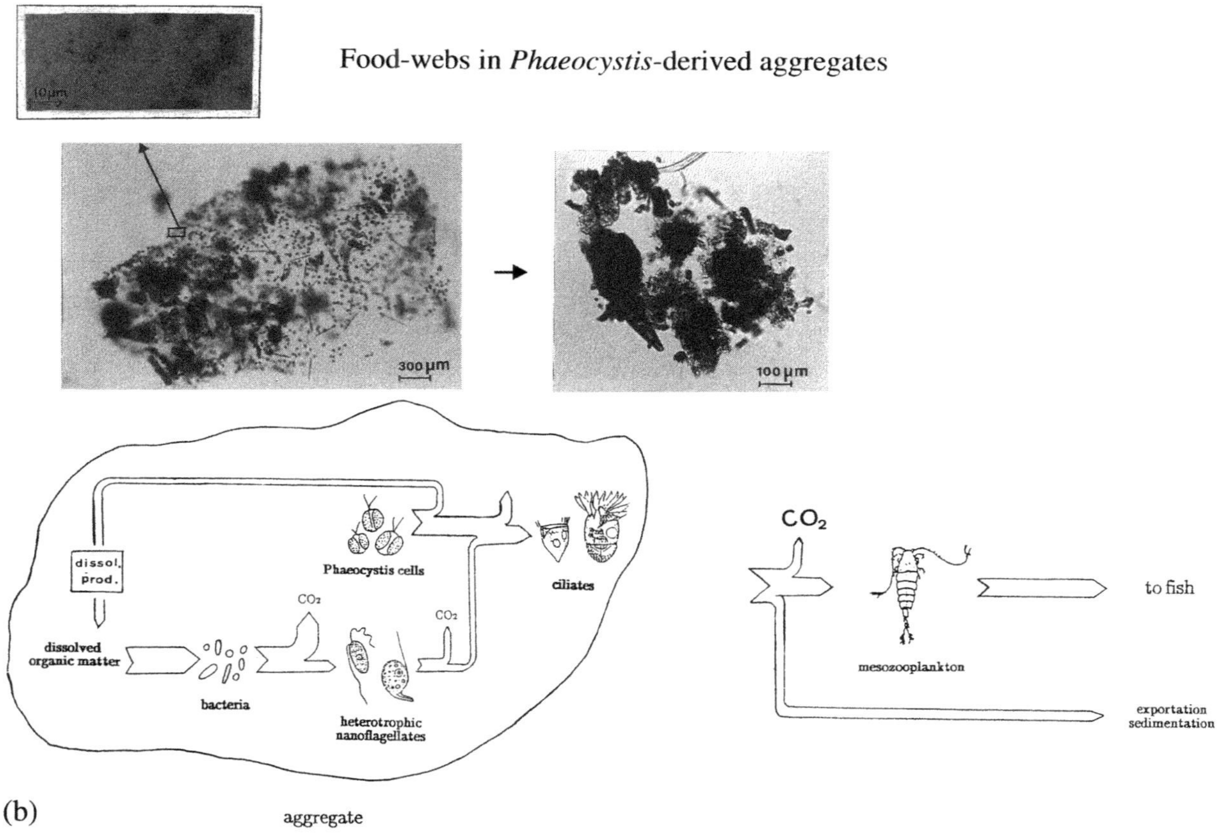

(b)

Fig. 12.3 Schematic representation of the structure of the microbial food-web in aggregates derived from *Phaeocystis* colonies in the southern North Sea, illustrating the progressive invasion of a senescent colony by various heterotrophic micro-organisms. (**a**) Invasion of a decaying *Phaeocystis* colony by protozoa (mostly ciliates) grazing on colony cells, and (**b**) subsequent development of complex and changing microbial food-webs involving *Phaeocystis* cells, bacteria, and bacterivorous and herbivorous protozoa. Photographs by V. Rousseau and S. Becquevort.

energetic gain depends on the colony size. Over an order of magnitude increase to a colony of 1 mm^3 (the average colony size at the height of a *Phaeocystis* bloom, Rousseau *et al.* 1990), the pool of energetic substrates available to the colony has been estimated to increase by a factor of 20.

On the other hand, experimental evidence suggests that the catabolism of colonial polysaccharides greatly modifies the chemical structure of the gel by providing intermediates of lower molecular weight inside the colonial matrix (Fig. 12.4; Veldhuis and Admiraal 1985). These chemical changes modify gel firmness and could indirectly be the cause of the progressive transformation of healthy homogeneous colonies to microbe-invaded decaying colonies. Indeed this transition mostly occurs at the height of the bloom when light is possibly limited due to the high *Phaeocystis* biomass.

The nutrient and trace element sequestering function of the mucilaginous matrix

Besides its structural role and energy storage function, the mucilaginous matrix of *Phaeocystis* colonies has been shown recently to act as a reservoir for phosphorus (Veldhuis *et al.* 1991) and trace elements, especially manganese (Davidson and Marchant 1987; Lubbers *et al.* 1990) and probably iron. These sequestration mechanisms result from a suite of abiotic chemical reactions, resulting from both the gel properties and the biological activity of colonial cells (Fig. 12.4; Lubbers *et al.* 1990). The colonial matrix constitutes a 3-dimensional network, embedding cells and seawater, and acts as diffusion barrier for solute molecules (Lubbers *et al.* 1990). Consequently, physico-chemical conditions (nutrients, pH, Eh) inside the colonies can be significantly different from those of the external medium due to the biological activity of colonial cells. For example, Lubbers *et al.* (1990) demonstrated that high pH conditions, corresponding to that of Mn/Fe oxyhydroxide precipitation, can be reached inside *Phaeocystis* colonies in culture when exposed to optimal light conditions. Manganese precipitation is strongly light-dependent and is slightly reversible under prolonged dark periods, making dissolved manganese available for colonial cells (Fig. 12.4; Lubbers *et al.* 1990). Deposits of Mn/Fe oxyhydroxides inside the mucilaginous matrix in turn drive the

(a) light metabolism and chemical reactions

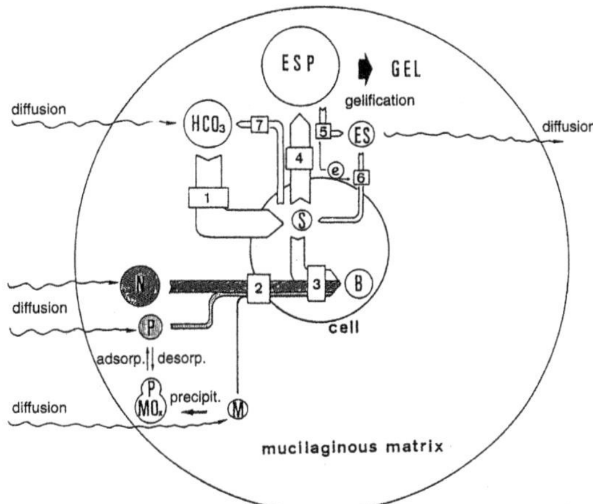

Phaeocystis colony

(b) dark metabolism and chemical reactions

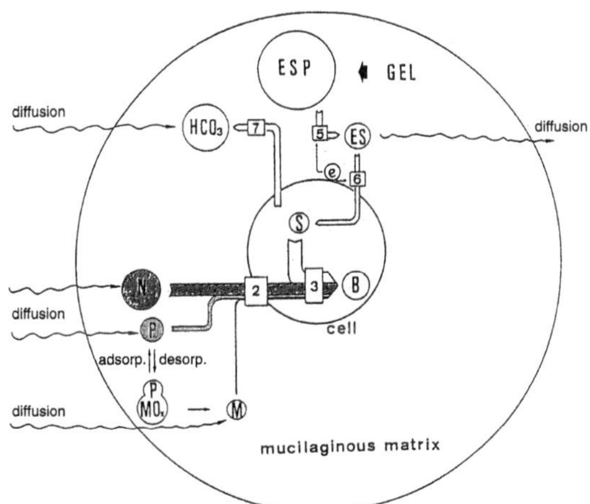

Phaeocystis colony

Fig. 12.4 Schematic representation of: (a) the light and (b) the dark metabolism of *Phaeocystis* colonies. *Intracellular and intracolonial pools*: B = cellular biomass; e = exoglucosidase; ES = extracellular oligosaccharide; ESP = extracellular polysaccharide; M = trace metals (mainly Mn and Fe); MO_x = trace metal oxy-hydroxide; N = inorganic nitrogen (nitrate and ammonium); P = phosphate; S = monomeric precursor. *Processes*: 1 = photosynthesis; 2 = nutrient uptake; 3 = cellular growth; 4 = polysaccharide secretion; 5 = polysaccharide hydrolysis; 6 = polysaccharide catabolism.

sequestering of other metals and of phosphate by adsorption from the enclosed seawater. This reaction is reversible, making phosphate available to colonial cells when phosphate in the external medium is depleted (Fig. 12.4; Veldhuis *et al.* 1991). Rough calculations, based on the potential amount of manganese precipitate inside a 1 mm^3 *Phaeocystis* colony (Lubbers *et al.* 1990), suggest that this nutrient storage mechanism can meet trace metal and phosphorus requirements of individual colonial cells. This makes colonial cells more competitive than free-living cells when ambient concentrations of these nutrients reach limiting values. Interestingly, nitrate and ammonium are not adsorbed on Mn/Fe oxy-hydroxides, and no mechanism of nitrogen sequestration in *Phaeocystis* colonies has yet been demonstrated.

The biological functioning of *Phaeocystis* colonies: its implication for the structure and functioning of *Phaeocystis*-dominated ecosystems

The near-complete dominance of colony forms in *Phaeocystis*-dominated ecosystems all over the world ocean indicates that similar mechanisms have developed both to decrease losses through grazing, sinking, and degradation, and to outcompete free-living *Phaeocystis* cells as well as other phytoplankters. Thus, the biological functioning of *Phaeocystis* colonies outlined above from southern North Sea data (Fig. 12.4) might be extended to the genus worldwide. Each *Phaeocystis* colony can be regarded as a biofilm in which several related and mutually-dependent biological and chemical processes are occurring for the benefit of the aggregated biological entity. Biogeographical and ecological features of *Phaeocystis*-dominated systems are here reappraised on this basis.

Our analysis suggests that the most remarkable physiological and ecological properties of *Phaeocystis* colonies are attributed to their capacity to synthesize gel-forming nutrient-deprived exopolysaccharides. Firstly, this mechanism leads either directly or indirectly to the building of significant supplementary intracolonial reserves of energetic substrates, phosphate, and trace elements for the benefit of colonial cells. The intracolonial energy reservoir might well explain the competitive edge *Phaeocystis* colonies have over free-living cells and other phytoplankters when energy-costly nitrates constitute the nitrogen source (Riegman *et al.* 1992). It may also explain the dominance of colonial forms in nitrate-enriched environments. To what extent the phosphate and micro-nutrient sequestrating mechanisms allow *Phaeocystis* to outcompete other algae by depriving them of essential elements is difficult to assess properly at present and will depend on the ambient chemistry of the marine system. Presumably, the role of trace metal sequestration by *Phaeocystis* colonies as mediators of species succession would be of less significance in the polluted coastal waters of the southern North Sea than in the Southern Ocean, which is typically characterized by low trace metal availability (Nolting *et al.* 1991; Westerlund and Ohman 1991). Much has yet to be known, however, about the regulation of Mn/Fe precipitation and dissolution inside the colony and the subsequent phosphate adsorption/desorption, and about the diffusion properties of the gelatinous matrix, but there is no doubt that *Phaeocystis* blooms, due to their importance and the dominance of colonial forms, play an important role in biogeochemical cycles. The fate of sequestrated trace elements is then strongly linked to the fate of *Phaeocystis* colonies.

Secondly, the nutrient-deprived mucous secretion, by rapidly increasing *Phaeocystis* colony size while lowering significantly its nutritional value, can be seen as causing the general unpalatibility of *Phaeocystis* colonies to most grazers (Weisse *et al.* 1994). Related to this, polysaccharides produced through colony lysis have been shown to be refractory to bacterial degradation (Thingstad and Billen 1994), causing accumulation of dissolved organic matter in the water column (Billen and Fontigny 1987).

Thirdly, the gel characteristics of the mucilaginous matrix and its rapid turnover rate considerably reduce the average density of the whole colony bringing it closer to that of surrounding seawater, with attendant benefits for suspension. The firmness of the gel is determined by the sugar residue composition of the exopolysaccharide and the Ca/Mg content of seawater (both possibly varying between *Phaeocystis*-dominated ecosystems). Simple calculation shows that this method of determining density would prevent *Phaeocystis* colonies from sinking in stratified seas where stratification is due to salinity differences. Such areas are typically coastal and polar seas, where massive blooms of *Phaeocystis* colonies have been observed (Davidson and Marchant 1992), stratification being due to river outflow and ice melting, respectively. On the other hand, gelation offers no particular advantage in the fully mixed tidal areas like the southern North Sea. Counteracting this mechanism, Mn/Fe oxyhydroxide deposits inside the colony modify the sinking characteristics of *Phaeocystis* colonies by significantly increasing colonial density. The present-day failure to appreciate properly how these opposite mechanisms regulate the density of *Phaeocystis* colonies might well explain the general confusion surrounding an appreciation of the sedimentation of *Phaeocystis* colonies in natural environments of contrasting hydrodynamics (Riebesell 1993; Wassmann 1994).

The sudden termination of *Phaeocystis* blooms, commonly characterized by the formation of senescent colonies and aggregates (Fig. 12.1) followed by their massive disappearance through either specific grazing on aggregates (Estep *et al.* 1990), sedimentation (Wassmann *et al.* 1990; Wassmann 1994), or dissolution in the water column (Billen and Fontigny 1987), while perceived as a perturbance, can also be seen as a consequence of

Phaeocystis physiology. Alteration of the chemical composition and structure of the mucilaginous matrix, due to *Phaeocystis* dark catabolism, constitutes one possible autogeneous mechanism triggering the decay of mature colonies and their subsequent colonization by various microorganisms through the modification of their attachment properties. The successive colonization of decaying colonies by attached auto- and heterotrophic microbial communities creates microenvironments based on regenerated production. This production, through nutrient regeneration, possibly enhances the bacterial degradation of the nitrogen-deficient polymers of mucus. Aggregate formation and transformation can thus considerably modify the density and food quality of the primary *Phaeocystis* colonies.

The subsequent fate of *Phaeocystis*-derived aggregates greatly varies between shallow and deep-sea environments, according to the physical structure of the marine habitat and the zooplankton present (Wassmann 1994). Trophodynamically, the formation of *Phaeocystis*-derived aggregates can be seen as a subtle mechanism to induce low grazing pressure on healthy colonies by copepods, and would constitute one explanation of the often high secondary biological production associated with *Phaeocystis*-dominated ecosystems. Mesozooplankton grazing on *Phaeocystis*-derived aggregates has, however, no impact on bloom regulation. Only the selective grazing by protozoa on *Phaeocystis* cells could be of significance for bloom regulation (Hansen and Boekel 1991). In this way, colony division and budding, as commonly observed during *Phaeocystis* bloom development (Verity *et al.* 1988; Rousseau *et al.* 1994; Fig. 12.2), might be seen as a subtle way of delaying aggregate formation and so escaping grazing by mesozooplankton. In addition, colony division, by providing smaller-sized healthy colonies, probably also reduces *Phaeocystis* colony losses by sedimentation. Thus, it is not only the ability to form gelatinous colonies of high biological competitivity that contributes to the success of *Phaeocystis* as a bloom-forming genus, but also the occurrence in their life history of various colony division events (Fig. 12.1) which are probably driven by physical conditions.

Acknowledgments

This work is a contribution to the EC research project on 'Modelling *Phaeocystis* blooms, their causes and consequences' (contract STEP-CT90-0062), and to the Belgian Impulse Programme 'Marine Sciences' (contract MS/11/070). We are greatly indebted to participants of the joint EC research project, especially, S. Becquevort, G. Billen, W. van Boekel, W. Gieskes, F. Hansen, S. Mathot, R. Riegman, M. Veldhuis, and T. Weisse, whose scientific contributions provided material for most ideas developed in this synthesis. We are grateful to C. Veth and P. Wassmann for their helpful discussions on *Phaeocystis* colony sedimentation. Finally, we thank A. Laffut for drawing the picture of the *Phaeocystis* life cycle.

References

Admiraal, W. and Venekamp, L. A. H. (1986). Significance of tintinnid grazing during blooms of *Phaeocystis pouchetii* (Haptophyceae) in Dutch coastal waters. *Netherlands Journal of Sea Research*, **20**, 61–6.

Al-Hasan, R. H., Ali, A. M., and Radwan, S. S. (1990). Lipids and their constituent fatty acids of *Phaeocystis* sp. from the Arabian Gulf. *Marine Biology*, **105**, 9–14.

Bätje, M. and Michaelis, H. (1986). *Phaeocystis pouchetii* blooms in the East Frisian coastal waters (German Bight, North Sea). *Marine Biology*, **93**, 21–7.

Baumann, M. E., Lancelot, C., Brandini, F. P., Sakshaug, E., and John, D. M. (1994). The taxonomic identity of the cosmopolitan prymnesiophyte *Phaeocystis*: a morphological and ecophysiological approach. *Journal of Marine Systems*, (In press).

Billen, G. and Fontigny, A. (1987). Dynamics of a *Phaeocystis*-dominated spring bloom in Belgian coastal waters. II. Bacterioplankton dynamics. *Marine Ecology Progress Series*, **37**, 249–57.

Boekel, W. H. M. van (1992). *Phaeocystis* colony mucus components and the importance of calcium ions for colony stability. *Marine Ecology Progress Series*, **87**, 301–5.

Buck, K. R. and Garrison, D. L. (1983). Protists from the ice-edge region in the Weddell Sea. *Deep Sea Research*, **30**, 1261–77.

Cadée, G. C. and Hegeman, J. (1990). Historical phytoplankton data of the Marsdiep. *Hydrobiological Bulletin*, **24**, 111–88.

Cadée G. C. and Hegeman, J. (1991). Phytoplankton primary production, chlorophyll and species composition, organic carbon and turbidity in the Marsdiep in 1990, compared with foregoing years. *Hydrobiological Bulletin*, **25**, 29–35.

Davidson, A. T. and Marchant, H. J. (1987). Binding of manganese by the mucilage of Antarctic *Phaeocystis pouchetii* and the role of bacteria in its release. *Marine Biology*, **95**, 481–7.

Davidson, A. T. and Marchant, H. J. (1992). The biology and ecology of *Phaeocystis* (Prymnesiophyceae). *Progress in Phycological Research*, **8**, 1–45.

Estep, K., Nejstgaard, J. C., Skjøldal, H. R., and Rey, F. (1990). Predation by copepods upon natural populations of *Phaeocystis pouchetii* as a function of the physiological state of the prey. *Marine Ecology Progress Series*, **67**, 235–49.

Fransz, H. G., Gonzalez, S. R., Cadée, G. C., and Hansen, F. H. (1992). Long-term change of *Temora longicornis* (Copepoda, Calanoida) abundance in a Dutch tidal inlet (Marsdiep) in relation to eutrophication. *Netherlands Journal of Sea Research*, **30**, 23–32.

Hansen, F. C. and Boekel, W. H. M. van (1991). Grazing pressure of the calanoid copepod *Temora longicornis* on a *Phaeocystis*-dominated spring bloom in a Dutch tidal inlet. *Marine Ecology Progress Series*, **78**, 123–9.

Kornmann, P. (1955). Beobachtung an *Phaeocystis*-kulturen. *Helgoländer Wissenschaft Meeresuntersuchungen*, **5**, 218–33.

Lancelot, C. (1983). Factors affecting phytoplankton extracellular release in the Southern Bight of the North Sea. *Marine Ecology Progress Series*, **12**, 115–21.

Lancelot, C. and Billen, G. (1985). Carbon-nitrogen relationship in nutrient metabolism of coastal marine ecosystems. In *Advances in aquatic microbiology*, vol. 3., (ed. H. W. Jannash and P. J. LeB. Williams), pp. 263–321. Academic Press, London.

Lancelot, C. and Mathot, S. (1985). Biochemical fractionation of primary production by phytoplankton in Belgian coastal waters during short- and long-term incubations with ^{14}C-bicarbonate. II. *Phaeocystis pouchetii* colonial population. *Marine Biology*, **86**, 227–32.

Lancelot, C. and Mathot S. (1987). Dynamics of a *Phaeocystis*-dominated spring bloom in Belgian coastal waters. I. Phytoplanktonic activities and related parameters. *Marine Ecology Progress Series*, **37**, 239–49.

Lancelot, C., Mathot, S., and Owens, N. J. P. (1986). Modelling protein synthesis, a step to an accurate estimate of net primary production: the case of *Phaeocystis pouchetii* colonies in Belgian coastal waters. *Marine Ecology Progress Series*, **32**, 193–202.

Lancelot, C., Billen, G., Sournia, A., Weisse, T., Colijn, F., Veldhuis, M., *et al.* (1987). *Phaeocystis* blooms and nutrient enrichment in the continental coastal zones of the North Sea. *Ambio*, **16**, 38–46.

Lancelot, C., Wassmann, P., and Barth, H. (1992). *Phaeocystis*-dominated ecosystems. *Marine Pollution Bulletin*, **24**, 56–7.

Liss, P. S., Malin, G., Turner, S. M., and Holligan, P. (1994). Dimethyl sulphide and *Phaeocystis*: a review. *Journal of Marine Systems*, (In press).

Lubbers, G. W., Gieskes, W. W. C., Del Castilho, P., Salomon, W., and Bril J. (1990). Manganese accumulation in the high pH microenvironment of *Phaeocystis* (Haptophyceae) colonies of the North Sea. *Marine Ecology Progress Series*, **59**, 285–93.

Martens, P. (1981). On the *Acartia* species in the northern Wadden Sea of Sylt. *Kieler Meeresforschungen Sonderhefte*, **5**, 153–63.

Nolting, R. F., de Baar, H. J. W., and van Bennekom, A. J. (1991). Cadmium, copper and iron in the Scotia Sea, Weddell Sea and Weddell/Scotia Confluence (Antarctica). *Marine Chemistry*, **35**, 219–43.

Pieters, H., Kluytmans, J. H., Zandee, D. I., and Cadée, G. C. (1980). Tissue composition and reproduction of *Mytilus edulis* in relation to food availability. *Netherlands Journal of Sea Research*, **14**, 349–61.

Riebesell, U. (1993). Aggregation of *Phaeocystis* during phytoplankton spring blooms in the southern North Sea. *Marine Ecology Progress Series*, **96**, 281–9.

Riegman, R., Noordeloos, A. A. M., and Cadée, G. C. (1992). *Phaeocystis* blooms and eutrophication of the continental coastal zone of the North Sea. *Marine Biology*, **106**, 479–84.

Rogers, S. I. and Lockwood, S. J. (1990). Observations on coastal fish fauna during a spring bloom of *Phaeocystis pouchetii* in the eastern Irish Sea. *Journal of the Marine Biological Association of the United Kingdom*, **70**, 249–53.

Rousseau, V., Mathot, S., and Lancelot, C. (1990). Calculating carbon biomass of *Phaeocystis* sp. from microscopical observations. *Marine Biology*, **197**, 305–14.

Rousseau, V., Vaulot, D., Casotti, R., Cariou, V., Lenz, J., Gunkel, J., *et al.* (1994). The life-cycle of *Phaeocystis* (Prymnesiophyceae): evidence and hypotheses. *Journal of Marine Systems*, (In press).

Smith, W. O., Codiposti, L. A., Nelson, D. M., Manley, T., Buskey, E. J., Niebauer, H. J. *et al.* (1991). Importance of *Phaeocystis* blooms in the high-latitude ocean carbon cycle. *Nature*, **352**, 514–16.

Thingstad, F. and Billen, G. (1994). Microbial degradation of *Phaeocystis* material in the water column. *Journal of Marine Systems*, (In press).

Veldhuis, M. J. W. and Admiraal, W. (1985). Transfer of photosynthetic products in gelatinous colonies of *Phaeocystis pouchetii* (Haptophyceae) and its effect on the measurement of excretion rate. *Marine Ecology Progress Series*, **26**, 301–4.

Veldhuis, M. J. W., Colijn, F., and Admiraal, W. (1991). Phosphate utilization in *Phaeocystis pouchetii* (Haptophyceae). *Marine Ecology*, **12**, 53–62.

Verity, P. G., Villareal, T. A., and Smayda, T. J. (1988). Ecological investigations of blooms of colonial *Phaeocystis pouchetii*. II. The role of life cycle phenomena in bloom termination. *Journal of Plankton Research*, **10**, 749–66.

Wassmann, P. (1994). Significance of sedimentation for the termination of *Phaeocystis* blooms. *Journal of Marine Systems*, (In press).

Wassmann, P., Vernet, M., Mitchell, B. G., and Rey, F. (1990). Mass sedimentation of *Phaeocystis pouchetii* in the Barents Sea. *Marine Ecology Progress Series*, **66**, 183–95.

Weisse, T. and Scheffel-Möser, U. (1990). Growth and grazing loss rates in single-celled *Phaeocystis* sp (Prymnesiophyceae). *Marine Biology*, **106**, 153–8.

Weisse, T., Tande, K., Verity, P., Hansen, F., and Gieskes, W. (1994). The trophic significance of *Phaeocystis* blooms. *Journal of Marine Systems*, (In press).

Westerlund, S. and Ohman, P. (1991). Iron in the water column of the Weddell Sea. *Marine Chemistry*, **35**, 199–217.

13. Mixotrophy in haptophytes

HARRIET L. J. JONES, B. S. C. LEADBEATER
School of Biological Sciences, University of Birmingham, Birmingham, UK

and J. C. GREEN
Plymouth Marine Laboratory, Citadel Hill, Plymouth, UK

Abstract

Mixotrophy, the uptake of particulate and dissolved organic carbon by photosynthetic algae, is widespread in haptophyte flagellates, in particular in the genus *Chrysochromulina.* Some species of *Chrysochromulina* have been shown to discriminate between particles on the basis of particle size and composition. Relatively large prey cells are selected by several species as well as, and sometimes in preference to, bacteria-sized particles. The digestion of prey has been studied in detail in *C. brevifilum* using transmission electron microscopy. Ingested food particles have been shown to be digested within food vacuoles. Some experiments have shown an increase in the rate of ingestion of particles under nutrient-limiting conditions though the crucial relationship with regard to mixotrophic behaviour is that between photosynthesis and phagotrophy. It is not yet known whether mixotrophic behaviour is governed by prey density or light intensity, and how this varies between species and between classes. The importance of phagotrophy to algal nutrition has shown that algae can be consumers as well as primary producers within the microbial loop. Evidence from grazing rates has also suggested that haptophyte flagellates may have a significant impact on the occurrence and density of their prey.

Introduction

Mixotrophy refers to the feeding behaviour of algae that obtain carbon both from the photosynthetic fixation of inorganic carbon and from the uptake of organic carbon. Organic carbon can be obtained as dissolved organic carbon (DOC) directly from the surrounding medium or by ingestion, digestion and assimilation of food particles. Most information

The Haptophyte Algae (ed. J. C. Green and B. S. C. Leadbeater), Systematics Association Special Volume No. 51, pp. 247–63. Clarendon Press, Oxford, 1994.

on mixotrophy in natural populations has come from studies of chryso-phytes, in particular *Dinobryon* spp. and *Ochromonas* spp. and at present this must be extrapolated to include mixotrophic haptophytes.

The uptake of DOC and the ingestion of particulate matter has been noted in many haptophyte flagellates. However, the implications of heterotrophic feeding on the growth rate and survival of mixotrophic species and the impact of heterotrophic feeding on particulate prey, has not been studied in detail. Conclusions on the ecological consequences of mixotrophy are, at present, mainly speculative (Boraas *et al.* 1988), but from information on grazing rates, determined for several species, it has been concluded that mixotrophs are of importance in the clearance of bacteria and other small micro-organisms from the water column (Estep *et al.* 1986; Porter 1988; Sanders *et al.* 1989).

The uptake of dissolved organic carbon

Some haptophytes can take up DOC compounds from their surrounding medium. Pintner and Provasoli (1968) examined several species of *Chrysochromulina* and found that no growth occurred in cultures maintained in darkness, and that cultures kept in the light with bacteria grew no better than the control cultures grown in the light, but without bacteria. However, the assimilation of a carbon source was clearly demonstrated in an axenic culture of *C. kappa*; by adding 1g l^{-1} glycerol, the final cell concentration was double that found in cultures without added glycerol. However, other species of *Chrysochromulina* were tested and showed little difference in their growth between controls and cultures with an added carbon source. Although no growth was obtained from cultures kept in the dark, all species tested, except *C. strobilus*, survived and glycerol seemed to be the best carbon source to ensure survival; it also gave the best rate of recovery when cultures were brought back into the light.

A glycerol concentration of 1 g l^{-1}, used in the experiments by Pintner and Provasoli (1968), was far in excess of what would be expected in the natural environment. However, the results do demonstrate the ability of species of *Chrysochromulina* both to take up and make use of DOC. The presence of glycine and L-valine (5.0 mM) and yeast extract (0.5 g l^{-1}) in culture media increased cell yields in *C. breviturrita* (Wehr *et al.* 1985) but did not increase cell division rates. Algae other than haptophytes also take up dissolved organic carbon either to sustain populations under light limiting conditions or to increase specific growth rates, for example the chlorophytes, *Ulva lactuca* (Markager and Sand-Jensen, 1990), *Chlorella regularis* (Endo *et al.* 1977), and *Scenedesmus falcatus* (Fingerhut *et al.* 1990), and the cryptophytes *Chroomonas salina* (=*Pyrenomonas salina*; Antia *et al.* 1973; Lewitus and Caron 1991; Lewitus *et al.* 1991).

The uptake of particulate matter

Particle selection

Many types of particles have been shown to be ingested by haptophytes (Table 13.1). A degree of particle selection has been demonstrated (Jones *et al.* 1993; Nygaard and Tobiesen 1993; Kawachi, personal communication) and many species will take up inert particles. For example, particle selection on the basis of particle size was found in three species of *Chrysochromulina* (Marchant, personal communication) using different sized microspheres. Kawachi *et al.* (1991) found organisms and particles from 0.3 μm to 5 μm diameter were ingested by *C. hirta,* but cells such as *Porphyridium marinum,* which exceeds 5 μm in diameter, were not ingested. Several species of unicellular green algae and three yeasts were presented to *C. brevifilum* to test for ingestion (Jones *et al.* 1993), but only one of the green algae, a pedinophyte, and two of the yeasts were ingested. When *C. brevifilum* was presented with a mixture of the pedinophyte and carmine particles it showed a preference for the small pedinophyte (Jones *et al.* 1993). Nygaard and Tobiesen (1993) suggested that some algal flagellates discriminate against heat killed bacteria; for example, both *C. polylepis* and *C. spinifera* ingested radiolabelled bacteria but did not ingest fluorescently labelled bacteria.

The uptake of relatively large food particles by species of *Chrysochromulina,* in one case the uptake of a diatom of 9 μm × 3 μm by *C. ericina,* 6–10 μm in diameter (Parke *et al.* 1956), challenges the theory that microorganisms preferentially prey on cells ten times smaller than themselves (Sheldon *et al.* 1972). In laboratory experiments *C. brevifilum,* approximately 5 μm in diameter, readily ingested a small pedinophyte of 3 μm diameter (Jones *et al.* 1993). *Chrysochromulina brevifilum* also showed a preference for this flagellate over bacteria (see Fig. 13.1 A, B). Results such as these suggest that the 10 : 1 ratio of size of predator to size of prey (Azam *et al.* 1983) may not apply strictly to mixotrophic haptophytes. Evidence suggests that many species of *Chrysochromulina* ingest a range of particles including detritus, inert particles, bacteria, yeasts, and small algal cells. These species could be opportunistic feeders that ingest any potential prey particles with which they come into contact. Under these conditions a 10 : 1 predator to prey size ratio would limit the predator in the range of particles it could ingest. It is interesting to note that natural samples have, on the whole, shown evidence for the uptake of small algal cells, whereas laboratory work has provided data on the uptake of bacteria sized particles.

Particle ingestion

Until recently it has been particle ingestion rather than mixotrophy that has been studied in haptophytes. Many of the early observations on

Table 13.1 Particle ingestion by haptophytes

Species	Particles ingested	Reference
Chrysochromulina acantha Leadbeater & Manton	Carmine	Jones *et al.* (1993)
Chrysochromulina alifera Parke & Manton	Graphite particles, bacteria	Parke *et al.* (1956)
Chrysochromulina bergenesis Leadbeater	Food vacuoles identified	Manton and Leadbeater (1974)
Chrysochromulina brevifilum Parke & Manton	Graphite, bacteria, algal cells	Parke *et al.* (1955)
	Carmine, yeast, small pedinophyte	Jones *et al.* (1993)
Chrysochromulina camella Leadbeater & Manton	Carmine	Jones *et al.* (1993)
Chrysochromulina campanulifera Manton & Leadbeater	Food vacuoles identified	Manton and Leadbeater (1974)
Chrysochromulina chiton Parke & Manton	Graphite, bacteria, scales	Parke *et al.* (1958)
	Carmine, small pedinophyte	Jones *et al.* (1993)
Chrysochromulina cymbium Leadbeater & Manton	Carmine, small pedinophyte	Jones *et al.* (1993)
Chrysochromulina ephippium Parke & Manton	Graphite, bacteria, algal cells	Parke *et al.* (1956)
Chrysochromulina ericina Parke & Manton	Graphite, bacteria, algal cells, incl. *Oicomonas* sp., *Stichococcus* sp., *Chlorella*	Parke *et al.* (1956)
	Carmine, small pedinophyte	Jones *et al.* (1993)

Table 13.1 (*cont.*)

Species	Particles ingested	Reference
Chrysochromulina herdlensis Leadbeater	Food vacuoles identified	Manton and Leadbeater (1974)
Chrysochromulina hirta Manton	*Nannochloropsis oculata, Micromonas pusilla, Coscinodiscus* sp.	Kawachi *et al.* (1991)
Chrysochromulina kappa Parke & Manton	Graphite, bacteria, algal cells	Parke *et al.* (1955)
	Carmine	Jones *et al.* (1993)
Chrysochromulina megacylindra Leadbeater	Diatoms, detritus	Manton (1972)
Chrysochromulina minor Parke & Manton	Graphite, bacteria, algal cells	Parke *et al.* (1955)
	Carmine	Jones *et al.* (1993)
Chrysochromulina polylepis Manton & Parke	Bacteria, microspheres	Nygaard and Tobiesen (1993)
Chrysochromulina pringsheimii Parke & Manton	Carmine	Jones *et al.* (1993)
Chrysochromulina spinifera (Fournier) Pienaar & Norris	Bacteria	Nygaard and Tobiesen (1993)
Chrysochromulina strobilus Parke & Manton	Graphite, bacteria, algal cells	Parke *et al.* (1959)
Coccolithus pelagicus (Wallich) Schiller	Graphite	Parke and Adams (1960)
Prymnesium parvum Carter	Bacteria	Nygaard and Tobiesen (1993)
Prymnesium saltans Massart	Small prey	Conrad (1941)

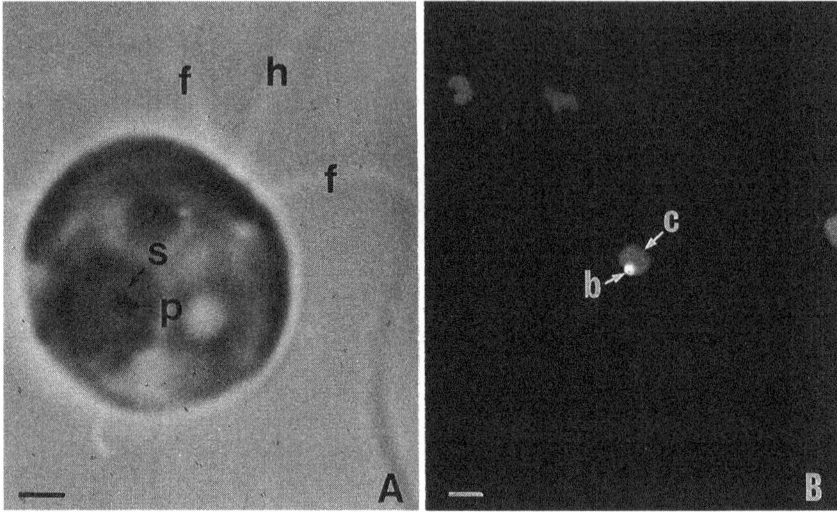

Fig. 13.1 Particle ingestion by *Chrysochromulina brevifilum*. (**A**) Light micrograph showing the ingestion of a small pedinophyte by *C. brevifilum*. Ingestion occurs at the non-flagellar pole (flagella, f; haptonema, h). The pyrenoid (p) and starch shell (s) of the pedinophyte can be identified and vesicular activity around the food particle can be seen. Scale bar = 1 μm. (**B**) Fluorescence micrograph showing the autofluorescent chloroplasts (c) of *C. brevifilum*. One cell has ingested bacteria (b) stained with the fluorescein DTAF. Scale bar = 5 μm.

particle ingestion in species of *Chrysochromulina* were carried out by Parke and Manton (Parke *et al.* 1955, 1956, 1959) who studied the phagotrophic behaviour of species of *Chrysochromulina*, including the uptake of inert particles, bacteria, and small pigmented microorganisms.

Suspended particles, bacteria, and small microorganisms up to 2.5 μm in diameter were found in cells of *C. kappa* (Parke *et al.* 1955). Ingestion was very rapid and particles that had recently been ingested could sometimes be seen in Brownian motion within a vacuole. The food was ingested at the non-flagellar pole of the cell; a colourless, slightly granular substance seemed to flow out from this region and surrounded the particles being ingested, which were then drawn quickly into the cell. Food particles were seen to move around within the cell and were then formed into a rounded food pellet. In *C. brevifilum* ingested food remained very close to the non-flagellar pole and did not move around inside the cell (Parke *et al.* 1955). In *C. ericina* (Parke *et al.* 1956) ingestion was also always at the non-flagellar pole, but after ingestion food particles were moved close to one of the pyrenoids. When cells with definite cell walls were ingested the cell contents were absorbed but the cell walls were egested.

Ingestion has been studied in more detail in *C. hirta* (Kawachi *et al.* 1991; Inouye and Kawachi, Chapter 4) using fluorescent microspheres and small microorganisms. They have shown fluorescent microspheres adhering to the haptonematal surface and then immediately moving down the haptonema towards the cell body. Particles aggregated at the base and were then moved towards the tip of the haptonema. As the aggregated particles arrived at the tip, the haptonema bent to position the particles on the cell surface and the aggregate was then ingested into a food vacuole towards the posterior end of the cell. This work showed that food capture and transport, involving the haptonema, was a sequence of events occurring in a particular order.

The ingestion of particles has also been studied in *C. spinifera* (Kawachi, personal communication). Particles adhered to the spines on the cell surface and the haptonema removed the particles from the spines and formed an aggregate in a similar manner to that seen in *C. hirta*. The aggregate was then positioned, by the haptonema, on a region of the cell surface that had a 'mouth-like' structure. *Chrysochromulina hirta* could ingest relatively large particles, up to 5 μm, since the particles attached directly to the haptonema. *Chrysochromulina spinifera* took up bacteria-sized particles because the food particles became attached to the spines before being transferred to the haptonema; the spines apparently in some way limiting the size of particle that could be ingested.

Digestion

The uptake of particles is not, in itself, evidence of phagotrophy (Porter 1988). It is therefore necessary to demonstrate that particles are digested, and that the products of digestion are assimilated before measurements of particle ingestion can be used to relate particle uptake to the impact of mixotrophs in the environment. The digestion of food particles in haptophytes has at present been studied in detail in only one species, *Chrysochromulina brevifilum*, in which the digestion of a small pedinophyte was investigated using transmission electron microscopy (Jones *et al.* 1993). The breakdown of the small pedinophyte took less than one hour. The pedinophyte was transformed into an electron dense, spherical food particle within a food vacuole in *C. brevifilum* and vesicular activity around the food vacuole was presumed to be involved in the transport of digesting enzymes to the food vacoule. The pyrenoid and surrounding starch shell of the pedinophyte prey were more resistant to digestive enzymes than other components of the prey cell and could still be identified when the rest of the cell had more or less disintegrated. This could also be observed using light microscopy (Fig. 13.1A). Eventually the pyrenoid and starch shell were digested and could no longer be identified in the food vacuole. In the chrysophyte *Ochromonas* sp., the process of digestion has been studied in a similar manner. The digestion of the blue-green alga *Anacystis nidulans*, by *Ochromonas* sp., was investigated using transmission electron microscopy. A two stage digestion process was described where the two stages occurred at different sites within the *Ochromonas* cell (Daley *et al.* 1973) in contrast to *C. brevifilum* where digestion occurred at only one site.

Acquisition of nutrients and vitamins through mixotrophy

Phagotrophy may be a means of obtaining limiting nutrients and vitamins (Caron *et al.* 1993), particularly through the ingestion of organisms, such as bacteria, that may have the potential to outcompete algae in the uptake of nutrients and vitamins from the surrounding medium (Boraas *et al.* 1988). The ingestion of potential competitors has been termed 'intraguild predation' (Polis *et al.* 1989). In many ways this is a more plausible explanation for mixotrophic behaviour than suggesting that it is for the acquisition of carbon compounds since carbon is rarely the growth-limiting factor. A possible exception to this may be during light limiting periods such as the dark months in polar regions, or at depths below the euphotic zone.

The ingestion of bacteria to gain phosphorus compounds has been demonstrated in *Chrysochromulina polylepis* and *Prymnesium parvum* (Nygaard and Tobiesen 1993). Cells were grown in batch culture with excess phosphate and in chemostats with reduced phosphate levels. *Chrysochromulina polylepis* and *P. parvum* both ingested more bacteria per hour in chemostats than in batch cultures. A close connection was found between a lack of phosphate in the algal flagellates and the degree of bacterivory. This agrees with results for *C. brevifilum* (Jones *et al.* 1993), where the population of prey cells declined more rapidly due to grazing when *C. brevifilum* had been grown in phosphate-reduced conditions than when it was grown in phosphate-rich conditions.

Pintner and Provasoli (1968) found that the addition of amino acids, nucleic acids and vitamins to other carbon sources, supplied to *C. brevifilum* and *C. strobilus* under light limiting conditions, resulted in better growth. The chrysophyte *Uroglena americana* (Kimura and Ishida 1985) acquired phospholipids by the ingestion of bacteria; they were found to be essential growth factors.

A novel method of feeding in *Chrysochromulina* has been suggested by Estep and MacIntyre (1989) in which *Chrysochromulina* might be able to extract essential nutrients from its prey. The hypothesis states that

Chrysochromulina releases a toxin that renders the cell membrane of the prey organism permeable to an efflux of nutrients, not necessarily killing the prey, but effecting a flow of nutrients from the prey into the surrounding medium, which *Chrysochromulina* could then take up and absorb. This method of inducing nutrient leakage would be most efficient under low nutrient conditions, when *Chrysochromulina* cells would be in greatest need of additional nutrition, such as during bloom conditions. They suggest that the scales covering the cell surface of *Chrysochromulina* serve to protect it from its own toxins.

Relationship between photosynthesis and phagotrophy

Most of the work published on the relationship between photosynthesis and phagotrophy has been on members of the Chrysophyceae, in particular *Poterioochromonas malhamensis* (Sanders *et al.* 1990). In this species light intensity had no effect on phagotrophy with approximately 90% of cells ingesting fluorescently labelled bacteria both at a light intensity of 200 $\mu E\ m^{-2}\ s^{-1}$ and at 25 $\mu E\ m^{-2}\ s^{-1}$. When bacteria were added to a previously phototrophic culture at 200 $\mu E\ m^{-2}\ s^{-1}$, the concentration of chlorophyll *a* decreased from 350 fg $cell^{-1}$ to 50 fg $cell^{-1}$ over a 24 hour period, and the chlorophyll concentrations increased again when the bacteria had been grazed to a concentration of approximately 10^6 cells ml^{-1}. Heterotrophy seems to be the primary mode of nutrition in this species, photoautotrophy being considered as a long-term survival strategy, particularly during periods of starvation.

The relationship between the availability of light and the abundance of potential food particles is crucial to understanding mixotrophy. It is not yet known whether phagotrophy is a response to low light intensity or the presence of suitable prey; mixotrophic behaviour could either be driven by available light or available prey. There is probably considerable variation between species regarding these two parameters. In *C. brevifilum* (Jones *et al.* 1993) it was found that as light intensity increased, the number of cells ingesting a small pedinophyte decreased. However, when *C. birgeri* Hällfors and Neimi formed a bloom under the ice in the Tvärminne Archipelago off the southern coast of Finland, no evidence of phagotrophy was found, and it was noted that the chloroplasts were very large, suggesting that nutrition was mainly, if not exclusively photoautotrophic (Hällfors and Niemi 1974).

A study has been carried out to investigate the response of *C. brevifilum* to a food source as it adapted to changes in light intensity (Durjun, personal communication). *Chrysochromulina brevifilum* was acclimated to a light intensity of 25 $\mu E\ m^{-2}\ s^{-1}$ and then moved to a light intensity of 100 $\mu E\ m^{-2}\ s^{-1}$. Each day a sample was taken and fed with a small pedinophyte that it was known to ingest (Jones *et al.* 1993), and chlorophyll *a* concentration per cell was also determined. Another set of *C. brevifilum* cultures was first acclimated at a light intensity of 100 μE m^{-2} s^{-1} and then moved to 25 μE m^{-2} s^{-1}, and samples were taken and treated in the same manner as above. Results demonstrated that as *C. brevifilum*, previously acclimated to the lower light intensity, became acclimated to the higher light intensity, the percentage of cells found with an ingested pedinophyte decreased from 16.0% to 0.3% over five days, as organic compounds gained photosynthetically did not need supplementing with phagotrophically gained substrates. In the reverse situation, as growth became limited by the low light intensity, the number of *C. brevifilum* cells with an ingested pedinophyte increased from 11.7% to 32.0% over five days. *Chrysochromulina brevifilum* grew considerably faster when fed with the prey flagellate, in comparison with control cultures without the prey flagellate (Jones *et al.* 1993), but chlorophyll levels remained constant (Durjun, personal communication). From these data it is suggested that in *C. brevifilum* photoautotrophy is the dominant mode of nutrition and that light intensity governs the degree of phagotrophy. This is in direct contrast to the balance between photoautotrophy and heterotrophy found in the chrysophyte *Poterioochromonas malhamensis* (Sanders *et al.* 1990), in which phagotrophy was found to be the dominant mode of nutrition.

Mixotrophy can be seen as an ideal survival mechanism. Algae can gain carbon photosynthetically when light intensities are high and not limiting to growth; when the light intensity falls, so that growth is carbon-limited, mixotrophs can feed phagotrophically and so retain their high growth rates. Mixotrophy requires the maintenance of two trophic systems, one for photosynthesis and one for phagotrophy. This could be energetically expensive and could be a reason for organisms to adapt and employ only one of those systems. If chloroplasts were gained through endosymbiosis, the organism gaining the chloroplast might then be mixotrophic and could either specialize in one of the two trophic states, or retain its ability to both feed and photosynthesise simultaneously. Alternatively the organism could have retained the ability, or the need, to use complex organic compounds, but only in the form of DOC.

Inter-species variation

Sanders *et al.* (1990) suggested that there may be a mixotrophic gradient ranging from species where heterotrophy is the dominant mode of nutrition to those where photoautotrophy is the dominant mode of nutrition. The chrysophyte *Poterioochromonas malhamensis* would be at the heterotrophic extreme of the scale, being an obligate heterotroph, relying on photosynthesis only when bacteria levels are low. *Cryptomonas* sp. would be near the autotrophic extreme, heterotrophy being employed only in low light conditions to maintain the cell and ensure survival until light no longer limits growth (Tranvik *et al.* 1989).

Haptophyte mixotrophs appear to be closer to the photoautotrophic end of the scale than chrysophytes such as *Ochromonas.* Two freshwater species at the photoautotrophic end of a mixotrophic gradient, *C. breviturrita* (Wehr *et al.* 1985) and *C. parva* (Parke *et al.* 1962), could not be induced to ingest any particulate matter, and *C. breviturrita* could be grown axenically. There is wide variation between haptophyte species in the amount of particles they ingest and there appears, therefore, to be a mixotrophic gradient within this group of algae. In an experiment to measure particle ingestion in *Chrysochromulina,* six different species were fed carmine particles (unpublished results). The results (Table 13.2) show that under the conditions of the experiment (20 °C, 16h L : 8h D cycle, 45 μE m^{-2} s^{-1}) *C. ericina* showed a greater tendency towards phagotrophy than the other species tested. It also demonstrates that results from experiments on single species cannot be extrapolated to incorporate all mixotrophic haptophytes.

There is some debate as to the relative importance of photoautotrophy and phagotrophy in the chrysophyte *Dinobryon cylindricum* with some studies showing phagotrophy to be dominant (Bird and Kalff 1987) and others suggesting photoautotrophy to be dominant (Caron *et al.* 1993). An explanation for this has been suggested in that there could be intra-species variation with some strains depending less on phagotrophy than others (Caron *et al.* 1993). Intra-species variation in feeding behaviour has been demonstrated in the heterotrophic flagellate *Paraphysomonas imperforata* (Dr J. Eccleston-Parry, personal communication) and could also be true for phagotrophic algae.

Table 13.2 Percentage of cells with ingested carmine particles, in selected *Chrysochromulina* species, four hours after the introduction of carmine particles to cultures (Jones, Leadbeater, and Green, unpublished data)

Species	% Cells with ingested carmine particles ± standard error
C. minor	0.25 ± 0.25
C. camella	0.48 ± 0.12
C. kappa	2.25 ± 0.63
C. chiton	12.00 ± 2.16
C. brevifilum	13.70 ± 3.48
C. ericina	63.00 ± 1.78

Ecological significance of mixotrophy

There is increasing evidence that phagotrophy is extensive in haptophtye flagellates. However, evidence from natural populations (Parke *et al.* 1956; Kawachi *et al.* 1991) suggests that the ingestion of 2 to 5 μm microorganisms predominates over the ingestion of bacteria in mixotrophic haptophytes. Studies on phagotrophy in haptophytes should not, therefore, use bacteria as the only source of prey, since this would not reflect the natural situation.

The full ecological significance of mixotrophy by haptophytes has yet to be resolved. Cells isolated from the wild have been found to contain food particles, mostly pigmented microorganisms, so confirming that prey is taken up in nature. Most laboratory studies confirm this ability to ingest prey but there is a lack of data on grazing rates. Nygaard and Tobiesen (1993) calculated the grazing rates of *Prymnesium parvum* and *Chrysochromulina ericina* using fluorescently labelled and radiolabelled bacteria (Table 13.3). Jones *et al.* (1993) have calculated a grazing rate for *C. brevifilum*, when feeding on a small pedinophyte, of 0.8 prey cells per *C. brevifilum* per hour. Although this grazing rate is lower than that found using fluorescently labelled bacteria in *C. ericina* (Nygaard and Tobiesen 1993), in terms of carbon gained per hour, 0.8 pedinophyte cells per hour is greater than 18.0 bacteria per hour.

During the bloom of *C. polylepis* in Scandinavian waters in 1988 (Moestrup, Chapter 14) it was noted that bacterial numbers were very low; this could have been explained by bacterial grazing by *C. polylepis* (Nielsen *et al.* 1990). However, results from laboratory studies on grazing rates using inert fluorescent particles did not agree with this assumption, and the production of an extracellular inhibitor by *C. polylepis* was found to

Table 13.3 Grazing rates for haptophyte flagellates

Species	Grazing Rate	Reference
Prymnesium parvum	5.8 flb cell^{-1} h^{-1}	Nygaard and Tobiesen (1993)
Chrysochromulina ericina	18.0 flb cell^{-1} h^{-1}	Nygaard and Tobiesen (1993)
Chrysochromulina brevifilum	0.8 ped cell^{-1} h^{-1}	Jones *et al.* (1993)
Chrysochromulina brevifilum	0.3 flb cell^{-1} h^{-1}	Jones, Leadbeater and Green (unpublished data)
Chrysochromulina hirta	8.3 fm cell^{-1} h^{-1}	Green (1991)
Chrysochromulina sp.	0.15 fm cell^{-1} h^{-1}	Green (1991)

flb = fluorescently labelled bacteria; fm = fluorescent microspheres; ped = pedinophyte have a deleterious effect on the bacterial population. This shows that certain haptophyte species could have an adverse impact on bacterial populations other than through phagotrophy.

In a study of mixotrophy in ice-covered lakes in the Pocono Mountains it was found that in Lake Lacawac, 48% of the algae were mixotrophic (Berninger *et al.* 1992). Bennett *et al.* (1990) found that mixotrophs made up the highest proportion of bacterial grazers in the winter and the lowest in midsummer. In mid-February, an average of 14% of the pigmented flagellates were mixotrophic. This suggests that mixotrophy could be a strategy to counteract the reduction in day length and available light, due to the low angle of the sun in the winter. By ingesting food particles, the mixotrophic species could maintain their growth rates at the levels they attain in the summer. Sanders *et al.* (1989) also found a seasonal variation in the degree of mixotrophy. In winter and early spring it was found that mixotrophs contributed up to 79% of grazing at a given depth. Nygaard and Tobiesen (1993) showed that bacterivory by mixotrophic flagellates may contribute up to 60% of total grazing on bacteria. Bennett *et al.* (1990) also found that, in February, mixotrophic flagellates were responsible for up to 60% of the flagellate grazing. Mixotrophs could, therefore, be a very important part of the microbial food web both as grazers and as primary producers.

Concluding remarks

Information on mixotrophic haptophytes is dominated by studies on particle uptake. It is surprising that it has taken over 30 years to relate the original observations of Parke *et al.* (1955, 1956, 1959) to the impact of algal grazing in the natural environment. Studies by Bird and Kalff (1986, 1987, 1989) assessed the ecological implications of mixotrophy in *Dinobryon cylindricum* (Chrysophyceae), and found that grazing rates were relatively high and could not be ignored in studies on the clearance rates of bacteria from the water column. It is now necessary to carry out research to investigate the effect mixotrophic haptophytes have on the populations of their respective prey, and so clarify their importance in the microbial food web. Studies on natural populations are important to ascertain the possible impact mixotrophs have on bacteria and small protists, and to find out how ingestion rates of mixotrophs compare with ingestion rates of non-pigmented phagotrophs of a similar size. The microbial food web now shows a more complex series of interactions between aquatic microorganisms than was first thought. The emergence of data on the importance of mixotrophy suggests that the microbial food web was, itself, too simple and that algae can now be considered as either primary producers or as consumers.

There is much potential in the research of mixotrophy in haptophyte flagellates. Grazing rates show that mixotrophs could have a significant impact on the population of prey organisms. The ecological factor governing the balance between phagotrophy and photoautotrophy is also of great importance and so far has only been studied in detail in the chrysophytes *Poterioochromonas malhamensis* (Sanders *et al.* 1990) and *Dinobryon cylindricum* (Bird and Kalff 1987; Caron *et al.* 1993), and the haptophyte *Chrysochromulina brevifilum* (Durjun, personal communication). These two groups of algae have been found to have a very different systems, the balance between phototrophy and phagotrophy in *P. malhamensis* being governed by prey density, and in *C. brevifilum* the balance is governed by light intensity; *D. cylindricum* requires further investigation. More species need to be investigated to ascertain whether this is purely a difference between species, or whether it can be extrapolated to suggest that mixotrophy in chrysophytes is determined by prey density and that mixotrophy in haptophytes is governed by light intensity.

Acknowledgements

We would like to thank Dr Masanobu Kawachi for allowing the use of his unpublished information on particle uptake, Dr Harvey Marchant for his unpublished information on particle selection, Dr Jackie Eccleston-Parry for the use of unpublished information and her comments on the manuscript, Miss Prema Durjun for providing data on phagotrophy and phototrophy in *C. brevifilum*, and Mr Steve Price for his help with photography.

References

Antia, N. J., Kalley, J. P., McDonald, J., and Bisalputra, T. (1973). Ultrastructure of the marine Cryptomonad *Chroomonas salina* cultured under conditions of photoautrophy and glycerol-heterotrophy. *Journal of Protozoology*, **20**, 377–85.

Azam, F., Fenchel, T., Field, J. G., Gray, J. S., Meyer-Reil, L. A., and Thingstad, F. (1983). The ecological role of water-column microbes in the seas. *Marine Biology Progress Series*, **10**, 257–63.

Bennett, S. J., Sanders, R. W., and Porter, K. G. (1990). Heterotrophic, autotrophic, and mixotrophic nanoflagellates: Seasonal abundances and bacterivory in a eutrophic lake. *Limnology and Oceanogrophy*, **35**, 1821–32.

Berninger, U. G., Caron, D. A., and Sanders R. W. (1992). Mixotrophic algae in three ice-covered lakes of the Pocono Mountains, U. S. A. *Freshwater Biology*, **28**, 263–72.

Bird, D. F. and Kalff, J. (1986). Bacterial grazing by planktonic algae. *Science*, **231**, 493–5.

Bird, D. F. and Kalff, J. (1987). Algal phagotrophy: regulating factors and importance relative to photosynthesis in *Dinobryon* (Chrysophyceae). *Limnology and Oceanography*, **32**, 277–84.

Bird, D. F. and Kalff, J. (1989). Phagotrophic sustenance of a metalimnetic phytoplankton peak. *Limnology and Oceanography*, **34**, 155–62.

Boraas, M. E., Estep, K. W., Johnson, P. W., and Sieburth, J. McN. (1988). Phagotrophic phototrophs: The ecological significance of mixotrophy. *Journal of Protozoology*. **35**, 249–52.

Caron, D. A., Sanders, R. W., Lim, E. L., Marrasé C., Amaral, L. A., Whitney, S. *et al.* (1993). Light-dependent phagotrophy in the freshwater mixotrophic chrysophyte *Dinobryon cylindricum*. *Microbial Ecology*, **25**, 93–111.

Conrad, W. (1941). Sur les Chrysomonadines à trois fouets. Aperçu synoptique. *Bulletin du Musée Royal d'Histoire Naturelle de Belgique*, **17**, 1–16.

Daley, R. J., Morris, G. P., and Brown, S. R. (1973). Phagotrophic ingestion of a blue-green alga by *Ochromonas*. *Journal of Protozoology*, **20**, 58–61.

Endo, H., Sansawa, H., and Nakajima, K. (1977). Studies on *Chlorella regularis*, heterotrophic, fast-growing strain. II. Mixotrophic growth in relation to light intensity and acetate concentration. *Plant and Cell Physiology*, **18**, 199–205.

Estep, K. W. and MacIntyre, F. (1989). Taxonomy, life cycle, distribution and dasmotrophy of *Chrysochromulina*: a theory accounting for scales, haptonema, muciferous bodies and toxicity. *Marine Ecology Progress Series*, **57**, 11–21.

Estep, K. W., Davis, P. G., Keller, M. D., and Sieburth, J. McN. (1986). How important are oceanic algal nanoflagellates in bacterivory? *Limnology and Oceanography*, **31**, 646–50.

Fingerhut, U., Groeneweg, J., and Soeder, C. J. (1990). Acetate utilization in *Scenedesmus falcatus*, an alga from high-rate ponds. *Algological Studies*, **60**, 57–64.

Green, J. C. (1991). Heterotrophy in prymnesiophyte flagellates. In *The biology of free-living heterotrophic flagellates*, Systematics Association Special Volume No. 45, (ed. D. J. Patterson and J. Larsen), pp. 401–14. Clarendon Press, Oxford.

Hällfors, G. and Niemi, A. (1974). A *Chrysochromulina* (Haptophyceae) bloom under the ice in the Tvärminne Archipelago, Southern Coast of Finland. *Memoranda Societatis pro Fauna et Flora Fennica*, **50**, 89–104.

Jones, H. L. J., Leadbeater, B. S. C., and Green, J. C. (1993). Mixotrophy in a marine species of *Chrysochromulina*: Ingestion and digestion of a small green flagellate. *Journal of the Marine Biological Association of the United Kingdom*, **73**, 283–96.

Kawachi, M., Inouye, I., Maeda, O., and Chihara, M. (1991). The haptonema as a food-capturing device: observations on *Chrysochromulina hirta* (Prymnesiophyceae). *Phycologica*, **30**, 563–73.

Kimura, B. and Ishida, Y. (1985). Photophagotrophy in *Uroglena americana*, Chrysophyceae. *Japanese Journal of Limnology*, **46**, 315–18.

Lewitus, A. J. and Caron, D. A. (1991). Physiological responses of phytoflagellates to dissolved organic substrate additions. 2. Dominant role of autotrophic nutrition in *Pyrenomonas salina* (Cryptophyceae). *Plant and Cell Physiology*, **32**, 791–801.

Lewitus, A. J., Caron, D. A., and Miller, K. R. (1991). Effects of light and glycerol on the organization of the photosynthetic apparatus in the facultative heterotroph *Pyrenomonas salina* (Cryptophyceae). *Journal of Phycology*, **27**, 578–87.

Manton, I. (1972). Observations on the biology and micro-anatomy of *Chrysochromulina megacylindrica* Leadbeater. *British Phycological Journal*, **7**, 235–48.

Manton, I. and Leadbeater, B. S. C. (1974). Fine-structural observations on six species of *Chrysochromulina* from wild Danish marine nanoplankton, including a description of *C. campanulifera* sp. nov. and a preliminary summary of the nanoplankton as a whole. *Det Kongelige Danske Videnskabernes Selskab Biologiske Skrifter*, **20**, 1–26.

Markager, S. and Sand-Jensen, K. (1990). Heterotrophic growth of *Ulva lactuca* (Chlorophyceae). *Journal of Phycology*, **26**, 670–3.

Nielsen, T. G., Kiorboe, T., and Bjornsen, P. K. (1990). Effects of a *Chrysochromulina polylepis* subsurface bloom on the planktonic community. *Marine Ecology Progress Series*, **62**, 21–35.

Nygaard, K. and Tobiesen, A. (1993). Bacterivory in algae: A survival strategy during nutrient limitation. *Limnology and Oceanography*, **38**, 273–9.

Parke, M. and Adams, I. (1960). The motile (*Crystallolithus hyalinus* Gaarder & Markali) and non-motile phases in the life history of *Coccolithus pelagicus* (Wallich) Schiller. *Journal of the Marine Biological Association of the United Kingdom*, **39**, 263–74.

Parke, M., Manton, I., and Clarke, B. (1955). Studies on marine flagellates. II. Three new species of *Chrysochromulina*. *Journal of the Marine Biological Association of the United Kingdom*, **34**, 579–606.

Parke, M., Manton, I., and Clarke, B. (1956). Studies on marine flagellates. III. Three further species of *Chrysochromulina*. *Journal of the Marine Biological Association of the United Kingdom*, **35**, 387–414.

Parke, M., Manton, I., and Clarke, B. (1958). Studies on marine flagellates. IV. Morphology and microanatomy of a new species of *Chrysochromulina*. *Journal of the Marine Biological Association of the United Kingdom*, **37**, 209–28.

Parke, M., Manton, I., and Clarke, B. (1959). Studies on marine flagellates. V. Morphology and microanatomy of *Chrysochromulina strobilus* sp. nov. *Journal of the Marine Biological Association of the United Kingdom*, **38**, 169–88.

Parke, M., Lund, J. W. G., and Manton, I. (1962). Observations on the biology and fine structure of the type species of *Chrysochromulina* (*C. parva* Lackey) in the English Lake District. *Archiv für Mikrobiologie*, **42**, 333–52.

Pintner, I. J. and Provasoli, L. (1968). Heterotrophy in subdued light of 3 *Chrysochromulina* species. Proceedings of the U. S.–Japan seminar on marine microbiology. *Bulletin of the Miaki Marine Biological Institute Kyoto University*, **12**, 25–31.

Polis, G. A., Myers, C. A., and Holt, R. D. (1989). The ecology and evolution of intraguild predation: potential competitors that eat each other. *Annual Review of Ecological Systematics*, **20**, 297–330.

Porter, K. G. (1988). Phagotrophic phytoflagellates in microbial food webs. *Hydrobiologia*, **156**, 89–97.

Sanders, R. W., Porter, K. G., and Caron, D. A. (1990). Relationship between phototrophy and phagotrophy in the mixotrophic chrysophyte *Poterioochromonas malhamensis*. *Microbial Ecology*, **19**, 97–109.

Sanders, R. W., Porter, K. G., Bennett, S. J., and DeBiase, A. E. (1989). Seasonal patterns of bacterivory by flagellates, ciliates, rotifers and cladocerans in a freshwater planktonic community. *Limnology and Oceanography*, **34**, 673–87.

Sheldon, R. W., Prakash, A., and Sutcliff, W. H. (1972). The size distribution of particles in the ocean. *Limnology and Oceanography*, **17**, 327–40.

Tranvik, L. J., Porter, K. G., and Sieburth, J. McN. (1989). Occurrence of bacterivory in *Cryptomonas*, a common freshwater phytoplankter. *Oecologia*, **78**, 473–6.

Wehr, J. D., Brown, L. M., and O'Grady, K. (1985). Physiological ecology of the bloom-forming alga *Chrysochromulina breviturrita* (Prymnesiophyceae) from lakes influenced by acid precipitation. *Canadian Journal of Botany*, **63**, 2231–9.

14. Economic aspects: 'blooms', nuisance species, and toxins

ØJVIND MOESTRUP
Department of Phycology, Botanical Institute, University of Copenhagen, Denmark

Abstract

Blooms of haptophytes have been described since the 1800s, the best known bloom-forming species belonging to *Emiliania*, *Phaeocystis*, and *Prymnesium*. More recently *Chrysochromulina* and *Corymbellus* were added to the list. Some species produce potent toxins with severe effects on the environment. Toxins have been identified in a few species of *Chrysochromulina* and *Prymnesium* as galactolipids, but the toxicity is expressed only under certain conditions, notably when the cells are in the stationary phase of growth. Considering that most species of *Chrysochromulina* and *Prymnesium* species remain poorly studied, it is likely that many more species are potentially toxic. Economic losses caused by *Phaeocystis* blooms (tourism, fishing) and blooms of *Chrysochromulina* and *Prymnesium* (mainly fishkills) can be substantial. These genera and most species are distributed worldwide.

Introduction

Haptophytes, also known as prymnesiophytes, occur in all seas and in fresh water, sometimes in toxic blooms that may cause considerable harm to the environment. Increased nutrient loading of many aquatic areas and increased aquacultural activity have resulted in blooms in many parts of the world, although the number of species known to be toxic is small. As postulated below, the list of potentially toxic species is likely to increase as the toxicology and autecology of haptophytes have, to date, received scant attention. Many haptophytes are known only from the original description, and many species remain undescribed.

Some of the economic aspects of haptophyte biology will be discussed below, with emphasis on the potentially toxic species.

The Haptophyte Algae (ed. J. C. Green and B. S. C. Leadbeater), Systematics Association Special Volume No. 51, pp. 265–85. Clarendon Press, Oxford, 1994.

Blooms

Blooms of haptophytes are not a new phenomenon. Savage (1930) suggested that masses of *Phaeocystis* affected the migration of commercially important fish like herring, and even before *Phaeocystis pouchetii* was formally described, Pouchet in June–July 1882 found the sea between the Lofoten Archipelago and the Varangerfjord in Norway to be filled with 'a new pelagic alga': 1–2 mm large gelatinous, translucent, weakly stained spheres (Pouchet 1892). Pouchet calculated that each cubic metre of seawater contained 10 cubic cm of *Phaeocystis.* He found the same organism in Torshavn, the Faroe Islands in August 1890.

Since then *Phaeocystis* has caused problems for the fishing industry and for tourism from New Zealand to the North Sea. In New Zealand reports of the 'Tasman Bay slime' (*Phaeocystis*) go back to the 1860s, the blooms destroying fish, choking fishing nets, and causing a variety of other problems (Hurley 1982). As a producer of dimethylsulfide (DMS), blooms of *Phaeocystis* have also been assumed to be responsible for increased precipitation of acid rain, both locally and globally (Lancelot *et al.* 1987).

Other bloom-forming haptophytes have been less frequently reported. The coccolithophorid *Emiliania huxleyi* caused massive milky-green discolouration in coastal waters off Norway in May–July 1955 (115×10^6 cells l^{-1}; Berge 1962). Smaller blooms were observed by Braarud (1937, 1940, 1945) from various parts of southern Norway from 1911 onwards, in some localities almost on an annual basis, and satellite remote sensing has now demonstrated coccolithophorid blooms to be a regular feature of the North Sea each summer from late May to August (Holligan *et al.* 1989).

Other coccolithophorids may form blooms in warmer waters: *Gephyrocapsa oceanica* in Jervis Bay, Australia in mid-December 1992 (18×10^6 cells l^{-1}, the bloom persisting for one month; Blackburn and Cresswell, personal communication), and in Northland, New Zealand 1992 (L. Rhodes, personal communication). Large numbers of *Umbilicosphaera sibogae* were recorded in sediment traps down to 3560 meters depth in the Panama Basin during June–July 1980 (Honjo 1982).

In April and May 1993 blooms of the colony-forming species *Corymbellus aureus* were detected in the open waters of the northern North Sea near the Fladen Ground (Gieskes and Kraay 1986). Like *Phaeocystis* the blooms succeeded the diatom spring bloom, reaching a maximum on 19 May 1983 with 9×10^6 cells l^{-1}. *Corymbellus* had not been found in the North Sea before. It was described in 1973 from the English Channel (Green 1976), and subsequently reported sporadically elsewhere (New Zealand: Moestrup 1979; Australia: Hallegraeff 1983; the Atlantic: Estep *et al.* 1984).

Widely publicized blooms of *Chrysochromulina polylepis* occurred in Scandinavian waters in 1988 (*e.g.* Kaas *et al.* 1991). These blooms had a tremendous impact on marine ecosystems, in some ways similar to those of *Prymnesium*, the only known member of the Haptophyceae known at the time to be toxic. In 1991 another serious toxic bloom of *Chrysochromulina* took place in Norwegian waters, this time a bloom of *C. leadbeateri*.

Prymnesium is a notorious fishkiller, mainly in brackish water, with reports going back to the turn of the century in Germany (Strodtmann 1898). A truly marine species has recently been found in New Zealand (Chang 1985). As in the case of *Chrysochromulina*, however, the toxic potential of most species of *Prymnesium* remains unknown.

Nuisance species

Although blooms of other haptophytes occur, as listed above, only species of *Chrysochromulina*, *Prymnesium*, and *Phaeocystis* may be termed 'nuisance species'. These three genera will be discussed separately below, and the species on which some information on the toxic potential has appeared are listed in Table 14.1.

The genus Chrysochromulina

Prior to the bloom of *Chrysochromulina polylepis* in Scandinavia in 1988 there was little indication of toxicity in any species of *Chrysochromulina* (47 species are presently named, but many remain undescribed). When describing new species of *Chrysochromulina* from 1955 onwards, Parke and Manton routinely tested cultures of their new species for toxicity to fish, but always with negative results. The first indication of possible toxicity appeared in a study of the bryozoan *Electra pilosa* by Jebram (1980), published in a German language zoological journal and therefore overlooked by most phytoplankton specialists.

Jebram used a wide range of algae as food for *Electra* and discovered that several species previously thought to be harmless were inhibitive or toxic when given to *Electra* as the main diet. This applied to four of six species of *Chrysochromulina* tested (Table 14.1), but only to old cultures of the algae. Jebram specifically stated with regard to *Chrysochromulina polylepis* that 'high concentrations of old cultures are toxic' (Jebram 1980, p. 383). Young cultures were excellent food.

It nevertheless came as a total surprise when massive blooms of *C. polylepis* caused havoc in Scandinavian waters in May–June 1988, affecting over 60 000 km^2 of open water (Dahl *et al.* 1989) with cell numbers reaching 200×10^6 l^{-1} (Kaas *et al.* 1991). *Chrysochromulina polylepis* had been previously found only sporadically in the area.

Table 14.1 Toxicological data on species of *Chrysochromulina* and *Prymnesium*

Name	Toxicity	Habitat	Selected references
C. acantha Leadbeater & Manton	Non-toxic to *Artemia*	Marine	Edvardsen and Paasche (1992)
C. alifera Parke & Manton	Non-toxic to fish	Marine	Parke *et al.* (1956)
C. brevifilum Parke & Manton	Non-toxic to fish Old cultures toxic to *Electra*	Marine	Parke *et al.* (1955) Jebram (1980)
C. breviturrita Nicholls	Undesirable odours in lakes; toxic to tadpoles?	Fresh water	Nicholls *et al.* (1982)
C. chiton Parke *et al.*	Non-toxic to fish	Marine	Parke *et al.* (1958)
C. ephippium Parke & Manton	Non-toxic to fish	Marine	Parke *et al.* (1956)
C. ericina Parke & Manton	Non-toxic to fish Non-toxic to *Artemia*	Marine	Parke *et al.* (1956) Edvardsen and Paasche (1992)
C. hirta Manton	Non-toxic to *Artemia*	Marine	Edvardsen and Paasche (1992)
C. kappa Parke & Manton	Non-toxic to fish; old cultures toxic to *Electra*	Marine	Parke *et al.* (1955) Jebram (1980)
C. leadbeateri Estep *et al.*	Toxic	Marine	Throndsen and Eikrem (1991)
C. minor Parke & Manton	Non-toxic to fish; non-toxic to *Electra*	Marine	Parke *et al.* (1955) Jebram (1980)
C. parva Lackey	Toxic to fish?	Fresh water	Hansen *et al.* (1993)

Table 14.1 (*cont.*)

Name	Toxicity	Habitat	Selected references
C. polylepis Manton & Parke	Toxic to a wide range of organisms	Marine	Jebram (1980), Rosenberg *et al.* 1988
C. pringsheimii Parke & Manton	Non-toxic to fish. Non-toxic to *Electra*	Marine	Parke and Manton (1962) Jebram (1980)
C. strobilus Parke & Manton	Non-toxic to fish. Old cultures toxic to *Electra*	Marine	Parke *et al.* (1959) Jebram (1980)
P. calathiferum Chang & Ryan	Toxic to fish shellfish?	Marine	Chang (1985)
P. parvum N. Carter	Toxic to fish and other organisms with gills	Brackish water	
P. patelliferum Green *et al.*	Toxic to fish (?)	Marine and brackish water	Valkanov (1964)

The same remarks apply to *C. leadbeateri*, found in small numbers by Leadbeater off Bergen in 1970 (Leadbeater 1972), as *Chrysochromulina* sp. It was later found in Australia (Hallegraeff 1983) and in the North Atlantic (Estep *et al.* 1984), always in small numbers until the bloom in the inner part of Vestfjorden near the Lofoten Archipelago in 1991, where cell numbers reached 3×10^6 l^{-1} (Heidal *et al.* 1991; Rey and Aure 1991; Throndsen and Eikrem 1991).

Deleterious effects resulting from species of *Chrysochromulina* are not confined to seawater, however. Only four species have been described from fresh water, and ill effects have been ascribed to two of these. In Ontario and New Hampshire Nicholls *et al.* (1982) reported obnoxious odours in five lakes between 1978 and 1980, caused by *C. breviturrita* (reaching 9×10^6 cells l^{-1}). Death of tadpoles may also have been caused by *C. breviturrita*. This is of interest since *Prymnesium parvum* is also toxic to tadpoles. Finally, in 1992, the type species of *Chrysochromulina*, *C. parva* occurred in enormous numbers in a Danish lake (600×10^6 cells l^{-1}), coinciding with substantial fishkills (Hansen *et al.*, 1993)

The genus Prymnesium

The first recorded fishkill caused by *Prymnesium* goes back to the Workum See in Holland, and was ascribed by Liebert and Deerns (1920) to an organism named by these authors 'the chrysomonad from Workum'. Periodically recurring fishkills in the Waterneverstorf Binnensee, a brackish water lagoon on the Baltic coast of Germany, were linked to the same organism (Lenz 1933), and Lenz has pointed out that fishkills had taken place in the area since the 1890s (Strodtmann 1898). The organism responsible was described as *Prymnesium parvum* by Carter (1937). In 1938 and 1939 two fishkills in Danish waters were caused by the same organism (Otterstrøm and Steemann Nielsen 1940), and this was followed by reports from many other parts of the world: England (Holdway *et al.* 1978), Norway (Kaartvedt *et al.* 1991), Finland (Lindholm and Virtanen 1992), the former Soviet Union (Krasnotschek and Abramowitsch 1971), Israel (Reich and Aschner 1947), and the USA (James and de la Cruz 1989). *Prymnesium parvum* is known also from British Columbia, Canada, and South Africa (Green *et al.* 1982), and from Japan (M. Kawachi, personal communication) but there have been no incidences of harmful effects.

Three additional species of *Prymnesium* are suspected fishkillers. The type species, *P. saltans*, occurred in numbers reaching 300×10^6 cells l^{-1} in slightly brackish waters (2–6‰) on the island of Rügen, Germany in April 1990 (Kell and Noack 1991), killing an estimated 2–300 tons of fish. However, the description of the causative species agreed more closely with *Prymnesium parvum* rather than with *P. saltans*. The main, if not the only, character by which *P. saltans* may be distinguished under the light microscope, the jumping movements which gave the species its name, were not described by Kell and Noack (1991). This feature has been seen only rarely since the original description by Massart (1920): Conrad (1941) recognized it in brackish water material from Belgium; and Heynig (1978) described it in material from Luppenau in eastern Germany in 1964, in a freshwater pond with relatively high concentrations of sulfate and chloride. Neither of these cases were associated with fishkills and there is, therefore, no documented case of a fishkill caused by *P. saltans*. Whether *P. saltans* and *P. parvum* are distinct species is another question, however. Material from the brackish water lake Selsø, Denmark, the cause of a fishkill in 1939, was identified as *Prymnesium parvum* by Otterstrøm and Steemann Nielsen (1940), but additional material from 1978 was identified by Ettl (1980) as *Prymnesium saltans*.

The cells involved in fishkills in the Varna lakes, Bulgaria, in August–October 1959 (Valkanov 1964; Petrova 1966) were identified as *P. parvum*, but Green *et al.* (1982) noted that the Varna material was probably *P. patelliferum*. The toxic potential of this species has now been documented, using material from Australia (Arlstad 1991). When tested against the dinoflagellate *Heterocapsa* it was found to be almost as toxic as *P. parvum*. *Prymnesium patelliferum* is known also from the Pacific coast of North America (Green *et al.* 1982), and from England (Green *et al.* 1982; Moestrup, unpublished data).

The most recently described species of *Prymnesium*, *P. calathiferum* from New Zealand, is truly marine (Chang and Ryan 1985; Chang 1985). Toxicity was suspected after fish and shellfish mortalities in the austral summer of 1983, and documented in the laboratory by exposing fish (*Gambusia*) to the cell free supernatant of a unialgal, dense culture of *P. calathiferum* (10^{11} cells l^{-1}).

The toxic potential of the few additional species of *Prymnesium* is unknown.

The genus Phaeocystis

The biology and ecology of *Phaeocystis* has recently been reviewed by Davidson and Marchant (1992; see also Lancelot and Rousseau, Chapter 12; Medlin *et al.*, Chapter 21).

One of the most urgent problems is resolution of the taxonomy of *Phaeocystis*. Nine species have been described (Sournia 1988), eight species between 1896 and 1930. The ninth species was found by Moestrup (1979) in plankton from New Zealand. The eight early species were described from light microscopical observations of the colonies, a highly variable character, and this variability led Kashkin (1963) to conclude that only a single species should be recognized. His advice was generally accepted, with the addition of *P. scrobiculata* (Moestrup 1979), but lately the discussion regarding the number of species has been brought up again by Jahnke and Baumann (1986, 1987) and Jahnke (1989). Two different taxa were identified in material collected from the North Sea, referred to as *P. pouchetii* and *P. globosa*. They differed in colony shape, in the arrangement of the cells in the colony, and in temperature preference. Cells of *P. pouchetii* were in groups of four in an often irregularly shaped colony, while cells of *P. globosa* were single and evenly spaced in a spherical colony. *Phaeocystis pouchetii* grew well between 0 and 14 °C and was therefore considered to be a coldwater species, while *P. globosa* grew between 4 and 22 °C.

Moestrup and Larsen (1992), using material from the Antarctic, found their material to resemble *P. globosa* in morphology, but *P. pouchetii* in its presence in cold water. This indicates the existence of yet another species, for which the names *P. antarctica* or *P. brucei* would be available. If these are identical, as suggested by Kashkin (1963), the oldest name takes preference. Alternatively, the eight 'old species' all belong to the same species. There is an urgent need for ultrastructural studies of the various isolates, to be combined with data on gene nucleotide sequences, isozymes, etc. If different ultrastructural and genetic entities exist, these may not necessarily reflect species differences, but may be recognized at the subspecific level.

Toxins

The toxins of haptophytes have not been fully analysed. *Phaeocystis* produces acrylic acid (Sieburth 1960), which has antibacterial properties, but no other toxins have been identified in *Phaeocystis*. The fishkill in Norway attributed to toxic *Phaeocystis* (Tangen, personal communication) requires further study.

Most attention has been paid to the toxins of *Prymnesium* (prymnesins). Shilo (1971) found that the main toxic component of *P. parvum* comprised a family of similar substances rather than a single compound. The biological spectrum of toxin activity was very broad and included cytotoxic, haemolytic, and neurotoxic effects (Shilo 1981). Ulitzur and Shilo (1970) suggested that the toxins were proteolipids, while Paster (1973) found glycolipids. This apparent discrepancy was examined by Kozakai *et al.* (1982) who, like Ulitzur and Shilo (1970), found six haemolytic compounds, using thin-layer chromatography. One of these (haemolysin 1) was present in higher quantities and was identified as a mixture of two galactolipids, the major fatty acid component being a $C_{18:4}$ and the minor a $C_{18:5}$ polyunsaturated fatty acid. All six fractions contained phosphorus, but this was not confirmed by Kozakai *et al.* (1982).

In *Chrysochromulina*, Yasumoto *et al.* (1990) found several hemolysins in *C. polylepis*. It was stated that these were 'not related' to prymnesins of *Prymnesium parvum*, but the authors found two galactolipids, in which the fatty acid components were identified as $C_{18:5}$ and $C_{20:5}$.

Both species of *Prymnesium* and *Chrysochromulina* are therefore capable of producing galactolipids with toxic properties. The findings of proteolipids and phosphate in *Prymnesium* need to be confirmed. It should be stressed, however, that only toxins of one species of *Prymnesium* (*P. parvum*) and one species of *Chrysochromulina* (*C. polylepis*) have been analysed.

Biological effects of the toxins

Prymnesium parvum toxins are known to affect the gills of fish (Ulitzur 1965) and invertebrates (the bivalves *Unio* and *Dreissensia*; Otterstrøm and Steemann Nielsen 1940). This led Conrad and Leloup (1939) to suggest that the stiff 'third flagellum' killed the fish by attaching to the gills and serving as a 'stylet inoculateur', injecting poison into the gills. The theory was discounted two years later (Otterstrøm and Steemann Nielsen 1940). Gill-breathing tadpoles are also affected, but not the non-gill breathing stages following metamorphosis. The main target of the toxins seems to be the cell membrane (Shilo 1981). Some organisms are able to exist together with *Prymnesium parvum*; for example, Otterstrøm and Steeman Nielsen (1940) found, in a bloom of 655×10^6 *Prymnesium* cells per litre, both the green alga *Scenedesmus* (1.9×10^6 cells l^{-1}), the bluegreen alga *Merismopedia* (1.5×10^6 colonies l^{-1}), and small diatoms (1.2×10^6 cells l^{-1}).

Prymnesium patelliferum was apparently responsible for the fishkills and extensive kills of other organisms in the Varna lakes, Bulgaria, in August–October 1959; protozoa, polychaetes, crustaceans, *Unio*, *Anodonta*, etc., perished. Tens of thousands of dead mussels of the latter two genera accumulated at the lake surface (Valkanov 1964).

Toxins of *Chrysochromulina* show many similarities to prymnesins. Rat hepatocytes lyse in the presence of *C. polylepis* (Aune 1989) and *C. leadbeateri* (Aune *et al.* 1991). Toxins of these species have also been found to interfere with transport of neurotransmitters into synaptosomes and synaptic vesicles, especially by inhibiting the uptake of L-glutamate (Meldahl and Fonnum 1991; Meldahl *et al.* 1993).

During the blooms of *C. polylepis* in Scandinavian waters in 1988 toxic effects were observed on bacteria, phytoplankton, zooplankton, benthic algae, and bottom dwelling invertebrates, conforming with the idea of the toxins being active on cell membranes. Bacteria were absent in the water layer containing *C. polylepis* (Nielsen *et al.* 1990), and some phytoplankton groups were considerably affected (Johnsen and Lømsland 1991): cryptomonads, *Ceratium*, and prasinophytes almost disappeared, while no negative effect was noted on chrysophytes and diatoms. According to Yasumoto *et al.* (1990) accumulation of the toxin takes place within mussels. When mussels were exposed to *C. polylepis* they exhibited five times higher haemolytic activity than non-contaminated mussels.

The bloom of *C. leadbeateri* in 1991 caused some damage to marine invertebrates such as sea urchins (Johannesen *et al.* 1991).

In culture experiments, the toxic effect of cell free extracts of both *Chrysochromulina polylepis*, *Prymnesium parvum*, and *Prymnesium patelliferum* on the phytoplankton dinoflagellate *Heterocapsa triquetra* was confirmed (Arlstad 1991), the latter being the first direct proof of toxicity in *P. patelliferum*.

Toxic algae as food

Evidence has been presented indicating that species of both *Chrysochromulina* and *Phaeocystis* can be used as food organisms and are grazed in nature. Jebram's findings that young cultures of *Chrysochromulina polylepis* provide excellent food for *Electra pilosa* have been mentioned above, and similar results were obtained by Tobiesen (1991) who provided *C. polylepis* as monofood to the heliozoan *Heterophrys marina*. Tobiesen found good growth of the heliozoan up to a food concentration of 2×10^6 cells l^{-1}. At higher concentrations of *C. polylepis* the growth rate of the heliozoan decreased and negative growth was seen in phosphate-limited cultures of *C. polylepis* above 75×10^6 cells l^{-1}. This almost certainly indicates that the alga is non-toxic, or nearly so, at lower concentrations. I have found no information on *Prymnesium* as a food organism, but Valkanov (1964) stated that protozoa were absent during the blooms in Bulgaria in 1959.

There is conflicting evidence regarding *Phaeocystis* as a food organism. According to some authors *Phaeocystis* is grazed readily. In the North Sea blooms of the dinoflagellate *Noctiluca scintillans* sometimes follow blooms of *Phaeocystis*, the dinoflagellate grazing on smaller colonies of *Phaeocystis*, thus limiting the extent of the bloom (Sørensen 1990). Other studies have indicated that *Phaeocystis* serves as an excellent food for copepods (Weisse 1983; Huntley *et al.* 1987). Jebram (1980), however, found *Phaeocystis* to be toxic to the bryozoan *Electra*, and several authors have found *Phaeocystis* to be incapable of supporting growth and reproduction of copepods (e.g. Verity and Smayda 1989).

For further references on this controversial subject, see Davidson and Marchant (1992).

Economic implications

The economic implications of haptophyte blooms can be substantial. In the case of *Phaeocystis* three problems occur. Fish are repulsed by the blooms and swim away resulting in reduced catches (Boalch 1987). In mariculture, Tangen (personal communication) attributed the death of farmed salmon, valued at 1 million NKK (*c.* 200 000 US dollars), in Norway in 1992 to *Phaeocystis*. Young oysters have been reported to be affected by *Phaeocystis* (Walne 1970). In New Zealand, massive blooms of *Phaeocystis*, known as Tasman Bay slime, have caused clogging of fishing nets and threatened the livelihood of fishermen, a phenomenon known in the Tasman Bay since last century (Hurley 1982).

The effect on tourism is less easily evaluated. The foam generated during a bloom, probably from decomposition of organic compounds, and blown on to the beaches is obviously a major nuisance for tourists, especially when the thickness of the foam reaches a meter or more. Such foams were apparently unknown before 1978 (Bätje and Michaelis 1986). In the austral summer of 1992/1993 foams were also reported from the Southern Hemisphere, in Christchurch, New Zealand (H. Chang, personal communication). As a producer of dimethylsulfide (DMS) *Phaeocystis* may have an impact on acid rainfall. The amount of DMS emitted from the sea to the atmosphere equals that from industrial sources, and part of the DMS is oxidized in the atmosphere to SO_2, resulting in acid rain (Charlson and Rodhe 1982). Oxidation into sulfate (SO_4) aerosols may influence the climate both locally and globally, the aerosols acting as cloud condensation centres which enhance cloud formation (Charlson *et al.* 1987; Bates *et al.* 1987). Numerous phytoplankton organisms have now been shown to produce DMS, but the Haptophyceae and Dinophyceae are the two most important classes (Keller *et al.* 1989). Both *Emiliania huxleyi*, *Prymnesium parvum*, and *Phaeocystis* are known producers of DMS (Keller *et al.* 1989; see also Malin *et al.*, Chapter 16).

Economic effects of *Chrysochromulina* species are presently restricted to Scandinavia. In 1988 it was estimated that about 900 tonnes of fish died during the *C. polylepis* bloom in Norway and Sweden (800 tonnes in Norway alone; Throndsen *et al.* 1994). The fish affected were mainly cod, salmon, and trout. In Norway, insurance companies paid out 60 million NKK in compensation (10 million US dollars). Smaller blooms, comprising a mixture of several species of *Chrysochromulina*, occurred in Denmark and Northern Germany in the spring of 1992 (Knipschildt 1992; Moestrup 1992), and losses from a single farm exceeded 35 tonnes of trout. The fishkill caused by *C. leadbeateri* in Northern Norway in 1991 killed 600 tonnes of fish (Eikrem and Throndsen 1993). The total loss of cultured fish in Norway in 1989–91, caused by *Prymnesium parvum*, reached 1200 tonnes, fish farms in Ryfylke losing 750 tons of salmon and trout in 1989, valued at 4.5 million US dollars (Johnsen and Lein 1989).

Losses caused by *Prymnesium* are, however, generally rather local. During the fishkill in the Varna lakes in Bulgaria the number of cells reached 110×10^6 cells l^{-1} and practically all fish in the lakes died within a few days (*c.* 300 tons) (Valkanov 1964). In a shallow carp pond on the island of Fehmarn, Germany, a concentration of 400×10^6 cells l^{-1} of *Prymnesium* in August–September 1975 threatened the carp population, which responded by taking up very little food (Hickel 1976). After two weeks, 50 kg $CuSO_4$ (5 g m^{-3}) was added to the pond to kill the algae, but the effect was the opposite: by the following morning nearly all fish had died (carp, eel, tench). The behaviour of the poisoned fish was striking, they lost their sense of balance and jumped on to the land. Also in Germany, on the island of Rügen, a concentration of 300 million cells of *Prymnesium* occurred in April 1990, an unusually early date, causing massive fishkills. Kell and Noack (1991) mention 500 kg dead fish per 100 m along the coast, the total probably in the range of 200–300 tons. In Texas, blooms of *Prymnesium parvum* in 1985, 1986, and 1988 were thought to be responsible for extensive fishkills in the Pecos River,

110 000 fish in 1985, 500 000 in 1986, and more than 1.5 million in 1988 (James and de la Cruz 1989). All losses took place in November and December. The identity of the organism responsible was not checked by electron microscopy, however, and the identification is therefore somewhat premature. James and de la Cruz (1989) give no information on the ecology of the river system and, if it is fresh water, *Prymnesium* is less likely to be the cause of the fish mortalities. The same remarks apply to another fishkill of 48 000 fish in November 1988 in Paint Creek, a tributary of the Brazos River, Texas (James and de la Cruz 1989).

In fresh water, few problems have been encountered with haptophytes. In Ontario and New Hampshire, the obnoxious odours caused by *Chrysochromulina breviturrita* in five lakes in 1978–1980 (Nicholls 1978; Nicholls *et al.* 1982) persisted for 4–6 weeks, but there was apparently little effect on the ecosystem. In one lake used for recreational purposes, $CuSO_4$ was added and this removed algae and odour in two days.

It is generally difficult to control the economic damage caused by phytoplankton blooms, an exception being smaller ponds or lakes with blooms of *Prymnesium.* This alga lyses when ammonia is added to the pond (Shilo 1981). Absence of light reduces toxin production, but the culture becomes toxic again when the cells enter the stationary phase of growth (Shilo 1971). Low phosphate concentrations are known to increase toxicity in both *Prymnesium parvum* and *Chrysochromulina polylepis* (Shilo 1971; Edvardsen *et al.* 1990; Tobiesen 1991).

Will this continue?

There is no doubt that haptophyte blooms have occurred for a long time. As mentioned above, blooms of *Emiliania* were reported around the turn of the century by Braarud, and *Phaeocystis* in the northern North Sea by Pouchet in the 1880s. These are probably normal, recurring phenomena. The large numbers of *Emiliania* in western Norway were thought to originate from the Atlantic, and the large blooms developed as a result of the wind-driven upwelling of nutrient-rich water (Berge 1962). While the blooms are probably natural phenomena, the extent of the blooms is no doubt affected by increased output of nutrients into the water. Thus the *Phaeocystis* blooms along the coasts of north-western Europe were mentioned by Smayda (1990) as an example indicating a long-term increase in both the frequency and extent of the blooms accompanying nutrient enrichment of coastal areas. Because of their worldwide distribution, similar blooms may be expected in almost any polluted marine area in the world.

The situation regarding *Prymnesium* is at present less clear. Nearly all fishkills caused by *Prymnesium* are in brackish water, typically at low salinities. The water may be almost fresh, as demonstrated by a bloom in northwestern Denmark in 1991, killing all fish in a fishpond where the salinity was only 0.8‰. The number of *Prymnesium* cells reached 146×10^6 cells l^{-1} (Bach and Jacobsen 1991). Many of the fishkills have taken place in waters polluted by aquaculture, and it is probably safe to predict fishkills by *Prymnesium* in almost any low-salinity, nutrient-rich area in temperate regions of both the Northern and the Southern Hemispheres. Data are lacking for the occurrence of *Prymnesium* in the tropics. In fully marine areas, fishkills caused by *Prymnesium* are exceptional, and at present known only from New Zealand.

The cause(s) of the *Chrysochromulina* blooms in Scandinavian waters have been much discussed and it is not clear whether such blooms have occurred before. In part of the area, the blooms took place near the pycnocline (8–10 m depth) leaving no visible sign of algae at the surface. Such blooms are therefore readily overlooked. The prerequisites for blooms, including high stability of the water column combined with high nutrient levels, are generally agreed upon, but the source of the nutrients remains uncertain. Edler (1984) calculated that nitrogen increased 4-fold and phosphorus 3 to 7-fold in the Kattegat in the period 1930–1980. Andersson and Rydberg (1988) found a doubling of inorganic nitrogen between 1971 and 1982. More directly, Haumann and Jørgensen (1989) found a surplus of nitrate in the pycnocline of Kattegat, when comparing the nitrate values for March 1988 (i.e. two months before the bloom was noted) and the previous year. An exceptional mixing of water from the Baltic and the North Sea has also been suggested as a likely cause (Kaas *et al.* 1991; Skjoldal and Dundas 1991), the latter perhaps containing increased concentrations of micronutrients (Dahl *et al.* 1989). Granéli and Moreira (1990) found that river water from forested areas promoted the growth of the dinoflagellate *Prorocentrum minimum*, and there has apparently been an increase of humic substances to coastal areas in Sweden during the last 15 years. Acid rain is thought to release the humic substances from the soil (Forsberg and Ahl, cited in Granéli and Moreira 1990).

Whatever the reason, there is little to explain why these particular species became dominant, although it is compelling to think of toxin production as a factor used to control the growth of competitor algae and potential grazers. Information on the autecology and physiology of individual species of marine phytoplankton is generally poor, allowing much speculation. Thus, it has been postulated that the coccolith cover on the cells of *Emiliania huxleyi* make this species well suited to occur at high light intensities which kill other algae (Steemann Nielsen and Hansen 1959; Berge 1962). While *Emiliania huxleyi* is considered to be non-toxic, some other coccolithophorids, e.g. *Ochrosphaera neapolitana* and *Hymenomonas* sp., are now thought to be toxic (Jebram 1980; Kobayashi *et al.* 1989). The unusual milky-green colour of *Emiliania* blooms is explained by reflection of light from the coccoliths, the cells themselves being yellow or yellow-brown (Braarud 1937).

Could the toxic blooms of *Chrysochromulina* occur again and could they occur elsewhere? There are indications that future blooms may be expected in many marine, brackish, or freshwater habitats worldwide. In a series of growth experiments, Arlstad (1991) showed that *Chrysochromulina polylepis, Prymnesium parvum,* and *P. patelliferum* grew well at all salinities between 10 and 30‰, while the division time increased at 5 and at 35‰. All three species proved toxic to the dinoflagellate *Heterocapsa triquetra* at all salinities, but cultures in the stationary phase of growth were more toxic than those in the exponential phase. In the exponential phase, *C. polylepis* was non-toxic to *Heterocapsa* at 35‰ and only very slightly toxic at 5, 10, and 30‰. This agrees well with Jebram's findings that only old cultures are toxic. Considering that Jebram (1980) found four of six tested species of *Chrysochromulina* to be toxic in old cultures, and bearing in mind that the number of *Chrysochromulina* species is probably not below 75, distributed in all the oceans of the world, the conclusion must obviously be that a toxic bloom of *Chrysochromulina* can occur almost anywhere. Attention should also be paid to other haptophytes, notably *Corymbellus,* which is very similar ultrastructurally to *Chrysochromulina* (Green 1976), differing mainly in its colonial habit.

References

Andersson, L. and Rydberg, L. (1988). Trends in nutrient and oxygen conditions within the Kattegat: effects of local nutrient supply. *Estuarine and Coastal Shelf Sciences,* **26**, 559–79.

Arlstad, S. (1991). Den allelopatiske effekt af *Prymnesium parvum, Prymnesium patelliferum, Chrysochromulina polylepis.* M.Sc. thesis, University of Copenhagen.

Aune, T. (1989). Toxicity of marine and freshwater algal biotoxins towards freshly prepared hepatocytes. In *Mycotoxins and phycotoxins '88,* (ed. S. Natori, K. Hashimoto, and Y. Ueno), pp. 461–8. Elsevier, New York.

Aune, T., Underdal, B., and Skulberg, O. M. (1991). Preliminary toxicological studies of the algal bloom in Vestfjorden, June 1991. In *The* Chrysochromulina leadbeateri *bloom in Vestfjorden, North Norway, May–June 1991,* Fisken og Havet No. 3, (ed. F. Rey), pp. 95–101. Havforskningsinstituttet, Bergen.

Bach, F. and Jacobsen, H. (1991). Giftige alger drœber hele fiskebestanden i vandhul i Vestjylland. *Vœkst,* **1991**. Det danske Hedeselskab.

Bates, T. S., Charlson, R. J., and Gammon, R. H. (1987). Evidence for the climatic role of marine biogenic sulfur. *Nature,* **329**, 319–21.

Bätje, M. and Michaelis, H. (1986). *Phaeocystis pouchetii* blooms in the East Frisian coastal waters (German Bight, North Sea). *Marine Biology,* **93**, 21–7.

Berge, G. (1962). Discoloration of the sea due to *Coccolithus huxleyi* 'bloom'. *Sarsia*, **6**, 27–40.

Boalch, G. T. (1987). Recent blooms in the western English Channel. *Rapports et Procès-Verbaux des Réunions, Conseil International pour Exploration de la Mer* **187**, 94–7.

Braarud, T. (1937). Om 'rødt vann', 'vannblomst' og lignende fenomener. *Naturen*, **1937** (2), 33–43.

Braarud, T. (1940). Grønnfargingen av Lenefjorden og Grønnsfjorden i Vest-Agder. *Naturen*, **1940** (2), 50–4.

Braarud, T. (1945). A phytoplankton survey of the polluted waters of inner Oslo Fjord. *Hvalrådets Skrifter*, **28**, 1–142.

Carter, N. (1937). New or interesting algae from brackish water. *Archiv für Protistenkunde*, **90**, 1–68.

Chang, F. H. (1985). Preliminary toxicity test of *Prymnesium calathiferum* n. sp. isolated from New Zealand. In *Toxic dinoflagellates*, (ed. D. M. Anderson, A. W. White, and D. G. Baden), pp. 109–12. Elsevier, New York.

Chang, F. H. and Ryan, K. G. (1985). *Prymnesium calathiferum* sp. nov. (Prymnesiophyceae), a new species isolated from Northland, New Zealand. *Phycologia*, **24**, 191–8.

Charlson, R. J. and Rodhe, H. (1982). Factors controlling the acidity of natural rainwater. *Nature*, **295**, 683–5.

Charlson, R. J., Lovelock, J. E., Andreae, M. O., and Warren, S. G. (1987). Oceanic phytoplankton, atmospheric sulfur, cloud albedo and climate. *Nature*, **326**, 655–61.

Conrad, W. (1941). Notes protistologiques. XXI. Sur les Chrysomonadines à trois fouets. Aperçu synoptique. *Bulletin du Musée Royal d'Histoire Naturelle de Belgique*, **17**, 1–16.

Conrad, W. and Leloup, E. (1939). Intoxications dues à des Flagellates autotrophes. *Annales de la Société Royale Zoologique de Belgique*, **69**, 243–6 (1938).

Dahl, E., Lindahl, O., Paasche, E., and Throndsen, J. (1989). The *Chrysochromulina polylepis* bloom in Scandinavian waters during spring 1988. In *Coastal and estuarine studies*, Vol. 35, *Novel phytoplankton blooms*, (ed. E. M. Cosper, V. M. Bricelj, and E. J. Carpenter), pp. 383–405. Springer-Verlag, Berlin.

Davidson, A. T. and Marchant, H. J. (1992). The biology and ecology of *Phaeocystis* (Prymnesiophyceae). In *Progress in phycological research*, Vol. 8, (ed. F. E. Round and D. J. Chapman), pp. 1–45. Biopress, Bristol.

Edler, L. (1984). Västerhavet. In *Eutrophication in marine waters surrounding Sweden. A review*, (ed. R. Rosenberg), pp. 74–111. Swedish Environmental Protection Board, Stockholm.

Edvardsen, B., Moy, F., and Paasche, E. (1990). Hemolytic activity in extracts of *Chrysochromulina polylepis* grown at different levels of selenite and phosphate. In *Toxic marine phytoplankton*, (ed. E. Granéli, B. Sundström, L. Edler, and D. M. Anderson), pp. 284–9. Elsevier, New York.

Edvardsen, B. and Paasche, E. (1992). Two motile stages of *Chrysochromulina polylepis* (Prymnesiophyceae): morphology, growth, and toxicity. *Journal of Phycology*, **28**, 104–14.

Eikrem, W. and Throndsen, J. (1993). Toxic prymnesiophytes identified from Norwegian coastal waters. In *Toxic phytoplankton blooms in the sea*, (ed. T. J. Smayda and Y. Shimizu), 687–92. Elsevier, Amsterdam.

Estep, K. W., Davis, P. G., Hargraves P. E., and Sieburth, J. McN. (1984). Chloroplast containing microflagellates in natural populations of North Atlantic nanoplankton, their identification and distribution; including a description of five new species of *Chrysochromulina* (Prymnesiophyceae). *Protistologica*, **20**, 613–34.

Ettl, H. (1980). Beitrag zur Kenntnis der Süsswasseralgen Dänemarks. *Botanisk Tidsskrift*, **74**, 179–223.

Gieskes, W. W. C. and Kraay, G. W. (1986). Analysis of phytoplankton pigments by HPLC before, during and after mass occurrence of the flagellate *Corymbellus aureus* during the spring bloom in the open northern North Sea in 1983. *Marine Biology*, **92**, 45–52.

Granéli, E. and Moreira, M. O. (1990). Effects of river water of different origin on the growth of marine dinoflagellates and diatoms in laboratory cultures. *Journal of Experimental Marine Biology and Ecology*, **136**, 89–106.

Green, J. C. (1976). *Corymbellus aureus* gen. et sp. nov., a new colonial member of the Haptophyceae. *Journal of the Marine Biological Association of the United Kingdom*, **56**, 31–8.

Green, J., Hibberd, D. J., and Pienaar, R. N. (1982). The taxonomy of *Prymnesium* (Prymnesiophyceae) including a description of a new cosmopolitan species, *P. patellifera* sp. nov., and further observations on *P. pavum* N. Carter. *British Phycological Journal*, **17**, 363–82.

Hallegraeff, G. M. (1983). Scale-bearing and loricate nanoplankton from the East Australian current. *Botanica Marina*, **26**, 493–515.

Hansen, L. R., Kristiansen, J., and Rasmussen, J. V. (1993). Potential toxicity of the freshwater *Chrysochromulina* species *C. parva* (Prymnesiophyceae). *Hydrobiologia*, (In press).

Haumann, L. and Jørgensen, L. A. 1989. Chemical conditions in comparison to previous years. In *The occurrence of* Chrysochromulina polylepis *in the Skagerrak and Kattegat in May–June 1988: an analysis of extent, effects and causes. Water Pollution Report*, Vol. 10, pp. 53–8. Commission of European Communities, Luxembourg.

Heidal, K., Skreslet, S., Mohus, Å., Eliassen, R., and Frogh, M. (1991). Distribution of the planktonalga *Chrysochromulina leadbeateri* and its environmental conditions in the Vestfjord, North Norway, 1991. In *The* Chrysochromulina leadbeateri *bloom in Vestfjorden, North Norway, May–June 1991*. Fisken og Havet No. 3, (ed. F. Rey), pp. 33–43. Havforskningsinstituttet, Bergen.

Heynig, H. (1978). *Prymnesium saltans* Massart (Chrysophyceae) in Gewässern des Bezirks Halle (DDR). *Archiv für Protistenkunde*, **120**, 222–8.

Hickel, B. (1976). Fischsterben in einem Karpfenteich bei einer Massenentwicklung des toxischen Phytoflagellaten *Prymnesium parvum* Carter (Haptophyceae). *Archiv für Fisch Wissenschaft*, **27**, 143–8.

Holdway, P. A., Watson, R. A., and Moss, B. (1978). Aspects of the ecology of *Prymnesium parvum* (Haptophyta) and water chemistry in the Norfolk Broads, England. *Freshwater Biology*, **8**, 295–311.

Holligan, P. M., Aarup, T., and Groom. S. B. (1989). The North Sea satellite colour atlas. *Continental Shelf Research*, **9**, 665–765.

Honjo, S. (1982). Seasonality and interaction of biogenic and lithogenic particulate flux at the Panama Basin. *Science*, **218**, 883–4.

Huntley, M., Tande, K., and Eilertsen, H. C. (1987). On the trophic fate of *Phaeocystis pouchetii* (Hariot). II. Grazing rates of *Calanus hyperboreus* (Kröyer) on diatoms and different size categories of *P. pouchetii*. *Journal of Experimental Marine Biology and Ecology*, **110**, 197–212.

Hurley, D. E. (1982). The 'Nelson Slime'. Observations on past occurrences. *New Zealand Oceanographic Institute Oceanographic Summary*, **20**, 1–11.

Jahnke, J. (1989). The light and temperature dependence of growth rate and elemental composition of *Phaeocystis globosa* Scherffel and *P. pouchetii* (Hariot) Lagerh. in batch cultures. *Netherland Journal of Sea Research*, **23**, 15–21.

Jahnke, J. and Baumann, M. (1986). Die marine Planktonalge *Phaeocystis globosa*. *Mikrokosmos*, **75**, 357–9.

Jahnke, J. and Baumann, M. (1987). Differentiation between *Phaeocystis pouchetii* (Har.) Lagerheim and *Phaeocystis globosa* Scherffel. *Hydrobiological Bulletin*, **21**, 141–7.

James, T. L. and de la Cruz, A. (1989). *Prymnesium parvum* Carter (Chrysophyceae) as a suspect of mass mortalities of fish and shellfish communities in Western Texas. *The Texas Journal of Science*, **41**, 429–30.

Jebram, D. (1980). Prospection for sufficient nutrition for the cosmopolitic marine bryozoan *Electra pilosa* (Linnaeus). *Zoologische Jahrbücher, Abteilung für Systematik, Ökologie und Geographie der Thiere*, **107**, 368–90.

Johannesen, T., Knutsen, J. A., and Paulsen, Ø. (1991). Effects of *Chrysochromulina leadbeateri* on reared salmon and natural fauna. In *The* Chrysochromulina leadbeateri *bloom in Vestfjorden, North Norway, May–June 1991*. Fisken og Havet No. 3, (ed. F. Rey), pp. 103–19. Havforskningsinstituttet, Bergen.

Johnsen, T. M. and Lein, T. E. (1989). *Prymnesium parvum* Carter (Prymnesiophyceae) in association with macroalgae in Ryfylke, southwestern Norway. *Sarsia*, **74**, 277–81.

Johnsen, T. M. and Lømsland, E. R. (1991). Growth experiments with *Chrysochromulina leadbeateri*. In *The* Chrysochromulina leadbeateri *bloom in Vestfjorden, North Norway, May–June 1991*. Fisken og Havet No. 3, (ed. F. Rey), pp. 85–8. Havforskningsinstituttet, Bergen.

Kaardtvedt, S., Johnsen, T. M., Aksnes, D. L., and Lie, U. (1991). Occurrence of the toxic phytoflagellate *Prymnesium parvum* and associated fish mortality in a

Norwegian fjord system. *Canadian Journal of Fisheries and Aquatic Sciences*, **48**, 2316–23.

Kaas, H., Larsen, J., Møhlenberg, F., and Richardson, K. 1991. The *Chrysochromulina polylepis* bloom in the Kattegat (Scandinavia) May–June 1988. Distribution, primary production and nutrient dynamics in the late stage of the bloom. *Marine Ecology Progress Series*, **79**, 151–61.

Kashkin, N. I. (1963). Data on the ecology of *Phaeocystis pouchetii* (Hariot) Lagerheim, 1893 (Chrysophyceae). II. Habitat and specification of biogeographical characteristics. *Okeanologia (Moscow)*, 3, 697–705. (Translated from the Russian).

Kell, V. and Noack, B. (1991). Fischsterben durch *Prymnesium saltans* Massart im Kleinen Jasmunder Bodden (Rügen) im April 1990. *Journal of Applied Ichthyology*, **7**, 187–92.

Keller, M. D., Bellows, W. K., and Guillard, R. R. L. (1989). Dimethyl sulphide production in marine phytoplankton. In *Biogenic sulfur in the environment*, American Chemical Society Symposium Series No. 393, (ed. E. S. Saltzman and W. J. Cooper), pp. 183–200. American Chemical Society, Washington DC.

Knipschildt, F. (1992). A new *Chrysochromulina* outbreak and fish kills in Danish coastal waters. *Harmful Algae News, UNESCO/IOC*, 2, 2.

Kobayashi, J., Ishibashi, M., Nakamura, H., and Ohizumi, Y. (1989). Hymenosulphate, a novel sterol sulphate with Ca-releasing activity from the cultured marine haptophyte *Hymenomonas* sp. *Journal of the Chemical Society Perkin Transactions*, **1**, 101–3.

Kozakai, H., Oshima, Y., and Yasumoto, T. (1982). Isolation and structural elucidation of hemolysin from the phytoflagellate *Prymnesium parvum*. *Agricultural and Biological Chemistry*, **46**, 233–6.

Krasnotschek, G. P. and Abramowitsch, L. S. (1971). Mass development of *Prymnesium parvum* Cart. in fish-breeding ponds. *Hydrobiological Journal*, **7**, 54–5.

Lancelot, C., Billen, G., Sournia, A., Weisse, T., Colijn, F., Veldhuis, M. J. W. *et al.* (1987). *Phaeocystis* blooms and nutrient enrichment in the continental coastal zones of the North Sea. *Ambio*, **16**, 38–46.

Leadbeater, B. S. C. (1972). Fine-structural observations on six new species of *Chrysochromulina* (Haptophyceae) from Norway with preliminary observations on scale production in *C. microcylindra* sp. nov. *Sarsia*, **49**, 65–80.

Lenz, F. (1933). Untersuchungen zur Limnologie von Strandseen. *Verhandlungen der internationalen Vereinigung für theoretische und angevandte Limnologie*, **6**, 166–77.

Liebert, F. and Deerns, W. M. (1920). Onderzoek naar de oorzak van een Vischsterfte in den Polder Workumer Nieuwland, nabij Workum. *Verhandlungen en Rapporten uitgegeven door Rijksinstituten voor Visscherijonderzoek*, **1**, 81–93.

Lindholm, T. and Virtanen, T. (1992). A bloom of *Prymnesium parvum* Carter in a small coastal inlet in Dragsfjärd, southwestern Finland. *Environmental Toxicology and Water Quality: an International Journal*, **7**, 165–70.

Manton, I. and Parke, M. (1962). Preliminary observations on scales and their mode of origin in *Chrysochromulina polylepis* sp. nov. *Journal of the Marine Biological Association of the United Kingdom*, **42**, 565–78.

Massart, J. (1920). Recherches sur les organismes inférieurs. VIII. Sur la motilité des Flagellates. *Bulletin de la Classe des Sciences, Académie Royale de Belgique*, Sér. 5, **6**, 116–41.

Meldahl, A-S., Edvardsen, B., and Fonnum, F. (1993). The effect of *Prymnesium*-toxin on neurotransmitter transport mechanisms: the development of a sensitive test method. In *Toxic phytoplankton blooms in the sea*, (ed. T. J. Smayda and Y. Shimizu), pp. 895–900. Elsevier, Amsterdam.

Meldah, A-S. and Fonnum, F. (1991). Toxin production in *Chrysochromulina leadbeateri*. In *The* Chrysochromulina leadbeateri *bloom in Vestfjorden, North Norway, May–June 1991*. Fisken og Havet No. 3, (ed. F. Rey), pp. 89–93. Havforskningsinstituttet, Bergen.

Moestrup, Ø. (1979). Identification by electron microscopy of marine nanoplankton from New Zealand, including the description of four new species. *New Zealand Journal of Botany*, **17**, 61–95.

Moestrup, Ø. (1992). Blooms of *Chrysochromulina* in Danish waters during April and May 1992. *Red Tide Newsletter*, **5** (1), 1–2.

Moestrup, Ø. and Larsen, J. (1992). Potentially toxic phytoplankton. 1. Haptophyceae (Prymnesiophyceae). ICES Identification Leaflets for Plankton, No. 179, pp. 1–11. International Council for the Exploration of the Sea, Copenhagen.

Nicholls, K. H. (1978). *Chrysochromulina breviturrita* sp. nov., a new freshwater member of the Prymnesiophyceae. *Journal of Phycology*, **14**, 499–505.

Nicholls, K. H., Beaver, J. L., and Estabrook, R. H. (1982). Lakewide odours in Ontario and New Hampshire caused by *Chrysochromulina breviturrita* Nich. (Prymnesiophyceae). *Hydrobiologia*, **96**, 91–5.

Nielsen, T. G., Kiørboe, T., and Bjørnsen, P.K. (1990). Effects of a *Chrysochromulina polylepis* subsurface bloom on the planktonic community. *Marine Ecology Progress Series*, **62**, 21–35.

Otterstrøm, C. V. and Steeman Nielsen, E. (1940). Two cases of extensive mortality in fishes caused by the flagellate *Prymnesium parvum*. *Report of the Danish Biological Station*, **44**, 1–24.

Parke, M., Manton, I., and Clarke, B. (1955). Studies on marine flagellates. II. Three new species of *Chrysochromulina*. *Journal of the Marine Biological Association of the United Kingdom*, **34**, 579–609.

Parke, M., Manton, I., and Clarke, B. (1956). Studies on marine flagellates. III. Three further species of *Chrysochromulina*. *Journal of the Marine Biological Association of the United Kingdom*, **35**, 387–414.

Parke, M., Manton, I., and Clarke, B. (1958). Studies on marine flagellates. IV. Morphology and microanatomy of a new species of *Chrysochromulina*. *Journal of the Marine Biological Association of the United Kingdom*, **37**, 209–28.

Parke, M., Manton, I., and Clarke, B. (1959). Studies on marine flagellates. V. Morphology and microanatomy of *Chrysochromulina strobilus* sp. nov. *Journal of the Marine Biological Association of the United Kingdom*, **38**, 169–88.

Parke, M. and Manton, I. (1962). Studies on marine flagellates. VI. *Chrysochromulina pringsheimii* sp. nov. *Journal of the Marine Biological Association of the United Kingdom*, **42**, 391–404.

Paster, Z. (1973). Pharmacognosy and mode of action of *Prymnesium*. In *Marine pharmacognosy*, (ed. D. F. Martin and G. M. Padilla), pp. 241–63. Academic Press, New York.

Petrova, V. J. (1966). Verbreitung und massenhafte Entwicklung der giftigen Chrysomonade *Prymnesium parvum* Carter in den Seen an der Bulgarischen Schwarzmeerküste. *Zeitschrift für Fischerei*, **14**, 9–14.

Pouchet, M. G. (1892). Sur une algue pélagique nouvelle. *Comptes Rendus Hebdomadaires des Séances et Mémoires de la Societé de Biologie*, **44**, 34–6.

Reich, K. and Aschner, M. (1947). Mass development and control of the phytoflagellate *Prymnesium parvum* in fish ponds in Palestine. *Palestine Journal of Botany*, **4**, 14–23.

Rey, F. and Aure, J. (1991). The *Chrysochromulina leadbeateri* bloom in the Vestfjord, North Norway, May–June 1991. Environmental conditions and possible causes. In *The* Chrysochromulina leadbeateri *bloom in Vestfjorden, North Norway, May–June 1991*, Fisken og Havet No. 3, (ed. F. Rey), pp. 13–32. Havforskningsinstituttet, Bergen.

Rosenberg, R., Lindahl, O., and Blanck, H. (1988). Silent spring in the sea. *Ambio*, **17**, 289–90.

Savage, R. W. (1930). The influence of *Phaeocystis* on the migration of the herring. *Fishery Investigations, London*, Series II, **12** (2); 5–14.

Sieburth, J. M. (1960). Acrylic acid, an 'antibiotic' principle in *Phaeocystis* blooms in antarctic waters. *Science*, **132**, 676–77.

Shilo, M. (1971). Toxins of Chrysophyceae. In *Microbial toxins*, Vol. 7, (ed. S. Kadis, A. Ciegler, and S. J. Ajl), pp. 67–103. Academic Press, New York.

Shilo, M. (1981). The toxic principles of *Prymnesium parvum*. In *The water environment. Algal toxins and health*, (ed. W. W. Carmichael), pp. 37–47. Plenum Press, New York.

Skjoldal, H. R. and Dundas, I. (1991). The *Chrysochromulina polylepis* bloom in Skagerrak and Kattegat in May–June 1988: environmental conditions, possible causes, and effects. *ICES Cooperative Research Report*, **175**, 1–59.

Smayda, T. J. (1990). Novel and nuisance phytoplankton blooms in the sea: evidence for a global epidemic. In *Toxic marine phytoplankton*, (ed. E. Granéli, B. Sundström, L. Edler, and D. M. Anderson), pp. 29–40. Elsevier, New York.

Sournia A. (1988). *Phaeocystis* (Prymnesiophyceae): how many species? *Nova Hedwigia*, **47**, 211–17.

Sørensen, H. M. (1990). Toksiske og potentielt toksiske algers økologi i danske farvande. In *Toksiske og potentielt toksiske alger i danske farvande*, (by T. Bjergskov,

J. Larsen, Ø. Moestrup, H. M. Sørensen, and P. Krogh), pp. 61–157. The Fish Inspection Service, Ministry of Fisheries, Copenhagen.

Steemann Nielsen, E. and Hansen, V. K. (1959). Light adaptation in marine phytoplankton populations and its interrelation with temperatures. *Physiologia Plantarum*, **12**, 353–70.

Strodtmann, S. (1898). Über die vermeintliche Schädlichkeit der Wasserblüte. *Forschungsberichte aus der Biologischen Station zu Plön*, **6**, 206–12.

Throndsen, J. and Eikrem, W. (1991). The biology of *Chrysochromulina leadbeateri* and other *Chrysochromulina* species. In *The* Chrysochromulina leadbeateri *bloom in Vest fjorden, North Norway, May–June 1991*, Fisken og Havet No. 3, (ed. F. Rey), pp. 63–71. Havforskningsinstituttet, Bergen.

Throndsen, J., Larsen, J., and Moestrup, Ø. (1993). Toxic algae: toxicity of *Chrysochromulina* with new ultrastructural information on *C. polylepis*. (In press).

Tobiesen, A. (1991). Growth rates of *Heterophrys marina* (Heliozoa) on *Chrysochromulina polylepis* (Prymnesiophyceae). *Ophelia*, **33**, 205–12.

Ulitzur, S. (1965). The mode of action of cofactors on the ichthyotoxic activity of *Prymnesium* toxin and other fish toxins. Ph.D. Thesis, Hebrew University, Jerusalem.

Ulitzur, S. and Shilo, M. (1970). Procedure for purification and separation of *Prymnesium parvum* toxins. *Biochimica Biophysica Acta*, **201**, 350–63.

Valkanov, A. (1964). Untersuchungen über *Prymnesium parvum* Carter und seine toxische Einwirkung auf die Wasserorganismen. *Kieler Meeresforschungen*, **20**, 65–81.

Verity, P. G. and Smayda, T. J. (1989). Nutritional value of *Phaeocystis pouchetii* (Prymnesiophyceae) and other phytoplankton for *Acartia* spp. (Copepoda): ingestion, egg production and growth of nauplii. *Marine Biology*, **100**, 161–71.

Walne, P. R. (1970). Studies on the food value of nineteen genera of algae to juvenile bivalves of the genera *Ostrea, Crassostrea, Mercenaria* and *Mytilus. Fishery Investigations, London*, **26**, 1–62.

Weisse, T. (1983). Feeding of calanoid copepods in relation to *Phaeocystis pouchetii* blooms in the German Wadden Sea. *Marine Biology*, **74**, 87–94.

Yasumoto, T., Underdal, B., Aune, T., Hormazabal, V., Skulberg, O., and Oshima, Y. (1990). Screening for hemolytic and ichthyotoxic components of *Chrysochromulina polylepis* and *Gyrodinium aureolum* from Norwegian coastal waters. In *Toxic marine phytoplankton*, (ed. E. Garnéli, B. Sundström, L. Edler, and D. M. Anderson), pp. 436–40. Elsevier, New York.

15. Haptophytes as feedstocks in mariculture

S. W. JEFFREY*, M. R. BROWN* and J. K. VOLKMAN†
*CSIRO Division of Fisheries and †CSIRO Division of Oceanography, Marine Laboratories, Hobart, Tasmania, Australia

Abstract

Mass-cultured microalgae, including haptophytes, are used extensively in mariculture as live feeds for all growth stages of molluscs, for the early larval stages of crustaceans and fish, and for brine shrimp and rotifers that are animal intermediates in mariculture food chains. Temperate strains of *Pavlova*, and temperate and tropical strains of *Isochrysis* have been used successfully worldwide for oyster and mussel farming since the 1950s. Microalgae must supply a balanced mixture of key nutrients. As part of a study of the nutritional (biochemical) composition of about 40 species of both classical and new isolates of microalgae from six algal classes, we have identified seven strains of haptophytes from the genera *Pavlova* and *Isochrysis* which show excellent nutritional profiles for larval animals. Grown under both standard laboratory and hatchery conditions they show good profiles of total protein, carbohydrate, and lipid (gross composition), they are rich in essential and non-essential amino acids, and they have significant concentrations of essential polyunsaturated fatty acids. Haptophytes are also rich in B, C, D, and K vitamins. The cells are easily assimilated by larval animals because of their small size (3–8 μm) and absence of a tough cell wall. Other attributes include fast growth rates, easy mass-culture, wide temperature and salinity tolerances, and absence of toxins. The nutritional attributes of haptophytes studied in our laboratories will be described in this paper.

Introduction

Microalgae in mariculture

Live microalgal diets are central to most marine farming activities, particularly those raising filter-feeding animals such as oysters and mussels (Fig. 15.1; Jeffrey *et al.* 1990). The adult grow-out stages of these animals are fed by the natural phytoplankton of local waters. In Australia, marine

The Haptophyte Algae (ed. J. C. Green and B. S. C. Leadbeater), Systematics Association Special Volume No. 51, pp. 287–302. Clarendon Press, Oxford, 1994.

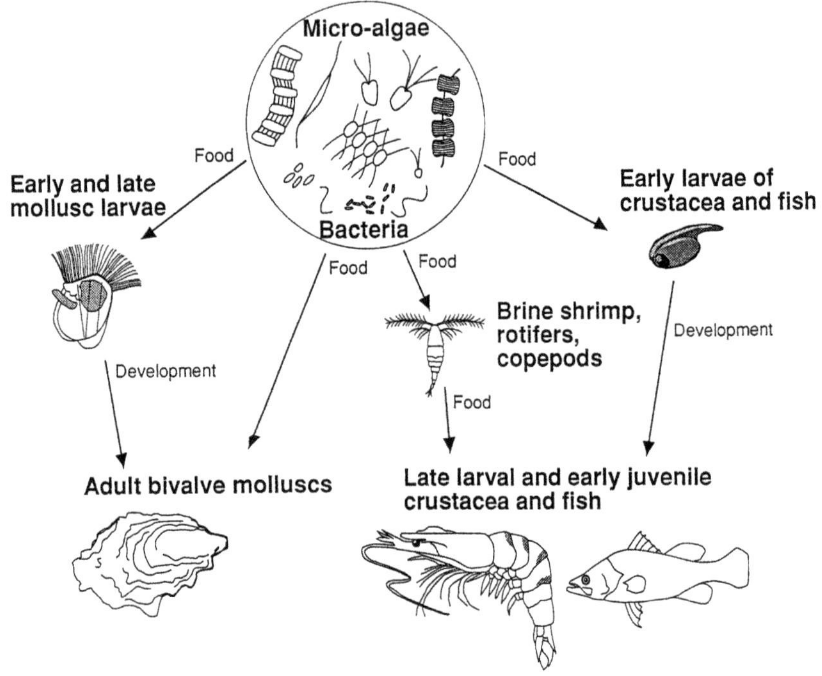

Fig. 15.1 The basic nutritional role of microalgae in mariculture.

farming is a new and small primary industry of immense potential (Jeffrey and Garland 1987; Garland 1988; O'Sullivan 1990) with pearl and edible oysters, mussels, trout, salmon, barramundi, and prawns being the most important animals farmed.

The success of microalgae as food species depends not only on their nutritional quality, but also on their tolerance to temperature, salinity, and light, especially if grown in outdoor tanks and ponds. Successful microalgal production for animal farming requires:

(1) knowledge of the optimal growth responses of desirable species under local conditions;
(2) knowledge of the nutritional value of microalgae matched to the nutritional requirements of animals at all stages of their life cycles; and
(3) development of new microalgal strains more suited to local conditions.

Microalgae suitable for mariculture are usually in the nanoplankton size range (2–20 μm), except when the animal larvae are raptorial feeders (e.g. prawn larvae) and need larger prey (e.g. chains of diatoms such as *Skeletonema costatum*) or grazers on algal turfs, composed of mats of 'sticky' pennate diatoms (e.g. abalone larvae). The biochemical (nutritional) profiles of a range of over 35 cultured species (diatoms, haptophytes, eustigmatophytes, chlorophytes, prasinophytes, cryptomonads, and rhodophytes) have now been analysed in our laboratories (Volkman *et al.* 1989; Brown 1991; Brown and Jeffrey 1992; Brown and Miller 1992; Dunstan *et al.* 1992; Volkman *et al.* 1993). In this paper we focus on the high nutritional quality of the haptophytes.

The role of haptophytes

The first haptophytes used in European mariculture were strains of *Isochrysis galbana* and *Pavlova lutheri* (Walne 1963, 1967, 1970) which were used for the culture of oyster (*Ostrea edulis*), clam (*Mercenaria* spp.), and mussel larvae (*Mytilus* spp.). Fertilized ponds and tanks were the early sources of these algal strains and they quickly proved useful feedstocks. Before long, culture collections of these organisms were established (see Guillard 1972), and a range of unialgal cultures of species suitable for mass culture and animal feeding was soon established.

Other haptophytes that have been used in overseas mariculture include strains of *Pleurochrysis* (*Cricosphaera*) *carterae, Dicrateria inornata, Emiliania huxleyi, Isochrysis galbana, Pavlova gyrans, Pavlova lutheri,* and *Pseudoisochrysis paradoxa* (Brown *et al.* (1989). *Prymnesium parvum* was unsuitable because of the presence of toxins. The success of filter-feeding molluscs raised on these diets are listed in Brown *et al.* (1989)

The three most frequently requested haptophyte species used in Australian mariculture are the two tropical species *Isochrysis* sp. (clone T.ISO), isolated from Tahiti and *Pavlova salina* (Sargasso Sea), and the temperate species *Pavlova lutheri*, isolated from Finland.

Methods

Algal cultures used in the present study

The haptophyte species studied in the present work, and their geographic sources, are shown in Table 15.1. The axenic strains, *Isochrysis* sp. strain T. ISO (CS-177) and *Pavlova lutheri* strain MONO (CS-182), were obtained from Dr R. R. L. Guillard, Culture Collection of Marine Phytoplankton, Bigelow Laboratory for Ocean Sciences, Maine, USA in 1986 and maintained in the CSIRO Algal Culture Collection (Jeffrey 1980). Cultures were grown in 1.2 l volumes as described in Volkman *et al.* (1989) and Brown (1991). Appropriate aliquots were taken for the analytical procedures as described above and in Brown and Miller (1992).

Table 15.1 Haptophyte species used in the present work, with culture code numbers, geographic source and mariculture use

Species	CSIRO culture code number	Geographic source	Present Australian mariculture use
Isochrysis galbana Parke	CS-22	unknown	This strain not yet used commercially
+*Isochrysis* sp.	CS-177	Tahiti	Widely used tropical strain
+*Pavlova lutheri* (Droop) Green	CS-23	unknown	This strain not yet used commercially
+*Pavlova lutheri* (Droop) Green	CS-182	Finland	Widely used temperate strain
Pavlova salina	CS-49	Sargasso Sea	Frequently used tropical strain
Pavlova sp.	CS-50	Sargasso Sea	This strain not yet used commercially
**Pavlova* sp.	CS-63	Port Phillip Bay, Australia	This strain not yet used commercially

All species were cultured in Guillard and Ryther's (1962) medium 'f' at half strength at 20 °C (or 25 °C for CS-49 and CS-177).

+Axenic culture

*Ultrastructural evidence suggests this species closely resembles *Pavlova gyrans* (Dr M. Vesk, unpublished data).

Analytical methods

Methods for sample preparation and analysis of the gross biochemical composition, amino acids, and sugars were as outlined in Brown (1991). Amino acids were analysed by high-performance liquid chromatography (HPLC) and sugars by gas chromatography (Brown 1991). The vitamins, ascorbic acid, and riboflavin were both assayed by reverse phase HPLC with fluorimetric detection (Shearer 1986; Brown and Miller 1992). Fatty acids, after transesterification to methyl esters, were analysed by capillary GC using polar and non-polar capillary columns and GC-mass spectrometry (Volkman *et al.* 1989). Lipid classes were determined by thin layer chromatography with flame ionization detection (Iatroscan TLC-FID) (Volkman and Nichols 1991).

Results and discussion

Gross biochemical composition of haptophytes

The gross biochemical composition (total protein, carbohydrate, lipid, and minerals) of haptophytes can vary substantially between species (Table 15.2). The range of values of haptophytes and other algal classes show little difference in composition except that diatoms have a higher content of mineral due to siliceous walls (not shown in Table 15.2; Parsons *et al.* 1961). For logarithmic phase cultures, protein is the major organic component of haptophytes (23–41%), followed by lipid (8.8–23%) and carbohydrate (6.0–12.9%). The composition is dependent on culture media (Wikfors *et al.* 1984), temperature (James *et al.* 1989), light intensity (Thompson *et al.* 1990), and photoperiod (Caron *et al.* 1988), but it is most sensitive to stage of harvest. Stationary phase cultures of the haptophytes *Pavlova lutheri* and *Isochrysis* sp. (T.ISO) contained approximately 3–4 times more carbohydrate, 30–50% less protein, and (for *P. lutheri* only) twice the lipid than logarithmic phase cultures (Brown *et al.* 1993*b*).

Differences in the gross composition of microalgae may contribute to differences in their nutritional quality, but in most cases specific nutrients such as fatty acids, sugars, and vitamins are the most important nutrients.

Amino acid composition

The amino acid content of over 35 species of microalgae is strikingly similar in composition, irrespective of algal class (Brown 1991; Brown and Jeffrey 1992; Volkman *et al.* 1993; Brown, unpublished data). Of 18 amino acids analysed, aspartate and glutamate occur in the highest concentrations (7.1–12.4%), cystine, methionine, tryptophan, and histidine in lowest concentrations (0.4–3.2%), with other amino acids ranging from 3.2–13.5%.

The nutritional value of a protein 'feed' is considered to be high if its composition of essential amino acids is close to that of the feeding animal (Webb and Chu 1983). The essential amino acids of all haptophyte algae analysed here were either equivalent to, or greater than, the levels in oyster larvae (*Crassostrea gigas*), indicating high nutritional value (Fig. 15.2). The protein quality of haptophytes also remained high under different culture conditions, e.g. stationary phase and logarithmic phase cultures were identical in amino acid composition (Brown *et al.* 1993*b*), as were cultures grown over a range of light intensities (Brown *et al.* 1993*a*).

Sugar composition

Glucose is usually the predominant sugar of microalgal polysaccharide, ranging from 21–87% of total sugars (Brown 1991). In the haptophytes, the principal sugars are glucose (43–81%) and galactose (4–19%) with

Table 15.2 The percentage dry weight of protein, carbohydrate, and lipid in microalgae

	Percentage dry weight		
Algal class/species	Protein	Carbohydrate	Lipid
Haptophytes (4)[1,2]			
Isochrysis galbana	29	12.9	23
Isochrysis sp. (T.ISO)	23	6.0	20
Isochrysis sp. (T.ISO)[a]	41	10.3	15
Pavlova lutheri	29	9.0	12
Pavlova lutheri[a]	28	7.8	8.8
Pavlova salina	26	7.4	12
Mean (range)	29 (23–41)	8.9 (6.0–12.9)	15 (8.8–23)
Diatoms (10)[1,3]			
Mean (range)	28 (12–38)	7.4 (4.7–12)	16 (7.2–20)
Chlorophytes (5)[1,4]			
Mean (range)	22 (15–30)	14 (5.9–23)	15 (8.5–21)
Prasinophytes (7)[1,4]			
Mean (range)	20 (6–31)	14.1 (12–17)	13 (9.5–17)
Eustigmatophytes (4)[1,2,5]			
Mean (range)	25 (18–34)	7.4 (6.0–9.0)	13 (7.8–18)
Cryptomonads (7)[6]			
Mean (range)	32 (17–47)	8.2 (5.3–12)	17 (12–26)
All species Mean (range)	26 (6.0–47)	9.8 (4.7–23)	15 (7.2–26)

All species were grown under the standard conditions of 20 °C (temperate species) or 25 °C (tropical species) at 70–80 $\mu E\ m^{-2} s^{-1}$ (12 : 12 h L : D cycles) in 2 l flasks containing 1.2 l of culture media, except [a] (80 l aerated bag cultures). The number of species analysed from each class is given in parentheses.

[1]Brown 1991;[2] Brown *et al.* 1993;[3] Brown and Jeffrey unpublished data;[4] Brown and Jeffrey 1992;[5] Volkman *et al.* 1993;[6] Jeffrey *et al.* unpublished data.

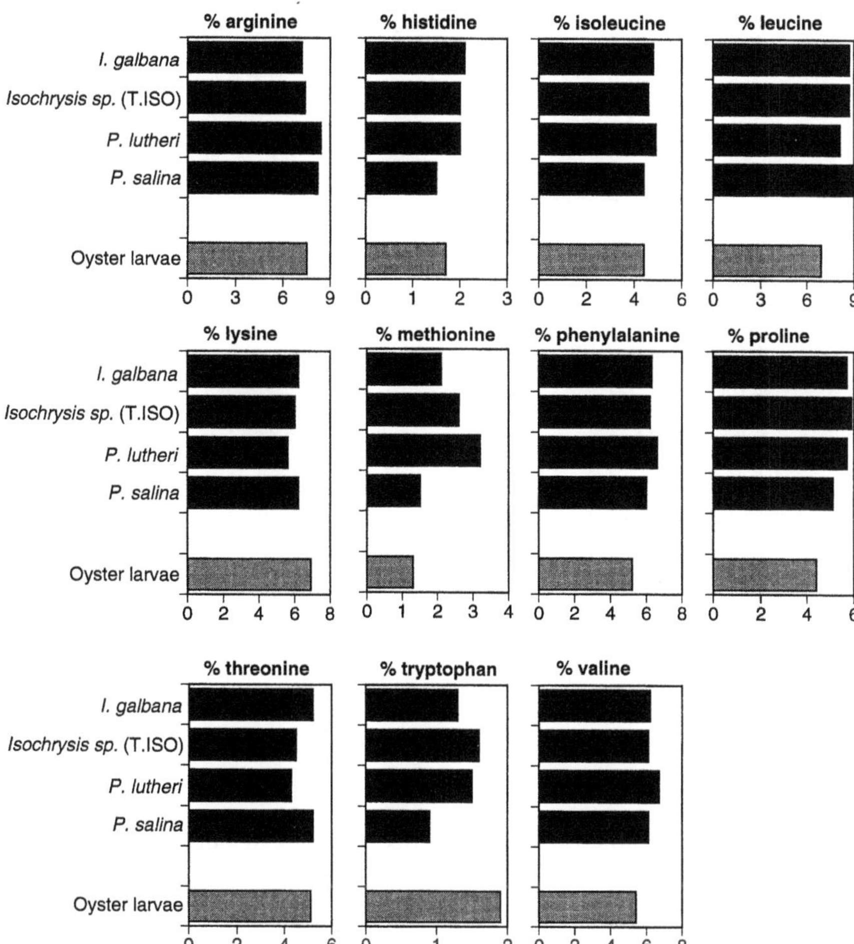

Fig. 15.2 The amino acid composition of haptophyte microalgae (■) compared with oyster larvae (▩).

lesser amounts of mannose, fucose, rhamnose, ribose, and xylose. Higher levels of arabinose (1.6–11.7%) are found in haptophytes compared with other classes of algae (0–2.4%; Fig. 15.3).

The nutritional significance of the algal sugar composition may be important to feeding animals because they digest different polysaccharide types with different efficiencies (Kristensen 1972; Onishi *et al.* 1985). Chrysolaminarin, a glucose-rich polysaccharide of haptophytes (Lee 1980), should be effectively digested by the amylase present in the digestive organs of molluscs and crustaceans (Kristensen 1972).

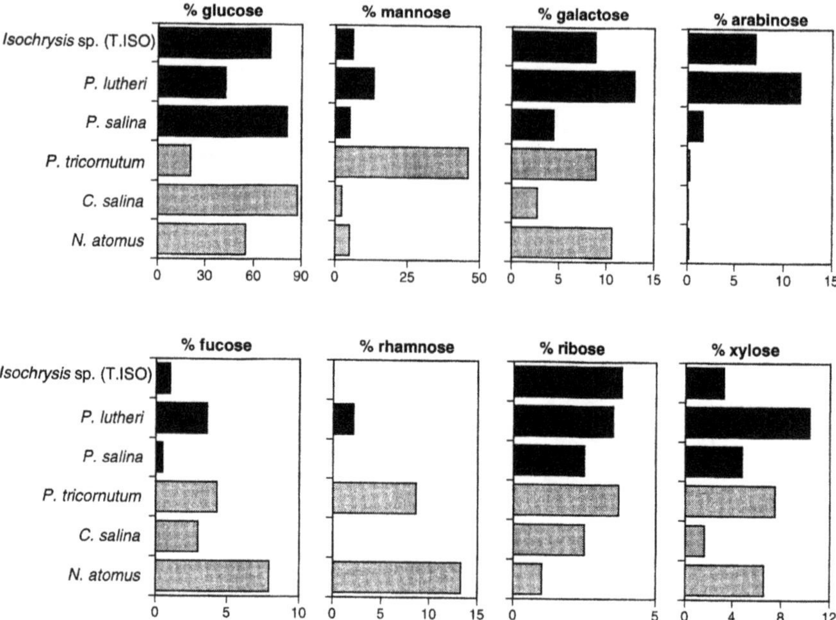

Fig. 15.3 The sugar composition of polysaccharide from haptophytes (■) and other microalgae (▩).

Vitamins

Few data are available on vitamin concentrations in microalgae, particularly those species used in mariculture. We have examined the ascorbic acid (vitamin C) composition of 11 species used for mariculture (Brown and Miller 1992), and the riboflavin (vitamin B_2) content of six species (Brown and Farmer 1994). Figure 15.4 shows that while the content of ascorbic acid in logarithmic phase cultures varied from 1.3 mg/g dry weight for *P. lutheri* to 16 mg/g for the diatom *Chaetoceros gracilis*, all species provided a rich source of both vitamins for maricultured animals, whose daily needs are much less (0.03–0.2 mg/g; Millikan 1982; Shigueno and Itoh 1988). Levels of riboflavin (20–40 μg/g) were also in excess in the dietary requirements of maricultured species (< 10 μg/g; Soliman and Wilson 1992).

DeRoeck-Holtzhauer *et al.* (1991) measured the content of 10 vitamins in five microalgae; thiamine (B_1), riboflavin (B_2), B_6, B_{12}, ascorbic acid (vitamin C), pyridoxyl phosphate, and the fat-soluble vitamins A, D, E, and K. Although all algae were rich in most vitamins, they also all had low concentrations of at least one or more. The two haptophytes *I. galbana* and *P. lutheri* were rich in thiamine, pyridoxyl phosphate, ascorbic acid,

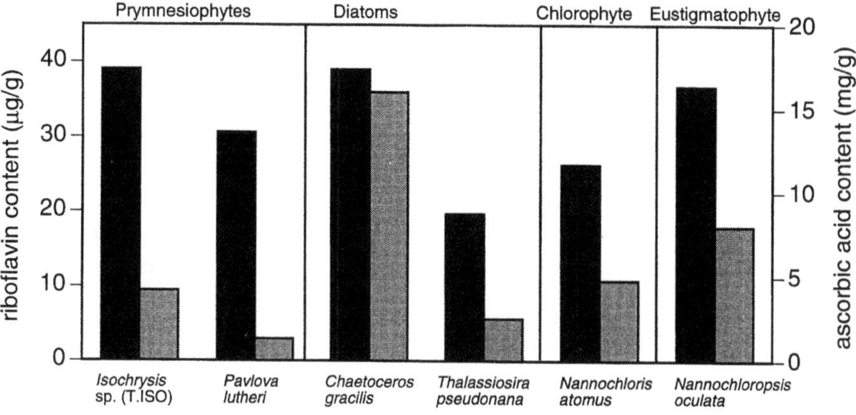

Fig. 15.4 Proportions of riboflavin (■) and ascorbic acid (▩) in microalgae.

and vitamins B_{12}, D, K, and (for *I. galbana* only) pyridoxine, but had low concentrations of vitamins A and E. The concentrations of ascorbic acid in microalgae reported by DeRoeck-Holtzhauer *et al.* (1991) were 2–100 times lower than our values (Brown and Miller 1992), probably due to their long storage at –30 °C prior to analysis.

Marine microalgae not only produce vitamins, but many species have requirements for specific vitamins. The haptophytes *Isochrysis galbana* and *Pavlova gyrans* require both thiamin and vitamin B_{12} (Turner 1979).

Differences in vitamin composition of microalgae could account for differences in their nutritional value. Carefully selected mixed algal diets would be needed to provide high concentrations of all needed vitamins for maricultured animals.

Lipid composition

Microalgae display a diversity of lipid compositions, both in terms of the amount and proportion of individual lipid classes, but also in the presence of unique lipids or characteristic distributions of compounds within a lipid class. Differences in sterol and fatty acid distributions often reflect taxonomic classifications (see Conte *et al.*, Chapter 19).

Total lipid content ranged from 8.8 to 23% of dry weight in *Isochrysis* and *Pavlova* species (Table 15.2). In *Pavlova*, this corresponds to 4.5–8.0 pg cell^{-1} (Volkman *et al.* 1991). Polar lipids (includes phospholipids, glycolipids, and others) predominated in most haptophytes (Volkman *et al.* 1989), but a few haptophytes such as *Emiliania huxleyi*, *Isochrysis* spp., *Chrysotila lamellosa*, and *Gephyrocapsa oceanica* also contain very long chain C_{37}-C_{39} unsaturated ketones and hydrocarbons (Volkman *et al.* 1980;

Marlowe *et al.* 1984; J. K. Volkman, unpublished data; Conte *et al.*, Chapter 19). The nutritional value of alkenones is unknown, but experiments have shown that they are not assimilated by the copepod *Calanus helgolandicus* when fed *Emiliania huxleyi* (Volkman *et al.* 1980).

Triacylglycerols are significant constituents in *P. lutheri*, which may enhance its nutritional value since triacylglycerols are readily utilized as energy reserves. This alga does not contain long-chain unsaturated ketones, but it does synthesize unusual 4-methyl sterols and steroidal diols (Ballantine *et al.* 1979; Volkman *et al.* 1990). Free fatty acids and hydrocarbons are relatively minor lipid constituents in these species.

Fatty acids

The fatty acid content and composition is a very important determinant of the nutritional value of an algal food. Many studies have shown that diets deficient in the long-chain essential polyunsaturated fatty acids 20:5(n-3) and 22:6(n-3) produce poor larval growth in a range of animal species (e.g. Enright *et al.* 1986; Kanazawa 1985). Haptophytes are generally a good source of these fatty acids compared with other microalgae (Fig. 15.5), although amounts do vary considerably between species.

The major fatty acids of *Isochrysis* sp. clone T.ISO are 18:4(n-3), 14:0, 18:1(n-9), 16:0, and 22:6(n-3) (Volkman *et al.* 1989). A distinctive feature of this fatty acid composition is the very low abundance of 20:5(n-3) and relatively high abundance of 22:6(n-3). Fatty acid analyses of the closely related species *Isochrysis galbana* have shown little agreement (reviewed by Volkman 1989; López Alonso *et al.* 1992). For example, the percentage of 20:5(n-3) and 22:6(n-3) reported for *I. galbana* ranges from 0 to 14.4% and 0 to 18.9% of the total fatty acids respectively. Such quantitative differences may account for the diversity of views concerning the value of *I. galbana* as a mariculture food (Helm and Laing, 1987). López Alonso *et al.* (1992) in a study of the fatty acid composition of 59 clones of *I. galbana* derived from a single clonal starter culture reported significant variations in the proportions of key fatty acids (e.g. 20:5(n-3) ranged from 13.2–31.9% and 22.6(n-3) from 4.3–13.4%). This suggests that genetic factors may also contribute the variations in biochemical composition.

The major fatty acids of *Pavlova lutheri* are 16:0, 16:1(n-7), 20:5(n-3), and 22:6(n-3) (Volkman *et al.* 1991). The high abundances of C_{20} and C_{22} PUFA make this a particularly useful food (Fig. 15.5). *Pavlova salina* and two tropical isolates of *Pavlova* sp. also have a similar fatty acid content (2.6–3.0 pg/cell) and distribution of fatty acids (Volkman *et al.* 1991). The tropical species grow well at 30 °C and thus might be useful replacements for the temperate *P. lutheri* in tropical hatcheries. Polyunsaturated fatty acids (PUFA) comprise more than half of the total fatty acids (50–64%). *Pavlova lutheri* contains less PUFA than the tropical *Pavlova* species

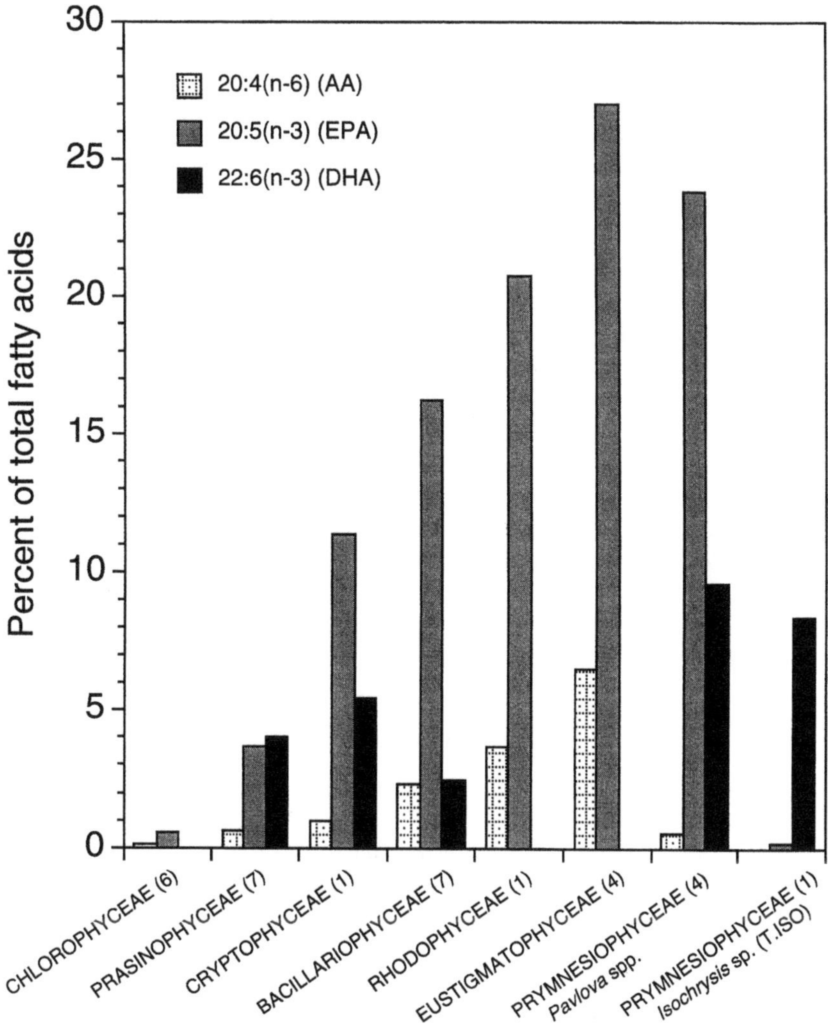

Fig. 15.5 Summary of the average percentage composition of the long-chain PUFAs 20 : 4(n-6), 20 : 5(n-3), and 22 : 6(n-3) in the total fatty acids of microalgal classes commonly used in mariculture. Values can vary across species of a class and by choice of culture conditions, and therefore should be used only as a guide to typical proportions found in each algal class. The number of species or strains examined within each class is given in parentheses.

(50.4%). Major PUFA include 18:4(n-3) (7–15%), 20:5(n-3) (21–28%), 22:5(n-6) (3–7%), and 22:6(n-3) (8–11%). *Pavlova* species apparently do not contain the unusual PUFA 18:5(n-3), which occurs in some haptophytes and dinoflagellates (Volkman *et al.* 1989).

The proportions of fatty acids and lipid content of *P. lutheri* vary with the age of the culture (Emdadi and Berland 1989). The concentration of 22:6(n-3) is reduced at very low light levels (Thompson *et al.* 1990), but the possible effects of temperature are less clear. Thompson *et al.* (1992) found that cell volume tended to increase with temperature, but lipid and carbohydrate content showed no consistent trends. Ackman *et al.*(1968) found little difference between the fatty acid composition of cultures grown at 10 °C and 20 °C, and Volkman *et al.*(1991) found only minor differences in *P. salina* grown at 20 °C and 27 °C.

Other haptophytes display a diversity of distributions with 16:0, 16:1(n-7), 18:1(n-9), 18:4(n-3), 20:5(n-3), and 22:6(n-3) fatty acids being most common (Holz 1981; Marlowe *et al.* 1984; Volkman *et al.* 1981; Conte *et al.*, Chapter 19). Such diversity of biochemical composition is unusual and probably reflects the heterogeneity of algal types that are included in this taxonomic grouping (Volkman *et al.* 1990).

Rotifers (*Brachionus plicatilus*) readily ingest *Pavlova* cells from which they can accumulate high concentrations of 20:5(n-3), 22:6(n-3), and other PUFA within a few hours (Nichols *et al.* 1989). After just three hours feeding on *Pavlova* the distribution of fatty acids in the rotifer becomes almost indistinguishable from its algal food. This provides a convenient means of transferring these acids to crustacean and fish larvae.

Concluding remarks

Haptophytes are excellent live microalgal feedstocks for use in mariculture. Species of *Isochrysis* and *Pavlova* are favoured because of their small size, fast growth rates, wide temperature ranges, absence of tough cell wall, and absence of toxins that could affect either the animal or the human consumer. Nutritionally they are rich in vitamins, contain high-quality protein rich in essential amino acids, as well as glucose-rich polysaccharides. The lipids have a high proportion of essential long-chain C_{20} and C_{22} polyunsaturated fatty acids. *Pavlova* species are an excellent source of both 20:5(n-3) and 22:6(n-3), whereas *Isochrysis* sp. (T.ISO) contains a high content of 22:6(n-3), but not 20:5(n-3). The lack of the latter acid apparently does not impair its nutritional value and is compensated by the high 22:6(n-3) content.

Acknowledgements

We thank Jeannie-Marie Leroi for growing the algal cultures; Kelly Miller, Suzanne Norwood, and Christine Farmer (CSIRO Division of Fisheries) for assistance with gross composition, sugar, amino acid, and vitamin analyses; Graeme Dunstan and Stephanie Barrett (CSIRO Division of Oceanography) for assistance with lipid and fatty acid analyses. This research was supported by Grants 1986/81, 1988/69, 1990/63, and 1991/59 from the Fishing Industry Research and Development Trust Account, Grant A 18831836 from the Australian Research Council, and a Rural Credits Development Grant.

References

Ackman, R. G., Tocher, C. S., and McLachlan, J. (1968). Marine phytoplankter fatty acids. *Journal of the Fisheries Research Board of Canada*, **25**, 1603–20.

Ballantine, J. A., Lavis, A., and Morris, R. J. (1979). Sterols of the phytoplankton-effects of illumination and growth stage. *Phytochemistry*, **18**, 1459–66.

Brown, M. R. (1991). The amino acid and sugar composition of 16 species of microalgae used in mariculture. *Journal of Experimental Marine Biology and Ecology*, **145**, 79–99.

Brown, M. R. and Farmer, C. (1994). Riboflavin content of six species of microalgae used in mariculture. *Journal of Applied Phycology*, (In press).

Brown, M. R. and Jeffrey, S. W. (1992). Biochemical composition of microalgae from the classes Chlorophyceae and Prasinophyceae. 1. Amino acids, sugars and pigments. *Journal of Experimental Marine Biology and Ecology*, **161**, 91–113.

Brown, M. R and Miller, K. A. (1992). The ascorbic acid content of eleven species of microalgae used in mariculture. *Journal of Applied Phycology*, **4**, 205–15.

Brown, M. R., Jeffrey, S. W., and Garland, C. D. (1989). Nutritional aspects of microalgae used in mariculture: a literature review. CSIRO Marine Laboratories Report 205.

Brown, M. R., Dunstan, G. A., Jeffrey, S. W., Volkman, J. K., Barrett, S. M., and LeRoi, J. M. (1993*a*). The influence of irradiance on the biochemical composition of the prymnesiophyte *Isochrysis* sp. (clone T-ISO). *Journal of Phycology*, **29**, 601–12.

Brown, M. R., Garland, C. D., Jeffrey, S. W., Jameson, I. D., and Leroi, J. M. (1993*b*). The gross and amino acid compositions of batch and semi-continuous cultures of *Isochrysis* sp. (clone T.ISO), *Pavlova lutheri* and *Nannochloropsis oculata*. *Journal of Applied Phycology*, **5**, 285–96.

Caron, L., Mortain-Bertrand, A. and Jupin, H. (1988). Effect of photoperiod on photosynthetic characteristics of two marine diatoms. *Journal of Experimental Marine Biology and Ecology*, **123**, 211–26.

Conte, M. H., Volkman, J. K., and Eglinton, G. (1993). Lipid biomarkers of the Prymnesiophyceae. (Chapter 19, this volume).

De Roeck-Holtzhauer, Y., Quere, I., and Claire, C. (1991). Vitamin analysis of five planktonic microalgae and one macroalga. *Journal of Applied Phycology*, **3**, 259–64.

Dunstan, G. A., Volkman, J. K., Jeffrey, S. W., and Barrett, S. M. (1992). Biochemical composition of microalgae from the green algal classes Chlorophyceae and Prasinophyceae. 2. Lipid classes and fatty acids. *Journal of Experimental Marine Biology and Ecology*, **161**, 115–34.

Emdadi, D. and Berland, B. (1989). Variation in lipid class composition during batch growth of *Nannochloropsis salina* and *Pavlova lutheri. Marine Chemistry*, **26**, 215–25.

Enright, C. T., Newkirk, G. F., Craigie, J. S., and Castell, J. D. (1986). Evaluation of phytoplankton as diets for juvenile *Ostrea edulis* L. *Journal of Experimental Marine Biology and Ecology*, **96**, 1–13.

Garland, C. D. (1988). Australian mariculture: the role of hatcheries in animal production. University of Tasmania, Hobart, Australia.

Guillard, R. R. L. (1972). Culture of phytoplankton for feeding marine invertebrates. In *The culture of marine invertebrate animals*, (ed. W. L. Smith and M. H. Chenley), pp. 26–60. Plenum, New York.

Helm, M. M. and Laing, I. (1987). Preliminary observations on the nutritional value of Tahiti *Isochrysis* to bivalve larvae. *Aquaculture*, **62**, 281–8.

Holz G. G. (1981) Non-isoprenoid lipids and lipid metabolism of marine flagellates. In *Biochemistry and physiology of protozoa*, (ed. M. Levandowsky and S. H. Hunter), pp. 301–32. Academic Press, New York.

James, C. M., Al-Hinty, S., and Salman, A. E. (1989). Growth and ω3 fatty acid and amino acid composition of microalgae under different temperature regimes. *Aquaculture*, **77**, 337–57.

Jeffrey, S. W. (1980). Cultivating uni-cellular marine plants. In *CSIRO Division of Fisheries and Oceanography Research Report*, 1977–1979, pp. 22–43.

Jeffrey, S. W. and Garland, C. D. (1987). Mass culture of microalgae essential for mariculture hatcheries. *Australian Fisheries*, **46**, 14–8.

Jeffrey, S. W. Garland, C. D., and Brown, M. R. (1990). Microalgae in Australian mariculture, In *The biology of marine plants*, (ed. M. N. Clayton and R. J. King), pp. 400–14. Longman Cheshire, Melbourne.

Kanazawa, A. (1985). Nutrition of penaid prawns and shrimps. In *Proceedings of the first international conference on the culture of penaid prawns/shrimps*, (ed. Y. Taki *et al.*), pp. 123–30. South East Asian Fisheries Development Center, Iloilo.

Kristensen, J. H. (1972). Carbohydrases of some marine invertebrates with notes on their food and on the natural occurrence of the carbohydrases studied. *Marine Biology*, **14**, 130–42.

Lee, R. E. (1980). Prymnesiophyceae. In *Phycologie*, (ed. R. E. Lee), pp. 155–72. Cambridge University Press.

López Alonso, D., Molina Grima, E., Sánchez Pérez, J. A., García Sánchez, J. L., and García Comarcho, F. (1992). Isolation of clones of *Isochrysis galbana* rich in eicosapentaenoic acid. *Aquaculture*, **102**, 363–71.

Marlowe, I. T., Green, J. C., Neal, A. C., Brassell, S. C., Eglinton, G., and Course, P. A. (1984). Long chain (n-C_{37}-C_{39}) alkenones in the Prymnesiophyceae. Distribution of alkenones and other lipids and their taxonomic significance. *British Phycological Journal*, **19**, 203–16.

Millikin, M. R. (1982). Qualitative and quantitative nutrient requirements of fishes: a review. *US National Marine Fisheries Service Fishery Bulletin*, **80**, 655–86.

Nichols, P. D., Holdsworth, D. G., Volkman, J. K., Daintith, M., and Allanson, S. (1989). High incorporation of essential fatty acids by the rotifer *Brachionus plicatilis* fed on the prymnesiophyte alga *Pavlova lutheri. Australian Journal of Marine and Freshwater Research*, **40**, 645–55.

O'Sullivan, D. (1990). Status and prospects for Australian aquaculture (1989–90). Special Report, DOSAQUA, Mowbray, Tasmania, Australia, 67 pp.

Onishi, T., Suzuki, M., and Kikuchi, R. (1985). The distribution of polysaccharide hydrolase activity in gastropods and bivalves. *Bulletin of the Japanese Society of Scientific Fisheries*, **51**, 301–8.

Parsons, T. R., Stephens, K., and Strickland, J. D. H. (1961). On the chemical composition of eleven species of marine phytoplankters. *Journal of the Fisheries Research Board of Canada*, **18**, 1001–16.

Shearer, M. J. (1986). Vitamins. In *HPLC of small molecules*, (ed. C. K. Lim), pp. 157–219. IRL Press, Oxford.

Shigueno, K. and Itoh, S. (1988) Use of Mg-L-ascorbyl-2-phosphate as a vitamin C source in shrimp diets. *Journal of the World Aquaculture Society*, **19**, 168–74.

Soliman, A. K. and Wilson, R. P. (1992). Water-soluble vitamin requirements of tilapia. 2. Riboflavin requirement of blue tilapia *Oreochromis aureus. Aquaculture*, **104**, 309–14.

Thompson, P. A., Harrison, P. J., and Whyte, J. N. C. (1990). Influence of irradiance on the fatty acid composition of phytoplankton. *Journal of Phycology*, **26**, 278–88.

Thompson, P. A., Guo, M.-X., and Harrison, P. J. (1992). Effects of variation in temperature. I. On the biochemical composition of eight species of marine phytoplankton. *Journal of Phycology*, **28**, 481–8.

Turner, M. F. (1979). Nutrition of some marine microalgae with special reference to vitamin requirements and utilization of nitrogen and carbon sources. *Journal of the Marine Biological Association of the United Kingdom*, **59**, 535–52.

Volkman, J. K. (1989). Fatty acids of microalgae used as feedstocks in aquaculture. In *Fats for the future*, (ed. R. C. Cambie), pp. 263–83. Ellis Horwood, Chichester.

Volkman, J. K. and Nichols, P. D. (1991). Applications of thin layer chromatography-flame ionization detection to the analysis of lipids and pollutants in marine and environmental samples. *Journal of Planar Chromatography*, **4**, 19–26.

Volkman, J. K., Eglinton, G., Corner, E. D. S., and Sargent, J. R. (1980). Novel unsaturated straight-chain C_{37}-C_{39} methyl and ethyl ketones in marine sediments and a coccolithophorid *Emiliania huxleyi*. In *Advances in organic geochemistry, 1979*, (ed. A. G. Douglas and J. R. Maxwell), pp. 219–27. Pergamon Press, Oxford.

Volkman, J. K., Smith, D. J., Eglinton, G., Forsberg, T. E. V., and Corner, E. D. S. (1981). Sterol and fatty acid composition of four marine haptophycean algae. *Journal of the Marine Biological Association of the United Kingdom*, **61**, 509–27.

Volkman, J. K., Jeffrey, S. W., Nichols, P. D., Rogers, G. I., and Garland, C. D. (1989). Fatty acid and lipid composition of ten species of microalgae used in mariculture. *Journal of Experimental Marine Biology and Ecology*, **128**, 219–40.

Volkman, J. K., Kearney, P. S., and Jeffrey, S. W. (1990) A new source of 4-methyl sterols and 5α(H)-stanols in sediments: prymnesiophyte microalgae of the genus *Pavlova. Organic Geochemistry*, **15**, 489–97.

Volkman, J. K., Dunstan, G. A., Jeffrey, S. W., and Kearney, P. S. (1991). Fatty acids from microalgae of the genus *Pavlova. Phytochemistry*, **30**, 1855–9.

Volkman, J. K., Brown, M. R., Dunstan, G. A., and Jeffrey, S. W. (1993). The biochemical composition of marine microalgae from the class Eustigmatophyceae. *Journal of Phycology*, **29**, 69–78.

Walne, P. R. (1963). Observations on the food value of seven species of algae to the larvae of *Ostrea edulis*. I. Feeding experiments. *Journal of the Marine Biological Association of the United Kingdom*, **43**, 767–84.

Walne, P. R. (1967). The food value of 13 species of unicellular algae to *Artemia salina*. ICES. C.M. 1967/E:5, 7 pp. (mimeo).

Walne, P. R. (1970). Studies on the food value of nineteen genera of algae to juvenile bivalves of the genera *Ostrea, Mercenaria*, and *Mytilus. Fishery Investigations, Ministry of Agriculture, Fisheries and Food*, Series 2, **25**(5), 1–62.

Webb, K. L. and Chu, F. E. (1983). Phytoplankton as a food source for bivalve larvae. In *Proceedings of the second international conference on aquaculture nutrition: Biochemical and physiological approaches to shellfish nutrition*, (ed. G. D. Pruder, C. J. Langdon, and D. E. Conklin), pp. 272–91. Louisiana State University, Baton Rouge.

Wikfors, G. H., Twarog, J. W., and Ukeles, R. (1984). Influence of chemical composition of algal food sources on growth of juvenile oysters, *Crassoestrea virginica. Biological Bulletin*, **167**, 251–63.

16. Dimethyl sulfide: production and atmospheric consequences

GILLIAN MALIN, PETER S. LISS and
SUZANNE M. TURNER
School of Environmental Sciences, University of East Anglia, Norwich UK

Abstract

The oceans are the major source of a number of volatile organic compounds which enter the atmosphere via air-sea exchange and play a significant role in biogeochemical cycles and global atmospheric chemistry. The focus of this chapter will be one such compound, dimethyl sulfide (DMS; $(CH_3)_2S$), its biogenic production by haptophytes and possible global effects.

Introduction

Research on DMS in the environmental context was stimulated by the pioneering work of Lovelock *et al.* (1972). Prior to 1972 there was already considerable evidence for the biogenic origin of marine DMS. In studies on biological methylation Challenger *et al.* (1957) demonstrated that DMS occurred in marine macroalgae, that its precursor was the tertiary sulfonium compound β-dimethylsulphoniopropionate (DMSP), and that DMSP reacted with cold aqueous alkali to produce DMS and acrylic acid in a 1:1 ratio (Challenger and Simpson 1948; Challenger *et al.* 1957). In the early 1960s, Sieburth found that bacterial sterility of penguin guts resulted from acrylic acid, released during the digestion of krill which had grazed on blooms of *Phaeocystis pouchetii* containing DMSP (Sieburth 1961). Subsequently, DMSP was detected in a range of unicellular marine phytoplankton species (Ackman *et al.* 1966; Ishida 1968).

The biological function of DMSP in marine phytoplankton is not fully understood. In addition to methylation and antibiotic potentials mentioned above, there are several lines of evidence suggesting that DMSP is a compatible solute: it accumulates with increasing salinity in several phytoplankton species (Vairavamurthy *et al.* 1985; Dickson and Kirst 1987); *in*

The Haptophyte Algae (ed. J. C. Green and B. S. C. Leadbeater), Systematics Association Special Volume No. 51, pp. 303–20. Clarendon Press, Oxford, 1994.

vitro studies show that it minimizes NaCl inhibition of malate dehydrogenase and improves the stability of some enzymes at high temperature (Gröne and Kirst 1991; Nishiguchi and Somero 1990), and it acts as a cryoprotectant in some Antarctic seaweeds and ice algae (Karsten *et al.* 1990*a*; Kirst *et al.* 1991). DMSP is synthesized from methionine (Greene 1962; Gröne and Kirst 1992), and Uchida *et al.* (1992) recently isolated methionine decarboxylase, which is a key enzyme in DMSP synthesis, from *Crypthecodinium cohnii.*

Field and culture studies

During the last 20 years field studies have shown that DMS is ubiquitous in surface seawater, although considerable spatial and seasonal variability in concentration is observed. DMS is now known to be the dominant volatile sulfur compound in seawater, and particulate and dissolved DMSP (DMSPp and DMSPd) can also be readily detected (Turner *et al.* 1988). Field surveys which have considered phytoplankton speciation in relation to DMS and/or DMSP concentration, and culture studies have identified dinoflagellates and haptophytes (including coccolithophorids and *Phaeocystis pouchetii*) as significant sources of DMS (e.g. Turner *et al.* 1988; Keller *et al.* 1989*a*).

Factors affecting the production of DMS and DMSP

Recently research has focussed increasingly on processes which affect DMSP production, and influence DMS formation and its subsequent air-sea exchange. We now consider factors which may influence DMSP levels in phytoplankton cells, and how transformations and interactions within the microbial food web might control the dissolved DMS/DMSP pool in seawater (see Fig. 16.1).

Phytoplankton

Phytoplankton speciation and abundance are the primary factors influencing the concentration of DMSP and DMS in seawater. Most publications related to DMS/DMSP in seawater and cultures have been concerned with haptophytes. Significant correlations have been found between *Phaeocystis* and DMS (Barnard *et al.* 1984; Holligan *et al.* 1987), and high DMS concentrations have been found in coastal blooms of *Phaeocystis* in Antarctica (Gibson *et al.* 1990) and the southern North Sea. For a review on *Phaeocystis* and DMS, see Liss *et al.* (1993). Holligan *et al.* (1987) recorded DMS, maxima at stations where a mixed coccolithophorid population, comprising *Emiliania huxleyi, Cyclococcolithus leptoporus* and

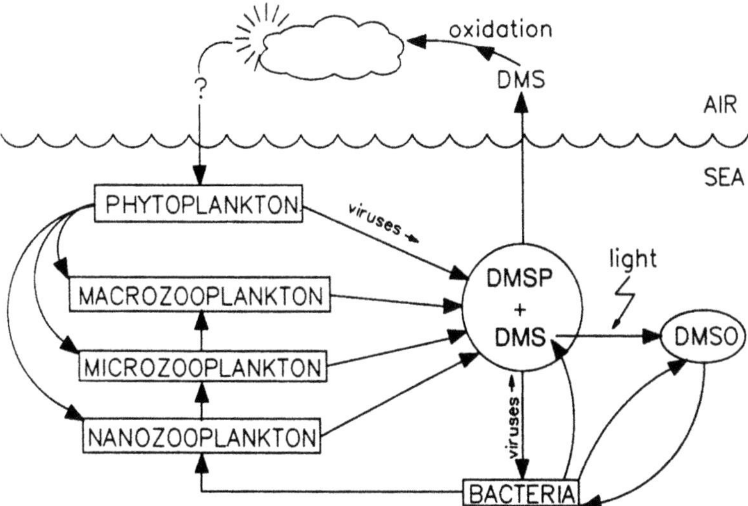

Fig. 16.1 Scheme to show interactions within the marine microbial food web leading to a pool of dissolved DMSP and DMS in seawater, and processes which influence how much DMSP is converted to DMS and the quantity of DMS entering the atmosphere by air-sea exchange. Production of DMSP and DMS and DMSP-containing phytoplankton cells can occur by exudation and natural senescence, but zooplankton grazing is likely to be a much more important route. Losses of DMS/DMSP via sedimentation of algal cells, zooplankton, and their faecal pellets are not shown. Viral lysis may represent a significant release pathway (→ viruses). Bacterioplankton can take up DMSP, DMS, and DMSO via different metabolic routes which may lead to release of DMS or DMSO into seawater, or incorporation into bacterial biomass. Bacterial uptake could possibly be reduced by viruses (→ viruses) and grazing on bacteria by nanozooplankton. DMSO is produced by photoxidation and bacterial DMS oxidase. The emission of DMS to the atmosphere is controlled by physical processes and, following atmospheric oxidation of DMS, leads to enhanced albedo of clouds. Charlson *et al.* (1987) hypothesized that the DMS/CCN/climate link might represent a climate regulating mechanism.

Gephyrocapsa, was dominant. Turner *et al.* (1988) subsequently observed a good correlation between DMS and chlorophyll *a* for samples in which *C. leptoporus* was abundant. Leck and Rodhe (1991) found a relationship between DMS and *Chrysochromulina* in the Baltic sea, and Keller *et al.* (1989*a*, *b*) noted high intracellular DMSP levels in a range of haptophytes including *E. huxleyi*, *Phaeocystis*, *Pleurochrysis carterae*, *Prymnesium parvum* and two *Chrysochromulina polylepis* clones isolated from a bloom in Swedish waters.

Evidence from satellite images shows that mesoscale blooms of *E. huxleyi* occur annually in several oceanic areas, including the north-east Atlantic,

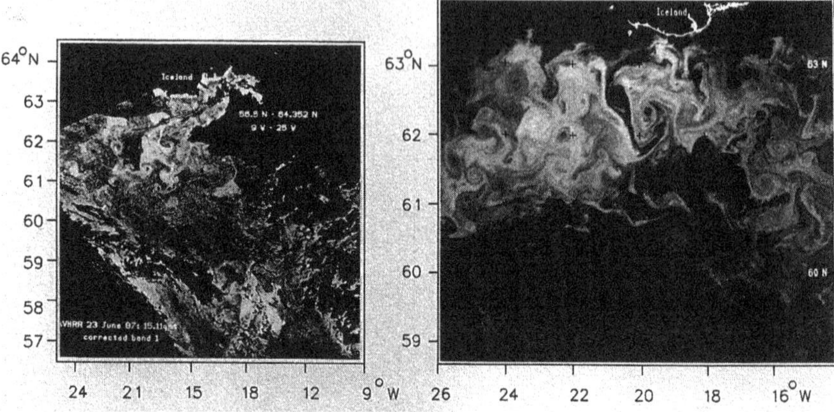

Fig. 16.2 Processed Advanced Very High Resolution Radiometer (AVHRR) satellite images for the north-east Atlantic for 23 June 1987 (*left*) and 17 June 1991 (*right*). Bright areas indicate high reflectance due to *Emiliania huxleyi* coccoliths. Land and areas of cloud cover appear black.

Gulf of Maine, and Argentine Continental Shelf (Balch *et al.* 1992). Our group participated in studies of *E. huxleyi* blooms in the north-east Atlantic in 1987 and 1991, and our intention here is briefly to compare results from the two cruises. For further details, see Malin *et al.* (1993) and Holligan *et al.* (1993). Satellite images received during each cruise, showing areas of high reflectivity indicative of advanced blooms of *E. huxleyi*, can be seen in Fig. 16.2. In 1987 the cruise track centred on the bloom around Rockall (58°00′N; 14°30′W) whereas in 1991 we focussed on the area south of Iceland. In 1987 and 1991 the average DMS concentrations for oceanic samples were 13.5 and 15.4 nmol dm^{-3} respectively, considerably higher than the average global concentration of DMS of 3 nmol dm^{-3} given by Andreae (1990). DMS, DMSPp, DMSPd, and chlorophyll *a* were spatially variable, but rather similar concentration ranges were observed in both studies (Table 16.1). Furthermore, correlations between DMS and chlorophyll *a*, and DMSPp and chlorophyll *a* were also similar (Fig. 16.3). This was surprising given that correlations between DMS and chlorophyll *a*, a taxon-insensitive indicator of biomass, are usually poor unless the data set is subdivided according to phytoplankton speciation and abundance (Turner *et al.* 1988). Spatially averaged flux calculations gave a value of 862 nmol sulphur m^{-2} h^{-1} for the oceanic area in 1987, and 1129 nmol m^{-2} h^{-1} for 1991. These values are somewhat higher than the value of 646 nmol m^{-2} h^{-1} we have calculated for the southern North Sea at the same time of year, and considerably higher than the range 208–416 nmol m^{-2} h^{-1} given for the western North Atlantic

Table 16.1 Surface seawater concentrations for DMS, particulate DMSP (DMSPp), dissolved DMSP (DMSPd, nmol dm^{-3}), and chlorophyll *a* (mg m^{-3}) for oceanic samples collected in the north-east Atlantic June–July 1987 and 1991

		n	average	minimum	maximum	$\sigma n-1$
1987	DMS	101	13.5	3.5	51.0	8.6
	DMSPp	95	107.1	10.8	279.9	49.7
	DMSPd	97	43.3	<1.3	199.0	31.8
	chl *a*	101	1.0	0.3	2.6	0.4
1991	DMS	142	15.4	0.8	59.7	14.4
	DMSPp	126	148.1	9.4	365.6	97.0
	DMSPd	128	30.4	<1.3	95.0	19.44
	chl *a*	140	1.1	0.3	2.7	0.6

(Berresheim *et al.* 1991), and the average value of 274 nmol m^{-2} h^{-1} computed by Bates *et al.* (1992) for the whole of the 65–50° N oceanic area. It seems very likely that large scale blooms of coccolithophorids and other haptophytes, play a significant role in the emission of DMS to the atmosphere.

Nutrients, light, and temperature

The structure of DMSP ($(CH_3)_2S^+(CH_2)_2COO^-$) is similar to that of glycine betaine (GBT, $(CH_3)_3N^+CH_2COO^-$) a quarternary ammonium compound with a well defined osmoregulatory role in a wide range of organisms (see Blunden and Gordon 1986). Since GBT contains nitrogen rather than sulfur, Andreae (1986) suggested that marine phytoplankton might favour DMSP production when nitrogen availability is low. There is some evidence to support this hypothesis: in a culture study, *E. huxleyi* grown in medium which was not supplemented with nitrate had a higher intracellular DMSP concentration throughout a two week period, compared with a culture with added nitrate; and secondly, samples from UK coastal and shelf waters showed higher particulate DMSP concentrations where nitrate was below 0.2 μM, than samples with a similar phytoplankton assemblage, but higher nitrate concentration (Turner *et al.* 1988). Furthermore, Gröne and Kirst (1992) found that intracellular DMSP in *Tetraselmis subcordiformis* cultures increased by 75% within 24 hours of nitrogen deficiency, and Kiene and Service (1993) detected elevated DMS levels in seawater spiked with GBT. They suggested that the

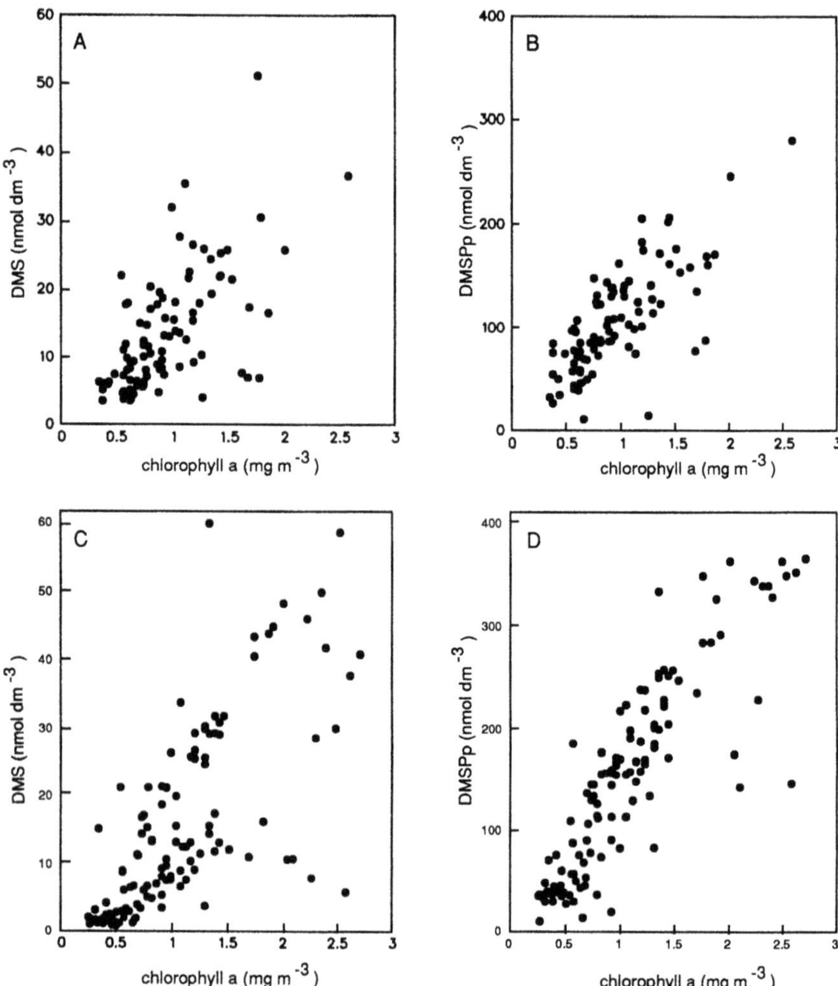

Fig. 16.3 Plots for DMS vs. chlorophyll *a* and particulate DMSP (DMSPp) vs. chlorophyll *a* for oceanic samples collected in the north-east Atlantic in 1987 (A and B) and 1991 (C and D). Regression data are as follows: (A) r = 0.601, signif. > 0.001, n = 101; (B) r = 0.741 signif. > 0.001, n = 95; (C) r = 0.695, signif. > 0.001, n = 139; (D) r = 0.860, signif. > 0.001, n = 123.

phytoplankton had assimilated GBT for use in osmoregulation and had broken down intracellular DMSP which was released as DMS.

Some loss of DMS occurs via photooxidation to dimethylsulfoxide (DMSO; Brimblecombe and Shooter 1986), but DMSPd seems resistant to chemical (Dacey and Blough 1987) and photochemical breakdown under abiotic conditions (our unpublished observations). It is also interesting to note that DMSP degradation to DMS has been observed in 0.2 μm filtered seawater (Turner *et al.* 1988). Enzymatic cleavage of DMSP has also been demonstrated with a partially purified extract of *Crypthecodinium cohnii* (Ishida 1968). Hence, extracellular algal DMSP lyase may be active in seawater.

There have been many studies on the effects of temperature and light on the growth of marine phytoplankton. However, we are not aware of any studies which have considered the effects of temperature, and only one study has been done on the effect of light, on the production of DMS or DMSP by marine phytoplankton. In studies on green macroalgae Karsten *et al.* (1990*b*, 1991, 1992) showed that DMSP concentration increased at low temperature and with increasing light intensity and daylength, and Vetter and Sharp (1993) found that the diatom *Skeletonema costatum* produced more DMS over an 18-day period at a light intensity of 380 $\mu E\ m^{-2}\ s^{-1}$ than at 38 $\mu E\ m^{-2}\ s^{-1}$, but DMS production per cell did not differ consistently between the light treatments during the experiment.

Zooplankton

Actively growing phytoplankton cultures show little DMS and DMSP release (Turner *et al.* 1988; Keller 1989*b*). However, Dacey and Wakeham (1986) demonstrated that dinoflagellate cultures grazed by copepods released more DMS than ungrazed cultures, and that DMS release continued when the zooplankton was moved into fresh medium without prey. Sedimentation of zooplankton faecal pellets may represent a significant pathway by which DMS/DMSP is lost from the euphotic zone. Leck *et al.* (1990) correlated copepod and total macrozooplankton biomass with DMS concentration in the Baltic Sea. Recent research has highlighted the importance of microzooplankton grazing in the natural environment, and this is particularly relevant for phytoplankton species, such as *Emiliania huxleyi* and species of *Chrysochromulina,* which are not effectively grazed by macrozooplankton. Belviso *et al.* (1990) studied filter-fractionated samples from the north-west Mediterranean and concluded that microzooplankton grazing was significant for DMS/DMSP release from phytoplankton cells. We carried out a study in collaboration with Drs Peter Burkill and Glen Tarran of Plymouth Marine Laboratory, using cultures of the heterotrophic dinoflagellate *Oxyrrhis marina* grazing on *E. huxleyi.* The results (Fig. 16.4) indicate that grazing can cause significant increases in DMS concentration. Experiments were also carried out in the coccolithophorid bloom in the north-east Atlantic in 1991, but although *E. huxleyi* was selectively grazed by microzooplankton (see Holligan *et al.* 1993), we were unable to detect concurrent changes in DMS concentration. This may have been due to rapid DMS and DMSP turnover by bacterioplankton.

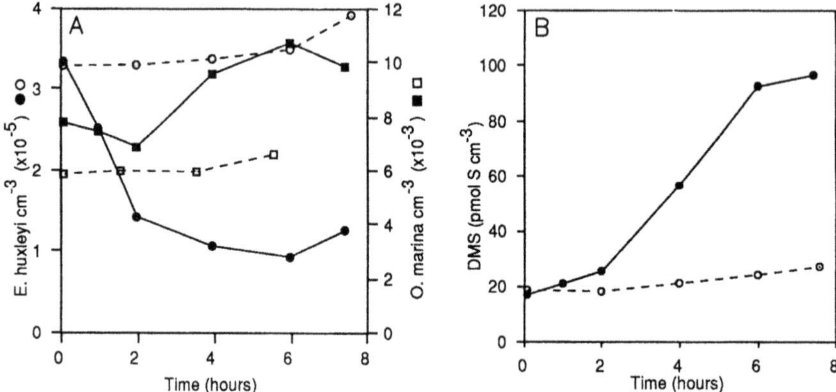

Fig. 16.4 Experiment to investigate the effect of microzooplankton grazing (*Oxyrrhis marina*) on production of DMS from a culture of a DMSP-containing phytoplankton species (*Emiliania huxleyi*). (A) Changes in cell density for the ungrazed *Emiliania huxleyi* control (--○--), and starved *Oxyrrhis marina* control (--□--), *E. huxleyi* in experimental sample in which it was the prey species (—●—), and *O. marina* grazing on *E. huxleyi* in the same sample (—■—). (B) Concomitant changes in DMS concentration in the *E. huxleyi* control (--○--), and the mixed culture in which *O. marina* was grazing on *E. huxleyi* (—●—). DMS was below the detection limit in the *O. marina* control.

It is interesting to note that many haptophyte species can ingest particles such as small phytoplankton cells and bacteria (see Sanders and Porter 1988; Jones *et al.*, Chapter 13). Furthermore, heterotrophic dinoflagellate species, e.g. *Crypthecodinium cohnii* (Kadota and Ishida 1968; Keller *et al.* 1989*a*) contain substantial levels of DMSP. Thus non- photosynthetic and mixotrophic species, may contribute to DMS and DMSP pools in the natural environment directly, and indirectly *via* grazing on other DMSP containing phytoplankton.

Bacteria

DMS, DMSO, and DMSP represent potential carbon and energy sources for marine bacteria, and bacterial production, transformation, and consumption processes have been identified as highly significant factors in the biogeochemical cycle of DMS in seawater (Fig. 16.1). Kiene and Bates (1990) found that bacterial consumption and/or transformation of DMS was between three and 430 times faster than DMS emission to the atmosphere in the Equatorial Pacific. There is considerable evidence of rapid bacterial turnover of the DMSPd, but a large fraction may be metabolized via demethylation yielding 3-methiolpropionate, a pathway which does not yield DMS (Kiene 1993).

Recent data obtained in the Pacific, using an improved gas chromatographic analysis for DMSO (Kiene, personal communication), have confirmed previous observations of a substantial DMSO pool in seawater (Andreae 1980; Gibson *et al.* 1990). Phototrophic bacteria can oxidize DMS to DMSO (Visscher and van Gemerden 1991), and oxidation of DMS *via* co-metabolism has been observed (Zhang *et al.* 1991). Isolates of marine bacteria can utilize DMSO as an electron acceptor in anaerobic respiration, resulting in production of DMS (Clarke and Ward 1988), and DMSO can also be used aerobically by *Hyphomicrobium* spp. as a carbon and energy source (Suylen 1988). Our recent results using antibodies to DMSO reductase suggests that respiratory DMSO reductase is also involved in DMSO assimilation pathways (Hatton *et al.* 1992). The overall significance of these pathways in seawater is not yet known.

Viruses

Viruses are highly abundant in marine aquatic environments, and viral infection is now recognized as an important mechanism for release of dissolved organic matter (see Fuhrman 1992). Infection is common in marine bacteria and cyanobacteria (Proctor and Fuhrman 1990), and intracellular virus-like particles (VLP) have been seen in most eukaryotic algal classes including the haptophytes (see van Etten *et al.* 1991). In a mesocosm study, viral lysis accounted for 25–100% of the mortality of *E. huxleyi* both when nutrient conditions were non-limiting and when nitrate was in short supply, but under phosphate-limiting conditions production of virus particles was restricted (Bratbak *et al.* 1993). Thus under certain conditions viruses may be responsible for the termination of phytoplankton blooms. Viral lysis could be an important mechanism for release of intracellular DMS and DMSP to seawater and we have indicated on Fig. 16.1 where viruses might be of particular significance in the biogeochemical DMS cycle.

Emission of DMS to the atmosphere

At present it is not possible to quantify directly the rate at which DMS is emitted to the atmosphere from seawater. Fluxes are calculated from field measurements of DMS in seawater and estimates of transfer velocity, a term which encompasses the physical processes affecting exchange. Current understanding of the processes controlling the air-sea exchange of gases such as DMS is discussed in Liss and Merlivat (1986). For recent global and regional scale natural sulfur emission estimates see Andreae (1990), and Bates *et al.* (1987, 1992).

Any compound transferred from the ocean to the atmosphere passes through the sea-surface microlayer, operationally defined as the top few microns (Liss 1975). Microbial activity and photochemical reactions occurring in this layer could affect air-sea exchange, e.g. microlayer bacteria could utilize a compound so that it is no longer available for transfer, or enhance or retard transfer by producing an end-product more or less volatile than the precursor. Nguyen *et al.* (1978) found that microlayer samples collected with a Garrett screen contained five times more DMS than subsurface samples. However, collection of microlayer samples for trace gas analysis is problematic, and in a comparison of different methods Turner and Liss (1989) concluded that there were only minor enrichments and depletions for DMS. They did observe DMS enrichment in Garrett screen samples during a phytoplankton bloom, but attributed these to inherent sampling artifacts arising from stressing of the organisms on contact with the screen.

In laboratory experiments, Goldman *et al.* (1988) found that high molecular weight surfactants could reduce gas exchange at the air–sea interface by reducing turbulent eddy energy. Furthermore, Frew *et al.* (1990) showed that the surface active compounds generated by phytoplankton cultures, including the haptophyte *Isochrysis galbana*, reduced oxygen exchange by between 5–50%. Surfactant production was more pronounced during stationary phase, but was independent of the type of nutrient limitation. It is unrealistic to extrapolate from these results to the situation in the environment because it is difficult to mimic oceanic turbulence in the laboratory. During blooms of *Phaeocystis* and *Emiliania huxleyi* the seawater surface can sometimes appear 'oily', suggesting that surfactants may be produced. However, current flux calculation methods cannot account for possible reduced gas exchange due to effects of biogenic surfactants at the air-sea interface.

Environmental change and DMS emissions

Environmental change resulting from anthropogenic perturbation may be significant for emissions of trace gases including DMS. In coastal areas eutrophication has led to increases in the levels of nitrate and phosphorus, whereas silicate concentrations have remained constant or declined. For the North Sea there is evidence for increases in algal biomass and the duration of blooms, and a shift in favour of flagellates rather than diatoms for the period 1950 to 1980 (see Lancelot *et al.* 1987). In mesocosm studies, Egge and Aksnes (1992) found that silicate regulated phytoplankton competition, in that diatoms outcompeted flagellates, including *Phaeocystis,* whenever silicate was available. Current efforts to limit nutrient inputs into areas such as the North Sea could reduce the prevalence and/or duration of coastal blooms of *Phaeocystis* and other flagellates, which could well reduce DMS emissions.

Stratospheric ozone depletion has led to a global increase in UV radiation and there is evidence to show that this is affecting marine phytoplankton. Worrest *et al.* (1978) reported that increased UV-B could lead to decreased primary productivity and changes in species composition. It is difficult to predict how increased UV-B might affect air-sea exchange of DMS, but it is interesting to note that diatoms, which generally contain low levels of DMSP, seem more susceptible to UV-B damage than other types of phytoplankton (Karentz *et al.* 1991). Furthermore, some dinoflagellates and *Phaeocystis pouchetii* produce compounds with UV-B screening properties (Carreto *et al.* 1990, Marchant *et al.* 1991). UV induced changes in community taxonomic structure favouring DMSP producing phytoplankton might lead to increased DMS emissions.

Global consequences of biogenic DMS emissions

The sulfur cycle and acidity

Following emission from seawater, DMS is subject to oxidation by hydroxyl and nitrate free radicals in the troposphere, which leads to formation of sulfur dioxide, methanesulphonic acid (MSA), sulfate, and other minor species (Plane 1989). A recent estimate of the flux of DMS from the oceans to the atmosphere of 35×10^{12} g sulfur per annum can be compared with an anthropogenic input of approximately 93×10^{12} g S (Brimblecombe *et al.* 1989). Hence, wet and dry deposition of DMS oxidation products plays an important role in the global sulfur cycle and significantly affects acidity in areas remote from anthropogenic influence (Liss 1991). Recent studies suggest that biogenic DMS production in the North Sea and north-east Atlantic contributes to sulfur deposition over areas such as western Scotland, Eire, and Scandinavia during spring and summer (e.g. Turner *et al.* 1988, 1993). Furthermore, data from stable sulfur isotope analyses on aerosol samples collected on the west coast of Ireland show that 25% of the sulfur in summer comes from a marine biogenic source (McArdle 1993).

DMS and climate

Charlson *et al.* (1987) put forward the hypothesis that the production of DMS by marine phytoplankton might represent a climate regulating mechanism. Their idea is that increased seawater temperature and/or light leads to increased DMS emissions, followed by atmospheric oxidation, production of CCN, and increased cloud albedo which would serve to counteract the initial temperature/light increase. Central to this idea is the assumption that DMS emissions from seawater are controlled by temperature or sea surface irradiance. As discussed earlier there are no reports on the effects of temperature on the production of DMS and

DMSP by marine phytoplankton, but a study using a diatom culture did show enhanced DMS production at high light intensity (Vetter and Sharp 1993). A few studies have considered DMSP production in relation to temperature and light in the Antarctic environment. These show that Antarctic macroalgae and ice microalgae increase intracellular DMSP levels as temperature decreases (Karsten *et al.* 1990*a*; Kirst *et al.* 1991), and green macroalgae had greater levels of DMSP with increasing light intensity and daylength (Karsten *et al.* 1990*b*, 1991, 1992).

Data for non-sea-salt (nss) sulfate and MSA in Antarctic ice cores suggest that atmospheric DMS concentrations were higher during cooler glacial periods (Legrand *et al.* 1991). Furthermore, ocean productivity is thought to have been higher during past glaciations (Sarnthein *et al.* 1987) due to a greater abundance of diatoms, but present day diatom species show relatively low DMSP levels. This evidence appears contrary to the Charlson *et al.* (1987) hypothesis, but ice core records should be interpreted with caution as they may reflect local variations in sulfur deposition rather than global changes (Holligan 1992).

The phytoplankton, DMS, and climate regulation aspect of the Charlson *et al.* (1987) hypothesis remains controversial, but evidence in support of biogenic DMS emissions affecting cloud albedo is more direct. Marine aerosols affect radiative transfer and climate directly by scattering and absorbing radiation, and indirectly by influencing the droplet size distribution and the albedo of clouds in the marine boundary layer (Fitzgerald 1991). Particles in clean marine air are composed predominantly of submicron sized nss sulfate and MSA derived from the atmospheric oxidation of biogenic DMS, which act as highly efficient cloud condensation nuclei (Savoie and Prospero 1989). Ayers *et al.* (1991) and Hegg *et al.* (1991) have shown a relationship between atmospheric DMS and CCN concentration. Furthermore, Falkowski *et al.* (1992) found correlations between satellite-derived chlorophyll *a* concentration, temperature, and low level cloud reflectance for the central North Atlantic south of Iceland. The albedo of clouds is related to CCN number concentration which is greatest in the shallow stratiform clouds common over some oceanic regions, and is therefore expected to contribute significantly to planetary albedo (Slingo 1989).

Despite the fact that we are still some way off establishing whether or not climate can be regulated by marine organisms, the hypothesis put forward by Charlson *et al.* (1987) has greatly stimulated research on DMS and DMSP production, transformation and consumption in seawater, air-sea exchange of DMS, and the atmospheric processes affecting DMS and its oxidation products.

Acknowledgements

We thank Patrick Holligan and Derek Harbour for help, advice, and encouragement during and following cruises in the north-east Atlantic; Peter Bayliss and Steve Groom for relaying and processing satellite data; Peter Burkill and Glen Tarran for use of unpublished data on microzooplankton grazing; Gary Creissen for drawing the figures; and Sheila Davies for photographic reproduction of Fig. 16.2. This research was supported by grants from the UK Natural Environment Research Council.

References

Ackman, R. G., Tocher, C. S., and McLachlan, J. (1966). Occurrence of dimethyl-β-propiothetin in marine phytoplankton. *Journal of the Fisheries Research Board of Canada*, **23**, 357–64.

Andreae, M. O. (1980). Dimethylsulfoxide in marine and freshwaters. *Limnology and Oceanography*, **25**, 1054–63.

Andreae, M. O. (1986). The ocean as a source of atmospheric sulfur compounds. In *The role of air-sea exchange in geochemical cycling*, (ed. P. Buat-Menard), pp. 331–62. Reidel, Dordrecht.

Andreae, M. O. (1990). Ocean-atmosphere interactions in the global biogeochemical sulfur cycle. *Marine Chemistry*, **30**, 1–29.

Ayers, G. P., Ivey, J. P., and Gillet, R. W. (1991). Coherence between seasonal cycles of dimethyl sulphide, methanesulphonate and sulphate in marine air. *Nature*, **349**, 404–6.

Balch, W. M., Holligan, P. M., and Kilpatrick, K. A. (1992). Calcification photosynthesis and growth of the bloom-forming coccolithophore, *Emiliania huxleyi*. *Continental Shelf Research*, **12**, 1353–74.

Barnard, W. R., Andreae, M. O., and Iverson, R. L. (1984). Dimethylsulfide and *Phaeocystis poucheti* in the southeastern Bering sea. *Continental Shelf Research*, **2**, 103–13.

Bates, T. S., Cline, J. D., Gammon, R. H., and Kelly-Hanson, S. R. (1987). Regional and seasonal variations in the flux of oceanic dimethylsulphide to the atmosphere. *Journal of Geophysical Research*, **92**, 2930–38.

Bates, T. S., Lamb, B. K., Guenther, A., Dignon, J., and Stoiber, R. E. (1992). Sulfur emissions to the atmosphere from natural sources. *Journal of Atmospheric Chemistry*, **14**, 315–37.

Belviso, S., Kim, S-K., Rassoulzadegan, F., Krajka, B., Nguyen, B. C., Milhalopoulos, N., and Buatemenard, P. (1990). Production of dimethylsufoniumpropionate (DMSP) by a microbial food web. *Limnology and Oceanography*, **35**, 1810–21.

Berresheim, H., Andreae, M. O., Iverson, R. L., and Li, S. M. (1991). Seasonal variations of dimethylsulfide emissions and atmospheric sulfur and nitrogen species over the western North Atlantic Ocean. *Tellus*, **43B**, 353–72.

Blunden, G. and Gordon, S. M. (1986). Betaines and their sulphonio analogues in marine algae. *Progress in Phycological Research*, **4**, 39–80.

Bratbak, G., Egge, J. K., and Heldal, M. (1993). Viral mortality of the marine alga *Emiliania huxleyi* (Haptophyceae) and termination of algal blooms. *Marine Ecology Progress Series*, **93**, 39–48.

Brimblecombe, P. and Shooter, D. (1986). Photo-oxidation of dimethylsulphide in aqueous solution. *Marine Chemistry*, **19**, 343–53.

Brimblecombe, P., Hammer, C., Rodhe, H., Ryaboshapko, A., and Bourton, C. F. (1989). Human influence on the sulphur Cycle. In *Evolution of the global biogeochemical sulphur cycle*, (ed. P. Brimblecombe and A. Yu Lein), pp. 77–121. Wiley, New York.

Carreto, J. I., Carignan, M. O., Daleo, G., and de Marco, S. G. (1990). Occurrence of mycosporine-like amino acids in the red-tide dinoflagellate *Alexandrium excavatum*: UV-photoprotective compounds? *Journal of Plankton Research*, **12**, 909–21.

Challenger, F. and Simpson, M. I. (1948). Studies on biological methylation. XII: A precursor of the dimethyl sulphide evolved by *Polysiphonia fastigata* dimethyl-2-carboxyethyl-sulphonium hydroxide and its salts. *Journal of the Chemical Society*, **3**, 1591–97.

Challenger, F., Bywood, R., Thomas, P., and Hayward, B. J. (1957). Studies on biological methylation. XVII. The natural occurrence and chemical reactions of some thetins. *Archives of Biochemistry and Biophysics*, **69**, 514–23.

Charlson, R. J., Lovelock, J. E., Andreae, M. O., and Warren, S. G. (1987). Oceanic phytoplankton, atmospheric sulphur, cloud albedo and climate. *Nature*, **326**, 655–61.

Clarke, G. J. and Ward, F. B. (1988). Purification and properties of trimethylamine-n-oxide reductase from *Shewenella* sp. NCMB 400. *Journal of General Microbiology*, **134**, 379–86.

Dacey, J. W. H. and Wakeham, S. G. (1986). Oceanic dimethylsulfide: production during zooplankton grazing on phytoplankton. *Science*, **233**, 1314–16.

Dacey, J. W. H. and Blough, N. V. (1987). Hydroxide decomposition of DMSP to form DMS. *Geophysical Research Letters*, **14**, 1246–9.

Dickson, D. M. J. and Krist, G. O. (1987). Osmotic adjustment in marine eukaryotic algae: the role of inorganic ions, quarternary ammonium, tertiary sulphonium and carbohydrate solutes. II. Prasinophytes and haptophytes. *New Phytologist*, **106**, 657–66.

Egge, J. K. and Aksnes, D. L. (1992). Silicate as regulating nutrient in phytoplankton competition. *Marine Ecology Progress Series*, **83**, 281–89.

Etten, J. L. van, Lane, L. C., and Meints, R. L. (1991). Viruses and viruslike particles of eukaryotic algae. *Microbiological Reviews*, **55**, 586–620.

Falkowski, P. G., Kim, Y., Kolber, Z., Wilson, C., Wirick, C., and Cess, R. (1992). Natural versus anthropogenic factors affecting low-level cloud albedo over the North Atlantic. *Science*, **256**, 1311–13.

Fitzgerald, J. W. (1991). Marine aerosols: a review. *Atmospheric Environment*, **25A**, 533–45.

Frew, D. C., Goldman, J. C., Dennett, M. R., and Johnson, A. S. (1990). Impact of phytoplankton-generated surfactants on air-sea exchange. *Journal of Geophysical Research*, **95**, 3337–52.

Fuhrman, J. A. (1992). Bacterioplankton roles in cycling of organic matter: the microbial food web. In *Primary productivity and biogeochemical cycles in the sea*, (ed. P. G. Falkowski and A. D. Woodhead), pp. 361–83. Plenum, New York.

Gibson, J. E. A., Garrick, R. C., Burton, H. R., and McTaggart, A. R. (1990) Dimethylsulphide and the alga *Phaeocystis pouchetti* in Antarctic coastal waters. *Marine Biology*, **104**, 339–46.

Goldman, J. C., Dennett, M. R., and Frew, N. M. (1988). Surfactant effects on air-sea exchange under turbulent conditions. *Deep-Sea Research*, **35**, 1953–70.

Greene, R. C. (1962). Biosynthesis of dimethyl-β-propiothetin. *Journal of Biological Chemistry*, **237**, 2251–54.

Gröne, T. and Kirst, G. O. (1991). Aspects of dimethylsulphoniopropionate effects on enzymes isolated from the marine phytoplankter *Tetraselmis subcordiformis* (Stein). *Journal of Plant Physiology*, **138**, 85–91.

Gröne, T. and Kirst, G. O. (1992). The effect of nitrogen deficiency, methionine and inhibitors of methionine metabolism on the DMSP contents of *Tetraselmis subcordiformis* (Stein). *Marine Biology*, **112**, 497–503.

Hatton, A. D., Bonnett, T. C., Benson, N., McEwan, A. G., Malin, G., and Liss, P. S. (1992). Immunochemical analysis of dimethylsulphoxide reductase in bacterial laboratory species and marine isolates. In *Abstracts of the Sixth International Symposium on Microbial Ecology*. Barcelona, 6–11 September, 1992. Abstract P2-10-05.

Hegg, D. A., Ferek, R. J., Hobbs, P. V., and Radke, L. F. (1991). Dimethylsulfide and cloud condensation nucleus correlations in the northeast Pacific Ocean. *Journal of Geophysical Research*, **96**, 13189–91.

Holligan, P. M. (1992). Do marine phytoplankton influence global climate? In *Primary productivity and biogeochemical cycles in the sea*, (ed. P. G. Falkowski and A. D. Woodhead), pp. 487–501. Plenum, New York.

Holligan, P. M., Turner, S. M., and Liss, P. S. (1987). Measurements of dimethyl sulphide in frontal regions. *Continental Shelf Research*, **7**, 213–24.

Holligan, P. M., Fernandez, E., Aiken, J., Balch, W. M., Boyd, P., Burkill, P. H., *et al.* (1993). A biogeochemical study of the coccolithophore, *Emiliania huxleyi*, in the North Atlantic. *Global Biogeochemical Cycles*, **7**, 879–900.

Ishida, Y. (1968). Physiological studies on evolution of dimethyl sulfide from unicellular marine algae. *Memoirs of the College of Agriculture, Kyoto University*, **94**, 47–82.

Kadota, H. and Ishida, Y. (1968). Effect of salts on enzymatical production of dimethyl sulfide from *Gyrodinium cohnii*. *Bulletin of the Japanese Society of Scientific Fisheries*, **34**, 512–18.

Karentz, D., Cleaver, E., and Mitchell, D. L. (1991). Cell survival characteristics and molecular responses of Antarctic phytoplankton to ultraviolet-B radiation. *Journal of Phycology*, **27**, 326–41.

Karsten, U., Wienke, C., and Kirst, G. O. (1990*a*). The β-dimethylsulphoniopropionate (DMSP) content of macroalgae from Antartica and southern Chile. *Botanica Marina*, **33**, 143–46.

Karsten, U., Wienke, C., and Kirst, G. O. (1990*b*). The effect of light intensity and daylength on the β-dimethylsulphoniopropionate (DMSP) content of marine green macroalgae from Antarctica. *Plant, Cell and Environment*, **13**, 989–93.

Karsten, U., Wienke, C., and Kirst, G. O. (1991). Growth pattern and β-dimethylsulphoniopropionate (DMSP) content of green macroalgae at different irradiances. *Marine Biology*, **108**, 151–55.

Karsten, U., Wienke, C., and Kirst, G. O. (1992). Dimethylsulphoniopropionate (DMSP) accumulation in green macroalgae from polar to temperate regions: interactive effects of light versus salinity and light versus temperature. *Polar Biology*, **12**, 603–7.

Keller, M. D., Bellows, W. K., and Guillard, R. R. L. (1989*a*). Dimethyl sulfide production in marine phytoplankton. In *Biogenic sulfur in the environment*, (ed. E. S. Saltzman and W. J. Cooper), pp. 183–200. American Chemical Society, Washington D. C.

Keller, M. D., Bellows, W. K., and Guillard, R. R. L. (1989*b*). Dimethylsulfide production and marine phytoplankton: an additional impact of unusual blooms. In *Novel phytoplankton blooms — causes and impacts of recurrent brown tides and other unusual blooms*, (ed. E. M. Cosper and E. J. Carpenter), pp. 101–15. Springer-Verlag, Berlin.

Kiene, R. P. (1993). Microbial sources and sinks for methylated sulfur compounds in the marine environment. In *Microbial growth on C_1 compounds*, (ed. J. C. Murrell and D. P. Kelly), pp. 15–33. Intercept, Andover.

Kiene, R. P. and Bates, T. S. (1990). Biological removal of dimethyl sulphide from seawater. *Nature*, **345**, 702–5.

Kiene, R. P. and Service, S. K. (1994). The influence of glycine betaine on dimethyl sulfide and dimethylsulfoniopropionate concentrations in sea water. In *The biogeochemistry of global change: radiatively active trace gases* (ed. R. S. Oremland). Chapman & Hall, New York (in press).

Kirst, G. O., Thiel, C., Wolff, H., Nothnagel, J., Wanzek, M., and Ulmke, R. (1991). dimethylsulphoniopropionate (DMSP) in ice-algae and its possible biological role. *Marine Chemistry*, **35**, 381–88.

Lancelot, C., Billen, G., Sournia, A., Weisse, T., Colyn, F., Veldhuis, M., *et al.* (1987). *Phaeocystis* blooms and nutrient enrichment in the continental coastal zones of the North Sea. *Ambio*, **16**, 38–46.

Leck, C., Larsson, U., Bagander, L. E., Johansson, S., and Hadju, S. (1990). DMS in the Baltic Sea — annual variability in relation to biological activity. *Journal of Geophysical Research*, **95**, 3353–63.

Leck, C. and Rodhe, H. (1991). Emissions of marine biogenic sulfur to the atmosphere of northern Europe. *Journal of Atmospheric Chemistry*, **12**, 63–86.

Legrand, M., Feinet-Saigne, C., Saltzman, E. S., Germain, C., Barkov, N. I., and Petrov, V. N. (1991). Ice-core record of oceanic emissions of dimethylsulphide during the last climate cycle. *Nature*, **350**, 144–6.

Liss, P. S. (1975). Chemistry of the sea surface microlayer. In *Chemical oceanography*, vol. 2, (ed. J. P. Riley and G. Scirrow), pp. 193–243. Academic Press, New York.

Liss, P. S. (1991). The sulphur cycle. In *Climate and global change*, (ed. J. C. Duplessy, A. Pons and R. Fantechi), pp. 75–89. CEC, Brussels.

Liss, P. S. and Merlivat, L. (1986). Air-sea exchange rates: introduction and synthesis. In *The role of air-sea exchange in geochemical cycling* (ed. P. Buat-Menard), pp. 113–27. Reidel, Dordrecht.

Liss, P. S., Malin, G., Turner, S. M., and Holligan, P. M. (1993). Dimethyl sulphide and *Phaeocystis*: a review. *Journal of Marine Systems*, (in press).

Lovelock, J. E., Maggs, R. J., and Rasmussen, R. A. (1972). Atmospheric sulphur and the natural sulphur cycle. *Nature*, **237**, 452.

Malin, G., Turner, S. M., Liss, P. S., Holligan, P. M., and Harbour, D. S. (1993). Dimethyl sulphide and dimethylsulphoniopropionate in the north east Atlantic during the summer coccolithophore bloom. *Deep-Sea Research I*, **40**, 1487–1508, (in press).

Marchant, H. J., Davidson, A. T., and Kelly, G. J. (1991). UV-B protecting compounds in the marine alga *Phaeocystis pouchetti* from Antarctica. *Marine Biology*, **109**, 391–5.

McArdle, N. C. (1993). The use of stable sulphur isotopes to distinguish between natural and anthropogenic sulphur in the atmosphere. Ph.D. thesis, University of East Anglia, UK.

Nishiguchi, M. K. and Somero, G. N. (1990). The role of dimethylsulphoniopropionate in protein conformation and stability. *EOS*, **71**, 104–5.

Nguyen, B. C., Gaudry, A., Bonsang, B., and Lambert, G. (1978). Re-evaluation of the role of dimethyl sulphide in the sulphur budget. *Nature*, **275**, 637–39.

Plane, J. M. C. (1989). Gas-phase atmospheric oxidation of biogenic sulphur compounds: a review. In *Biogenic sulfur in the environment*, (ed. E. S. Saltzman and W. J. Cooper), pp. 404–23. American Chemical Society, Washington DC.

Proctor, L. M. and Fuhrman, J. A. (1990). Viral mortality of marine bacteria and cyanobacteria. *Nature*, **343**, 60–2.

Sanders, R. W. and Porter, K. (1988). Phagotrophic phytoflagellates. In *Advances in microbial ecology*, Vol. 10, (ed. K. C. Marshall), pp. 167–92. Plenum Press, New York.

Sarnthein, M., Winn, K., and Zahn, R. (1987). Palaeoproductivity of oceanic upwelling and the effect on anthropogenic CO_2 and climatic change. In *Abrupt climate change*, (ed. W. H. Berger and L. D. Labeyrie), pp. 311–37. Reidel, Dordrecht.

Savoie, D. L. and Prospero, J. M. (1989). Comparison of oceanic and continental sources of non-sea-salt sulphate over the pacific Ocean. *Nature*, **339** 685–7.

Sieburth, J. McN. (1961). Antibiotic properties of acrylic acid, a factor in the gastrointestinal antibiosis of polar marine animals. *Journal of Bacteriology*, **82**, 72–9.

Slingo, A. (1989). Sensitivity of the earth's radiation budget to changes in low clouds. *Nature*, **343**, 49–51.

Suylen, G. M. H. (1988). Microbial metabolism of dimethyl sulphide and related compounds. Ph.D. thesis, University of Technology, Delft, The Netherlands.

Turner, S. M., Malin, G., Liss, P. S., Harbour, D. S., and Holligan, P. M. (1988). The seasonal variation of dimethyl sulfide and dimethylsulfonio-propionate concentrations in near-shore waters. *Limnology and Oceanography*, **33**, 364–75.

Turner, S. M. and Liss, P. S. (1989). A cryogenic technique for sampling the sea surface microlayer for trace gases. *Chemical Oceanography*, **10**, 379–92.

Uchida, A., Ooguri, T., Ishida, T., and Ishida, Y. (1992). Distribution of dimethylsulphide and biosynthesis of its precursor (Dimethylthiopropanoic acid). In *Abstracts of the Sixth International Symposium on Microbial Ecology*. Barcelona, 6–11 September, 1992. Abstract P2-10-18.

Vairavamurthy, A., Andreae, M. O., and Iverson, R. L. (1985). Biosynthesis of dimethylsulfide and dimethylpropiothetin by *Hymenomonas carterae* in relation to sulfur source and salinity variations. *Limnology and Oceanography*, **30**, 59–70.

Vetter, Y.-A. and Sharp, J. H. (1993). The influence of light intensity on dimethylsulfide production by a marine diatom. *Limnology and Oceanography*, **38**, 419–25.

Visscher, P. T. and van Gemerden, H. (1991). Photoautotrophic growth of *Thiocapsa roseopersicina* on dimethyl sulphide. *FEMS Microbiology Letters*, **81**, 247–50.

Worrest, R. C., van Dyke, H., and Thomson, B. E. (1978). Impact of enhanced simulated solar ultraviolet radiation upon a marine community. *Photochemistry and Photobiology*, **27**, 471–8.

Zhang, L., Kunyoshi, I., Hirai, M., and Shoda M. (1991). Oxidation of dimethylsulphide by *Pseudomonas acidovorans* DMR-11 isolated from a peat biofilter. *Biotechnology Letters*, **13**, 223–8.

17. *Emiliania huxleyi* as a key to biosphere-geosphere interactions

PETER WESTBROEK
Department of Chemistry, University of Leiden, The Netherlands

JAN E. VAN HINTE
Faculty of Earth Sciences, Free University, Amsterdam, The Netherlands

GEERT-JAN BRUMMER, MARCEL VELDHUIS
NIOZ, 1790 AB den Burg, The Netherlands

COLIN BROWNLEE
Marine Biological Association, Citadel Hill, Plymouth, UK

J. C. GREEN, ROGER HARRIS
Plymouth Marine Laboratory, Prospect Place, West Hoe, Plymouth, UK

BERIT RIDDERVOLD HEIMDAL
Department of Fisheries and Marine Biology, University of Bergen, Norway

---

Abstract

Cells of *Emiliania huxleyi* are surrounded by 'coccoliths', minute, elegant scales of calcium carbonate. In all the oceans, particularly at mid-latitudes, this species forms gigantic blooms, readily visualized by satellite imagery. Coccolith-bearing organisms, of which *E. huxleyi* is by far the most abundant representative, are major contributors to the ocean floor limestone sediments, and this in turn is the largest long-term sink of inorganic carbon on earth. In addition to coccoliths and organic carbon, *E. huxleyi* blooms emit vast amounts of dimethyl sulfide (DMS), a gas which, on oxidation, is a dominant source of cloud nuclei. The profusion of *E. huxleyi* and its intimate involvement with the global biogeochemical cycles of both carbon and sulfur make it a key component of the greenhouse effect (CO_2), natural acid rain, and albedo regulation.

The Haptophyte Algae (ed. J. C. Green and B. S. C. Leadbeater), Systematics Association Special Volume No. 51, pp. 321–34. Clarendon Press, Oxford, 1994.

We propose to investigate and model the *Emiliania huxleyi* system. This system is to be regarded as a key to the biological involvement in open ocean-climate interactions. Advantages of this species-centred approach are: (1) that a high degree of integration of the research is possible, including biochemical and molecular biological, physiological, field and geological studies, and satellite imagery; and (2) that the coupling between the production of particulate inorganic carbon, particulate organic carbon, and DMSP (the precursor of DMS) can be immediately studied.

Introduction

Today's massive increase of the world's population density and industrial activity raises public anxiety for imminent climate change. Predictions based on General Circulation Models of the ocean/atmosphere system play a key role in the climate debate. It is felt, however, that current model representations are too simple for the predictions to be reliable. The global climate system is complex, with multiple physical, chemical, biological, and cultural forces interacting, and our poor understanding of these interactions precludes a fair judgement of the situation.

To illustrate the complex role of biosphere-geosphere interactions in climate regulation, this chapter briefly discusses the importance of biological calcification in the ocean. Massive amounts of calcium carbonate are removed from the ocean surface waters by biological systems, and this activity is thought to affect the global carbon cycle and climate. We propose a strategy whereby the behaviour of a single prominent calcifying organism is investigated as an introduction to a more general study of climate forcing by planktonic organisms. In general, in-depth studies of selected representative biological systems are useful in the implementation of realistic models of biosphere-geosphere interactions.

Ocean supersaturation and biological calcification

Dissolved calcium and carbonate are steadily flowing into the oceans by runoff, deep sea seepage, and hydrothermal activity (Wilkinson and Algeo 1989). As a result, the ocean tends to be supersaturated with respect to calcium carbonate. On geological time scales, a steady state is maintained, whereby the fluxes of calcium and carbonate entering the ocean equal the amounts leaving the system as calcium carbonate sediment. If no biota were present, the solubility product of calcium carbonate would be the major factor controlling production (and redissolution) of calcium carbonate in the ambient sea water. Encrustration by spontaneous calcium carbonate crystallization presents a continuous hazard to the marine biota, and this has elicited the evolution of powerful mechanisms whereby calcification is prevented. Mucilages (Borman *et al.* 1982) and humic acids (Berner *et al.* 1978) are some of the major crystal poisons responsible for the maintenance of a supersaturated condition of the ocean waters. Low concentrations of such polyanionic macromolecules are known to inhibit crystallization efficiently by binding to crystal nuclei and dislocations. As a result, immediate physico-chemical precipitation from the bulk ocean water is almost entirely suppressed, and biologically-regulated calcification remains as the only major mechanism by which the sedimentation is initiated. Robbins and Blackwelder (1992) gave new support to the scenario of bulk precipitation inhibition and biological regulation of calcification with the observation that even 'whitings' are biologically formed, i.e., by picoplankton calcification. These clouds of aragonite needles, frequently occurring in tropical lagoonal waters, used to be considered as typical products of inorganic precipitation. The implication of this scenario is that no immediate coupling exists between bulk ocean supersaturation and calcium carbonate formation.

A considerable subset of the biota is capable of calcification; in these systems dedicated microenvironments are created where the process takes place. In microbial ecosystems this may be achieved by local degradation of the inhibitory macromolecules. Eukaryotic organisms are capable of more organized skeleton production whereby the supra-macromolecular structure of an organic matrix subtly interacts with growing crystallites (Lowenstam and Weiner 1983, 1989; Simkiss and Wilbur 1989). Crystal poisons are kept away from the growing crystal surfaces; polyanionic macromolecules, attached to the solid matrix substratum may either induce crystallization epitaxially, or bring it to a halt (Westbroek *et al.* 1984). Thus, polyanionic macromolecules play a major role in both the prevention and the regulation of calcification.

The oceanic calcium carbonate sink

One important consequence of the overwhelming biological mediation of calcium carbonate production in the oceans is the fact that, instead of being kept in check by bulk supersaturation, this process behaves in a highly non-linear fashion as it depends on the well-being of the calcifying biota. On the regional scale, slight changes in nutrient availability, illumination, hydrography, grazing, etc., may cause exponential growth or sudden collapse of the calcifying factory. Two environmental factors are particularly important in the stimulation of calcification: (1) moderately oligotrophic conditions favour calcification over biomass production, as the nutritional requirements for calcification are modest; and (2) the bulk of calcification takes place in the energy-rich euphotic zone, where coupling with phototrophic activity represents a major stimulus, according to the reactions:

$$2HCO_3^- + Ca^{++} \rightarrow CaCO_3 + CO_2 + H_2O \quad (17.1)$$
$$CO_2 + H_2O \rightarrow CH_2O + O_2 \quad (17.2)$$

The weight of the mineral forces many calcifying organisms to assume a benthic life style. This, together with the requirements of illumination and oligotrophy, makes the shallow peripheral zone of the tropical oceans, and in particular those areas where upwelling or runoff are absent, as favoured regions of calcium carbonate production. When, in addition, these regions are subsiding through geological time, extensive 'carbonate platforms' may be generated: biologically generated 'solar collectors', often more than 1000 m thick, with the calcifying tissue covering the flat, submerged top (Westbroek 1991).

In the present ocean, the platforms and other shallow water accumulations account for the removal of about 30% of the calcium carbonate introduced into the ocean. The remaining 70% accumulates in the deep sea and results from calcification by pelagic organisms in the euphotic zone. The unicellular coccolithophorid algae are the most important contributors to this latter portion of the carbonate sink, followed by foraminifera and small molluscs. The pelagic life style of these organisms makes unusual demands on the regulation of calcification: to maintain boyancy, the organisms must produce delicate, lightweight skeletons. In several coccolithophorid species, calcification is known to take place inside a specialized Golgi-derived intracellular vacuole and the coccoliths are only extruded after their synthesis is completed. No wonder that pelagic calcification was a late evolutionary innovation: the coccolithophorids appeared on the scene in the Late Triassic (some 200 Ma ago) and it was not until well into the Jurassic, and particularly the Middle Cretaceous (100 Ma ago), that they became important contributors to the oceanic calcium carbonate sink. Prior to that time, carbonate accumulation remained largely restricted to the peripheral parts of the ocean, although deep-sea limestone deposition is known to have occurred in Devonian and Carboniferous times. It appears that the evolution of a specialized vacuole triggered a major geochemical event which affected the global carbon cycle and climate.

Pelagic carbon cycle

The pelagic carbon cycle of the ocean is schematically represented in Fig. 17.1, with some quantitative estimates added with respect to the stocks and fluxes. Reservoir and standing stock sizes (gigatons of carbon) are indicated in bold and flux sizes (gigatons of carbon per year) in italics. Two biological carbon pumps from the surface to the deep ocean are

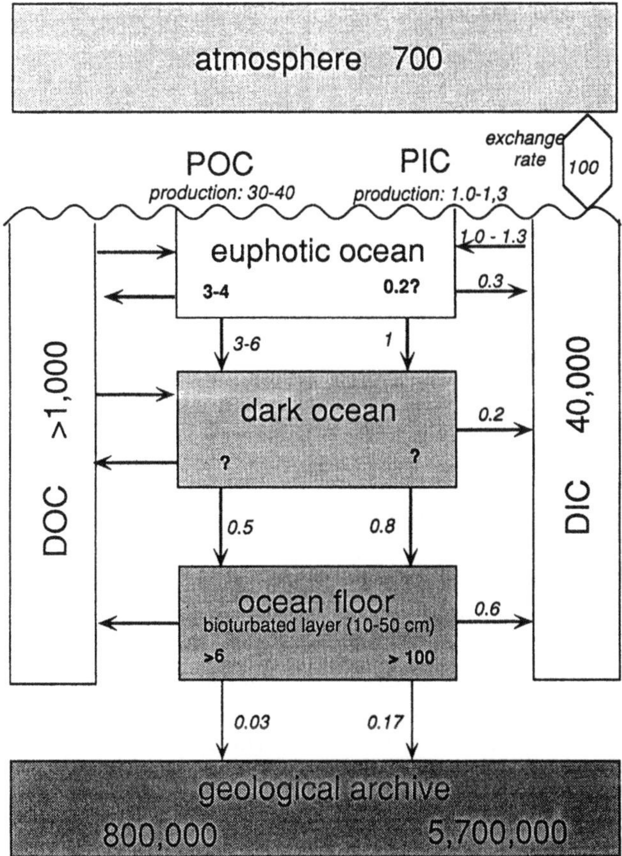

Fig. 17.1 Carbon cycle in the pelagic ocean. **Reservoirs** and **standing stocks** in gigatons of carbon; *fluxes* in gigatons of carbon per year; for further explanation, see text.

distinguished: the carbonate pump (on the right of Fig. 17.1) and the organic carbon pump (on the left).

The annual production of carbonate (particular inorganic carbon — PIC) in the euphotic zone is between 1.0 and 1.3 gigatons per year, only 0.17 gigaton of which accumulates in the deep-sea sediment. The fact that carbonate production depends on biological calcification, and is not immediately coupled with bulk seawater saturation, results in a fourfold overproduction in the pelagic sea. The remainder is re-dissolved in the water column and in the upper 10–50 cm of the ocean sediments.

In the organic carbon pump, between 30 and 40 gigatons of particulate organic carbon (POC) are produced annually in the euphotic zone. Only 3–6 gigatons are exported in particulate form to the dark ocean and 0.03 gigatons remain in the sedimentary record. The 'lost' organic carbon is respired and accumulates as CO_{2aq}, mainly in the deep water. As a net result of this process, dissolved CO_{2aq} is pumped out of the surface water down to the deep ocean, where it accumulates. After a delay of about 500 years it is returned to the surface by upwelling.

The effect of the carbonate and the organic carbon pumps on the concentration of CO_{2aq} are opposite: CO_{2aq} is released upon calcification and absorbed by dissolution. In addition, the carbonate pump affects alkalinity, while the organic pump does so only to a much lesser extent. Note, incidentally, that equation 17.1 does not take into account pH and alkalinity effects, and therefore inadequately reflects this relationship. In particular, less CO_{2aq} is released into the ocean-atmosphere system by calcification than the equation suggests. The two pumps are coupled: not only photosynthesis and calcification, but also respiration and carbonate dissolution tend to stimulate each other. We suggest that the release of acidic respiration and fermentation products of bacterial origin in organic-rich microenvironments, such as faecal pellets or marine snow, may be an important factor in the dissolution of pelagic carbonate particles while they sink down to the ocean floor. The resulting pH-buffering by carbonate particles may in turn enhance the degradation of organic material, thereby subtly influencing the operation of the organic carbon cycle in the oceans.

Note that about 85% of the total carbon stored in the sedimentary sink is in the form of calcium carbonate, while only 15% is organic carbon. Thus, although carbonate transport out of the euphotic ocean is dwarfed by the export of POC, the ultimate sink of carbonate carbon is far greater than of organic carbon. It has been concluded that the carbonate pump exerts a minor climatic effect on short time scales (less than 200 years), and only becomes the dominant factor over geological time (Morse and MacKenzie 1990). If, however, the intimate coupling between the various climatic forcing functions of pelagic calcifying organisms is taken into consideration, a different picture emerges. To demonstrate these relationships we focus on a single dominant calcifying species, *Emiliania huxleyi.*

The *Emiliania* system

Emiliania huxleyi is a marine unicellular alga (Haptophyta, Coccosphaerales) surrounded by 'coccoliths', oval, delicately structured particles, each consisting of a radial array of calcite crystal elements with a highly complex shape (van Emburg, 1989; Young *et al.* 1992; de Vrind-de Jong *et al.*, Chapter 8). Clearly, in the formation of coccoliths, crystallization of the calcite mineral is strictly organized by the cell, the coccoliths being produced intracellularly in a specialized vacuole (Holligan *et al.* 1983; Westbroek *et al.* 1984). After termination of the calcification event, the coccoliths are extruded and incorporated in the extracellular coccosphere.

About 200 extant coccolithophorid species are known, but *E. huxleyi* is by far the most abundant. In all the world's oceans, particularly at mid-latitudes above the spring and early summer thermocline, this species forms huge blooms which can be visualized by satellite imagery. The high reflectance of the blooms, which makes them easily detectable, is caused by backscattering of light by the coccoliths. Typical maximum densities of cells and detached coccoliths are 5×10^3 and 10^5 ml^{-1}, respectively. For a 20 m surface mixed layer, the estimated standing crops of coccoliths are in the order of 4×10^{12} m^{-2}, representing 20 g $CaCO_3$ m^{-2} (Holligan *et al.* 1983; Balch *et al.* 1991). Massive blooming has also been detected in tropical upwelling regions and in the Black Sea. In the main tropical ocean, *E. huxleyi* is a common, but less dominant component of algal communities living at the base of the euphotic zone.

E. huxleyi blooms typically follow the early spring diatom blooms, which appear to be terminated by silica starvation of the upper mixed water layer. Limitation in CO_{2aq} following these preceding blooms would favour calcifying phototrophs (equations 17.1 and 17.2), in particular *E. huxleyi* with its unusually short generation time of about one day. Indeed, *E. huxleyi* is capable of direct uptake of CO_{2aq} from the ambient medium for photosynthesis. Physiological experiments strongly suggest that for calcification HCO_3^- is utilized, so that the CO_{2aq} liberated in that process can be immediately captured by the enveloping chloroplast (Fig. 17.2) (Nimer and Merrett 1992). The assumption of a strong coupling between calcification and photosynthesis in *E. huxleyi* is further corroborated by the observations that the former process is strictly light-dependent (Linschooten *et al.* 1991) and that, in contrast to other algal blooms, those of *E. huxleyi* do not give rise to a local decrease in CO_{2aq} (Robinson, personal communication).

Agglutination of the cells (Cadée 1985), viral and bacterial infection, and grazing by micro-and macro-zooplankton are thought to be the major factors whereby a bloom of *E. huxleyi* is terminated. The contribution of *E. huxleyi* to the ocean's organic carbon flux may be considerable. Cellular remains and coccoliths are transported to the deep sea as faecal pellets or marine snow, and at the ocean floor above the calcite compensation depth the accumulated coccoliths form a long-term sink which may be used as an archive of the evolution of the system. In addition, *E. huxleyi*

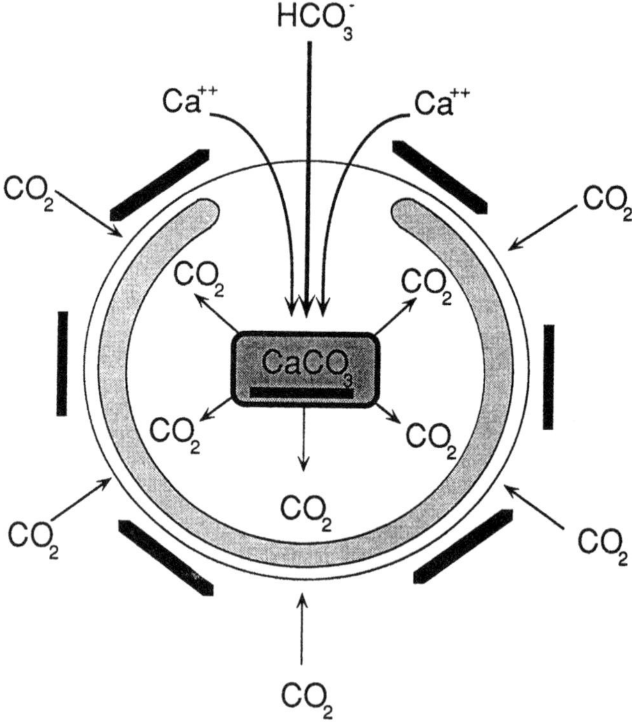

Fig. 17.2 *Emiliania huxleyi.* Cross section through cell with external coccoliths, chloroplast lining the interior of the cell, and a coccolith-synthesizing vesicle in the middle. CO_2 for photosynthesis is taken up directly from the medium. An additional source is bicarbonate which is converted into CO_3^- for calcification and CO_2 for photosynthesis. The latter mechanism is thought to be advantageous during CO_2 shortage in the surrounding seawater.

produces long-chain (C_{37}–C_{38}) alkenones as biomarkers that can be preserved at any depth in the ocean and are used to create a record of sea surface palaeotemperatures (Eglinton *et al.* 1992; Conte *et al*, Chapter 19).

A sulfur-containing osmoregulator, dimethylsulfoniopropionate or DMSP, is another product of *E. huxleyi*. While the cells die off and the blooms disintegrate, this solute decomposes and the smelly gas dimethyl sulphide (DMS) is set free. In the atmosphere, the DMS is oxidized and an aerosol rich in sulphuric acid is generated, thought to be an ideal site for the nucleation of white, highly reflectant clouds (Charlson *et al.* 1987; see also Malin *et al.*, Chapter 16). Thus, *E. huxleyi* influences the ocean carbon cycle by pumping carbonate and organic carbon from the surface water towards the ocean floor, while the earth's albedo appears to be affected by its DMS-generated clouds (Fig. 17.3).

The concept of *Emiliania huxleyi* as a distinct forcing function of the world's climate has some interesting implications. Firstly, it underscores the idea of extreme nonlinearity of the climate system, as this is influenced by the recurrent, but irregular blooming and busting behaviour of

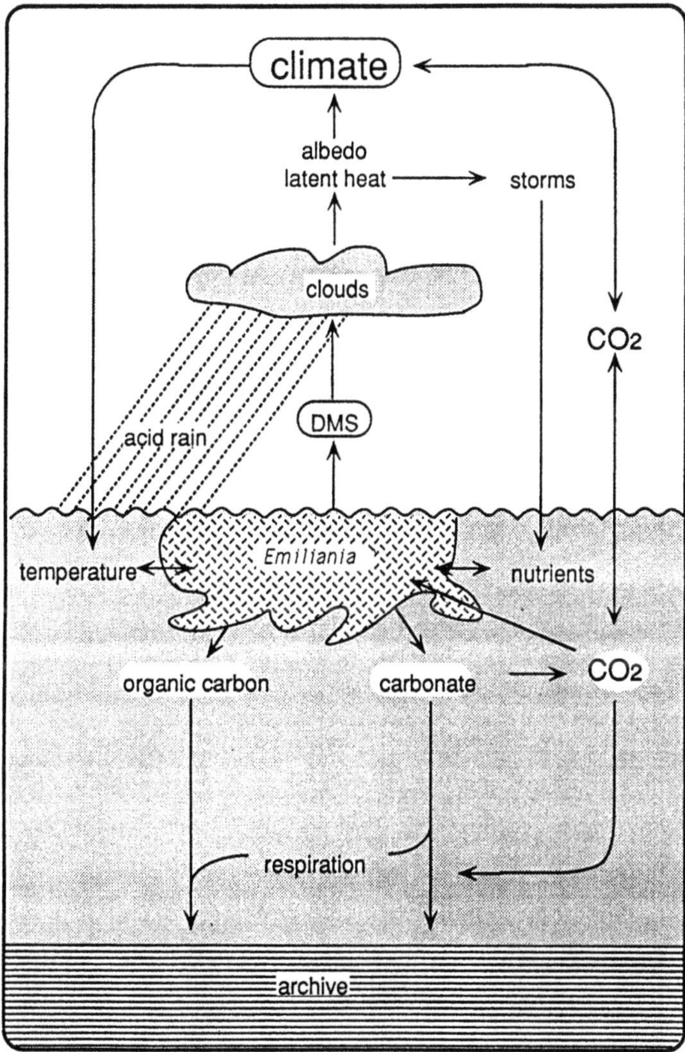

Fig. 17.3 Overview of major oceanic and atmospheric processes associated with the *Emiliania huxleyi* phenomenon.

this and other pelagic organisms. This complex behaviour is also apparent on longer time scales, as *E. huxleyi* originated 278 000 years ago and only adopted its dominant, blooming life-style in the last 85 000 years (Thierstein *et al.* 1977).

Secondly, it shows that three major processes by which pelagic algae can influence climate, the production of organic carbon, carbonate mineral, and DMSP, do not operate in isolation, but that they are intimately coupled. It is carbonate production by *E. huxleyi* that stimulates its biomass production and DMSP formation. The nature of the coupling depends on precise fit of the idiosyncratic cellular organisation of this remarkably successful organism and prevailing oceanic and climatic conditions. It should be borne in mind, however, that the ultimate fate of these products, and hence their climatic effects, not only depends on the productive machinery of *E. huxleyi* itself, but also on subsequent processing by grazers, bacteria, etc., and on physical and chemical factors.

Many organisms in the open ocean are involved in the production of PIC, POC, and DMS-precursors and in most studies only bulk productivity is estimated. We advocate a radically different approach. If *Emiliania huxleyi* is selected as a model organism it may be the key to the entire system. Instead of estimating mean values for the various fluxes, a reliable in-depth picture is pursued of the precise working of the system in so far as *E. huxleyi* is concerned. This approach ensures that the complex, non-linear behaviour of the oceanic calcifying activity is apparent from the start, as well as the coupling of calcification with other biological functions of the climate system. As an additional advantage, an integrated experimental and numerical modelling study becomes feasible, ranging in scale from the molecular to the global level and back through geological time.

While in the past the scientific interest in *Emiliania huxleyi* was restricted to the work of a few laboratories, its popularity has dramatically increased in recent years. The contributions by Brownlee *et al.* (Chapter 7) and de Vrind-de Jong *et al.* (Chapter 8) relate to the numerous studies on the cellular organization of *E. huxleyi*. Holligan *et al.* (1993) is a recent example of fieldwork dealing with the relation between ambient conditions in the ocean, *E. huxleyi* blooming, and its PIC, POC, and DMSP productivity. Further information about the dynamics and distribution of the blooms is provided by remote sensing studies using satellites (e.g., Brown and Yoder 1993). The latter technology also permits interactive planning of navigation itineraries. Zooplankton specialists and microbiologists investigate the termination of the blooms and the degradation of the cells and coccoliths while they sink to the ocean floor (e.g., Bratbak, *et al.* 1993), and geologists measure sedimentation rates and reconstruct the evolution of the system with deep sea cores (e.g., Thierstein *et al.* 1977).

Specific fossil biomarkers are an additional tool for reconstructing the dynamics of the *E. huxleyi* system, as well as climate fluctuations in the past (e.g., Eglinton *et al.* 1992; Conte *et al.*, Chapter 19).

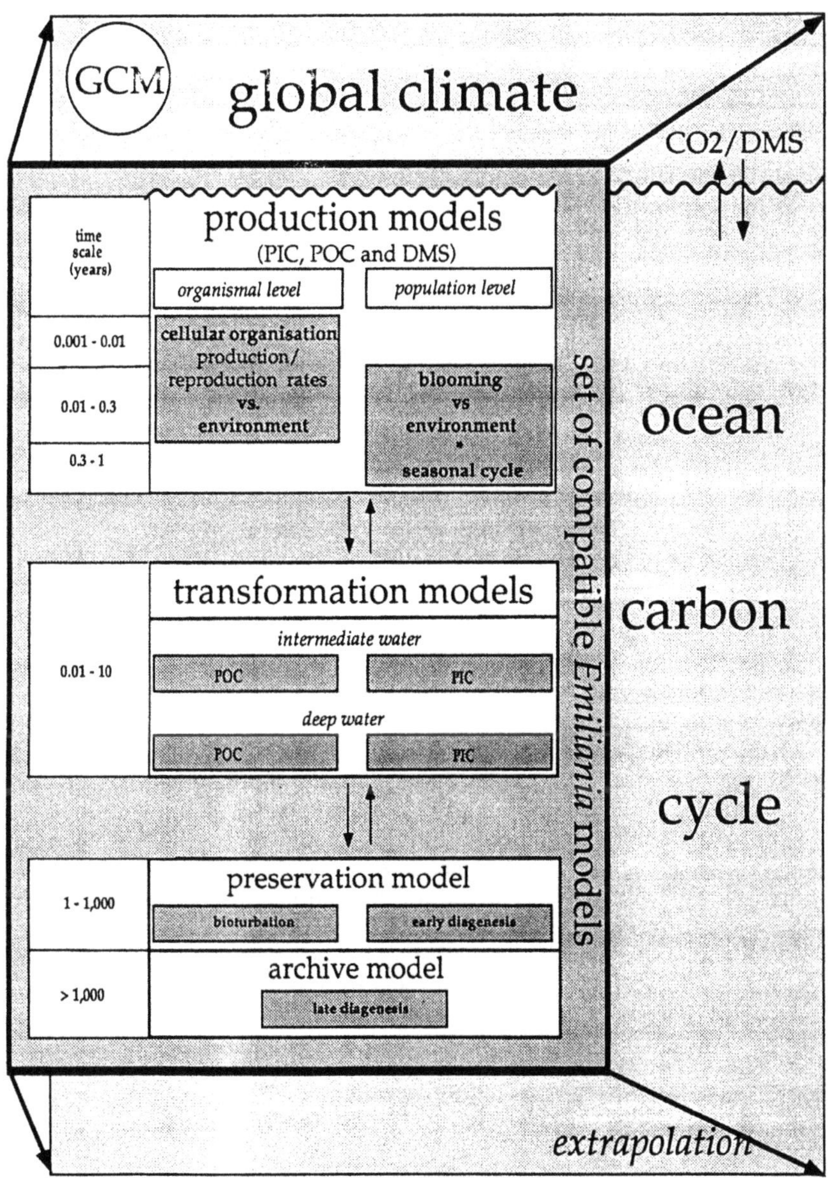

Fig. 17.4 Strategy for modelling the *Emiliania huxleyi* phenomenon and introduction into General Circulation Model (GCM).

To integrate this growing body of information and to direct the research towards the elucidation of biosphere-geosphere interactions, computer modelling is an essential tool. Modelling can best be approached from two ends (Fig. 17.4). In the first, organismic, approach the organisation of the *E. huxleyi* cell is the starting point and the goal is to predict the organism's behaviour in the ocean. This should lead to an enhanced understanding of the causes of blooming and of the spectacular success of this species; also, quantitative predictions of PIC, POC, and DMSP production should be envisaged. Modules simulating the subsequent transformation of these products in the ocean as well as sedimentation and diagenesis are required for estimating the contribution of *E. huxleyi* to global biogeochemical cycling and climate. As the relevant processes operate on different time scales, as indicated in Fig. 17.4, a coherent system of compatible modules is more feasible than a single all-embracing model of the *E. huxleyi* system. This modular system will then be the key to interpreting the geological record. Validation will depend on ongoing experimental and observational studies. Auxiliary investigations and literature data on complementary organisms will then be required for expanding the models to embrace the total interaction between the pelagic biota and the world climate, and for the incorporation of these effects in General Circulation Models.

The second, 'global climate centered', approach adopted for modelling would be to start from existing 3-D climate models and to home in on the *E. huxleyi* system. This would further elucidate the role of this organism in the global climate system and serve as an extra calibration for the organismic module system.

Acknowledgements

We thank the participants of the Global *Emiliania* Modelling Initiative (GEM) for their vital role in formulating the strategy for this research. GEM was created during three successive international meetings at the Château de Blagnac, near Bordeaux, France. Generous support for this international research collaboration was provided by the EC programme MAST II, Norwegian Government agencies, the Netherlands Organization for Scientific Research (NWO), National Research Programme for Global Climate (NOP), the Richard Lounsbery Foundation (New York), and the Natural Environmental Research Council (NERC–UK).

References

Balch, W. M., Holligan, P. M., Ackleson, S. G., and Voss, K. J. (1991). Biological and optical properties of mesoscale coccolithophore blooms in the Gulf of Maine. *Limnology and Oceanography*, **36**, 629–43.

Berner, R. A., and Lasaga, A. C. (1989). Modeling the geochemical carbon cycle. *Scientific American*, **260**, 74–81.

Berner, R. A., Westrich, J. T., Graber, R., Smith, J., and Martens, C. S. (1978). Inhibition of aragonite precipitation from supersaturated seawater: a laboratory and field study. *American Journal of Science*, **278**, 816–37.

Borman, A. H., De Vrind-de Jong, E. W., Huizinga, M., Kok, D., Westbroek, P., and Bosch, L. (1982). The role in $CaCO_3$ crystallization of an acid Ca^{++}-binding polysaccharide associated with coccoliths of *E. huxleyi*. *European Journal of Biochemistry*, **129**, 179–83.

Bratbak, G., Egge, J. K., and Heldal, M. (1993). Viral mortality of the marine alga *Emiliania huxleyi* (Haptophyceae) and termination of an algal bloom. *Marine Ecology Progress Series*, **93**, 39–48.

Brown, C. W. and Yoder, J. A. (1993). Distribution pattern of coccolithophorid blooms in the global ocean. *Continental Shelf Research*, **14**, 175–97.

Cadée, G.C. (1985). Macroaggregates of *Emiliania huxleyi* in sediment traps. *Marine Ecology Progress Series*, **24**, 193–6.

Charlson, R. J., Lovelock, J. E., Andreae, M. O., and Warren, S. G. (1987). Oceanic phytoplankton, atmospheric sulphur, cloud albedo and climate. *Nature*, **326**, 655–61.

Eglinton, G., Bradshaw, S. A., Rosell, A., Sarnthein, M., Pflaumann, U., and Tiedemann, R., (1992). Molecular record of secular sea surface changes on 100-year timescales for glacial terminations I, II and IV. *Nature*, **356**, 423–6.

Emburg P. R. van, (1989). Coccolith formation in *Emiliania huxleyi*. PhD thesis, Leiden University.

Holligan, P. M., Viollier, M., Harbour, D. S., and Champagne-Philipe, M. (1983). Satellite and ship studies of coccolithophore production along a continental shelf-edge. *Nature*, **304**, 339–42.

Holligan, P. M., Fernández, E., Aiken, J., Balch, W. M., Boyd, P., Burkill, P.H., *et al.* (1993). A biogeochemical study of the coccolithophore *Emiliania huxleyi* in the north Atlantic. *Global Biogeochemical Cycles*, **7**, 879–900.

Linschooten, C., van Bleijswijk, J. D. L., van Emburg, P. R., de Vrind, J. P. M., Kempers, W. S., Westbroek, P. *et al.* (1991). Role of the light-dark cycle and medium composition on the production of coccoliths by *Emiliania huxleyi* (Haptophyceae). *Journal of Phycology*, **27**, 82–6.

Lowenstam, H. A. and Weiner, S. (1983). Mineralization by organisms and the evolution of biomineralization. In '*Biomineralization and biological metal accumulation. Biological and geological perspectives*', (ed. P. Westbroek and E. W. De Vrind-de Jong), 191–203. Reidel, Dordrecht.

Lowenstam, H. A. and Weiner, S. (1989). *On biomineralization*. Oxford University Press, New York.

Morse, J. W. and MacKenzie, F. T. (1990). *Geochemistry of sedimentary carbonates*. Elsevier, Amsterdam.

Nimer, N. A. and Merrett, N. J. (1992). Calcification and utilization of inorganic carbon by the coccolithophorid *Emiliania huxleyi* Lohmann. *New Phytologist*, **121**, 173–7.

Robbins, L. L. and Blackwelder, P. L. (1992). Biochemical and ultrastructural evidence for the origin of whitings: a biologically induced calcium carbonate precipitation mechanism. *Geology*, **20**, 464–8.

Simkiss, K. and Wilbur, K. M. (1989). *Biomineralization, cell biology and mineral deposition.* Academic Press, New York.

Thierstein, H. R., Geitzenauer, K. R., Molfino, B., and Shackleton, N. J. (1977). Global synchroneity of late Quarternary coccolith datum levels: validation by oxygen isotopes. *Geology*, **5**, 400–4.

Westbroek, P., de Vrind-de Jong, E. W., van der Wal, P., Borman, A. H., de Vrind, J. P. M., Kok, D., *et al.* (1984). Mechanism of calcification in the marine alga *Emiliania huxleyi. Philosophical Transactions of the Royal Society, London,* Series B, **304**, 435–44.

P. Westbroek, (1991). *Life as a geological force.* W. W. Norton, New York.

Wilkinson, B. H. and Algeo, T. J. (1989). Sedimentary carbonate record of calcium-magnesium cycling. *American Journal of Science*, **289**, 1158–94.

Young, J. R., Didymus, J. M., Bown, P. R., Prins, B., and Mann, S. (1992). Crystal assembly and phylogenetic evolution in heterococcoliths. *Nature*, **356**, 516–18.

18. Coccolithophorid biocoenosis: production and fluxes to the deep sea

KOZO TAKAHASHI

Department of Marine Sciences and Technology, School of Engineering, Hokkaido Tokai University, Sapporo, Japan

Abstract

Coccolithophorids are one of the more important pelagic autotrophic calcium carbonate-producing plankton groups involved in the global carbon cycle. Their calcitic scales are produced in the euphotic layers and eventually become incorporated into sinking particles. Practically all of the coccoliths recovered from the deep sea have been rapidly transported via aggregates, such as marine snow and faecal pellets. Since the coccoliths in the deep sea were transported within a protective organic membrane, intact coccolithophorids can be recovered from deep sediment traps deployed below the calcium carbonate compensation depth. Coccoliths represent approximately two-thirds of the pelagic calcium carbonate production, followed by planktonic foraminifera and pteropods. Coccolithophorid standing stock and production levels, as well as seasonal flux amplitude, diminish towards the northern Polar region, suggesting that there is greater production in subarctic and lower latitudes. *Emiliania huxleyi*, a cosmopolitan coccolithophorid species, is the dominant form in most parts of the world oceans. In the northern high latitudes in the Atlantic, however, a cold water species, *Coccolithus pelagicus*, becomes seasonally more important than *E. huxleyi* during spring and summer. This is especially evident when the average coccolith mass of each species is incorporated in an assessment of the carbonate flux contribution: a placolith of *C. pelagicus* weighs two orders of magnitude more than a placolith of *E. huxleyi*.

Introduction

Coccolithophorids, members of the Haptophyta, constitute one of the most abundant autotrophic plankton groups in the upper ocean. They are

The Haptophyte Algae (ed. J. C. Green and B. S. C. Leadbeater), Systematics Association Special Volume No. 51, pp. 335–50. Clarendon Press, Oxford, 1994.

potentially immensely important in the global environment, particularly in global mass balance and climate change. There are two fundamental reasons why coccolithophorids are important. Firstly, their scales are made of calcite calcium carbonate which contains carbon. Together with the organic carbon in their cytoplasm, the carbon in the calcitic scales can make a substantial contribution to the global carbon cycle. Major producers of calcium carbonate in the pelagic realm include coccolithophorids, planktonic foraminifera (both calcite), and pteropods (aragonite) in descending order. Secondly, in the vast areas where oligotrophic conditions prevail (e.g. the central Pacific gyre), the relative abundance and production of coccolithophorids is generally greater than their siliceous counterparts (i.e., diatoms). Thus, the total coccolithophorid production from such a vast area is large compared with that of the diatoms which are important in limited areas, e.g. upwelling and high latitude regions.

One of the better ways to monitor coccolithophorid production in the upper ocean is to deploy sediment traps at different depths in the oceans (Honjo and Doherty 1988). The coccolithophorid production eventually turns into a sidocoenosis, a sinking assemblage (Takahashi 1993). Although some of the organic matter photosynthesized by autotrophs, such as coccolithophorids, is remineralized in the upper ocean (Fowler and Knauer 1986), the bulk of the calcite scales make their way down to the sea-floor. Since most of the upper layer of the ocean is saturated with respect to calcium carbonate (T. Takahashi 1975), practically no significant calcite dissolution occurs in the upper ocean. Even below the calcium carbonate compensation depth (CCD), intact coccoliths and coccospheres (hereafter simply referred to as coccoliths, wherever appropriate) can be recovered from sediment traps deployed in bathypelagic layers (Steinmetz 1991) because they are protected from dissolution in aggregates (Takahashi 1986).

In this paper, the production and distribution of coccolithophorids in the upper ocean and their fluxes to the deep sea are reviewed. Also included is the historical emergence and significance of sediment trapping, which arose from the need to address the question of how fine particles such as coccoliths sink to the deep-sea floor.

Coccolithophorid biocoenosis and the development of sediment trap studies

Basin-wide biogeographic studies carried out in the Atlantic (McIntyre and Be 1967; Okada and McIntyre *et al.* 1979) and Pacific Oceans (McIntyre *et al.* 1970; Okada and Honjo 1973), on both coccolithophorid biocoenosis (living assemblages) and thanatocoenosis (sedimented assemblages), significantly advanced our understanding of oceanic particle production

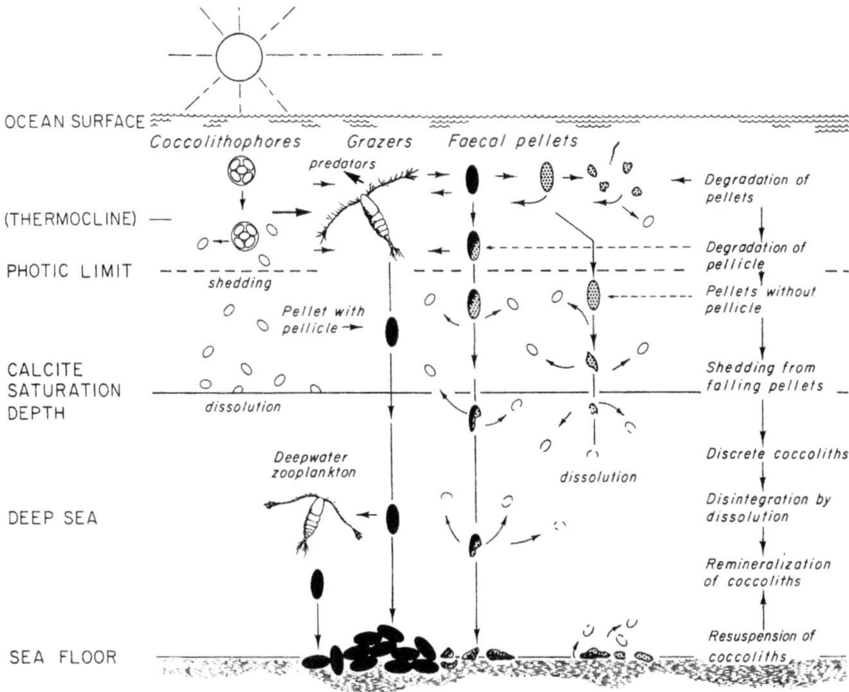

Fig. 18.1 A conceptual model for coccolithophorid production in the upper ocean and coccolith faecal pellet transport to deep sea (Honjo 1976).

and sedimentation. Distributions of coccoliths in the upper ocean and in surficial sediments coincide reasonably well (McIntyre and Bé 1967) without obvious wide lateral dispersion, indicating rather rapid transport of fine particles, such as coccoliths.

The idea of faecal pellet packaging and vertical transport dates back to the late 1960s (Smayda 1969, 1970). Furthermore, during the early 1970s faecal pellets were considered to be the major mechanism responsible for the accelerated sinking of fine particles (Fig. 18.1) (Honjo 1976). These ideas eventually led to the deployment of sediment traps to collect faecal pellets quantitatively in the deep sea (Wiebe *et al.* 1976; Honjo 1978). In the middle to late 1970s the time was ripe for the development of large-scale field programmes for sediment trapping. Although the idea of sediment trapping already existed in the 1960s, and some field work was carried out in the marine environment (Berger 1967), no significant advance was achieved as the importance of sediment trapping was not fully recognized.

Coccolithophorid fluxes

Vertical transport mechanism

The idea that fine particles such as coccoliths could be transported as faecal pellets led to the development of sediment trapping (see above). This resulted in major advances in diverse fields rather than simply proving that faecal pellets were a major pathway for the transport of fine particles. The expected key role of faecal pellets in the vertical transport of fine particles has been found to be limited to certain regions such as the Antarctic Ocean (Wefer *et al.* 1988; Honjo 1990) where specific organisms, such as krill, dominate the ecosystems. In many parts of the pelagic ocean the contribution of faecal pellet transport is in the order of 10% or less of the total flux (Dunber and Berger 1981; Pilskaln and Honjo 1987), suggesting that other sinking particles, for example marine snow, are numerically more important. In the high latitude eastern North Pacific, uniform sinking rates of a variety of particles ranging in size from approximately 10 μm to a few mm were observed, indicating accelerated sinking *via* amorphous aggregates (e.g. marine snow) rather than *via* faecal pellets (Takahashi 1986, 1989). The importance of coccolith transport in amorphous aggregates, if it really occurs, must be assessed over much larger areas of the pelagic domain in future investigations. The role of salp faecal pellets (Pfannkuche and Lochte 1993) also needs to be clarified. It is hoped that the sinking modes of particles including coccoliths, and the assessment of flux within ecosystems, will soon be determined in major parts of the world oceans by employing time-series sediment traps. This will certainly provide essential data for global carbon mass balance calculations.

Early coccolith flux studies

One of the earliest coccolith flux studies was carried out around 1980 using low latitude sediment trap material, and its results were eventually published a decade later (Steinmetz 1991). This study was based on collections from a single-cup sediment trap achieved by the PARFLUX (Particle Flux) Programme at three equatorial locations (western equatorial Atlantic, central Pacific, and Panama Basin) for 61 to 112 days duration (Honjo 1980). The contribution of discrete coccolith carbonate in < 63 μm size-fractions ranged, on average, from 9% (western equatorial Atlantic site) to 43% (central equatorial Pacific site; data from the upper two traps were not included due to swimmer problems: see Takahashi 1991*a*). The discovery of well-preserved coccoliths at great depths clearly supported the rapid transport hypothesis of Honjo (1976) (Fig. 18.1). The main occurrence of coccoliths as free-coccoliths suggested that faecal pellets or oceanic aggregates break down within or just above the traps.

The species diversity of coccoliths depended on coccolith flux levels; the correlation coefficient between coccolith flux and the number of coccolith species at three sediment trap locations is r = –0.97. The Panama Basin, a relatively highly productive hemipelagic area, had only 25 species whereas the western equatorial Atlantic, which experienced the lowest coccolith flux of the three regions, showed the highest diversity of 50 species. It is important to note that this inverse correlation of diversity with coccolith productivity can be used to infer levels of palaeoproductivity.

Coccolith flux levels

Coccolith fluxes recorded at three tropical stations ranged from 125 to 1180×10^6 individuals m^{-2} d^{-1} (Steinmetz 1991). Coccolith flux at 4000 m in the Japan Trench was 3400×10^6 m^{-2} d^{-1} (Okada 1989). This was of the same order of magnitude as the Panama Basin data of Steinmetz (1991). In a northern high latitude station, the mean coccolith flux level was much lower, for example, 40×10^6 m^{-2} d^{-1} (range: 0 to 137×10^6 m^{-2} d^{-1}), than those observed at the mid- and low latitude stations (Samtleben and Bickert 1990). Although further flux measurements are required for refinement, we now have basic data on coccolith flux spanning two orders of magnitude, enabling a variety of preliminary global mass balance calculations to be made.

Temporal variability

Seasonality

Seasonal changes in the abundance of coccolithophorid species constitute one of the most important issues in interpreting environmental data. Okada and McIntyre (1979) carried out a detailed characterization of seasonal changes in coccolithophorid biocoenosis in the North Atlantic. This study was based on suspended particulate material collected in collaboration with the United States Coast Guard along a north (56°N) to south (32°N) transect of North Atlantic weather stations, from north to south Stations BRAVO, CHARLIE, DELTA, HOTEL, ECHO, and BERMUDA. The biogeographic zones covered by the study ranged from subarctic to subtropical.

According to Okada and McIntyre (1979), average monthly coccolithophorid standing stock levels were a few tens of thousand of cells per litre and they did not change drastically in all the climatic regions studied. The standing stocks at 100 m showed an increasing trend southwards. The study clearly indicated the general predominance among coccolithophorids of *Emiliania huxleyi*, a cosmopolitan species (Fig. 18.2) (Westbroek *et al.*, Chapter 17). An exception is that at Station BRAVO, a northern high latitude site where *Coccolithus pelagicus*, a cold water species (Okada

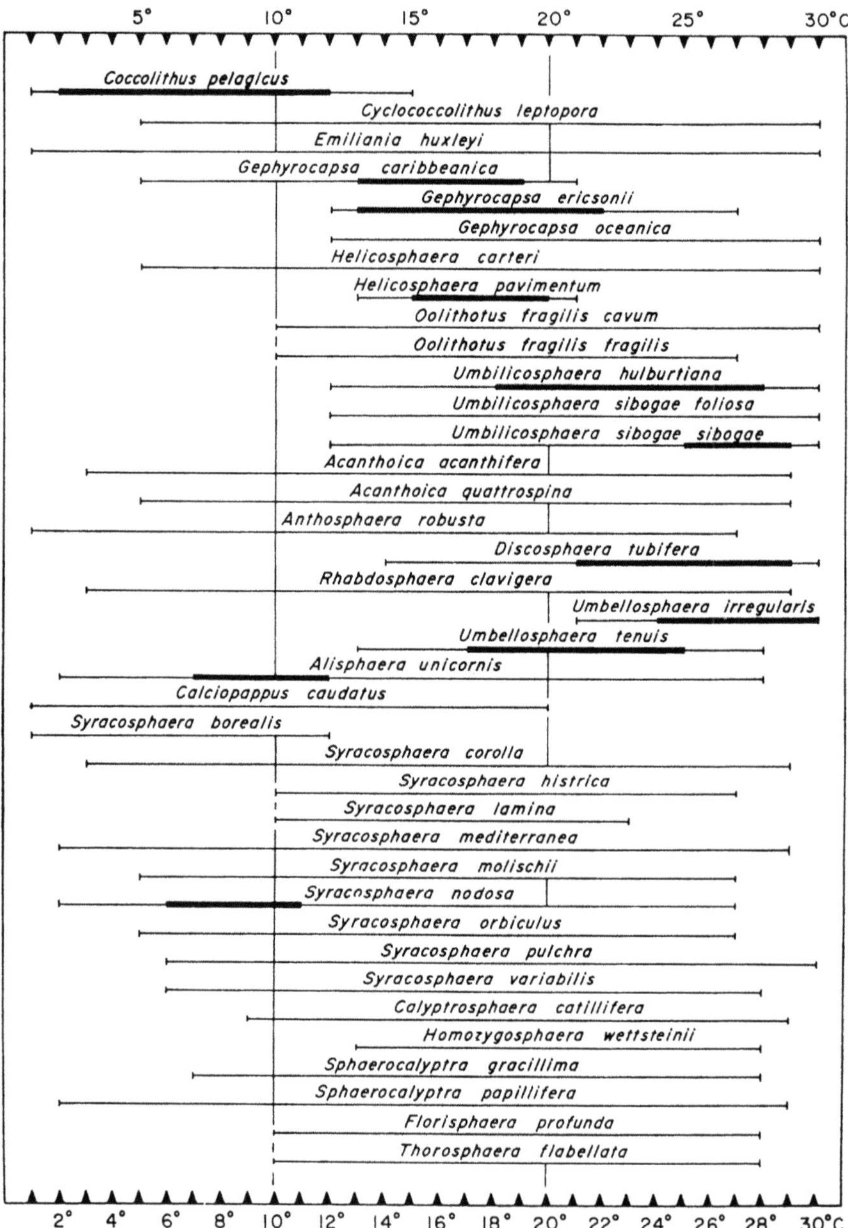

Fig. 18.2 Maximum (*thick bars*) and optimum (*thin lines*) temperature ranges for coccolithophorid production based on the data from the North Atlantic and the Pacific Oceans. From Okada and McIntyre (1979).

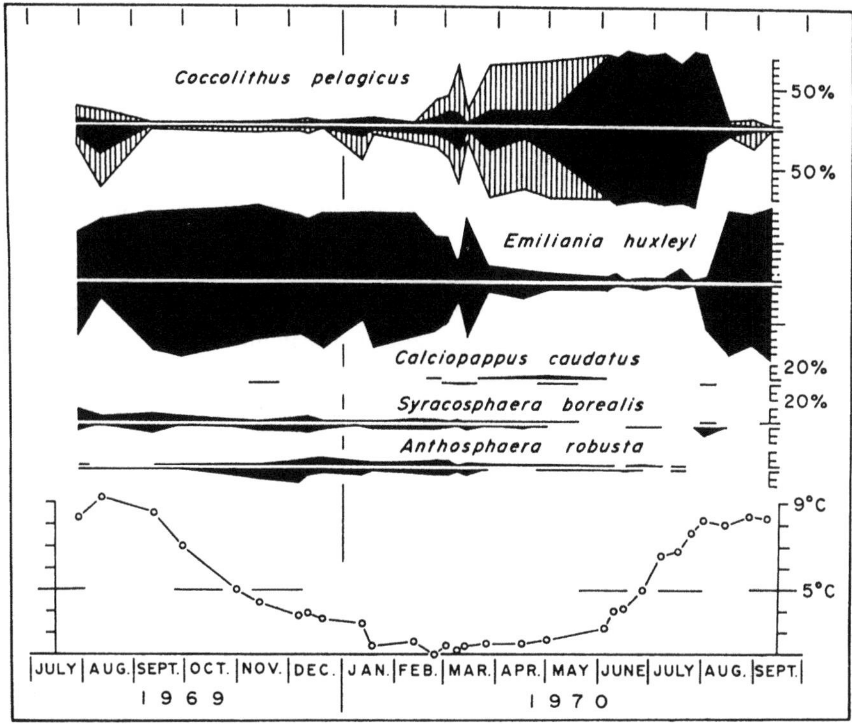

Fig. 18.3 Seasonal distribution of quantitatively important coccolithophorid species and water temperature at surface and 20 m depths at Station BRAVO (56°30′N; 51°00′W) in the subarctic North Atlantic. Black areas above base lines indicate % of each species at surface, and black areas below base lines indicate % of each species at 20 m. For *Coccolithus pelagicus*, hatched areas represent motile-phase cells, and black areas non-motile-phase cells. From Okada and McIntyre (1979).

and McIntyre 1979; Fig. 18.2), was seasonally more abundant. At this station, *C. pelagicus* had a high relative abundance during spring and early summer (March–August) whereas *E. huxleyi* showed a high relative abundance during summer and early spring (August–March; Fig. 18.3). *Coccolithus pelagicus* maintained its high relative abundance for a longer period at 100 m than at the surface and 20 m depth. It was noted that abundant motile-phase *C. pelagicus* in spring preceded a bloom of non-motile cells during summer. Additional seasonal studies are discussed below under separate headings.

Periodic coccolithophorid blooms

Periodic blooms of coccolithophorids have been known for a long time in coastal waters, such as the Oslo Fjord. Satellite observations are effective in documenting geographic coverage of coccolith blooms in the pelagic and hemipelagic realms (Hay 1987). In the Panama Basin, a bloom of *Umbilicosphaera sibogae* was observed using a time-depth series of sediment traps deployed for a year (Honjo 1982). This was the first time that time-series (2-month increment) sediment trapping had been employed for documenting temporal changes of coccolithophorid fluxes. According to Honjo, a flux of approximately 1.6 grams of calcium carbonate m^{-2} d^{-1}, two orders of magnitude more than during adjacent sampling intervals, was recorded at 890 m depth in one sample increment of two months (June–July) in 1980 (Fig. 18.4). This sample consisted primarily of *U. sibogae*, a temperate to warm water coccolithophorid. The decreasing flux of *U. sibogae* with depth indicated a relatively small scale (e.g., meso) coccolithophorid bloom patch in the upper water. A high flux level of such short duration has not been reported elsewhere (Honjo 1990). The importance of monitoring such a rare event cannot be stressed too strongly.

Steinmetz (1991) also recognized a nearly monospecific assemblage of *U. sibogae* in the 4280 m single-cup sediment trap in the central Pacific where the coccolith flux was 6.1 mg $CaCO_3$ m^{-2} d^{-1}. By comparing samples from single-cup traps deployed above and below this depth, it was concluded that the assemblage at 4280 m was already descending in the mid-water column when the trap mooring was deployed. An almost exclusively *Emiliania huxleyi* flux was recognized in early May in the North Sea using a time-series sediment trap (Cadee 1985).

Time-series flux measurements and continuous satellite observations of coccolithophorid blooms should be combined to delineate comprehensive views of phenomena which are of temporally short duration, but may be crucial to global mass balance.

Norwegian Sea fluxes

One of the most comprehensive seasonal coccolith flux studies was carried out in the Norwegian Sea (Samtleben and Bickert 1990). As stated earlier, the coccolith flux levels in the Norwegian Sea were relatively low compared to those at mid- and low latitudes. A substantial portion of the calcium carbonate flux was due to coccoliths. The contribution of coccoliths to total carbonate fluxes ranged from 36 to 86% except from January to April when no coccolith flux was observed. A major contribution of coccoliths to total carbonate mainly occurred from May to September. An increased contribution by planktonic foraminifera occurred from October

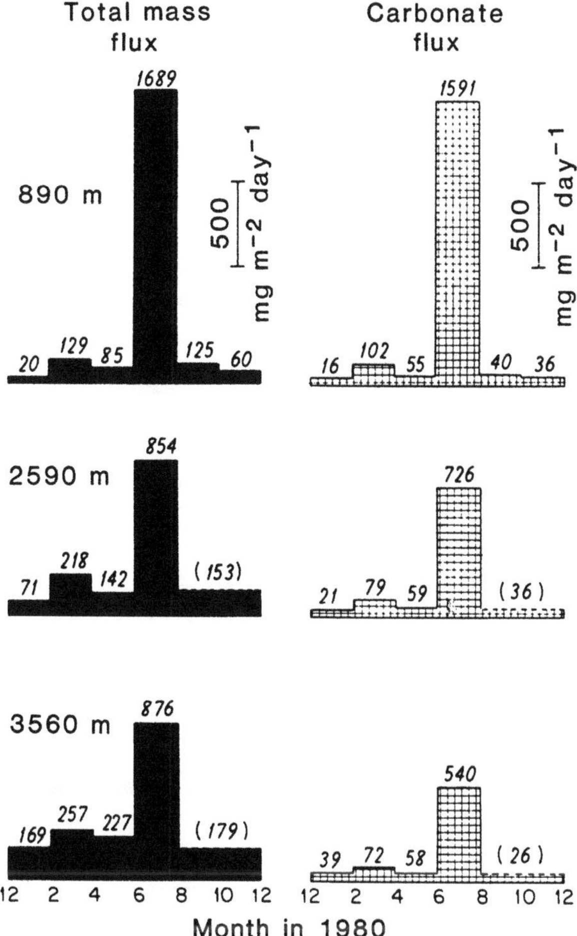

Fig. 18.4 Seasonal fluxes of total mass and calcium carbonate during 1980 in the Panama Basin. The flux peak of June–July consisted primarily of a monospecific assemblage of *Umbilicosphaera sibogae*. Modified from Honjo (1982).

to December. The carbonate flux data in the Norwegian Sea were initially published by Honjo *et al.* (1987).

Samtleben and Bickert (1990) clearly demonstrated that only two species of coccolithophorids were important in the mass balance of carbon in their study regions; namely *Emiliania huxleyi* and *Coccolithus pelagicus*. Generally, the *E. huxleyi* flux in terms of coccolith number surpassed that of *C. pelagicus* at all three Norwegian stations (Fig. 18.5). However, when the carbonate mass flux due to different species of coccoliths was com-

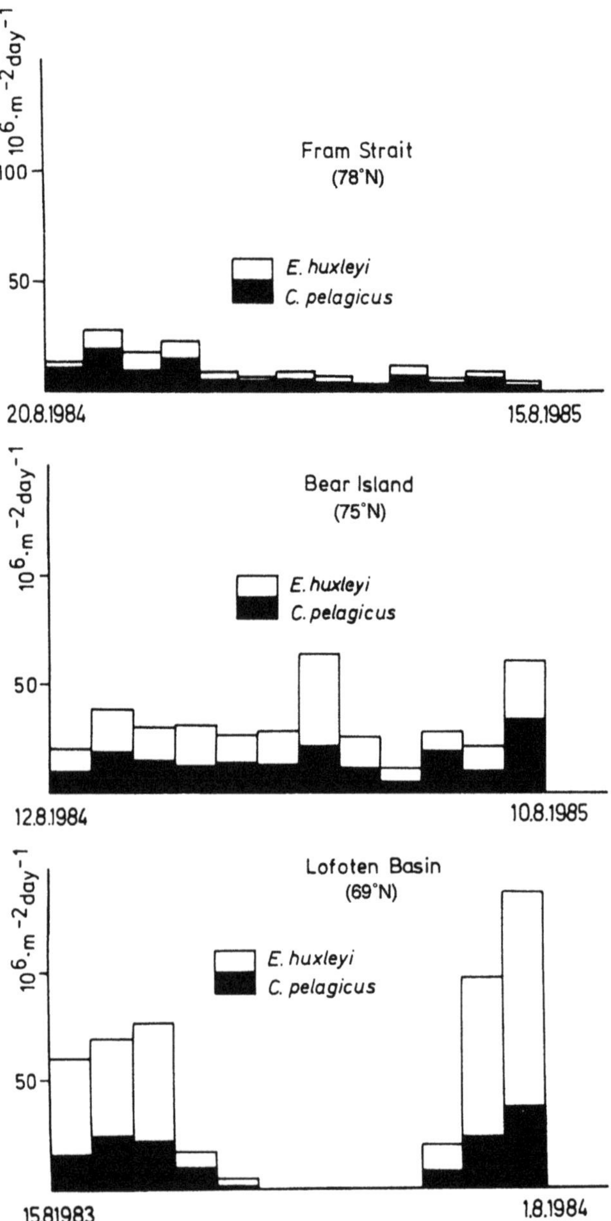

Fig. 18.5 Seasonal fluxes of two dominant species of coccolithophorids in the Norwegian Sea during 1983–4: *Emiliania huxleyi* and *Coccolithus pelagicus*. From Samtleben and Bickert (1990).

puted (see below), the contribution by *C. pelagicus* became greater than that of *E. huxleyi.*

A decreasing flux of coccoliths occurred with increasing latitude. The seasonal flux variability, which is clearly represented at the Lofoten Basin (69°N) Station, becomes rather vague at the Bear Island Station (75°N), and is lost almost completely further north in the Fram Strait (78°N) (Fig. 18.5).

Estimates of individual coccolith mass

Estimates of individual coccolith (e.g. placolith) mass for each taxa are extremely desirable. The information is, essential in order to derive the coccolith contribution to total calcium carbonate flux, for example, since it constitutes the basis for various mass balance computations. Paasche (1962) first estimated the calcite mass of an individual placolith of *E. huxleyi,* employing the ^{14}C method: 5.5×10^{-13} g. Geometric considerations based on SEM observations have provided revised information concerning the average weight of placoliths (Table 18.1) (Honjo, 1976, 1977). Using this method the mass of *E. huxleyi* coccoliths was found to be 8×10^{-12} g which is approximately one order of magnitude greater than that obtained by Paasche (1962). Similarly Samtleben and Bickert (1990) estimated the average weight for the coccolith of both *E. huxleyi* and *C. pelagicus* in the Norwegian Sea (Table 18.1). The most frequent species next to *E. huxleyi* in the middle to low latitude is *Gephyrocapsa oceanica* (McIntyre *et al.* 1970; Okada and Honjo 1973; Okada and McIntyre 1979). An estimate of coccolith mass in this species is 1.8×10^{-11} g since placoliths of this taxon should weigh approximately six times as much as those of *E. huxleyi* (Okada, personal communication 1993).

Table 18.1 Average coccolith weights derived from geometric and other considerations

Taxon	Weight (g)	Reference
Emiliania huxleyi	5.5×10^{-13}	Paache 1962
Emiliania huxleyi	8.0×10^{-12}	Honjo 1977
Emiliania huxleyi	3.0×10^{-12}	Samtleben and Bickert 1990
Gephyrocapsa oceanica	1.8×10^{-11}	Okada, personal communication 1993
Coccolithus pelagicus	1.3×10^{-10}	Samtleben and Bickert 1990
Coccolithus neohelis	6.0×10^{-12}	Honjo 1977
Coccolithus leptopora	3.3×10^{-10}	Honjo 1977

It is important to note that large-sized coccoliths are orders of magnitude heavier than their smaller-sized counterparts. For instance, an average *C. pelagicus* placolith weighs roughly two orders of magnitude more than one of *E. huxleyi* (Table 18.1). As mentioned above, a significant difference in coccolith mass resulted in a greater flux contribution by a less abundant taxon (*C. pelagicus*), than by the more common *E. huxleyi* at a northern high latitude site. Although the currently available coccolith mass information may be adequate for preliminary construction of a carbon mass balance, eventually more accurate estimates will be required for comprehensive computations. Perhaps better methods for measuring individual coccoliths will become available in the future which will lead to accurate mass balance estimates for carbon and carbonates.

Fractionation of carbonate fluxes by coccoliths

Although numerical knowledge of calcium carbonate plankton standing stocks was available in the early 1970s (Lisitzin 1972), it was not until the late 1970s that fractionation of carbonate production among the major carbonate plankton groups was attempted, employing sediment trap data (Honjo 1977). By far the greatest contribution by coccoliths to total carbonate flux was clearly documented in the Sargasso Sea by the author (Table 18.2). Deuser and Ross (1989) obtained similar results for carbonate flux fractionation derived from their Sargasso Sea trap data (Table 18.2). They assumed that all calcium carbonate in < 37 μm fraction was due to coccoliths, and that carbonate in < 37 μm fraction, plus planktonic foraminifera (weight measured), plus pteropods (not weight measured) equals total calcium carbonate (100%). A similar results was obtained from Station Papa in the Gulf of Alaska (Table 18.2) (Fabry 1989). Her calculations involved various assumptions including 20 coccoliths per cell, 8×10^{-9} mg $CaCO_3$ per coccolith and reproductive rates for coccolitho-

Table 18.2 Contribution of $CaCO_3$ flux by major pelagic plankton groups

	% of $CaCO_3$ flux		
	Honjo (1978)	Deuser and Ross (1989)	Fabry (1989)
Coccoliths	59	62	59–77
Planktonic foraminifera	29	22	18–30
Pteropods	7	16	4–13
Thoracosphaera	5	NA	NA

phorids of 0.5 to 1.0 divisions per day. Here again the predominance of coccoliths in total calcium carbonate flux was demonstrated from an ecological and climatic region very different to that of the Sargasso Sea.

Ideally, the actual fluxes of all three or four (*Thoracosphaera* included) major carbonate producing plankton groups should be accurately measured to yield more accurate fractionation data. Characterization of calcium carbonate fractionation by the three main plankton groups should be followed by organic carbon fractionation in future work. These data can be combined with fractionation data of organic carbon and biogenic silica contributed by diatoms, silicoflagellates and radiolarians (Takahashi 1991*b*). This will provide a more accurate understanding of the cycling of important elements including carbon.

Environmental proxy

Coccoliths have long been recognized as excellent stratigraphic markers (e.g., Perch-Nielsen 1985), but their value for environmental interpretation has only explored recently (see Young *et al.* Chapter 20). There have been only a few palaeoceanographic studies accomplished thus far: they include Molfino *et al.* (1982), Gartner (1988), Matsuoka and Okada (1989), and Molfino and McIntyre (1990). Matsuoka and Okada (1989), for example, explored time transgressive morphological change in coccolith species such as *Gephyrocapsa oceanica* in the Pleistocene sedimentary record. Rather significant changes with time in the relative abundance of different morphotypes of *G. oceanica* were recognized, suggesting an excellent possibility for an environmental proxy record. Furthermore, information of relevance for environmental reconstructions may be derived from detailed studies of coccoliths in sediment trap materials.

Conclusions

The sediment trap method has proved to be an effective way of reconstructing recent coccolithophorid production events in the upper ocean. The coccolith contribution is approximately two-thirds of total calcium carbonate flux, indicating the fundamental importance of coccolithophorids in the global carbon cycle. Further detailed flux measurements in all the major calcium carbonate producing plankton groups should be carried out in various oceanic environments in order to provide accurate data on carbonate flux fractionation by each group. Improved mass estimates of average coccoliths are the key to an accurate assessment of fractionation. Coupled with satellite observations, coccolith fluxes should be monitored in order to document rare but potentially important coccolith bloom events.

Sinking aggregates are the dominant mechanism for the vertical transport of coccoliths. Intact coccolithophorids and coccoliths shed from aggregates can be found in deep-sea traps deployed even below CCD because of protective transport. Although significant faecal pellet transport cannot be totally dismissed, and does occur in some regions, the major transport mode in other regions seems to be via amorphous aggregates such as marine snow. Only in the Antarctic Ocean, where krill is important, do faecal pellets become rather significant in particle sinking to the sea-floor. Except for the Antarctic Ocean, faecal pellets constitute in the order of 10% or less to the total flux in the regions studied. Further information from sediment trapping will help to clarify these two different modes of vertical particle transport in the world oceans.

Acknowledgments

I am grateful to Dr Hisatake Okada of the Yamagata University who contributed to this paper in providing me with information concerning individual weights of various coccolith species. An anonymous reviewer kindly provided constructive criticisms which improved the paper. Numerous colleagues, too many to specifically list, contributed to this paper in many ways through casual discussions. This work has been funded by US NSF OCE-8800665 and Scientific Research on Priority Area Project No. 05216210, Ministry of Education, Science and Culture of Japan.

References

Berger, W. H. (1967). Planktonic foraminifera: field experiment on production rate. *Science*, **156**, 1495–7.

Cadee, G. C. (1985). Macroaggregates of *Emiliania huxleyi* in sediment traps. *Marine Ecology Progress Series*, **24**, 193–6.

Deuser, W. G. and Ross, E. H. (1989). Seasonally abundant planktonic foraminifera of the Sargasso Sea: succession, deep-water fluxes, isotopic compositions, and paleoceanographic implications. *Journal of Foraminiferal Research*, **19**, 268–93.

Dunber, R. B. and Berger, W. H. (1981). Fecal pellet flux to modern bottom sediment of Santa Barbara Basin (California) based on sediment trapping. *Geological Society of America Bulletin*, Part. I, **92**, 212–18.

Fabry, V. J. (1989). Aragonite production by pteropod mollusks in the subarctic Pacific. *Deep-Sea Research*, **36**, 1735–51.

Fowler, S. W. and Knauer, G. A. (1986). Role of large particles in the transport of elements and organic compounds through the oceanic water column. *Progress in Oceanography*, **16**, 147–94.

Gartner, S. (1988). Paleoceanography of the mid-Pleistocene. *Marine Micropaleontology*, **13**, 23–46.

Hay, B. J. (1987). Sediment trap accumulation in the central western Black Sea over the past 5100 years. *Paleoceanography*, **3**, 491–508.

Honjo, S. (1976). Coccoliths: production, transportation and sedimentation. *Marine Micropaleontology*, **1**, 65–79.

Honjo, S. (1977). Biogenic carbonate particles in the ocean; do they dissolve in the water column? In *The fate of fossil fuel CO_2 in the oceans*, (ed. N. R. Andersen and A. Malahoff), pp. 269–94. Plenum, New York.

Honjo, S. (1978). Sedimentation of materials in the Sargasso Sea at a 5,367 m deep station. *Journal of Marine Research*, **36**, 469–92.

Honjo, S. (1980). Material fluxes and modes of sedimentation in the mesopelagic and bathypelagic zones. *Journal of Marine Research*, **38**, 53–97.

Honjo, S. (1982). Seasonality and interaction of biogenic and Lithogenic particulate flux at the Panama Basin. *Science*, **218**, 883–4.

Honjo, S. (1990). Particle fluxes and modern sedimentation in the polar oceans. In *Polar oceanography, Part B: chemistry, biology, and geology*, (ed. W. O. Smith) pp. 687–739. Academic Press, New York.

Honjo, S. and Doherty, K. W. (1988). Large aperture time-series sediment traps; design objectives, construction and application. *Deep-Sea Research*, **35**, 133–49.

Honjo, S., Manganini, S. J., Karowe, A., and Wooding, B. L. (1987). Particle fluxes, north-eastern Nordic Seas: 1983–1986. *Woods Hole Oceanographic Institution Technical Report WHOI-87-17*, pp. 1–84.

Lisitzin, A. P. (1972). Sedimentation in the World Ocean. *Society of Economic Paleontologists, Special Publication No. 17*, pp. 1–218.

Matsuoka, H. and Okada, H. (1989). Quantitative analysis of Quaternary nannoplankton in the subtropical Northwestern Pacific Ocean. *Marine Micropaleontology*, **14**, 97–118.

McIntyre, A. and Bé, A. W. H. (1967). Modern Coccolithophoridae of the Atlantic Ocean. I. Placoliths and Cyrtoliths. *Deep-Sea Research*, **14**, 561–97.

McIntyre, A., Bé, A. W. H., and Roche, M. B. (1970). Modern Pacific Coccolithophorida: a paleontological thermometer. *Transactions of the New York Academy of Sciences*, Series II, **32**, 720–31.

Molfino, B. and McIntyre, A. (1990). Precessional forcing of nutricline dynamics in the equatorial Atlantic. *Science*, **249**, 766–9.

Molfino, B., Kipp, N. G., and Morley, J. J. (1982). Comparison of foraminiferal, coccolithophorid, and radiolarian paleotemperature equations: assemblage coherency and estimate concordancy. *Quaternary Research*, **17**, 279–313.

Okada, H. and Honjo, S. (1973). The distribution of oceanic coccolithophores in the Pacific. *Deep-Sea Research*, **20**, 355–74.

Okada, H. (1989). Calcareous nannoplankton fluxes in JT-02 samples. *Kaiyo Monthly*, **21**, 228–31 [in Japanese].

Okada, H. and McIntyre, A. (1977). Modern coccolithophores of the Pacific and North Atlantic Oceans. *Micropaleontology*, **23**, 1–55.

Okada, H. and McIntyre, A. (1979). Seasonal distribution of modern coccolithophores in the Western North Atlantic Ocean. *Marine Biology*, **54**, 319–28.

Paasche, E. (1962). Coccolith formation. *Nature*, **193**, 1094–5.

Perch-Nielsen, K. (1985). Cenozoic calcareous nannofossils. In *Plankton stratigraphy*, (ed. H. M. Bolli, J. B. Saunders, and K. Perch-Nielsen), pp. 427–554. Cambridge University Press.

Pfannkuche, O. and Lochte, K. (1993). Open ocean pelago-benthic coupling: cyanobacteria as tracers of sedimenting salp faeces. *Deep-Sea Research*, **40**, 727–37.

Pilskaln, C. H. and Honjo, S. (1987). The fecal pellet fraction of biogeochemical particle fluxes to the deep sea. *Global Biogeochemical Cycles*, **1**, 31–48.

Samtleben, C. and Bickert, T. (1990). Coccoliths in sediment traps from the Norwegian Sea. *Marine Micropaleontology*, **16**, 39–64.

Smayda, T. J. (1969). Some measurements of the sinking rate of fecal pellets. *Limnology and Oceanography*, **14**, 621–5.

Smayda, T. J. (1970). The suspension and sinking of phytoplankton in the sea. *Oceanography and Marine Biology Annual Review*, **8**, 353–414.

Steinmetz, J. C. (1991). Calcareous nannoplankton biocoenosis: sediment trap studies in the equatorial Atlantic, central Pacific, and Panama Basin. In *Ocean biocoenosis*, Series No. 1. (ed. S. Honjo), pp. 1–85. Woods Hole Oceanographic Institution Press, Woods Hole, Mass.

Takahashi, K. (1986). Seasonal fluxes of pelagic diatoms in the subarctic Pacific, 1982–1983. *Deep-Sea Research*, **33**: 1225–51.

Takahashi, K. (1989). Silicoflagellates as productivity indicators: evidence from long temporal and spatial flux variability responding to hydrography in the northeastern Pacific. *Global Biogeochemical Cycles*, **3**, 43–61.

Takahashi, K. (1991*a*). Radiolaria: flux, ecology, and taxonomy in the Pacific and Atlantic. In *Ocean biocoenosis*, Series No. 3, (ed. S. Honjo), pp. 1–303. Woods Hole Oceanographic Institution Press, Woods Hole, Mass.

Takahashi, K. (1991*b*). Mineral flux and biogeochemical cycles of marine planktonic protozoa. In *Protozoa and their role in marine processes*, (ed. P. C. Reid, C. M. Turley, and P. H. Burkill), pp. 347–60. Springer-Verlag, Berlin.

Takahashi, K. (1993). Sidocoenosis: a sinking assemblage mediating rapid particle transport to deep sea. *Marine Micropaleontology*, (In press).

Takahashi, T. (1975). Carbonate chemistry of sea water and the calcite compensation depth in the oceans. In *Dissolution of deep sea carbonate*, (ed. W. V. Sliter, A. W. H. Bé, and W. H. Berger), pp. 11–26. Special Publication No. 13, Cushman Foundation for Foraminiferal Research, Sharon, Mass.

Wefer, G., Fischer, G., Fuetterer D., and Gersonde, R. (1988). Seasonal particle flux in the Bransfield Strait, Antarctica. *Deep-Sea Research*, **35**, 891–8.

Wiebe, P. H., Boyed, S. H., and Winget, C. (1976). Sedimentological trap for use above the sea-floor with preliminary results of its use in the Tongue of Ocean, Bahamas. *Journal of Marine Research*, **34**, 341–54.

19. Lipid biomarkers of the Haptophyta

MAUREEN H. CONTE*

Biogeochemistry Research Centre, School of Chemistry, University of Bristol, Bristol, UK

**Present address: Department of Chemistry and Geochemistry, Woods Hole Oceanographic Institute, Woods Hole, MA 02543, USA*

JOHN K. VOLKMAN

CSIRO Division of Oceanography, Hobart, Tasmania, Australia

and GEOFFREY EGLINTON

Biogeochemistry Research Centre, School of Chemistry, University of Bristol, Bristol, UK

Abstract

Lipid biomarker composition (fatty acids, sterols, alkenones, alkenoates, alkenes) of the Haptophyta is reviewed, and new data are presented on strain variability in the coccolithophorid *Emiliania huxleyi*. Groupings in both fatty acid and sterol profiles are consistent with taxonomic divisions. The Pavlovales are distinct in their higher relative concentrations of the polyunsaturated acid 20:5ω3 and in the presence of 4-methyl sterols. Biomarker profiles indicate significant genetic variability within the *Isochrysis* and *Phaeocystis* genera. Eleven strains of *Emiliania huxleyi*, cultured at 15 °C, had largely similar alkenone and alkenoate compositions and cellular abundances, although tetraunsaturated alkenones were present in higher concentrations in some coastal strains. In contrast, alkene composition and cellular abundance differed greatly, especially in the presence or absence of C_{37} and C_{38} alkenes. Field studies indicate biogeographical variability in the primary production of alkenones and alkenoates in cold surface waters, suggesting genetic differences in *E. huxleyi* populations. The closely related *Gephyrocapsa oceanica* also contains alkenones and alkenoates and may be an important source in warm water regions.

Introduction

This paper reviews fatty acid and sterol data for haptophyte species examined since 1968, and recent culture and field data on haptophyte-derived alkenones, alkenoates, and alkenes. Earlier studies of haptophytes are reviewed in Ackman *et al.* (1968) and Erwin (1973). We have used the classification scheme of the recent revision of the Haptophyta (Jordan and Green 1994).

The Haptophyte Algae (ed. J. C. Green and B. S. C. Leadbeater), Systematics Association Special Volume No. 51, pp. 351–77. Clarendon Press, Oxford, 1994.

Although the Haptophyta, with one class, the Prymnesiophyceae, have been studied more than many other algal groups, our knowledge of their biomarker composition is extremely limited; only 10% of the species listed in Jordan and Green (in press) have been examined. The orders Isochrysidales and Pavlovales, which contain species of economic interest as food sources in mariculture, are the most thoroughly studied. The coccolithophorids are the least well studied, with only two of approximately 60 genera examined. Most studies have concentrated on fatty acid and/or sterol composition. However, a number of species have been examined specifically for the presence of long-chain alkenones, and the related alkenoates and alkenes (Marlowe *et al.* 1984*b*). These compounds, synthesized most notably by the coccolithophorid *Emiliania huxleyi* (Volkman *et al.* 1980*a*, *b*), are of interest due to their importance as palaeo sea surface temperature proxies in the sedimentary record.

Fatty acids

Fatty acid composition is of primary interest in studies of ecosystem energetics and food webs (e.g. Sargent and Whittle 1981) and in assessment of the nutritional quality of species for aquaculture (e.g. Volkman *et al.* 1989). Systematic variations in fatty acid profiles are found among different phytoplankton classes (Ackman *et al.* 1968; Chuecas and Riley 1969; Erwin 1973), making these compounds useful indicators of taxonomic affiliations.

Table 19.1 gives the composition of selected fatty acids in haptophyte algae examined since 1968. Studies prior to 1980 (e.g. Ackman *et al.* 1968; Chuecas and Riley 1969; DeMort *et al.* 1972) should be interpreted with caution as the methods used generally could not resolve closely eluting compounds (e.g. 18:4ω3 and 18:5ω3 fatty acids), and may have resulted in losses of labile polyunsaturated fatty acids (PUFAs). Several features distinguish haptophytes from other algal classes. In general, C_{16} PUFAs and 18:0 concentrations are lower than in other classes. Many species have high 18:4ω3 and/or 18:5ω3 concentrations. (*Prymnesium parvum* and *Phaeocystis* sp. are exceptions.) The 20:5ω3/22:6ω3 ratio is also generally < 1.0 except in the families Pavlovaceae and Prymnesiaceae.

Five types of fatty acid profile, which correspond roughly to family divisions, are apparent:

1. The first group, comprising the Pavlovaceae, is characterized by 14:0/16:0 ratios between 1.0–1.5, 16:0/16:1 ratios < 1.2, and high 20:5ω3 concentrations. The 20:5ω3/22:6ω3 ratio is > 2 in *Pavlova*, whereas this ratio is generally < 1 for all other families except the Prymnesiaceae. Individual fatty acid abundances in *Pavlova* species,

Table 19.1 Selected fatty acids in microalgae of the Haptophyta (strains analyzed, when identified, are given in parentheses)

SPECIES (strain[c])	Fatty acid[a]												
	Family b	14:0	16:0	16:1	16 PUFA	18:0	18:1 ω9	18:2 ω6	18:3 ω3	18:4 ω3	18:5 ω3	20:5 ω3	22:6 ω3	Ref.
Pavlova lutheri[d]	I	12	15	24	4	+[e]	1	1	+	5[f]	–	22	11	(1)
Pavlova lutheri[d] (SMBA# 60)	I	9	10	20	23	+	6	2	–	+[f]	–	19	–	(2)
Pavlova lutheri[d]	I	14	29	21	1	+	7	–	1	1[f]	–	10	4	(3)
Pavlova lutheri	I	11	23	26	+	2	3	3	+	4	–	14	8	(4)
Pavlova lutheri (CS#182)	I	12	23	26	–	1	2	1	+	4	–	17	8	(5)
Pavlova lutheri (CS#182)	I	12	21	17	+	1	2	2	2	6	–	20	9	(6)
Pavlova lutheri (CS#182)	I	13	18	14	1	+	3	2	2	8	–	22	11	(7)
Pavlova lutheri (NEPCC#5)[g]	I	12	17	24	+	–	1	3	+	6	–	17	10	(8)
Pavlova salina (CS#49)[h]	I	15	15	5	1	+	+	2	1	14	–	27	11	(7)
Pavlova sp. (CS#50)[h]	I	18	13	10	1	+	+	1	2	11	–	24	9	(7)
Pavlova sp. (CS#63)[h]	I	22	12	10	+	+	+	2	1	9	–	23	9	(7)
Dicrateria inornata (PML#13)	II	1	16	8	4	+	17	5	13	20[f]	–	8	–	(2)
Isochrysis galbana (PML#I)	II	1	16	5	10	–	15	11	14	17[f]	–	3	–	(2)
Isochrysis galbana	II	11	22	16	+	2	14	2	+	8[f]	–	7	4	(9)
Isochrysis galbana	II	11	15	14	3	3	13	2	2	1[f]	–	14	11	(10)
Isochrysis galbana	II	8	9	+	1	1	7	4	11	9	–	4	19	(11)
Isochrysis galbana	II	10	21	22	2	3	6	1	1	4[f]	–	12	6	(3)
Isochrysis galbana	II	23	14	1	2	1	13	5	8	10	3	2	14	(12)
Isochrysis galbana (UTEX#LB987)	II	22	9	5	–	11	20	20	5	3	–	–	3	(13)
Isochrysis galbana[i]	II	12	15	16	1	1	+	+	–	8	–	23	8	(14)

Table 19.1 (*cont.*)

SPECIES (strain[c])	Family b	14:0	16:0	16:1	16 PUFA	18:0	18:1 ω9	18:2 ω6	18:3 ω3	18:4 ω3	18:5 ω3	20:5 ω3	22:6 ω3	Ref.
	Fatty acid[a]													
Isochrysis galbana (T-Iso)	II	19	12	4	6	+	14	4	6	19	–	–	8	(15)
Isochrysis galbana (T-Iso)	II	nd	nd	nd	nd	nd	nd	7	4	9	–	+	11	(16)
Isochrysis galbana (T-Iso) (CS#177)	II	16	15	4	3	+	20	3	4	17	3	+	8	(7)
Isochrysis galbana (T-Iso) (NEPCC#601)[g]	II	18	10	1	+	–	22	6	4	9	4	+	14	(8)
Isochrysis sp. (UTEX#2307)	II	13	12	6	–	–	15	4	6	17	–	2	13	(17)
Pseudoisochrysis paradoxa	II	24	18	5	1	1	18	4	2	2[f]	–	+	3	(3)
Emiliania huxleyi[j] (PML#92)	III	6	17	28	12	1	10	2	1	1[f]	–	17	–	(2)
Emiliania huxleyi	III	35	5	+	+	+	15	2	7	8	10	+	11	(12)
Emiliania huxleyi[k]	III	14	5	1	+	+	4	4	5	11	22	1	28	(18)
Prymnesium parvum	IV	4	11	2	1	4	19	7	7	1[f]	–	4	–	(10)
Prymnesium parvum (SMBA# 65)	IV	6	16	10	4	–	25	18	11	2[f]	–	4	–	(2)
Phaeocystis pouchetti[l]	V	10	20	11	–	1	1	1	1	21	7	7	13	(19)
Phaeocystis pouchetti[m]	V	22	31	4	+	3	30	1	1	1	1	+	+	(20)
Phaeocystis pouchetti (Al-3)[h,n]	V	21	36	+	–	12	29	–	1	1	–	–	–	(21)
Phaeocystis pouchetti (DE10)[h,o]	V	38	14	8	–	3	14	3	2	3	–	1	5	(21)
Phaeocystis pouchetti (Al-4)[h,n]	V	46	16	7	–	4	6	1	1	4	–	1	2	(21)
Phaeocystis sp.[p]	V	nd	46	+	+	+	48	+	+	1	5	–	–	(22)
Coccolithus pelagicus[q]	VI	11	25	+	4	+	4	5	5	16	7	5	10	(12)

Table 19.1 (*cont.*)

SPECIES (strain[c])	Fatty acid[a]													
	Family b	14:0	16:0	16:1	16 PUFA	18:0	18:1 ω9	18:2 ω6	18:3 ω3	18:4 ω3	18:5 ω3	20:5 ω3	22:6 ω3	Ref.
Pleurochrysis carterae[r]	VII	+	22	2	5	1	7	10	8	24[f]	–	4	9	(1)
Pleurochrysis carterae[s] (PML#156)	VII	9	9	21	25	+	3	3	–	2[f]	–	20	–	(2)
Pleurochrysis carterae[t]	VII	+	26	2	8	1	7	10	13	20	–	4	4	(12)
Pleurochrysis elongata[u] (SMBA# 62)	VII	7	10	21	23	–	2	2	+	+[f]	–	28	–	(2)

[a] fatty acid notation is carbon number: number of double bonds, ω indicates position of first double bond from terminal methyl carbon. 16 PUFA = 16 : 2 + 16 : 3 + 16 : 4. [b] Families: I: Pavlovaceae; II: Isochrysidaceae; III: Noelaerhabdaceae; IV: Prymnesiaceae; V: Phaeocystaceae; VI: Coccolithaceae; VII: Pleurochrysidaceae.

[c] Strain designations: PML = Plymouth Marine Laboratory (UK); UTEX = University of Texas (USA); CS = CSIRO Marine Laboratories (Australia); NEPCC = Northeast Pacific Culture Collection (Canada); SMBA = Scottish Marine Biological Association Laboratory Collection, Millport (now at Culture Collection of Algae and Protozoa, Dunstaffnage Marine Laboratory, Oban); Caen = University of Caen (France).

[d] as *Monochrysis lutheri*; [e] + = minor component (<1%); [f] 18 : 4 ω3 + 18 : 5ω3; [g] at 17.5° C; [h] mean of two analyses; [i] average of 59 clones; [j] as *Coccolithus huxleyi*; [k] average of 7 strains; [l] surface slick of gelantinous colonies, Balsfjorden, northern Norway; [m] <22 μm size fraction containing single cells, Irish Sea bloom; [n] predominantly colonial stage, Antarctica isolate; [o] colonial and flagellate stages, Antarctica isolate; [p] colonial stage, Arabian sea bloom; [q] as *Crystallolithus hyalinus*; [r] as *Syracosphaera carterae*; [s] as *Cricosphaera carterae*, [t] as *Hymenomonas carterae*; [u] as *Cricosphaera elongata*.

References cited: (1) Ackman *et al.* (1968); (2) Chuecas and Riley (1969); (3) Chu and Dupuy (1980); (4) Langdon and Waldcock (1981); (5) Nichols *et al.* (1989); (6) Volkman *et al.* (1989); (7) Volkman *et al.* (1991); (8) Thompson *et al.* (1992); (9) Watanabe and Ackman (1974); (10) DeMort *et al.* (1972); (11) Waldock and Nascimento (1979); (12) Volkman *et al.* (1981); (13) Ben-Amotz *et al.* (1987); (14) López Alonso *et al.* (1992b); (15) Pillsbury (1985); (16) Helm and Laing (1987); (17) Ben-Amotz *et al.* (1985); (18) M. H. Conte and J. C. Green, unpublished data; (19) Sargent *et al.* (1985); (20) Claustre *et al.* (1990); (21) Nichols *et al.* (1991); (22) Al-Hasan *et al.* (1990).

analyzed by a number of investigators and under a range of culture conditions, generally vary by less than a factor of two. An exception is *P. salina*, which contained higher concentrations of PUFAs, especially 18:4ω3, and lower concentrations of 16:1 than other *Pavlova* species cultured under the similar conditions (Volkman *et al.* 1991).

2. The second group, comprising the Isochrysidaceae, generally has 16:0/16:1 ratios >1.2, and high 18:1ω9, 18:4ω3 and 22:6ω3 concentrations. The 'Tahitian' *Isochrysis* strain (T-Iso) is distinct from *I. galbana* in its low 20:5ω3/22:6ω3 ratio (< 0.2) and high 18:4ω3 concentration. Duplicate analyses of the T-Iso strain are generally in close agreement. However, duplicate analyses of *I. galbana* show pronounced differences, especially in PUFA concentrations; this variability has been previously noted (Volkman *et al.* 1989). López Alonso *et al.* (1992*a*, *b*) also reported high variability in fatty acid profiles among 59 clones obtained from a single *I. galbana* culture and also among 40 isolates made from a single clone. These results may indicate greater genetic diversity within *I. galbana* than in *Pavlova lutheri*, or in the T-Iso strain of *Isochrysis.*

3. The third group, comprising only *Emiliania huxleyi*, has high 14:0/16:0 and 16:0/16:1 ratios, very high 18:4ω3, 18:5ω3 and 22:6ω3 concentrations, but negligible 20:5ω3 concentrations. The occurrence of high concentrations of 18:5ω3 sets *E. huxleyi* apart from other families and is of interest as 18:5ω3 is generally abundant only in dinoflagellates (e.g Nichols *et al.* 1984).

4. The fourth group, comprising the Phaeocystaceae, is distinguished by very low PUFA concentrations, particularly 20:5ω3 and 22:6ω3. However, different analyses of *Phaeocystis* species in blooms show discrepancies which suggest significant biochemical differences exist among populations in different regions. Blooms of *Phaeocystis* species in Antarctica (Nichols *et al.* 1991), the Irish Sea (Claustre *et al.* 1990), and the Arabian Sea (Al-Hasan *et al.* 1990) contained few PUFAs, whereas a bloom of *Phaeocystis* in Balsfjorden, Norway contained abundant PUFAs (Sargent *et al.* 1985). Wieße (1983) found that zooplankton feeding on blooms of *Phaeocystis* in the North Sea had elevated 18:4ω3 concentrations, suggesting that North Sea *Phaeocystis* populations also contain high C_{18} PUFA concentrations. Differences in pigment profiles similarly suggest significant biochemical variability within *Phaeocystis.* In Antarctic strains, 19′-hexanoyloxyfucoxanthin predominated (Nichols *et al.* 1991), whereas in a temperate strain of *Phaeocystis* 19′-butanoyl- and 19′-hexanoyloxyfucoxanthins predominated (Wright and Jeffrey 1987). In contrast, 19′-acyloxyfucoxanthins

were not major components of *Phaeocystis* blooms in the North and Irish Seas (Gieskes and Kraay 1986; Claustre *et al.* 1990). Jeffrey and Wright (Chapter 6) report pigment variations in cultured *Phaeocystis* strains, and Vaulot *et al.* (in press) report differences in DNA content and cell size, as well as pigments in cultured strains from the Atlantic, Mediterranean, North Sea, and Southern Ocean. These differences in biomarker profiles accord well with dissimilarities found in rRNA sequences between northern and southern hemisphere in populations of *Phaeocystis* (Medlin *et al.*, Chapter 21).

5. The fifth group, comprising the Coccolithaceae and Pleurochrysidaceae, has low 14:0/16:0 and high 16 : 0/16 : 1 ratios. Concentrations of 18 : 4ω3 are generally higher but concentrations of C_{20} and C_{22} PUFAs are lower than in other species. However, this group is poorly defined as the number of species analyzed in these families is limited.

While general groupings in fatty acid profiles are noteworthy, rigorous assessment of chemotaxonomic affinities is not possible. Early analyses need to be repeated using modern methods. Furthermore, fatty acid composition, especially with regard to PUFAs, is highly sensitive to growth stage and culture conditions (e.g. Ben-Amotz *et al.* 1985; Emdadi and Galois 1987; Thompson *et al.* 1990, 1992), and a wide range of culture conditions were used in the studies reported here.

Sterols

Sterols, and their sedimentary diagenetic products the sterenes, steranes, and diasteranes, are used frequently as biomarkers of the origins of organic matter in sediments (e.g. Mackenzie *et al.* 1982; Volkman 1986). Sterol profiles for 21 haptophyte species are shown in Table 19.2. Data from early studies have been omitted where more recent analyses of the same species are available. Although this data base is limited, it is rather more extensive than those available for many other classes. Haptophytes generally have quite simple sterol profiles, with a single sterol often comprising more than 75% of the total. Four major groupings can be recognized:

1. The first group contains a very simple sterol distribution dominated by cholest-5-en-3β-ol and 24-ethylcholest-5-en-3β-ol (*Chrysochromulina polylepis, Prymnesium patelliferum, Ochrosphaera neapolitana,* and *O. verrucosa*). Small amounts of C_{28} and C_{29} $\Delta^{5,22}$ sterols, including the uncommon sterol 23,24-dimethylcholesta-5, 22E-dien-3β-ol, also occur in the *Ochrosphaera* species. The close similarity of the *Chrysochromulina*

Table 19.2 Sterols in microalgae of the Haptophyta (strains analyzed, when identified, are given in parentheses). Strain designations given in Table 19.1.

SPECIES (strain)	Family b	Sterol[a]								
		A	B	C	D	E	F	G	H	Ref
Pavlova lutheri[c] (PML#75)	I	27	18	23	I	31	–	–	–	(1)
Pavlova lutheri	I	+[h]	16	73	–	10	–	–	–	(2)
Pavlova lutheri (CS#182)	I	–	12	24	–	30	–	1	33	(3)
Pavlova salina (CS#49)	I	+	4	5	–	43	–	1	47	(3)
Pavlova sp. (CS#50)	I	5	–	15	–	46	–	14	19	(3)
Pavlova sp. (CS#63)	I	9	1	5	–	50	–	20	15	(3)
Pavlova gyrans (PML#93)	I	+	2	5	–	45	–	11	35	(4)
Isochrysis galbana	II	4	–	–	96	–	–	–	–	(5)
Isochrysis galbana	II	1	+	2	97	–	–	–	–	(2)
Isochrysis sp.[d]	II	2	–	–	98	–	–	–	–	(2)
Isochrysis sp.	II	1	–	–	97	–	–	–	–	(6)
Isochrysis sp. (T-Iso)	II	–	–	–	99	–	–	–	–	(7)
Chrysotila lamellosa (PML#353)	II	4	+	+	86	10	–	–	–	(8)
Chrysotila lamellosa (PML#528)	II	13	–	–	78	6	–	1	–	(8)
Chrysotila lamellosa (Caen#17)	II	–	6	–	49	44	–	–	–	(9)
Chrysotila stipitata (PML#377)	II	3	–	+	50	47	–	–	–	(8)
Dicrateria inornata	II	–	–	–	93	–	–	–	–	(10)
Emiliania huxleyi (PML#92D)	III	2	–	–	98	+	–	–	–	(5)
Emiliania huxleyi[e]	III	1	–	–	99	–	–	–	–	(11)
Gephyrocapsa oceanica (JB-02)	III	2	–	–	98	–	–	–	–	(7)
Chrysochromulina polylepis (PML#200)	IV	74	–	26	–	–	–	–	–	(8)
Prymnesium patelliferum (PML#527)	IV	94	–	6	–	–	–	–	–	(2)
Phaeocystis pouchetti (A1-3)	V	9	–	–	91	–	–	–	–	(12)
Phaeocystis pouchetti (A1-4)	V	–	–	–	100	–	–	–	–	(12)
Umbillicosphaera sibogae	VI	–	–	–	44	56	–	–	–	(7)
Coccolithus pelagicus[f]	VI	–	–	–	88	12	–	–	–	(5)

Table 19.2 (*cont.*)

SPECIES (strain)	Family b	Sterol[a] A	B	C	D	E	F	G	H	Ref
Pleurochrysis carterae [g]	VII	–	–	–	54	33	13	–	–	(5)
Pleurochrysis carterae (Cocco II)	VII	–	–	–	40	30	30	–	–	(9)
Pleurochrysis placolithoides	VII	23	–	11	25	22	12	–	–	(13)
Ochrosphaera neapolitana (PML#162)	VIII	59	–	1	12	12	12	4	–	(8)
Ochrosphaera verrucosa (PML#466)	VIII	36	–	18	8	35	3	–	–	(8)

[a] Sterols are as follows: A: cholest-5-en-3β-ol; B: 24-methylcholest-5-en-3β-ol; C: 24-ethylcholest-5-en-3β-ol; D: 24-methylcholesta-5,22E-dien-3β-ol; E: 24-ethylcholesta-5,22E-dien-3β–ol; F: 23,24-dimethyl-cholesta-5,22E-dien-3β-ol; G: 5α(H)-stanols; H: 4-methylsterols.

[b] Families: I: Pavlovaceae; II: Isochrysidaceae; III: Noëlaerhabdaceae; IV: Prymnesiaceae; V: Phaeocystaceae; VI: Coccolithaceae; VII: Pleurochrysidaceae; VIII: Hymenomonadaceae.

[c] as *Monochrysis lutheri*: [d]as *Pseudoisochrysis paradoxa*; [e]average of 7 strains; [f]as *Crystallolithus hyalinus*; [g]as *Cricosphaera carterae*; [h]+ minor components (<1%).

References cited: (1) Ballantine *et al.* (1979); (2) Lin *et al.* (1982); (3) Volkman *et al.* (1990); (4) Gladu *et al.* (1991); (5) Volkman *et al.* (1981); (6) Berenberg and Patterson (1981); (7) J. K. Volkman, unpublished data; (8) Marlowe *et al.* (1984*b*); (9) Raederstorff and Rohmer (1984); (10) Gladu *et al.* (1990); (11) M. H. Conte and J. C. Green, unpublished data; (12) Nichols *et al.* (1991); (13) J. C. Dauger and J. Fresnel (1989, unpublished data).

and *Prymnesium* sterol distributions accords well with their close taxonomic relationship.

2. The second group appears to be restricted to the genus *Pavlova* and contains a much more diverse range of 4-desmethyl sterols including 24-methylcholest-5-en-3β-ol, 24-ethylcholest-5-en-3β-ol and 24-ethylcholest-5,22E-dien-3β-ol, as well as uncommon 5α(H)-stanols (Volkman *et al* 1990). A novel feature is the occurrence of 4-methyl sterols. These sterols have not been found in any other haptophytes analyzed to date, nor are they common in other classes (an exception is their occurrence in certain *Porphyridium* species; Beastall *et al.* 1974), which agrees well with suggestions that the Pavlovales form a distinct group within the Haptophyta at the subclass or class level. While Lin *et*

al. (1982) did not report 4-methyl sterols (possibly due to analytical problems), and Ballantine *et al.* (1979) found them only in stationary phase cells, more recent work (Volkman *et al.* 1990) has shown that they are present in both exponential and stationary growth phases. 4-Methyl sterols are also found in many dinoflagellates, but the major constituent is almost always dinosterol (4α,23,24-trimethyl-5α-cholest-22E-en-3β-ol), often considered to be characteristic of this class (e.g. Nichols *et al.* 1984). Dinosterol has not been found in haptophytes, but the side-chain of dinosterol does occur in one 4-desmethyl sterol (sterol F) in a few species. Some *Pavlova* species also contain unusual compounds thought to be steroidal diols (Ballantine *et al.* 1979). Volkman *et al.* (1990) proposed that the major diol present is 3,4-dihydroxy-4,24-dimethylcholestane, which has been confirmed by G. W. Patterson and coworkers (G. W. Patterson, personal communication).

3. The third, and perhaps more typical, distribution of sterols consists almost exclusively of 24-methylcholesta-5,22E-dien-3β-ol, found in *Emiliania huxleyi*, the closely related species *Gephyrocapsa oceanica*, *Phaeocystis pouchetii*, and *Isochrysis* species. This sterol is also the single most abundant sterol in many diatoms (Volkman 1986). The stereochemistry of the methyl group at position 24 is predominantly 24α (i.e. epibrassicasterol) which is the same stereochemistry as that found in diatoms (Maxwell *et al.* 1980; Gladu *et al.* 1990). Many of the species in this group also synthesize unusual long-chain alkenones.

4. The fourth group contains in addition to large amounts of 24-methylcholesta-5,22E-dien-3β-ol, significant amounts of 24-ethylcholesta-5,22E-dien-3β-ol. A number of genera are represented here, including *Chrysotila*, *Umbilicosphaera*, *Pleurochrysis*, and *Coccolithus*. Species of *Pleurochrysis* also synthesize the unusual sterol 23,24-dimethylcholesta-5,22E-dien-3β-ol.

Although at least four chemotaxonomic groups can be recognized, it should be pointed out that all of the 4-desmethyl sterol distributions can arise from a single biosynthetic pathway (Goodwin 1973). Thus, differences between genera may simply reflect the absence or low activity of enzymes needed to carry out the various biosynthetic steps. This in turn may be related to how recently the group evolved. While variations in the sterol profiles may occur among different strains (e.g. Nichols *et al.* 1991), or between algae grown under different culture conditions (e.g. Ballantine *et al.* 1979; Nichols *et al.* 1991), these variations are usually much smaller than differences between the main taxonomic groups.

Alkenones, alkenoates, and the related long-chain alkenes

1. Background

Long-chain unsaturated ketones (alkenones, Fig. 19.1) were first observed in Black Sea sediments (Boon *et al.* 1978). At the same time, a biological source from the cosmopolitan haptophyte *Emiliania huxleyi* was recognized (Volkman *et al.* 1980*a*). In addition to alkenones, long-chain C_{36} fatty acid methyl and ethyl esters (alkenoates) were also identified, Fig. 19.1 (Volkman *et al.* 1980*a*). Alkenones and, in most cases, alkenoates are now known to be globally distributed in marine sediments, extending back into pre-Pleistocene sediments (Marlowe *et al.* 1990). The similarity between alkenone and alkenoate distributions in *E. huxleyi* and that observed in particulate matter and sediments strongly suggests that *E. huxleyi* is the predominant source in the present-day ocean.

Alkenones and alkenoates have attracted enormous interest following the discovery that the unsaturation ratio of the alkenones reflects the organism's growth temperature (Marlowe 1984). Brassell *et al.* (1986) first noted that C_{37} alkenone unsaturation (U^{k}_{37}) in sediments was correlated with overlying surface water temperature and hypothesized that these compounds preserved a record of palaeo sea surface temperature (SST). This hypothesis has been confirmed by a number of detailed studies of U^{k}_{37} stratigraphy (e.g. Poynter *et al.* 1989; Farrimond *et al.* 1990; Ten Haven and Kroon 1990; Eglinton *et al.* 1992).

The application of these compounds as palaeo SST proxies has enormous potential for improving our knowledge of ocean circulation and palaeoclimate. However, the molecular signal preserved in the sediments must be carefully calibrated with temperature. Current alkenone-based estimates of palaeo SST use a simple linear temperature calibration of U^{k}_{37} derived from a single culture experiment using a NE Pacific strain of *E. huxleyi* (Prahl *et al.* 1988). While temperatures estimated using this calibration are qualitatively reasonable (Prahl and Wakeham 1987), recent detailed surveys of alkenone and alkenoate distributions in surface waters have found biogeographical variability in the alkenone/temperature calibration, suggesting genetic differences in the source organisms' biochemical response to growth temperature (section 3).

2. Occurrence in the Haptophyta

In addition to *Emiliania huxleyi*, alkenones have been identified in other species of the Isochrysidales (Table 19.3) but, to date, not in any other order. Not all members of the Isochrysidales contain alkenones; they were absent in *Chrysotila stipata, Dicrateria inornata,* and *Imantonia rotunda,* and in two species of *Ochrosphaera* (Marlowe *et al.* 1984*b*). The occurrence of

Table 19.3 Long-chain alkenones in microalgae from the Order Isochrysidales (Strains analyzed are given in parentheses). Structures of alkenones given in Fig. 19.1. All strains were cultured at 15 °C (except *Gephyrocapsa oceanica*, which was cultured at 20 °C) and harvested in exponential phase. All *Emiliania huxleyi* strains were calcifying except where noted in footnote[a].

	Alkenone																
SPECIES (strain[a])	33:3 Me	33:2 Me	34:2 Et	37:4 Me	37:3 Me	37:2 Me	38:4 Et	38:4 Me	38:3 Et	38:3 Me	38:2 Et	38:2 Me	39:3 Et	39:3 Me	39:2 Et	39:2 Me	Ref.
Chrysotila lamellosa (PML#353)	–	–	–	30	54	4	–	–	6[b]	–	3[b]	–	3	–	–	–	(1)
Chrysotila lamellosa (PML#528)	–	–	–	37	49	2	–	–	8[b]	–	2[b]	–	2	–	–	–	(1)
Isochrysis galbana (PML#I)	2	12	5	22	48	2	–	–	8[b]	–	2[b]	–	–	–	–	–	(1)
Isochrysis sp. (PML#507)	–	–	–	10	68	10	–	–	5[b]	–	+[b]	–	6	–	–	–	(1)
Isochrysis sp. (T-Iso) (PML#506a)	–	–	–	–	74	17	–	–	5[b]	–	2[b]	–	1	–	+	–	(1)
Gephyrocapsa oceanica (JB-02)	–	–	–	+	20	17	–	–	15	5	40	+	1	–	2	–	(2)
Emiliania huxleyi (PML#92)	–	–	–	–	32	14	–	–	26	7	14	2	3	–	2	–	(1)
E. huxleyi (PML#92d)	–	–	–	4	47	16	1[b]	–	13	3	11	3	2	–	+	–	(1)
E. huxleyi (PML#92d)[c]	–	–	–	4	37	14	1	+	15	11	11	4	2	+	2	+	(3)
E. huxleyi (PML#G1779Ge)[c]	–	–	–	1	27	16	+	+	16	11	15	5	4	1	3	1	(3)

Table 19.3 (*cont.*)

SPECIES (strain[a])	Alkenone																Ref.
	33:3 Me	33:2 Me	34:2 Et	37:4 Me	37:3 Me	37:2 Me	38:4 Et	38:4 Me	38:3 Et	38:3 Me	38:2 Et	38:2 Me	39:3 Et	39:3 Me	39:2 Et	39:2 Me	
E. huxleyi (CCMP#1A1)[c]	–	–	–	1	38	17	–	+	15	10	12	3	2	+	1	+	(3)
E. huxleyi (PML#DW53/74/6)[c]	–	–	–	+	36	14	–	–	21	9	12	3	3	+	1	+	(3)
E. huxleyi (PML#S. Africa)[c]	–	–	–	+	32	12	+	+	12	9	15	5	2	+	2	+	(3)
E. huxleyi (PML#B21)[c]	–	–	–	1	28	20	+	–	14	10	16	6	3	+	3	+	(3)
E. huxleyi (CCMP#88Cocco)	–	–	–	2	31	16	1	+	16	12	13	5	3	+	2	+	(3)
E. huxleyi (CCMP#88E)	–	–	–	2	31	14	–	1	19	10	14	3	3	+	1	+	(3)
E. huxleyi (VAN55)	–	–	–	1	27	15	+	–	17	11	16	6	3	+	2	+	(3)
E. huxleyi (VAN556)	–	–	–	1	27	17	+	–	14	11	17	6	3	+	3	+	(3)

[a] Collection locations, date of *E. huxleyi* strains: 92 — English Channel, 1950; 92d — English Channel, 1975 (poorly calcifying); G1779Ge — Iceland Basin, 1989; 1A1 — Sargasso Sea, 1987; DW53/74/6 — subtropical NE Atlantic, 1990; S. Africa — SW Indian Ocean, 1983; B21 — Norwegian fjord, 1992; 88Cocco, 88E — Gulf of Maine, 1988 (both poorly calcifying); VAN55, VAN556 — NE Pacific (both noncalcifying).
[b] includes Et + Me compounds; [c] average of 2 culture experiments
References cited: (1) Marlowe *et al.* (1984*b*); (2) J. K. Volkman, unpublished data; (3) M. H. Conte and J. C. Green, unpublished data.

alkenones in both coccolith-bearing and uncalcified haptophytes indicates that they are not restricted to the coccolithophorids, and provides a biochemical link between these different types of organism. Recently, alkenones and alkenoates have also been identified in the coccolithophorid *Gephyrocapsa oceanica* (Volkman, Blackburn and Barrett, unpublished data). The marine stratigraphic record of alkenones predates the appearance of *E. huxleyi*, and it is believed that *E. huxleyi* inherited its alkenone/alkenoate biochemistry from a presumed gephyrocapsid ancestor (Marlowe *et al.* 1990).

Alkenoates (Fig. 19.1, Table 19.4) and long-chain alkenes (Table 19.5) also occur in alkenone-containing species, but not in species lacking alkenones. The chain length and degree of unsaturation of the alkenoates is similar to that of the alkenones, and alkenoate and alkene distributions also vary with growth temperature (Marlowe 1984; Conte and Eglinton 1993). These observations suggest that alkenones, alkenoates, and alkenes are biochemically linked (Volkman *et al.* 1980*b*).

Alkenones have also been found in freshwater sediments. Cranwell (1985) identified C_{37}-C_{39} methyl and C_{38}-C_{40} ethyl alkenones, but no alkenoates, in English lake sediments. Their abundance was correlated with that of microalgae of the orders Ochromonadales and Prymnesiales. However, lakes containing only Ochromonadales species lacked alkenones, suggesting their source may possibly be an unidentified species of the Prymnesiales.

Alkenones were also found in sediments of a saline lake in Antarctica (Volkman *et al.* 1988). The distribution of alkenones in this lake is quite unusual and shows a very strong predominance of the tetra-unsaturated C_{37} alkenone, whereas alkenones from waters of the Southern Ocean show almost none of this alkenone (Sikes and Volkman 1993). Coccolithophorids have not been observed in the lake, so it seems likely that the alkenone source is a coastal coccolith-free microalga (as yet undescribed) which became isolated when the lake separated from the sea during isostatic uplift of the Antarctic continent.

3. Temperature control on synthesis and biogeographical variation in the alkenone/alkenoate vs. temperature relationship in natural waters

Marlowe (1984) first reported that alkenone, alkenoate, and alkene compositions in both *Emiliania huxleyi* and *Isochrysis galbana* varied with growth temperature. Prahl *et al.* (1988) and Prahl and Wakeham (1987) demonstrated that C_{37} alkenone unsaturation (U^{k}_{37}) was linearly correlated with growth temperature. Conte and Eglinton (1993) showed that alkenone and alkenoate distribution in North Atlantic surface waters was primarily controlled by temperature, and was largely unaffected by nutrient concentration, bloom status, or other surface water variables.

I Diunsaturated Compounds:

$(CH_2)n — C(=O)R$

II Triunsaturated compounds

$(CH_2)n — C(=O)R$

III Tetraunsaturated compounds

$(CH_2)n — C(=O)R$

ALKENONES (n = 5 - 7; R = Me, Et)

37:4Me - heptatriaconta-8E,15E,22E,29E-tetraen-2-one	
37:3Me - heptatriaconta-8E,15E,22E-trien-2-one	
37:2Me - heptatriaconta-15E,22E-dien-2-one	
38:4Me - octatriaconta-9E,16E,23E,30E-tetraen-2-one	**38:4Et** - octatriaconta-9E,16E,23E,30E-tetraen-3-one
38:3Me - octatriaconta-9E,16E,23E-trien-2-one	**38:3Et** - octatriaconta-9E,16E,23E-trien-3-one
38:2Me - octatriaconta-16E,23E-dien-2-one	**38:2Et** - octatriaconta-16E,23E-dien-3-one
39:3Me - nonatriaconta-10E,17E,24E-trien-2-one	**39:3Et** - nonatriaconta-10E,17E,24E-trien-3-one
39:2Me - nonatriaconta-17E,24E-dien-2-one	**39:2Et** - nonatriaconta-17E,24E-dien-3-one

ALKENOATES (n = 5, 6; R = OEt, OMe)

36:4FAME - methyl hexatriaconta-7E,14E,21E,28E-tetraenoate	
36:3FAME - methyl hexatriaconta-7E,14E,21E-trienoate	**36:3FAEE** - ethyl hexatriaconta -7E,14E,21E-trienoate
36:2FAME - methyl hexatriaconta -14E,21E-dienoate	**36:2FAEE** - ethyl hexatriaconta -14E,21E-dienoate
37:2FAME - methyl heptatriaconta -15E,22E-dienoate	

Fig. 19.1 Structures and shorthand notation for C_{37}–C_{39} alkenones, and C_{36} and C_{37} alkenoates. The double bonds are in the biologically rare *trans* configuration (Recka and Maxwell 1988). Structures confirmed by synthesis for the C_{37} di- and tri-unsaturated alkenones (de Leeuw *et al.* 1980); structures for other compounds are assumed to be analogous.

Similarly, alkenone and alkenoate distribution in a bloom of *E. huxleyi* in a Norwegian fjord was nearly identical to that observed in an experimental mesocosm at the same water temperature, even though cell counts were 1.5 orders of magnitude lower in the mesocosm and NO_3^- concentrations an order of magnitude higher (Conte *et al.*, 1994). Dunstan *et al.* (1993) observed that light intensity had little effect on alkenone distributions in

Table 19.4 Long-chain alkenoates in microalgae from the order Isochrysidales. (Strains analyzed are given in parentheses. Strain details and culturing conditions given in Table 19.3. Structures of alkenoates given in Fig. 19.1)

SPECIES (strain)	Alkenoate							Ref.
	32:2 FAEE	36:4 FAEE	36:3 FAEE	36:3 FAME	36:2 FAEE	36:2 FAME	37:2 FAME	
Chrysotila lamellosa (PML#353)	–	36	36	–	28	–	–	(1)
Chrysotila lamellosa (PML#528)	–	42	43	–	15	–	–	(1)
Isochrysis galbana (PML#I)	3	67	27	4	–	–	–	(1)
Isochrysis sp. (PML#507)	–	–	15	71	8	7	–	(1)
Isochrysis sp. (T-Iso) (PML#506a)	–	–	–	–	–	100	–	(1)
Gephyrocapsa oceanica (JB-02)			–	+[b]	++[c]	+	–	(2)
Emiliania huxleyi (PML#92)	–	–	–	11	57	33	–	(1)
E. huxleyi (PML#92d)	–	–	–	8	67	25	–	(1)
E. huxleyi (PML#92d)[a]	–	–	–	17[b]	–	79[b]	4[d]	(3)
E. huxleyi (PML#G1779Ge)[a]	–	–	–	15[b]	–	82[b]	4[d]	(3)
E. huxleyi (CCMP#1A1)[a]	–	–	–	19[b]	–	77[b]	4[d]	(3)
E. huxleyi (PML#DW53/74/6)[a]	–	–	–	14[b]	–	83[b]	2[d]	(3)
E. huxleyi (PML#S. Africa)[a]	–	–	–	12[b]	–	85[b]	3[d]	(3)
E. huxleyi (PML#B21)[a]	–	–	–	13[b]	–	85[b]	2[d]	(3)
E. huxleyi (CCMP#88Cocco)	–	–	–	18[b]	–	80[b]	2[d]	(3)
E. huxleyi (CCMP#88E)	–	–	–	17[b]	–	81[b]	2[d]	(3)
E. huxleyi (VAN55)	–	–	–	14[b]	–	77[b]	8[d]	(3)
E. huxleyi (VAN556)	–	–	–	14[b]	–	77[b]	8[d]	(3)

[a] average of 2 culture experiments; [b] includes ethyl as well as methyl alkenoates; [c] predominate compound; [d] possibly includes compounds originally present as ethyl as well as methyl esters.

References cited: (1) Marlowe *et al.* (1984*b*); (2) J. K. Volkman, unpublished data; (3) M. H. Conte and J. C. Green, unpublished data.

Isochrysis (T-Iso), although total abundances per cell varied slightly. While the above studies show that temperature exerts the primary control on the synthesis of these compounds, we still do not know their biochemical function(s) or location within the algal cell.

Recent surveys of alkenone and alkenoate distributions in surface waters indicate geographic variation in the alkenone *versus* temperature relationship, suggesting genetic differences in the source populations inhabiting these regions. Conte and Eglinton (1993) surveyed alkenone and alkenoate distributions in North Atlantic surface waters along a 20°W transect extending from the subpolar waters of the Iceland Basin to subtropical, oligotrophic waters off Africa. In the cold water region (< 15 °C) of the North Atlantic, where *Gephyrocapsa* species are absent, alkenoate abundance (AA_{36}) was strongly inversely correlated with temperature, while alkenone unsaturation was only weakly correlated. Conversely, in warmer (> 15 °C) waters, alkenoate abundance remained low but uncorrelated with temperature, while the unsaturation ratio of *both* the C_{37} and C_{38} methyl alkenones (U^k_{37} and U^k_{38Me}, respectively) was strongly correlated with temperature. The slopes of both regressions of U^k_{37} and U^k_{38Me} versus temperature were identical, yet were 50% greater than the slope of the Prahl and Wakeham (1987) field calibration. Sikes and Volkman (1993) made a similar survey of U^k_{37} in surface waters along a transect extending from Tasmania to the Antarctica ice shelf. Above the polar frontal boundary, temperatures spanned a similar range as the northern section of the North Atlantic transect. In contrast to that transect, U^k_{37} in subAntarctic waters varied linearly with temperature with a slightly higher slope than that of the Prahl and Wakeham (1987) calibration. In Antarctic waters below the polar frontal boundary, U^k_{37} remained low but was poorly correlated with temperature.

The alkenone/alkenoate versus temperature relationship was also studied over a similar temperature range (7–15 °C) in experimental mesocosms in a Norwegian fjord (Conte *et al.* 1994). In contrast to both the Conte and Eglinton (1993) and Sikes and Volkman (1993) findings, tetraunsaturated alkenones showed the strongest correlation with temperature while both AA_{36} and U^k_{37} remained nearly constant. This discrepancy is especially interesting as *Emiliania huxleyi* strains isolated from North Atlantic (G1779Ge) and fjord waters (B21) during the respective field studies have indistinguishable Type A morphology (J. C. Green, personal communication). Freeman and Wakeham (1992) have also reported U^k_{37} values for surface waters of the Black Sea which are inconsistent with the Prahl and Wakeham (1987) or Conte and Eglinton (1993) calibrations. It is noteworthy that in both the Black Sea and Norwegian fjords, tetraunsaturated alkenones occur in significantly higher abundances than in open waters at similar temperatures.

In addition to genetic variability in *Emiliania huxleyi*, production of alkenones and alkenoates by *Gephyrocapsa* species may also affect empirical temperature calibrations in temperate and subtropical waters where these species occur. Both culture studies of *Emiliania* and *Gephyrocapsa* and additional surveys of alkenone/alkenoate surface water distributions are needed to assess the effect of genetic and/or species variations on empirical temperature calibrations in natural waters.

4. Species and strain-related variations in biomarker profiles

Table 19.3 shows the alkenone profiles for alkenone-containing species which have been cultured at 15 °C under similar conditions at Plymouth Marine Laboratory. Both *Chrysotila lamellosa* strains have a similar predominance of 37:4 Me and 37:3Me. In contrast, alkenones in the *Isochrysis* strains vary greatly in both chain length and degree of unsaturation, *Isochrysis galbana* (PML strain I) also contains C_{33} and C_{34} alkenones absent in other species. *Emiliania huxleyi* strains, recently isolated from locations ranging from the subpolar North Atlantic to be subtropical Indian Ocean, show largely similar alkenone profiles when cultured at 15 °C, although significantly higher abundance of tetraunsaturated alkenones occur in English Channel (92D), Gulf of Maine (88E, 88Cocco), and Norwegian fjord (B21) strains. The closely related *Gephyrocapsa oceanica* (cultured at 20 °C) is distinguished from *E. huxleyi* by significantly higher abundances of C_{38} ethyl alkenones, especially 38 : 2Et.

Table 19.4 shows the alkenoate profiles for the same strains. Alkenoate profiles in both *Chrysotila lamellosa* strains are again very similar, while in *Isochrysis* species they differ significantly. A C_{32} alkenoate is present only in *Isochrysis galbana* (PML Strain I); this corresponds to the presence of C_{33} alkenones in this strain. Only slight variations in alkenoate profiles are observed among *Emiliania huxleyi* strains, although 37 : 2FAME concentrations in north-east Pacific strains (Van55, Van556) are approximately twice that in other strains.

Overall cellular concentrations of alkenones and alkenoates show greater variability between strains than does relative composition (Fig. 19.2). Total alkenones and alkenoates comprised 4–9% of the total organic cell carbon for these analyses, slightly lower than that reported by Prahl *et al.* (1988) for their north-east Pacific strain. Cellular concentrations averaged 0.359 ± 0.172 pg cell^{-1}, with alkenoates comprising between 10–14% of the total. Alkenoate concentrations strongly covaried with that of the alkenones, supporting the assumption that these compounds are biochemically linked.

Both alkene profiles and cellular composition varied greatly among the species and *Emiliania huxleyi* strains (Table 19.5). Both *Chrysotila* and *Isochrysis* species have a very simple alkene distribution, with a single 31 : 2

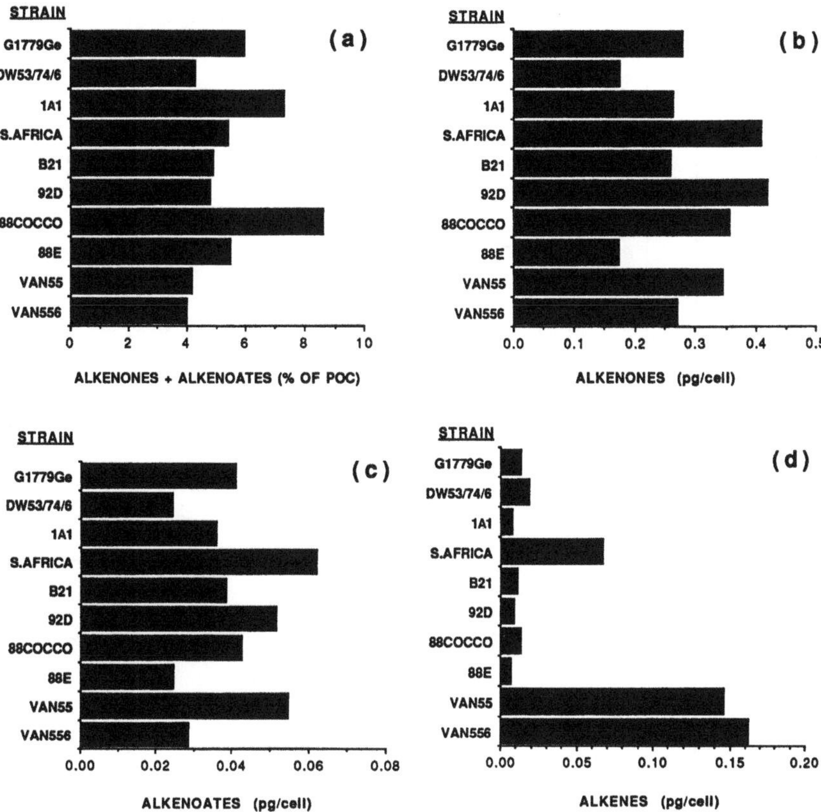

Fig. 19.2 Strain variations in alkenone, alkenoate, and alkene concentrations in *Emiliania huxleyi* cultured at 15 °C (M. H. Conte and J. C. Green, unpublished data). Location, collection date, and culture conditions given in Table 19.3 (a) Total alkenone and alkenoate concentration as percentage of cellular organic carbon. (b) Total C_{37}–C_{39} alkenones (pg cell^{-1}). (c) Total C_{36} and C_{37} alkenoates (pg cell^{-1}). (d) Total C_{31}, C_{33}, C_{37}, and C_{38} alkenes (pg cell^{-1}).

alkene predominating. In contrast, a range of C_{31}, C_{33}, C_{37} and C_{38} alkenes occur in *E. huxleyi.* The alkene profiles in the *E. huxleyi* strains show three significantly different groupings. In the first group, comprising strain 92 (English Channel, 1950) and both north-east Pacific strains (VAN55, VAN556), the 37:3 and 38:3 alkenes predominate. These compounds are absent in other strains. In the second group, comprising the English Channel (92D), Norwegian fjord (B21), and Gulf of Maine (88E, 88Cocco) strains, significant concentrations of 33 : 4 alkenes are present. These alkenes are minor constituents in other strains. It is noteworthy that the second group consists entirely of coastal strains having enhanced

Table 19.5 Long-chain alkenes in microalgae from the order Isochrysidales (Strains analyzed are given in parentheses. Strain details and culture conditions as in Table 19.3); The positions of double bonds in the alkyl chain have not yet been determined — isomers are listed in order of elution on 50m × 0.25 μm Chrompack CPSil5CB GC column.

	Alkene[a]													
SPECIES (strain)	25:2	27:2	29:2	31:2	31:2	31:2	33:4	33:4	33:3	33:3	33:2	37:3	38:3	Ref.
Chrysotila lamellosa (PML#353)	–	–	–	100	–	–	–	–	–	–	–	–	–	(1)
Chrysotila lamellosa (PML#528)	–	–	–	100	–	–	–	–	–	–	–	–	–	(1)
Isochrysis galbana (PML#I)	+[b]	1	1	96	–	–	–	–	1	–	1	–	–	(1)
Isochrysis sp. (PML#507)	–	–	–	100	–	–	–	–	–	–	–	–	–	(1)
Isochrysis sp. (T-Iso) (PML#506a)	–	–	–	100	–	–	–	–	–	–	–	–	–	(1)
E. huxleyi (PML#92)	–	–	–	+	n.a.[c]	1	n.a.	n.a.	n.a.	n.a.	–	83	16	(1)
E. huxleyi (PML#92d)	–	–	–	27	4	53	1	1	9	4	–	–	–	(1)
E. huxleyi (PML#92d)[d]	–	–	–	21	5	41	9	20	5	16	–	–	–	(2)
E. huxleyi (PML#G1779Ge)[d]	–	–	–	+	7	54	–	–	13	28	–	–	–	(2)
E. huxleyi (CCMP#1A1)[d]	–	–	–	+	11	80	–	–	4	5	–	–	–	(2)
E. huxleyi (PML#DW53/74/6)[d]	–	–	–	27	33	40	–	–	+	+	–	–	–	(2)
E. huxleyi (PML#S. Africa)[d]	–	–	–	19	8	74	–	–	–	–	–	–	–	(2)
E. huxleyi (PML#B21)[d]	–	–	–	11	+	46	7	12	–	23	–	–	–	(2)
E. huxleyi (CCMP#88Cocco)	–	–	–	7	+	26	18	21	–	24	–	–	–	(2)
E. huxleyi (CCMP#88E)	–	–	–	6	+	46	12	16	–	21	–	–	–	(2)
E. huxleyi (PML#VAN55)	–	–	–	+	1	5	+	1	2	7	–	68	15	(2)
E. huxleyi (PML#VAN556)	–	–	–	+	+	3	+	+	1	5	–	74	16	(2)

[a] Notation is carbon number: number of double bonds; [b] + = minor component (<1%); [c] not analyzed; [d] average of 2 culture experiments.
References cited: (1) Marlowe (1984); (2) M. H. Conte and J. C. Green, unpublished data.

concentrations of tetraunsaturated alkenones. In the third group, comprised of the Iceland Basin (G1779Ge), Sargasso Sea (1A1), subtropical eastern North Atlantic (DW53/74/6), and subtropical Indian Ocean (South Africa) strains, 31:2 alkenes predominate although lesser quantities of 33:3 alkenes also occur. This third group consists entirely of open ocean strains. Cellular concentrations of alkenes also showed extreme differences between the *E.huxleyi* strains (Fig. 19.2). In particular, the alkene concentration in the north-east Pacific strains (VAN55, VAN556) averaged approximately 0.15 pg cell^{-1}, nearly an order of magnitude greater than that found in most other strains. The South African strain also has enhanced alkene concentration, averaging 0.06 pg cell-1 for two separate cultures.

These biomarker variations are consistent with other evidence for genetic variability in *Emiliania huxleyi.* Two distinct *E.huxleyi* morphotypes (Type A and B), which also differ in immunological response, coexist in the North Atlantic region (van Bleijswiijk *et al.* 1991). Variations in growth rates among *E.huxleyi* isolates have also been reported (Brand 1982). Pigment profiles of many of these same strains differ in abundances of 19′-hexanoylfucoxanthin and fucoxanthin (R. F. C. Mantoura, unpublished data; Jeffrey and Wright, chapter 6). However, gene sequences for both the small subunit rRNA and spacer region between the Rubisco genes are identical for strains 92D, G1779Ge, 1A1, and DW53/74/6 (Medlin *et al.* Chapter 21), suggesting that the observed biochemical differences reflect variations on a subspecies rather than higher taxonomic level.

Summary

1. Lipid biomarker profiles of haptophyte algae are distinct from those of other major algal classes and show groupings consistent with current taxonomic divisions. The Pavlovales differ from other orders in both fatty acid and sterol profiles, particularly in the occurrence of unusual 4-methyl sterols. The biomarker composition of most species has not been examined, and there is a particular need for data on coccolithophorid and freshwater species.

2. Biomarker profiles indicate significant genetic variability within the *Isochrysis* genera. Fatty acid profiles of species of *Phaeocystis* in blooms and cultures also indicate geographical differences, in agreement with recent molecular genetic evidence for genetic dissimilarity in *Phaeocystis.*

3. Both culture and field studies indicate biochemical differences among *Emiliania huxleyi* strains, although no consistent division between Type

A and Type B morphotypes in found. Enhanced concentrations of tetraunsaturated alkenones are present in some coastal strains. Alkene profiles showed the greatest variability, e.g. C_{37} and C_{38} alkenes are present in high concentrations in some strains but totally absent in others.

4. Chemotaxonomy is a sensitive tool for assessing phylogenetic affinities and is complementary to morphological and molecular genetic studies. However, because biomarker composition can be affected by growth conditions, standardized culturing, and analytical methodology must be developed so that results from different laboratories are comparable. Cellular concentrations as well as class composition of biomarkers may show significant variations among taxonomic divisions and should be reported.

Acknowledgements

We thank L. A. S. Madureira, R. Jordan, R. Harris, and R. Head for assistance with *Emiliania huxleyi* isolations on BOFS cruises, and A. Thompson, S. Brown, and R. Green for assistance with laboratory analyses. This work was supported by NERC grants GST/02/386, GR3/8139 and GR9/787, and the mass spectrometry facilities by GR3/3758. Funding to J. Volkman from FRDC grant 91/59 is gratefully acknowledged. BOFS contribution No. 154. Thanks are also due to R. Andersen (Bigelow Laboratory), R. N. Pienaar (University of Witwatersrand), and F. J. R. Taylor (University of British Columbia) who kindly made cultures of *Emiliania huxleyi* available to us.

References

Ackman R. G., Tocher, C. S., and McLachlan, J. (1968). Marine phytoplankter fatty acids. *Journal of the Fisheries Research Board of Canada*, **25**, 1603–20.

Al-Hasan R. H., Ali, A. M., and Radwan, S. S. (1990). Lipids, and their constituent fatty acids, of *Phaeocystis* sp. from the Arabian Gulf. *Marine Biology*, **105**, 9–14.

Ballantine, J. A., Lavis, A., and Morris, R. J. (1979). Sterols of the phytoplankton-effects of illumination and growth stage. *Phytochemistry*, **18**, 1459–66.

Beastall, G. H., Tyndall, A. M., Rees, H. H., and Goodwin, T. W. (1974). Sterols in *Porphyridum* series. 4α-Methylcholesta-5,22-dien-3β-ol and 4,24-dimethyl-5α-cholesta-8,22-dien-3β-ol: two novel sterols from *Porphyridium cruentum*. *European Journal of Biochemistry*, **41**, 301–9.

Ben-Amotz, A., Tornabene, T. G., and Thomas, W. H. (1985). Chemical profile of selected species of microalgae with emphasis on lipids. *Journal of Phycology*, **21**, 72–81.

Ben-Amotz, A., Fishler, R., and Schneller, A. (1987). Chemical composition of dietary species of marine unicellular algae and rotifers with emphasis on fatty acids. *Marine Biology*, **95**, 31–6.

Berenberg, C. J. and Patterson, G. W. (1981) The relationship between dietary phytosterols and the sterols of wild and cultivated oysters. *Lipids*, **16**, 276–8.

Bleijswijk, J. van, van der Wal, P., Kempers, R., Velduis, M., Young, J. R., Muyzer, G. *et al.* (1991). Distribution of two types of *Emiliania huxleyi* (Prymnesiophyceae) in the northeast Atlantic region as determined by immunofluorescence and coccolith morphology. *Journal of Phycology*, **27**, 556–70.

Boon J. J., van der Meer, F. W., Schuyl, P. J. W., de Leeuw, J. W., Schenck, P. A., and Burlingame, A. L. (1978). Organic geochemical analysis of core samples from Site 362, Walvis Ridge, DSDP leg 40. In *Initial reports of the Deep-Sea Drilling Project*, Vol. 40, supplement, (ed. H. M. Bolli and W. F. B. Ryan), pp. 627–37. US Government Printing Office, Washington.

Brand, L. E. (1982) Genetic variability and spatial patterns of genetic differentiation in the reproductive rates of the marine coccolithophorids *Emiliania huxleyi* and *Gephyrocapsa oceanica. Limnology and Oceanography*, **27**, 236–45.

Brassell, S. C., Eglinton, G., Marlowe, I. T., Pflaumann, U., and Sarnthein, M. (1986). Molecular stratigraphy: a new tool for climatic assessment. *Nature*, **320**, 129–33.

Chu, F-L. E. and Dupuy, J. L. (1980). The fatty acid composition of three unicellular algal species used as food sources for larvae of the American oyster (*Crassostrea virginica*). *Lipids*, **15**, 356–64.

Chuecas, L. and Riley, J. P. (1969). Component fatty acids of the total lipids of some marine phytoplankton. *Journal of the Marine Biological Association of the United Kingdom*, **49**, 97–116.

Claustre, H., Poulet, S. A., Williams, R., Marty, J.-C., Coombs, S., Ben Mlih, F., *et al.* (1990). A biochemical investigation of a *Phaeocystis* sp. bloom in the Irish Sea. *Journal of the Marine Biological Association of the United Kingdom*, **70**, 197–207.

Conte, M. H. and Eglinton, G. (1993). Alkenone and alkenoate distributions within the euphotic zone of the eastern North Atlantic: correlation with production temperature. *Deep-Sea Research*, **40**, 1935–61.

Conte, M. H., Thompson, A., and Eglinton, G. (1994). Primary production of lipid biomarker compounds by *Emiliania huxleyi*: Results from an experimental mesocosm study in fjords of southern Norway. *Sarsia*, (in press).

Cranwell. P. A. (1985). Long-chain unsaturated ketones in recent lacustrine sediments. *Geochimica et Cosmochimica Acta*, **49**, 1545–51.

DeMort, C. L., Lowry, R., Tinsley, I., and Phinney, H. K. (1972). The biochemical analysis of some estuarine phytoplankton species. I. Fatty acid composition. *Journal of Phycology*, **8**, 211–16.

Dunstan, G. A., Volkman, J. K., Barrett, S. M., and Garland, C. D. (1993). Changes in the lipid composition and maximization of the polyunsaturated fatty acid content of three microalgae grown in mass culture. *Journal of Applied Phycology*, **5**, 71–83.

Eglinton G., Bradshaw, S. A., Rosell, A., Sarnthein, M., Pflaumann, U., and Tiedemann, R. (1992). Molecular record of secular sea surface changes on 100-year timescales for glacial terminations I, II and IV. *Nature*, **356**, 423–6.

Emdadi, D. and Galois, R. (1987). Influence de la température et de la salinité sur la croissance et la composition lipidique de *Pavlova lutheri* (Prymnesiophycée) en fonction de l'âge de la culture. *Océanis*, **13**, 495–504.

Erwin, J. A. (1973). Comparative biochemistry of fatty acids in eucaryotic microorganisms. In *Lipids and biomembranes of eucaryotic microorganisms*, (ed. J. A. Erwin), pp. 41–143. Academic Press, New York.

Farrimond, P., Poynter, J. G., and Eglinton, G. (1990). A molecular stratigraphic study of Peru margin sediments, Hole 686B, Leg 112. In *Proceedings of the Ocean Drilling Program, Scientific Results*, Vol. 112, (ed. W. E. Dean, K. - C. Emels, E. Suess, and R. von Huene), pp. 547–53. US Government Printing Office, Washington.

Freeman, K. H. and Wakeham, S. G. (1992). Variations in the isotopic compositions and concentrations of alkenones in Black Sea particles. In *Advances in organic geochemistry 1991*, (ed. C. B. Eckardt, J. R. Maxwell, S. R. Larter, and D. A. C. Manning), *Organic Geochemistry*, **19**, 277–86.

Gieskes, W. W. C. and Kraay, G. W. (1986). Analysis of phytoplankton pigments by HPLC before, during and after mass occurrence of the microflagellate *Corymbellus aureus* during the spring bloom in the open northern North Sea in 1983. *Marine Biology*, **92**, 45–52.

Gladu, P. K., Patterson, G. W., Wikfors, G. H., Chitwood, D. J., and Lusby, W. R. (1990). The occurrence of brassicasterol and epibrassicasterol in the Chromophyta. *Comparative Biochemistry and Physiology*, **97B**, 491–4.

Gladu, P. K., Patterson, G. W., Wikfors, G. H., and Lusby, W. R. (1991). Free and combined sterols of *Pavlova gyrans*. *Lipids*, **26**, 656–9.

Goodwin, T. W. (1973). Comparative biochemistry of sterols in eucaryotic microorganisms. In *Lipids and biomembranes of eucaryotic microorganisms*, (ed. J. A. Erwin), pp. 1–39. Academic Press, London.

Helm, M. M. and Laing, I. (1987). Preliminary observations of the nutritional value of 'Tahiti *Isochrysis*' to bivalve larvae. *Aquaculture*, **62**, 281–8.

Jordan, R. W. and Green, J. C. (1994) A check-list of the extant Haptophyta of the world. *Journal of the Marine Biological Association of the United Kingdom*, **74**, 149–74.

Langdon, C. J. and Waldock, M. J. (1981). The effect of algal and artificial diets on the growth and fatty acid composition of *Crassostrea gigas* spat. *Journal of the Marine Biological Association of the United Kingdom*, **61**, 431–48.

Leeuw J. W., de, van der Meer, F. W., Rijpstra, W. I. C., and Schenck, P. A. (1980). On the occurrence and structural identification of long chain unsaturated ketones and hydrocarbons in sediments. In *Advances in organic geochemistry 1979*, (ed. A. G. Douglas and J. R. Maxwell), pp. 211–17. Pergamon Press, New York.

Lin, D. S., Ilias, A. M., Conner, W. E., Caldwell, R. S., Cory, H. T., and Davies, G. D. (1982). Composition and biosynthesis of sterols of marine phytoplankton. *Lipids*, **17**, 818–24.

López Alonzo, D., Molina Grima, E., Sánchez Pérez, J. A., García Sánchez, J. L. and García Camacho, F. (1992*a*). Fatty acid variation among different isolates of a single strain of *Isochrysis galbana*. *Phytochemistry*, **31**, 3901–4.

López Alonzo, D., Molina Grima, E., Sánchez Pérez, J. A., García Sánchez, J. L., and García Camacho, F. (1992*b*). Isolation of clones of *Isochrysis galbana* rich in eicosapentaenoic acid. *Aquaculture*, **102**, 363–71.

Mackenzie, A. S., Brassell, S. C., Eglinton, G., and Maxwell, J. R. (1982). Chemical fossils: the geological fate of steroids. *Science*, **217**, 491–504.

Marlowe, I. T. (1984). Lipids as palaeoclimatic indicators. Unpublished D. Phil. thesis. University of Bristol.

Marlowe, I. T., Brassell, S. C., Eglinton, G., and Green, J. C. (1984*a*). Long chain unsaturated ketones and esters in living algae and marine sediments. In *Advances in organic geochemistry 1983*, (ed. P. A. Schenk, J. W. de Leeuw, and G. M. W. Lijmbach). *Organic Geochemistry*, **6**, 135–41.

Marlowe, I. T., Green, J. C., Neal, A. C., Brassell, S. C., Eglinton, G., and Course, P. A. (1984*b*). Long chain (n-C_{37}-C_{39}) alkenones in the Prymnesiophyceae. Distribution of alkenones and other lipids and their taxonomic significance. *British Phycology Journal*, **19**, 203–16.

Marlowe, I. T., Brassell, S. C., Eglinton, G., and Green, J. C. (1990). Long-chain alkenones and alkyl alkenoates and the fossil coccolith record of marine sediments. *Chemical Geology*, **88**, 349–75.

Maxwell, J. R., MacKenzie, A. S., and Volkman, J. K. (1980). Configuration at C-24 in steranes and sterols. *Nature*, **286**, 694–7.

Nichols, P. D., Jones, G. J., de Leeuw, J. W., and Johns, R. B. (1984). The fatty acid and sterol composition of two marine dinoflagellates. *Phytochemistry*, **23**, 1043–7.

Nichols, P. D., Holdsworth, D. G., Volkman, J. K., Daintith, M., and Allanson, S. (1989). High incorporation of essential fatty acids by the rotifer *Brachionus plicatilis* fed on the prymnesiophyte alga *Pavlova lutheri*. *Australian Journal of Marine and Freshwater Research*, **40**, 645–55.

Nichols, P. D., Skerratt, J. H., Davidson, A., Burton, H., and McMeekin, T. A. (1991). Lipids of cultured *Phaeocystis pouchetii*: signatures for food-web, biogeochemical and environmental studies in Antarctica and the Southern Ocean. *Phytochemistry*, 30, 3209–14.

Pillsbury, K. S. (1985). The relative food value and biochemical composition of five phytoplankton diets for the queen conch, *Strombus gigas* (Linne) larvae. *Journal of Experimental Marine Biology and Ecology*, **90**, 221–31.

Poynter, J. G., Farrimond, P., Brassell, S. C., and Eglinton, G. (1989). A molecular stratigraphic study of sediments from Holes 658A and 660A, Leg 108. In *Proceedings of the Ocean Drilling Program, Scientific Results*, Vol. 108, (ed. J. Baduaf, G. R. Heath, W. F. Ruddiman, and M. Sarnthein), pp. 387–94. US Government Printing Office, Washington.

Prahl, F. G. and Wakeham, S. G. (1987). Calibration of unsaturation patterns in long-chain ketone compositions for palaeotemperature assessment. *Nature*, **330**, 367–9.

Prahl, F. G., Muelhausen, L. A., and Zahnle, D. L. (1988). Further evaluation of long-chain alkenones as indicators of paleoceanographic conditions. *Geochimica et Cosmochimica Acta*, **52**, 2303–10.

Raederstorff, D. and Rohmer, M. (1984). Sterols of the unicellular algae *Nemato-chrysopsis roscoffensis* and *Chrysotila lamellosa*: isolation of 24(E)-24-*n*-propylidenecholesterol and 24-*n*-propylcholesterol. *Phytochemistry*, **23**, 2835–8.

Reckha, J. A. and Maxwell, J. R. (1988). Characterisation of alkenone temperature indicators in sediments and organisms. In *Advances in organic geochemistry 1987*, (ed. L. Mattavelli and L. Novelli), pp. 727–34. Pergamon Press, Oxford.

Sargent, J. R. and Whittle, K. J. (1981). Lipids and hydrocarbons in the marine food web. In *Analysis of marine ecosystems*, (ed. A. R. Longhurst), pp. 491–533. Academic Press, London.

Sargent J. R., Eilertsen, H. C., Falk-Petersen, S., and Taasen, J. P. (1985). Carbon assimilation and lipid production in phytoplankton in northern Norwegian fjords. *Marine Biology*, **85**, 109–16.

Sikes, E. L. and Volkman, J. K. (1993). Calibration of long-chain alkenone unsaturation ratios for paleotemperature estimation in cold polar waters. *Geochimica et Cosmochimica Acta*, **57**, 1883–9.

Ten Haven, H. L. and Kroon, H. (1990). Late Pleistocene sea surface temperature variations off Oman as revealed by the distribution of long-chain alkenones. In *Proceedings of the Ocean Drilling Program, Scientific Results*, Vol. 117, (ed. W. L. Prell, N. Niitsuma *et al.*), pp. 445–52. US Government Printing Office, Washington.

Thompson, P. A., Harrison, P. J., and Whyte, J. N. C. (1990). Influence of irradiance on the fatty acid composition of phytoplankton. *Journal of Phycology*, **26**, 278–88.

Thompson, P. A., Guo, M-X., Harrison, P. J., and Whyte, J. N. C. (1992). Effects of variation in temperature. II. On the fatty acid composition of eight species of marine phytoplankton. *Journal of Phycology*, **28**, 488–97.

Vaulot, D., Birrien, J. L., Marie, D., and Casotti, R. (1994). *Phaeocystis* spp: DNA content, cell size and pigment composition of cultured strains. *Journal of Phycology* (in press).

Volkman, J. K. (1986). A review of sterol markers for marine and terrigenous organic matter. *Organic Geochemistry*, **9**, 83–99.

Volkman J. K., Eglinton, G., Corner, E. D. S., and Sargent, J. R. (1980*a*). Novel unsaturated straight-chain C_{37}-C_{39} methyl and ethyl ketones in marine sediments and a coccolithophorid *Emiliania huxleyi*. In *Advances in organic geochemistry 1979*, (ed. A. G. Douglas and J. R. Maxwell), pp. 219–27. Pergamon Press, Oxford.

Volkman J. K., Eglinton, G., Corner, E. D. S., and Sargent, J. R. (1980*b*). Long-chain alkenes and alkenones in the marine coccolithophorid *Emiliania huxleyi*, *Phytochemistry*, **19**, 2619–22.

Volkman, J. K., Smith, D. J., Eglinton, G., Forsberg, T. E. V., and Corner, E. D. S. (1981). Sterol and fatty acid composition of four marine haptophycean algae. *Journal of the Marine Biological Association of the United Kingdom*, **61**, 509–27.

Volkman, J. K., Burton, H. R., Everitt, D. A., and Allen, D. I. (1988). Pigment and lipid compositions of algal and bacterial communities in Ace Lake, Vestfold Hills, Antarctica. *Hydrobiologia*, **165**, 41–57.

Volkman J. K, Jeffrey, S. W., Nichols, P. D., Rogers, G. I., and Garland, C. D. (1989). Fatty acid and lipid composition of 10 species of microalgae used in mariculture. *Journal of Experimental Marine Biology and Ecology*, **128**, 219–40.

Volkman, J. K., Kearney, P., and Jeffrey, S. W. (1990). A new source of 4-methyl sterols and 5α(H)-stanols in sediments: prymnesiophyte microalgae of the genus *Pavlova*. *Organic Geochemistry*, **15**, 489–97.

Volkman J. K. Dunstan, G. A., Jeffrey, S. W., and Kearney, P. S. (1991). Fatty acids from microalgae of the genus *Pavlova*. *Phytochemistry*, **30**, 1855–9.

Volkman, J. K., Barrett, S. M., Dunstan, G. A., and Jeffrey, S. W. (1993). Geochemical significance of the occurrence of dinosterol and other 4-methyl sterols in a marine diatom. *Organic Geochemistry*, **20**, 7–15.

Waldock, M. J. and Nascimento, I. A. (1979). The triacylglycerol composition of *Crassostrea gigas* larvae fed on different algal diets. *Marine Biology Letters*, **1**, 77–86.

Watanabe, T. and Ackman, R. G. (1974). Lipids and fatty acids of the American (*Crassostrea virginica*) and European flat (*Ostrea edulis*) oysters from a common habitat, and after one feeding with *Dicrateria inornata* or *Isochrysis galbana*. *Journal of the Fisheries Research Board of Canada*, **31**, 403–9.

Weiße, T. (1983). Feeding of calanoid copepods in relation to *Phaeocystis pouchetti* blooms in the German Wadden Sea area off Sylt. *Marine Biology*, **74**, 87–94.

Wright, S. W. and Jeffrey, S. W. (1987). Fucoxanthin pigment markers of marine phytoplankton analyzed by HPLC and HPTLC. *Marine Ecology Progress Series*, **38**, 259–66.

20. Palaeontological perspectives

JEREMY R. YOUNG
The Natural History Museum, Cromwell Road, London, UK

PAUL R. BOWN and JACKIE A. BURNETT
Department of Geology, University College, London, UK

Abstract

Haptophytes are represented in the fossil record by a phenomenal abundance of coccoliths and cryptic nannoliths. The biological information available from this record is outlined here. In particular palaeontological perspectives on coccolith based taxonomy and phylogenetic inferences are described. It is shown that the geological record of coccoliths is selective, but very complete for the selected species. Research on exceptionally preserved fossil coccoliths, the Cretaceous/Tertiary boundary, palaeoceanography, and Milankovitch cyclicity is outlined.

Introduction

Coccoliths are readily preserved in sediments and as a result haptophyte algae have one of the most abundant fossil records of any division in terms of numbers of preserved individuals, volume of sediment, or percentage of individuals which have been fossilized. Moreover, this fossil record is continuous from their first occurrence in the Late Triassic to the Present. Their abundance is best displayed by the Late Cretaceous white chalk facies which occurs across most of northern Europe with equivalents around the world, and which is essentially composed of coccoliths and debris from them (Fig. 20.1). The Chalk was unusual in being deposited from extensive shelf seas, but similar coccolith dominated sediments are forming today below very large areas of the deep ocean, and have formed through most of the history of coccoliths.

Despite their importance in the earth's biogeochemical cycles, the prime geological interest in the group has been in biostratigraphy, i.e. the determination of the age of sediments from the fossils contained in them. Coccoliths are exceptionally good for biostratigraphy since they are

The Haptophyte Algae (ed. J. C. Green and B. S. C. Leadbeater), Systematics Association Special Volume No. 51, pp. 379–92. Clarendon Press, Oxford, 1994.

(a) (b)

Fig. 20.1 The chalk of the Late Cretaceous. This deposit covered most of northwest Europe and similar deposits occurred around the world. Coccoliths constitute about 80% of it. (a) Culver Bay, Isle of Wight. Scanning electron micrograph of a typical chalk surface showing numerous coccoliths, rod-shaped nannoliths, and finer debris from coccoliths. Scale bar = 10 μm.

common, evolved fast, and show low endemicity. Also their small size means they can be studied from minute rock chips, which is of particular value in the oil industry and scientific drilling. The detailed biostratigraphic knowledge of coccoliths combined with their exceptional geological record gives haptophyte algae special potential as a group for palaeobiological research; including study of evolutionary processes, and environmental change.

Definite coccoliths are usually found together with *nannoliths* which are calcareous fossils of similar size but uncertain affinities. Most nannoliths probably were formed by haptophytes; they and coccoliths are usually studied together and referred to in the literature as *calcareous nannofossils* or *nannoplankton.*

State of research

Up to the 1950s only a very few workers studied fossil coccoliths — notably Deflandre and Kamptner — and they probably described < 10% of the fossil species now known. From the 1950s the biostratigraphical potential of coccoliths began to be recognized and transmission electron microscopy was applied to their study, even though this required elaborate carbon-replication preparation techniques. During the late 1960s there was a rapid acceleration of taxonomical and stratigraphical work due to a number of factors, including application of coccolith biostratigraphy in the oil industry, inception of the Deep-Sea Drilling Project in 1968, and the application of scanning electron microscopy. Over the past decade this descriptive work has declined, whilst biostratigraphical studies are evolving into a multidisciplinary high resolution science, and broader aspects of the relationship between coccolithophorids and the palaeoenvironment are being addressed. Electron microscopy is used for research on coccolith morphology and taxonomy. For routine work, however, light microscopy using strew mounts is always used since it is much faster and can be applied even with very poorly preserved samples.

Fossil coccoliths are now being actively used and studied by many (perhaps 300) workers across the world in both academic and industrial institutions, all studying essentially the same set of taxa (there are < 1000 routinely cited species). So detailed data is available for most taxa, including, particularly, distribution in time and space. However, there is a shortage of authoritative monographs, and species level taxonomy is often confused and ill-disciplined. This in turn makes it difficult to synthesise available non-taxonomic data. The most useful existing synthesis of both taxonomy and biostratigraphy is Perch-Nielsen (1985*a*, *b*). We have tried here to cite a broad selection of interesting recent papers, so that the references would form a useful introductory reading list.

Palaeontological taxonomy

It is sometimes argued that taxonomy of fossil coccoliths is inevitably artificial due to special features of coccolithophorid biology: coccoliths are only isolated components of a total exoskeleton; some species have life cycles with alternation of holococcolith and heterococcolith stages (Billard, Chapter 9); some coccospheres have more than one discrete type of coccolith; and coccolith morphology is affected by non-taxonomic processes. However, equivalent problems can be found in most other fossils, from mammal teeth to dinoflagellate cysts. Tests such as internal consistency of the classification, and correlation with living equivalents, indicate that at least the general outline of heterococcolith classification is sound, though there are problems in detail (see also Aubry 1989).

Species concepts

Genetic recombination during sexual reproduction underlies the concept of the species, but in practice even biologists can rarely apply it, so alternative working definitions are needed. For extant coccolithophorids the usual criterion is morphological continuity; within a species morphology is continuously variable, whilst between species there should be no overlap in morphology. Evolution causes morphology to change through time so

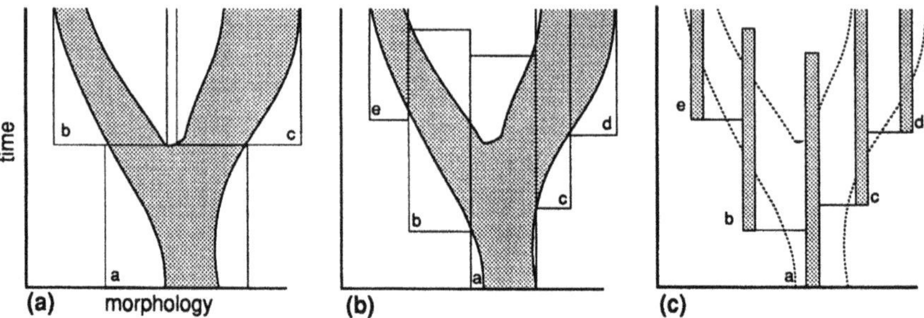

Fig. 20.2 Species definition and representation. (a) Subdivision of a lineage into three species based on time of evolutionary divergence. (b) A practical biostratigrapher's subdivision of the same lineage with finer resolution and morphology as the basis of species definitions. (c) A conventional representation of this pattern.

simple application of this criterion to fossils inevitably causes problems. A philosophically attractive solution is to define species in terms of phyletic branching events (Fig. 20.2a). However, this means that identification of a morphotype is often dependent on the age of the sediment it occurs in, which for biostratigraphy is inverted logic. For biostratigraphical work it is much more practical to define species so as to include all occurrences of a given morphotype irrespective of their age, so a species is *a consistently recognisable morphotype which has a discrete stratigraphic range.* This can give rise to a very different subdivision of the same evolutionary system (Fig. 20.2b). Nannofossil species are often based on arbitrary criteria in this way, and within single samples individual species may demonstrably intergrade. Nonetheless, lineages are conventionally illustrated by diagrams that suggest discontinuous evolution (Fig. 20.2c), consequently such diagrams need to be interpreted with care (see also Pearson 1992).

Higher taxa

Genera and families are the two main higher taxonomic ranks used for nannofossils. Within nannofossil genera individual species are typically variations on a morphological theme — often separated only by size or shape. Nannofossil families are sets of genera with similar structure, especially rim structure, as reflected by light microscopy effects and electron microscope study. Individual genera may differ in central area structure, or in details of rim structure such as relative development of different cycles.

Work on coccolith ultrastructure and biomineralization has produced a more analytical approach based on the use of crystallographic orientation to identify homologous structures (Romein 1979; Aubry 1989; Young *et al.* 1992; Young 1992). This approach should allow us to characterize families better and to investigate phylogenetic relationships between families. However, the basic classification has developed from a large amount of observational work and is likely to prove fundamentally sound.

Macroevolutionary history

The development of nannofloras through geological time has been discussed by Aubry (1992), Bown *et al.* (1991), Green *et al.* 1990, Roth (1989), Perch-Nielsen (1985*a*, *b*). We give only a brief overview here.

Mesozoic nannofloras

The earliest known calcareous nannofossils are from the Late Triassic (Fig. 20.3). They may extend somewhat further back but, despite searching, no convincing Palaeozoic coccoliths have been found; and there are no significant Palaeozoic pelagic carbonates. Definite dinoflagellates appear at about the same time as nannofossils, whilst diatoms and silicoflagellates do not occur until the Cretaceous. Thus the modern phytoplankton is essentially Mesozoic in origin and Palaeozoic oceans may have had very different primary productivity regimes.

The first calcareous nannofossils are an heterogeneous group including small calcispheres, enigmatic nannoliths, and a couple of coccoliths (Bown 1987). Only the coccoliths continue into the Early Jurassic, but these undergo a strong radiative evolution and gradually produce the diverse coccolith types of the Mesozoic families (Fig. 20.3; Prins 1969; Bown 1987). During the Middle and Late Jurassic the coccoliths are relatively conservative. During the Cretaceous there is a more or less steady appearance of new taxa. Extinction rates are rather low, most taxa are long-lived, and there is a gradual increase in diversity through the Late Cretaceous with ultimately about 140 species in 16 families. This is also the time of maximum abundance of coccoliths and a period of relatively high modal coccolith size (5–10 μm vs. 2–6 μm at the present day). It is not clear whether this acme of coccolith success was due to favourable conditions or to evolutionary success relative to other groups.

Among the nannoliths the Polycyclolithaceae and Microrhabdulaceae show similar distribution patterns to typical coccoliths and a heterococcolith type ultrastructure of complex crystal units. They were probably derived from coccoliths. The nannoconids have distribution patterns that are independent of coccoliths and have quite different ultrastructure (Busson and Noël 1991; van Niel 1992). It is possible that they are entirely unrelated. They are most important in Early Cretaceous Tethyan peri-continental sediments where they are often the main rock-forming component.

Cretaceous/Tertiary (K/T) boundary extinctions

All the important Late Cretaceous families become extinct at or just after the Cretaceous/Tertiary (K/T) boundary (Fig. 20.3). This event coincides with the extinction of dinosaurs, ammonites and various other groups, and massive turnovers in many more, including the planktonic foraminifera and silicoflagellates, but excluding diatoms and dinoflagellates (Hallam and Perch-Nielsen 1990). This event has been the subject of intensive research, particularly following the suggestion that the extinctions may have been caused by a meteorite impact. The nannofossil record has been studied in particular detail and many sections are now known which give a detailed record of the boundary interval, e.g. Romein (1979), Thierstein (1981), Hallam and Perch-Nielsen (1990), Pospichal and Bralower (1992).

About 80% of the Cretaceous species, including all the common ones, become extinct. A limited group of nannofossils survive, several of which show relative abundance peaks ('blooms') during this interval. Subsequently a new set of species evolve, these can mostly be traced back to the surviving species but develop into a quite distinct group of new families. The pattern thus appears to be: catastrophic extinction of the main nannoflora, survival of a few opportunistic species, and subsequent radiation of a new nannoflora.

The magnitude and rapidity of the event supports the more catastrophic explanations of the K/T boundary — massive volcanic eruptions or meteorite impact — over the more gradualistic — ecological collapse due to severe climate or sea-level change. There are, however, still problems of interpretation. For instance, the Cretaceous species which occur abundantly just above the K/T boundary might be either survivors or reworked specimens. As a result it is unclear whether, the extinctions occurred through mass mortality of virtually the entire nannoplankton over a period of at most a few years, or by progressive diminution of the nannoflora over a few tens of thousands of years.

Tertiary nannofossils and origins of the present day flora.

After the K/T boundary event extinctions there was a rapid evolutionary radiation during the Palaeocene and early Eocene (Romein 1979). This lead to the development of many new forms, including most of the families of the modern nannoflora. Compared to the Early Jurassic radiation, species appearance rates are much higher (> 8 spp./Ma vs 1–2 spp./Ma) and individual species in most families have short ranges.

Subsequently diversity decreases in the late Eocene and into the Oligocene before recovering in the Miocene. This pattern is similar to that shown by planktonic foraminifera and is thought to be related to Antarctic ice development in the Oligocene. The onset of Pleistocene

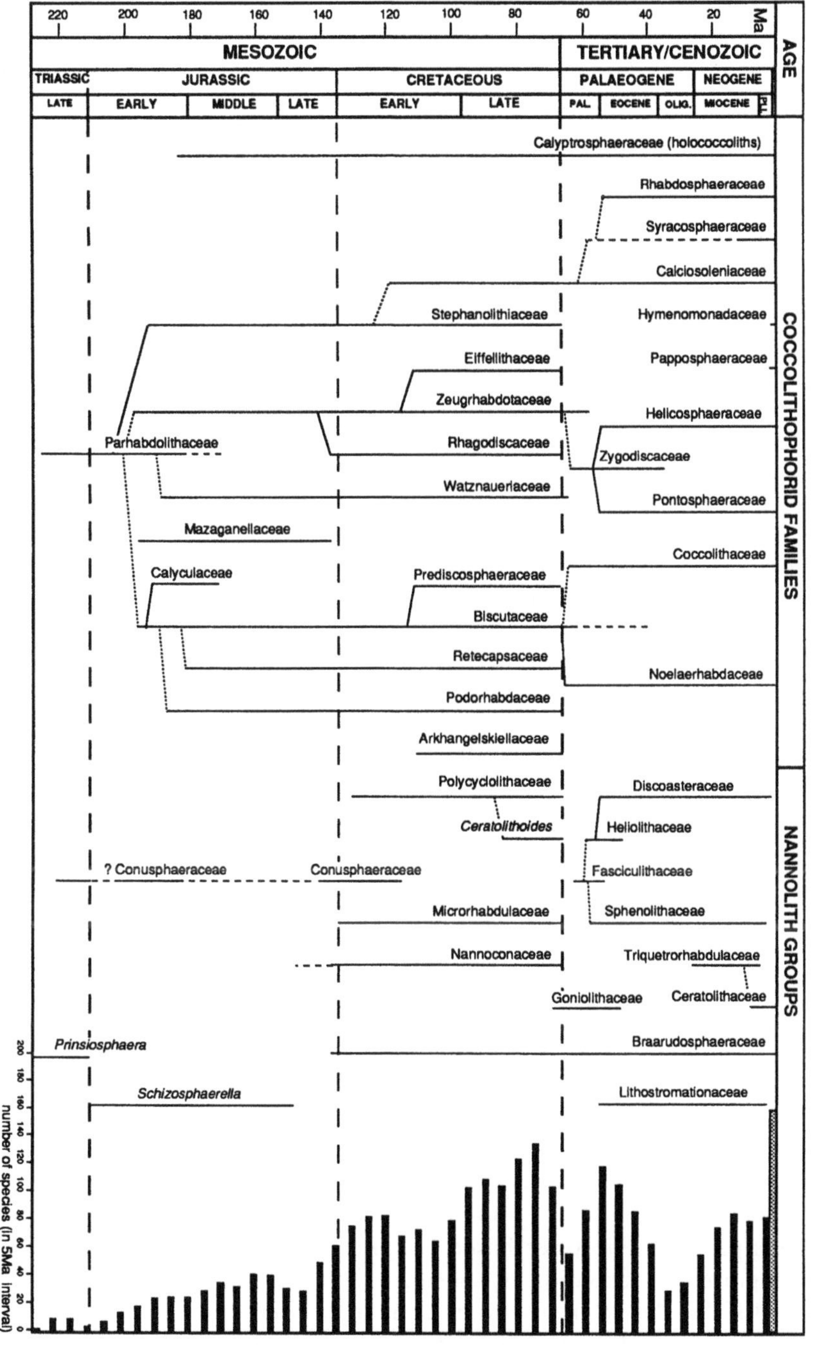

Fig. 20.3 Coccolith phylogeny-synthesis of family level relationships, emended from Bown *et al.* (1991). Excluding most individual genera *incertae sedis*.

glaciation sees some extinctions, notably of discoasters, but without any strong effect on the total diversity.

The modern flora includes about 200 species of coccolithophorids, but only about 40 of them are known from the latest Quaternary fossil record. The causes of this are partly preservational and partly observational, both of which are strongly size selective. With light microscopy it is difficult to identify coccoliths < 3.5 μm long, and virtually impossible even to observe coccoliths < 2 μm long. Moreover, preservation processes during both sedimentation and lithification strongly favour larger crystals over smaller, so electron microscopy does not usually reveal many additional small species. As a result, the Papposphaeraceae and Hymenomonadaceae have no known fossil record whilst the Syracosphaeraceae, Rhabdosphaeraceae, and Calyptrosphaeraceae are represented by only a few larger species without the enormous diversity of smaller living species. In contrast, in the Helicosphaeraceae, Pontosphaeraceae, Coccolithaceae, and Noëlaerhabdaceae most of the modern species have coccoliths larger than about 3 μm and are known as fossils. Overall the fossil record of the coccolithophorids is patchy, being extremely good for some families, limited for others, and non-existent for a few.

The fossil record suggests that the modern coccolithophorids are essentially the product of two radiations, the first in the Early Jurassic, the second in the Palaeocene (Fig. 20.3). The close relationship suggested for the Coccolithaceae and Noëlaerhabdaceae (including *Emiliania*) is interesting since features of *Emiliania*, such as absence of true a base-plate scale or motile phase holococcoliths, separate the Noëlaerhabdaceae from other coccolithophorids (Green *et al.* 1990). However, the fossil evidence from K/T boundary sections is good, although not incontrovertible. This suggests that the differentiating features of the Noëlaerhabdaceae are derived rather than primitive characters.

Microevolutionary studies

Microevolution is the process of evolution within and between species; as opposed to macroevolution, the origins and relations of higher taxa. Pelagic sediments often give a continuous record of individual nannofossil species through both time and space so there is enormous potential for rigorous microevolutionary work. There have, however, been relatively few detailed studies — most apparently microevolutionary studies are best-guess phylogenetic trees based on knowledge of the morphology and stratigraphic distribution of species (e.g. Theodoridis 1984; Bown 1987; Varol 1992). These are useful information syntheses but are of little value as studies of evolutionary pattern and process. Several studies have used biometrics to quantify microevolutionary change (e.g. Romein 1979;

Young 1990), however, it is difficult to find independent biometric parameters to measure on coccoliths so most studies are monoparametric and contain unresolvable ambiguities (Young 1990). One exception is work on *Gephyrocapsa* by Samtleben (1980), Rio (1982), Matsuoka and Okada (1990), and Sato *et al.* (1991). Here three independent parameters are available — coccolith length, central area opening, and bridge angle. This allows sensitive discrimination of species. Matsuoka and Okada (1990) showed that a succession of distinctive larger species repeatedly differentiated from the continuously present complex of small *Gephyrocapsa* spp. through the Pleistocene. The other studies revealed essentially the same pattern, and calibration from magnetic and oxygen isotope chronologies show that these microevolutionary changes were synchronous to within a few 1000 years across the Pacific, Indian, and Atlantic Oceans. This very strongly supports the use of nannofossils in biostratigraphy and suggests that the nannoplankton, or at least *Gephyrocapsa*, exists as a single well-mixed population on this timescale.

Obtaining palaeobiological data

Exceptional preservation, laminated deposits, and blooms

The study of exceptionally preserved fossil deposits — *Lagerstätten* — is immensely useful throughout palaeontology; obvious examples include the Burgess Shale, and Solenhofen Limestone. Unusual conditions of original deposition and subsequent fossilization can allow delicate species and anatomical details to be preserved which are normally lost.

For calcareous nannofossils a number of deposits are now known which might be characterized as *Lagerstätten*. Typically these are laminated organic-rich sediments, which indicate anoxic bottom conditions (Noël *et al.* 1987). Under these conditions material reaching the sediment is unlikely to be broken down or disrupted. So providing that coccoliths are supplied to the sediment, and that subsequent diagenesis is low, they are likely to be preserved in their original associations, and as pellets or coccospheres. Anoxic bottom conditions are most common in eutrophic basins with restricted circulation and so many, though not all, such deposits contain abnormal assemblages, and often monospecific assemblages probably recording blooms.

Although deposits of this type are rare, study of them has provided a disproportionately large amount of data including: biological data on, the arrangement of coccoliths on coccospheres, dimorphism, and ontogeny (e.g. Covington 1985; Lambert 1986; Young and Bown 1991); phylogenetic data from occurrences of species usually too small or delicate to be preserved (Goy 1981; Bown 1993; see Fig. 20.4); ecological data from the

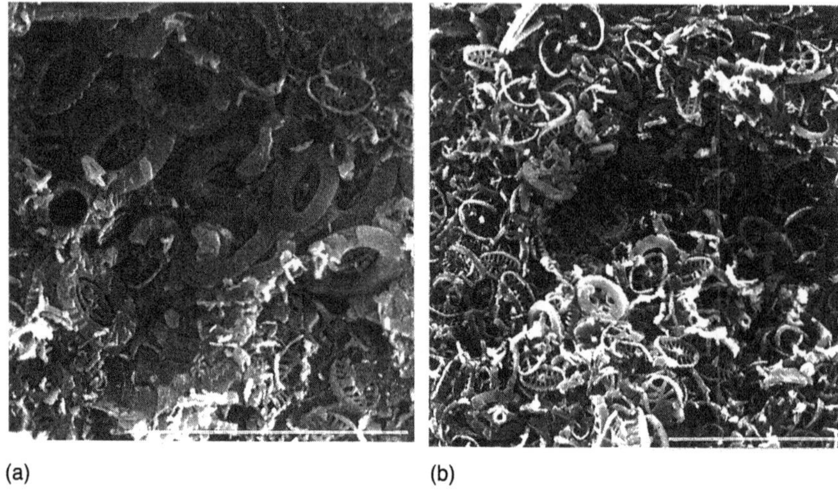

Fig. 20.4 Exceptional preservation of coccoliths, Lower Jurassic laminated sediment from France. This material has been described by Goy (1981), and by Bown (1993) (a) Collapsed coccosphere with coccoliths at various growth stages. (b) Surface showing abundant delicately preserved small coccoliths (compare with Fig. 20.1). Scale bars = 10 μm.

study of undisturbed associations recording single blooms and annual cycles (Thomsen 1989).

Palaeobiogeography, Palaeoceanography, and Palaeoclimatology

This is a very large and very active field characterized by multidisciplinary studies using geochemical data (stable isotope studies, element ratios, organic biomarkers, etc.) combined with quantitative palaeontological data (from planktonic foraminifera, dinoflagellates etc.), with modelling becoming increasingly important. Major themes include the development of thermohaline circulation, ice-caps, and current systems. Interesting nannofossil studies include Thierstein (1981), Bralower (1988), Wei and Wise (1990), Aubry (1992), and Pujos (1992). This research field creates a very strong palaeontological interest in the large scale controls on the distribution of coccolithophorids, and phytoplankton in general. To date, however, there have not been enough well constrained studies to gain clear answers from the nannofossil record as to the pattern or causes of species distribution, and many studies have tended to over-invoke single physical controls, particularly temperature.

A particularly promising new research area is Milankovitch cyclicity. Milankovitch (1941) predicted that changes in the precession, eccentricity, and obliquity of the earth's orbit would produce regular climatic changes on 20 000, 40 000 and 200 000 year frequencies. This proposition was not immediately accepted, but study of oxygen isotope derived palaeotemperature signals from fossil planktonic foraminifera revealed that such cycles control the timing of Pleistocene glaciations. Subsequently, Milankovitch-scale cyclicity has been recognized throughout the geological column and is currently the subject of intense research. It has great potential for providing a high resolution calibration of the geological time scale, and for understanding the responses to ecological change of particular environments and fossil groups. Typically it is thought that rather weak primary changes in solar radiation, caused by the orbital changes, are amplified by climate change and oceanic circulation change to produce a relatively strong primary productivity signal, which produces changes in, for instance, clay/carbonate ratios and so lithology.

The nannoplankton almost certainly plays a key role both as phytoplankton in the driving mechanism, and as carbonate producers in the production of the visible sedimentary record of Milankovitch cyclicity. The study of nannofossils is, therefore, especially relevant to the subject. Interesting recent studies include Erba *et al.* (1992), and work showing strong signals from *Florisphaera* in the Quaternary (Molfino and McIntyre 1990) and from *Discoaster* in the Pliocene (Chepstow-Lusty *et al.* 1992). *Florisphaera* is a deep-dwelling form and *Discoaster* may have been; and in both cases high relative abundances of these genera probably reflect intervals of lower surface productivity.

Conclusion

Palaeontological studies of coccoliths have traditionally been dominated by biostratigraphic work whose primary purpose was to date sediments. These produced an enormous literature, taxonomy, and distribution database, but this was difficult to use and yielded few results of direct biological value. This situation is changing fast due to: critical re-examination of phylogeny; detailed microevolutionary studies; rapidly developing palaeoceanographic studies, including work on Milankovitch cyclicity; and study of exceptionally preserved deposits.

References

Aubry, M.-P. (1989). Phylogenetically based calcareous nannofossil taxonomy: implications for the interpretation of geological events. In *Nannofossils and their applications*, (ed. J. A. Crux and S. E. van Heck), pp. 21–40. Ellis Horwood, Chichester.

Aubry, M.-P. (1992). Late Paleogene calcareous nannoplankton evolution: a tale of climatic deterioration. In *Eocene-Oligocene climatic and biotic evolution*, (ed. D. R. Prothero, and W. A. Berggren), pp. 272–309. Princeton University Press.

Bown, P. R. (1987). Taxonomy, evolution and biostratigraphy of Late Triassic–Early Jurassic calcareous nannofossils. *Special Papers in Palaeontology*, **38**, 1–118.

Bown, P. R. (1993). New holococcoliths from the Toarcian-Aalenian (Jurassic) of northern Germany. *Senckenbergiana Lethaia*, **73**, 407–19.

Bown, P. R., Burnett, J. A., and Gallagher, L. (1991). Critical events in the evolutionary history of calcareous nannoplankton. *Historical Biology*, **5**, 279–90.

Bralower, T. J. (1988). Calcareous nannofossil biostratigraphy and assemblages of the Cenomanian-Turonian boundary interval: implications for the origin and timing of oceanic anoxia. *Paleoceanography*, **3**, 275–316.

Busson, G. and Noël, D. (1991). Nannoconids as paramount environmental recorders of Late-Jurassic Early-Cretaceous oceans and epeiric seas. *Oceanologica Acta*, **14**, 333–56.

Chepstow-Lusty, A., Shackleton, N. J., and Backman, J., (1992). Upper Pliocene *Discoaster* abundance variations from the Atlantic, Pacific and Indian Oceans: the significance of productivity pressure at low latitudes. *Memorie di Scienze Geologiche*, **43**, 1–17.

Covington, M. (1985). New morphologic information on Cretaceous nannofossils from the Niobrara Formation (Upper Cretaceous) of Kansas. *Geology*, **13**, 683–6.

Erba, E., Castradori, D., Guasti, G., and Ripepe, M. (1992). Calcareous nannofossils and Milankovitch cycles: the example of the Albian Gault Clay Formation (southern England). *Palaeogeography, Palaeoclimatology, Palaeoecology*, **93**, 47–69.

Goy G. (1981). *Nannofossiles calcaires des schistes carton (Toarcien Inférieur) du Bassin de Paris*. Éditions du Bureau de Recherches Géologiques et Minières, 1–86.

Green, J. C., Perch-Nielsen, K., and Westbroek, P. (1990). Phylum Prymnesiophyta. In *Handbook of Protoctista*, (ed. L. Margulis *et al.*), pp. 293–317. Jones, and Bartlett, Boston.

Hallam, A. and Perch-Nielsen, K. (1990). The biotic record of events in the marine realm at the end of the Cretaceous: calcareous, siliceous and organic-walled microfossils and macroinvertebrates. *Tectonophysics*, **171**, 347–57.

Lambert, B. (1986). La notion d'espèce chez le genre *Braarudosphaera* Deflandre 1947. Mythe et réalité. *Revue de Micropaléontologie*, **28**, 255–64.

Matsuoka, H. and Okada, H. (1990). Time-progressive morphometric changes of the genus *Gephyrocapsa* in the Quaternary sequence of the tropical Indian Ocean, Site 709. *Proceedings of the Ocean Drilling Project, Scientific Results*, **115**, 255–70.

Milankovitch M. (1941). *Kanon der Erdbestrahlung*. Éditions specials du Academie Royale Serbe, 133, pp. 1–633.

Molfino, B., and McIntyre, A. (1990). Precessional forcing of nutricline dynamics in the equatorial Atlantic. *Science*, **249**, 766–9.

Niel, B. van (1992): New observations on the morphology of *Nannoconus*. In *Nannoplankton research, I, General topics, Mesozoic biostratigraphy*, (ed. B. Hamrsmid and J. R. Young). Knihovnicka ZPN, **14a**, 73–85.

Noël, D., Bréhéret J. G., and Lambert B. (1987). Enregistrement sédimentaire de floraisons phytoplanctoniques calcaires en milieu confiné. Synthèse de données sur l'actuel et observations géologiques. *Bulletin du Societé Géologique de France*, **8/3**, 1097–106.

Pearson, P. N. (1992). Survivorship analysis of fossil taxa when real-time extinction rates vary: the Palaeogene planktonic foraminifera. *Paleobiology*, **18**, 115–31.

Perch-Nielsen, K. (1985*a*). Mesozoic calcareous nannofossils. In *Plankton stratigraphy*, (ed. H. M. Bolli, J. B. Saunders, and K. Perch-Nielsen), pp. 329–426. Cambridge University Press.

Perch-Nielsen, K. (1985*b*). Cenozoic calcareous nannofossils. In *Plankton stratigraphy*, (ed. H. M. Bolli, J. B. Saunders, and K. Perch- Nielsen), pp. 427–554. Cambridge University Press.

Pospichal, J. J. and Bralower, T. J. (1992). Calcareous nannofossils across the Cretaceous/Tertiary boundary, Site 761, northwest Australian margin. *Proceedings of the Ocean Drilling Project, Scientific Results*, **122**, 735–51.

Prins, B. (1969). Evolution and stratigraphy of coccolithinids from the lower and middle Lias. In *Proceedings I International Conference on Planktonic Microfossils, Geneva*, (ed. P. Brönniman and H. H. Renz), pp. 547–59. E. J. Brill, Leiden.

Pujos A., (1992): Calcareous nannofossils of Plio-Pleistocene sediments from the north western margin of tropical Africa. In *Upwelling systems: evolution since the early Miocene*, (ed. C. P. Summerhayes *et al.*), pp. 343–58. Geological Society of London Special Publication, 64.

Rio, D. (1982). The fossil distribution of coccolithophore genus *Gephyrocapsa* Kamptner and related Plio–Pleistocene chronostratigraphic problems. *Initial Reports of the Deep Sea Drilling Project*, **68**, 325–43.

Romein, A. J. T. (1979). Lineages in early Palaeogene calcareous nannoplankton. *Utrecht. Micropalaeontological Bulletins*, **22**, 1–231.

Roth, P. H. (1989). Oceanic circulation and calcareous nannoplankton evolution during the Jurassic and Cretaceous. *Palaeogeography, Palaeoclimatology and Palaeoecology*, **74**, 111–26.

Samtleben, C. (1980). Die Evolution der Coccolithophoriden-Gattung *Gephyrocapsa* nach Befunden im Atlantik. *Paläontologisches Zeitschrift*, **54**, 91–127.

Sato, T., Kameo, K., and Takayama, T. (1991). Coccolith biostratigraphy of the Arabian Sea. *Proceedings of the Ocean Drilling Project, Scientific Results*, **117**, 37–54.

Theodoridis, H. R. (1984). Calcareous nannofossil biostratigraphy of the Miocene and revision of the helicoliths and discoasters. *Utrecht Micropalaeontological Bulletins*, **32**, 1–271.

Thierstein, H. R. (1981). Late Cretaceous nannoplankton and the change at the Cretaceous-Tertiary boundary. *Society of Economic Paleontologists and Mineralogists Special Publication*, **32**, 355–94.

Thomsen, E., (1989). Seasonal variation in boreal Early Cretaceous calcareous nannofossils. *Marine Micropalaeontology*, **15**, 123–52.

Varol, O. (1992). *Sullivania* a new genus of Palaeogene coccoliths. *Journal of Micropalaeontology*, **11**, 141–50.

Wei, W. and Wise, S. W. (1990). Biogeographic gradients of middle Eocene-Oligocene calcareous nannoplankton in the South Atlantic Ocean. *Palaeogeography, Palaeoclimatology and Palaeoecology*, **79**, 29–61.

Young, J. R. (1990). Size variation of Neogene *Reticulofenestra* coccoliths from Indian Ocean DSDP Cores. *Journal of Micropalaeontology*, **9**, 71–86.

Young, J. R. (1992). The description and analysis of coccolith structure. In *Nannoplankton Research, I General topics, Mesozoic biostratigraphy*, (ed. B. Hamrsmid and J. R. Young), *Knihovnicka ZPN*, **14a**, 35–71.

Young, J. R. and Bown, P. R., (1991). An ontogenetic sequence of coccoliths from the late Jurassic Kimmeridge Clay of England. *Palaeontology*, **34**, 843–50.

Young, J. R., Didymus, J. M., Bown, P. R., Prins, B., and Mann, S. (1992). Crystal assembly and phylogenetic evolution in heterococcoliths. *Nature*, **356**, 516–8.

21. Molecular biology and systematics

L. K. MEDLIN*, G. L. A. BARKER†, M. BAUMANN*,
P. K. HAYES† and M. LANGE*

*Alfred Wegener Institute for Polar and Marine Research, Bremerhaven, Germany
†Department of Botany, University of Bristol, Bristol, UK

Abstract

Sequence data from both the nuclear and plastid encoded small subunit ribosomal RNA (ssu rRNA) gene have been used to infer the phylogenetic position of the haptophyte host cell and its plastid. Both are in distinct lineages that do not share a recent evolutionary history with their chromophyte counterparts. Within the haptophyte lineage ssu rRNA sequence data support the recognition of at least three colony forming species within the ubiquitous genus, *Phaeocystis*. In contrast, the two distinct morphotypes of the cosmopolitan bloom-forming coccolithophorid, *Emiliania huxleyi*, are identical in the sequence of both their ssu rRNAs and the spacer regions between the plastid encoded large and small Rubisco subunits. The sensitive technique of RAPD fingerprinting has revealed the presence of genotypic variation between geographically isolated clones of *E. huxleyi*. The impact of these findings on the taxonomy of *E. huxleyi* is discussed.

Introduction

Molecular biological techniques are being used increasingly in algal systematics and ecology. Examples of such techniques, although not all algal based, are DNA fingerprinting (Lynch 1988), plastid DNA Restriction Fragment Length Polymorphisms (Goff and Coleman 1988), species specific probes (Amann *et al.* 1990), randomly amplified polymorphic DNA (RAPD) analysis (Welsh and McClelland 1990), and macromolecular sequencing (Bhattacharya *et al.* 1992; Woese 1987). The last two techniques, when used for large-scale phylogenetic and taxonomic analysis, depend on the polymerase chain reaction technique (PCR) to amplify either a specific gene sequence or, in the RAPD analysis, the sequence between closely spaced inverted repeats. These techniques are applied at various taxonomic levels to provide an objective approach to taxonomy, phylogeny, or biogeographical distribution; not all are quantitative. None

The Haptophyte Algae (ed. J. C. Green and B. S. C. Leadbeater), Systematics Association Special Volume No. 51, pp. 393–411. Clarendon Press, Oxford, 1994.

is expected to replace traditional methods of taxonomy or ecology, but instead should provide another measure of inter- and intraspecific variation and a framework upon which further investigations should be based.

Nucleic acids sequencing has been applied most widely to taxonomic (Medlin *et al.* 1991; Maggs *et al.* 1992), phylogenetic (Chapman and Buchheim 1991; Bhattacharya *et al.* 1992; Bird *et al.* 1992; Saunders and Druehl 1992), and biogeographical problems (Bakker *et al.* 1992; van Oppen *et al.* 1993) within the algae. If a nucleic acid sequence is to be phylogenetically informative it must meet certain criteria (Woese 1987). It must have the same origin, evolutionary history, and function in each organism, and it should accumulate base substitutions randomly throughout the molecule but in a 'clock-like' fashion over time. For the reconstruction of both close and distant phylogenetic relationships, the sequence should contain several domains that vary in their rate of base substitution. Regions meeting most, but not necessarily all, of these requirements are: (1) the ribosomal RNA genes and their spacer regions; (2) *rbc*L, *rbc*S, and their spacer region if both are plastid encoded; (3) *psa*A and *psa*B; (4) actin genes; and (5) *tuf*A.

In this contribution we report the use of PCR based techniques to determine the phylogenetic position of the haptophyte host cell and its plastid, and to assess the species concept in two ecologically sensitive haptophyte genera, *Phaeocystis* and *Emiliania.*

Phylogenetic position of the haptophyte host cell

The phylogenetic position of the Prymnesiophyceae (division: Haptophyta) has been the subject of several morphological and cladistic investigations (Cavalier-Smith 1987 and Chapter 22; Andersen 1991). Although it was removed from the Class Chrysophyceae by Christensen (1962), some still believe that the group is closely related to the Chromophyta/Oomycophyta (Cavalier-Smith 1986 and Chapter 22; Andersen 1991). Sequence data from the large (lsu) and small (ssu) subunits of the ribosomal RNA (rRNA) molecules have been used to provide an independent assessment of the phylogenetic position of the Haptophyta. Analysis of the complete sequence of the ssu rRNA (Fig. 21.1A) from *E. huxleyi* (Bhattacharya *et al.* 1992) and *Phaeocystis* (Medlin *et al.* 1994), and a partial sequence of the lsu rRNA from *Prymnesium parvum* and *Cricosphaera roscoffensis* (Perasso *et al.* 1989), indicate that the Haptophyta is a distinct eukaryotic lineage separate from the Chromophyta/Oomycophyta (Fig. 21.1A). The Cryptophyta and Dinoflagellata are also outside the chromophyte/oomycete lineage

Fig. 21.1 Phylogenetic representation of the position of the ssu rRNA gene from the haptophyte nucleus (A) and its plastid (B). Tree A is the most parsimonious

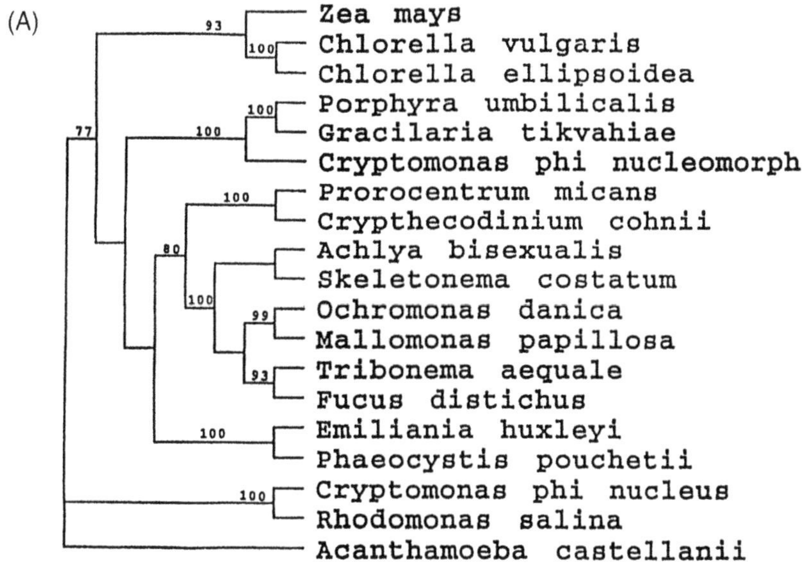

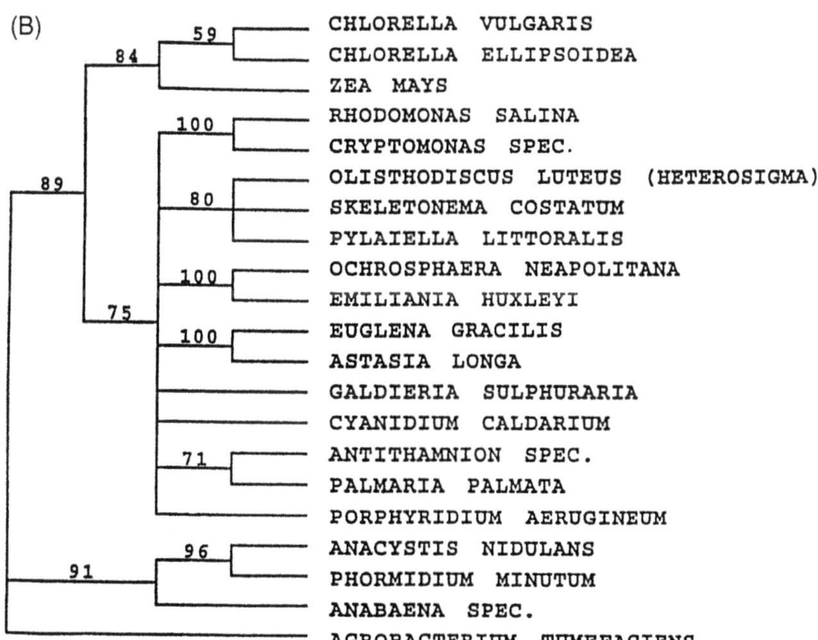

cladogram using the heuristic branch swapping algorithm (PAUP). Tree B is a consensus tree using the same treeing algorithm. Bootstrap values for 100 replications are shown at the branch nodes of both trees.

(Fig. 21.A). These four major algal lineages are on average 80% similar to one another (Medlin *et al.* 1994) and represent a major radiation within the eukaryotes. High bootstrap values at the branch nodes of our parsimony tree, which is rooted using *Acanthamoeba castellanii* as an outgroup, support this branching order (Fig. 21.1A), and high values within each lineage indicate that each are monophyletic.

Phylogenetic position of the haptophyte plastid

The endosymbiotic origin of plastids has been the subject of many investigations during the last few years. Recent evidence suggests that some, but not all plastids have a composite phylogenetic origin. The Rhodophyta, Chromophyta, Cryptophyta, and Haptophyta have plastid encoded RNA genes suggesting a cyanobacterial ancestor (Markowicz and Loiseaux de Goër 1991), but their Rubisco large and small subunits are related to those of β-purple bacteria (Fujiwara *et al.* 1993; Assasli *et al.* 1990; Douglas *et al.* 1990; Valentin and Zetsche 1990). Chlorophyte plastids have both rRNA and Rubisco genes related to cyanobacterial ancestors (see references in Markowicz and Loiseaux de Goër 1991). In the Euglenophyta the rRNA genes are organized like those of the Rhodophyta and Chromophyta (Markowicz and Loiseaux-de Goër 1991), while the Rubisco genes are related to those of the Chlorophyta (Martin *et al.* 1992).

Molecular evidence has supported, without question, the theory of Schimper (1883) and Mereschkowsky (1905, 1910) that plastids are originally derived from the primary endosymbiosis of a cyanobacterial ancestor. However, once this primary endosymbiotic event occurred, there are conflicting theories as to how plastids in the various algal groups were derived (Sagan 1967; Raven 1970; Cavalier-Smith 1982). Studies using ssu rRNA genes show that there is a deep divergence between the green algal and land plant lineage, and the lineage bearing all other plastid types thus far sampled (Douglas and Turner 1991; Markowicz and Loiseaux-de Goër 1991). A secondary endosymbiotic event in which a non-photosynthetic phagotroph engulfed a photosynthetic eukaryote (Gibbs 1981) is generally accepted to have occurred within this second lineage. Primary morphological evidence for this comes from the presence of three to four membranes surrounding the plastids of cryptophytes, haptophytes, dinoflagellates, and chromophytes (Gibbs 1981). Ultrastructural studies predicted that this secondary endosymbiosis event involved a red alga (Gibbs 1981); unequivocal molecular evidence documenting it was published recently (Douglas *et al.* 1991). The nucleus of *Cryptomonas* was related to the amoeboid protozoan *Acanthamoeba*, while its nucleomorph (vestigial nucleus contained within its plastid) was related to the red algae.

Valentin and Zetsche (1990) also showed that the Rubisco genes of the Chromophyta and Cryptophyta originated from an endosymbiotic unicellular red alga.

Cavalier-Smith (1982 and Chapter 22) proposed that this secondary endosymbiotic event occurred only once and that the diversity of the plastid types arose by the modification of the original endosymbiont. The main question is, therefore: does the molecular evidence support his hypothesis?

Sequence data for the ssu rRNA molecule from 21 taxa were used in our analyses. Both distance and parsimony trees inferred from these data show a deep divergence for the two plastid lineages from their cyanobacterial ancestors (Fig. 21.1). The trees are not completely congruent, and differ primarily in the branching order within the lineage containing the rhodophyte plastid and the plastids surrounded by 3-4 membranes. This is not surprising in view of the short branch lengths separating each major sub-lineage in the distance tree. The ssu rRNA gene may be unable to resolve correctly the branching order (Fig. 21.2A, B). However, it is significant that both trees depict lineages that correspond to extant algal groupings. Distinct lineages are shown for the Cryptophyta and Haptophyta. Members of the Florideophycidae are grouped together, but the Bangiophycidae appear polyphyletic. *Cyanidium* and *Galdieria* are sister taxa to the Chromophyta and Euglenophyta. *Porphyridium aerugineum*, a 'blue-greenish' unicellular red alga, lies at the base of a Cryptophyta/Haptophyta/Chromophyta lineage in the parsimony tree. Significantly, *Cyanidium* and *Galdieria* do not appear before the Rhodophyceae as predicted by Cavalier-Smith (1987), nor do the Chromophyta diverge from the Cryptophyta (Cavalier-Smith 1982). In the distance tree, the Haptophyta share a recent evolutionary history with the Cryptophyta. They are separated from the Chromophyta and other lineages in both the distance and parsimony tree; this separation is supported by the differences in their plastid structure (Hibberd 1976). The long branch length leading to the divergence of the two haptophytes indicates that the lineage was well established before it diversified. A 50% bootstrap consensus tree collapses most of the branches (Fig. 21.1B); however, even in the collapsed tree, the Haptophyta remain in lineage distinct from the Chromophyta.

Our molecular data do not contradict the hypothesis that the second major plastid lineage, containing the 3–4 membrane-bound plastids, is the result of a secondary endosymbiotic event. The number of events is still an open question because the branching order cannot be resolved with certainty, nevertheless, there are some significant points that can be made; (1) the Bangiophycidae appear polyphyletic, and (2) the Haptophyta form a distinct lineage outside the Chromophyta. If, with further

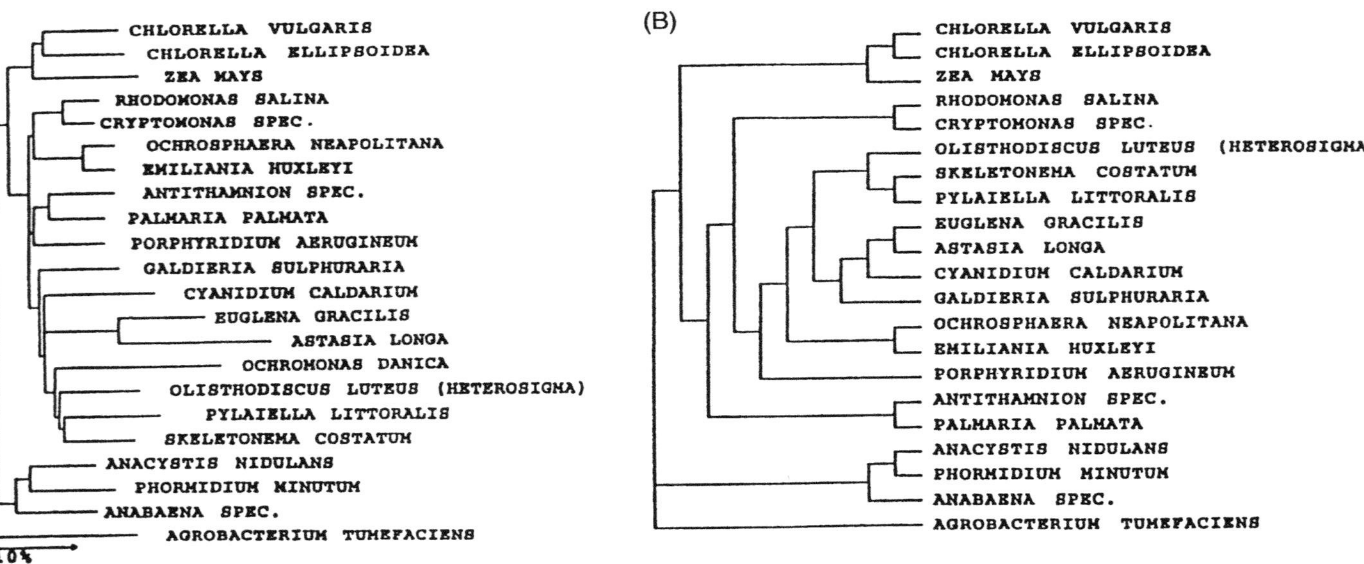

Fig. 21.2 Plastid phylogeny based on distance matrix (A) and maximum parsimony analysis (B) from 16S rRNA genes. The distance corresponding to 10 changes per 100 nucleotide positions is placed below the distance tree and is reflected in the horizontal separation of taxa in the tree. The parsimony tree is the most parsimonious one found using the heuristic branch swapping algorithm (PAUP).

sampling, the red algae remain polyphyletic, then this may be evidence for multiple secondary endosymbiotic events.

The only chlorophyll *c* bearing group insufficiently sampled for plastid ssu rRNA sequence data is the dinoflagellates. These protozoans have a varied plastid structure and pigment composition (Dodge 1989), and well-documented ultrastructural evidence for an photosynthetic eukaryotic endosymbiont (Tomas and Cox 1973). In view of our data, if the plastids within this group are polyphyletic then we would expect to see the various dinoflagellate plastid lineages diverging from within the various groups of ancestral photosynthetic organisms.

Case studies in systematics in the Haptophyta

The genus Phaeocystis

The genus *Phaeocystis* is recognized as both a nuisance alga and an ecologically important phytoplankton species (see review in Lancelot *et al.* 1987; Lancelot and Rousseau, Chapter 12; Moestrup, Chapter 14). It has a polymorphic life cycle with colonial and flagellated cells (Kornmann 1955); both can reach bloom proportions (Booth *et al.* 1982; Cadée and Hegeman 1986).

Phaeocystis was erected by Lagerheim (1893) to accommodate the colonial state of the alga originally described as *Tetraspora poucheti* by Hariot in Pouchet (1892). *Phaeocystis pouchetii*(its correct orthography) is a cold-water species forming globular, cloud-like colonies with cells arranged in packets of four (see Jahnke and Baumann 1987, for illustrations). *Phaeocystis globosa* (Scherffel 1900), a warm water species, forms spherical colonies with cells arranged homogeneously within the gelatinous matrix (Jahnke and Baumann 1987). Early workers separated both taxa based on their different distributions and different colonial morphologies until Kornmann (1955) demonstrated, in his life cycle studies on *Phaeocystis,* that *Ph. globosa* cell types were juvenile forms of *Ph. pouchetii.* Since that time, colony morphology has been judged an unreliable specific character.

Sournia (1988) reviewed the diagnostic features of *Phaeocystis* and examined the validity of its nine species. He recognized only two: *Phaeocystis scrobiculata,* known only from the flagellated state (Moestrup 1979) and *Ph. pouchetii,* which included *Ph. globosa* as a later synonym. Upon Sournia's (1988) recommendation, most marine ecologists report *Phaeocystis* colonies as *Ph. pouchetii* (the older name), or as *Phaeocystis* sp. to avoid confusion. Thus, reports of the colonial *Phaeocystis* record it as cosmopolitan with areas of massive blooms in Antarctic and Arctic waters, and extensions into north temperate regions where areas of high nutrient enrichment occur (Lancelot *et al.* 1987; Davidson and Marchant 1990).

Jahnke and Baumann (1987) reinvestigated the morphology of juvenile and older colonies of both *Ph. globosa* and *Ph. pouchetii* and found them to be stable (Table 21.1). Their clones also exhibited differences in temperature tolerance (Table 21.1; Jahnke 1989) and differences in features of the flagellated cells (Table 21.2). These results re-emphasize that *Ph. globosa* and *Ph. pouchetii* should be kept as separate species. Baumann *et al.* (1993) also recognized a third colonial species from Antarctic waters, which (a) shares features of colony morphology with *Ph. globosa*, (b) exhibits temperature tolerances similar to *Ph. pouchetii* (Table 21.1), but (c) expresses a different pigment spectrum (Buma *et al.* 1991; Vaulot *et al.* personal communication).

Table 21.1 Colony morphology and temperature tolerance of *Phaeocystis globosa*, *P. antarctica*, and *P. pouchetii*, from Jahnke and Baumann (1987) and Baumann *et al.* (1994)

Criteria	*P. globosa*	*P. antarctica*[1]	*P. pouchetii*
Colony morphology			
Maximum size	*c.* 8–9 mm	at least 9 mm ?	1.5–2 mm
Shape	spherical and numerous derived forms		spherical up to Ø 0.1 mm, above: cloud-like
Cell distribution	evenly in the periphery	evenly in the periphery	only in the curves of the clouds, and mostly regular: 4 cells form a square, cell free mucilage in between
Mucilage	solid	solid	delicate
Physiology			
Growth range	4 to 22 °C	–1.6[2] to 14 °C	–2 to 12 °C
Growth optimum	16 °C	4.5 °C	8 °C
Temperature tolerance	–0.6[3] to 22 °C	< –2 to 14 °C	< –2 to 14 °C

[1] Baumann *et al.* (1994)
[2] No deeper temperature was tested so far
[3] Cadée (1991)

Table 21.2 Features of the motile cell of *Phaeocystis* which are commonly assumed to be species specific after Baumann *et al.* (1994)

	Phaeocystis globosa	*P. scrobiculata*	*P. pouchetii*
Size	3 to 8 μm[1]	8 μm[2]	Ø *c.* 5 μm[4]
Threads[1,2,3]	pentagonal figure (length up to 20 μm, Ø 0.05 μm)[1]	nine ray figure (four pairs plus one) (length can exceed 50 μm, Ø 0.1 μm)[2]	pentagonal figure[4]
Scales[1,2,3]	two different scale types: both show a pattern of radiating ridges, visible on both surfaces	two different scale types: both show on the ventral side a pattern of ridges which radiate from a plain centre, the dorsal side is without visible patterning	two different scale types
	– larger scales are most circular flat plates with vertically upstanding rims, usually exactly 48 ridges, radiating from an approximately rectangular plain centre (0.18×0.19 μm)[1] (Ø: 0.25 μm)[3]	– larger scales are oval with a peripheral upstanding rim which shows no distinct pattern (0.60×0.45 μm)[1] (0.41×0.3 μm)[3]	
	– smaller scales are oval plates with strongly inflexed rims, 30 ridges, radiating from an oval plain centre (0.10×0.13 μm)[1] (Ø: 0.12 μm)[3]	– smaller scales are circular oval with a dorsal patternless rim (0.19 to 0.21 μm)[1] (0.1 μm)[3]	
Flagella[2]	about 10 to 15 μm long[2]	23 to 30 μm long[2]	8 μm long[4]
Haptonema[2]	non coiling type, slightly shorter than that of *P. scrobiculata*[2]	non coiling type, 5 μm long[2]	3 μm long[4]

[1] Parke, Green and Manton (1971) [2] Moestrup (1979) [3] Hallegraeff (1983) [4] Baumann and Jahnke (1986)

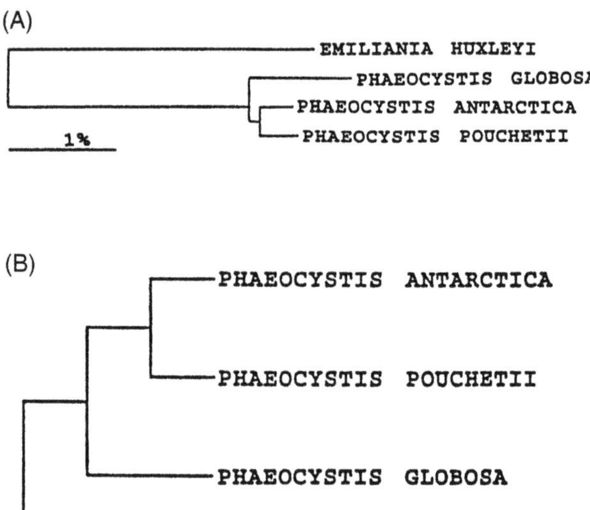

Fig. 21.3 Separation of *Phaeocystis* spp. using distance matrix (A) and maximum parsimony methods (B). The distance corresponding to 1 change per 100 nucleotide positions is placed below the distance tree and is reflected in the horizontal separation of taxa in the tree. The parsimony tree is the most parsimonious one found using the heuristic branch swapping algorithm (PAUP).

We have investigated the validity of the three colony-forming species recognized by Baumann *et al.* (1994) using sequence data from the nuclear-encoded ssu rRNA gene. Our analysis substantiates their interpretation (Fig. 21.3 and Table 21.3; Medlin *et al.* 1994), and we recognize the Antarctic cold water form as *Phaeocystis antarctica* (Karsten 1905) and the traditional species, *Ph. globosa* and *Ph. pouchetii.* This separation is further supported by differences in DNA content and pigment spectra (Vaulot *et al.*, personal communication), and by differences in the morphology of the colonial stage (Chretiennot-Dinet *et al.*, personal communication).

Similarity values and absolute number of base substitutions (Table 21.3) between these three *Phaeocystis* species are comparable to species differences within genera of diatoms (Medlin *et al.* 1991), protozoa (Sogin *et al.* 1986), and green algae (Huss and Sogin 1991). A distance tree inferred from these data show that *Ph. globosa* diverges first followed by a separation of the two cold-water forms, *Ph. antarctica* and *Ph. pouchetii* (Fig. 21.3A); parsimony analysis shows the same divergence (Fig. 21.3B). *Phaeocystis* appears to have originated as a warm water genus spreading into colder waters. *Phaeocystis antarctica* retained the morphology of the

Table 21.3 Percent similarity between small subunit ribosomal RNA sequences of *Emiliania huxleyi* and *Phaeocystis* spp. (*upper grouping*) and absolute number of nucleotide differences between these sequences excluding ambiguous nucleotides and gaps (*lower grouping*)

	Percent similarity to:			
Organism	*E. hux.*	*Ph. ant.*	*Ph. glob.*	*Ph. pouch.*
E. huxleyi		0.949	0.945	0.948
Ph. antarctica	81		0.989	0.996
Ph. globosa	88	22		0.987
Ph. pouchetii	82	10	22	

warm water form, while that of *Ph. pouchetii* diverged. Medlin *et al.* (submitted) have documented climatic changes and tectonic events in the middle to late Cenozoic that support this direction of change.

More *Phaeocystis* species may be erected based on their ssu rRNA genes. Similarly, many existing *Phaeocystis* records may be invalid because colonies with a *Phaeocystis*-like colonial stage release flagellate stages that can be assigned to other genera (Marchant and Thomsen, Chapter 11). The perplexing problem will be whether sufficient morphological markers can be identified to aid ecologists in their routine identification where different genotypes are known to overlap in their distribution.

The species Emiliania huxleyi

Emiliania huxleyi (Lohm.) Hay et Mohl., the most abundant representative of the Haptophyta, can be found in oceanic and neritic waters from subpolar to tropical latitudes. Annual North Atlantic blooms may cover up to half a million square kilometres, with cell densities of up to 10^5 ml^{-1} (see review in Aiken *et al.* 1992; and Chapters 8 and 16–20).

Morphometric, physiological, biochemical, and immunological studies support the division of *E. huxleyi* into two morphotypes, designated A and B (van Bleijswijk *et al.* 1991; Young and Westbroek 1991; Fig. 21.4). Morphotype B cells are generally larger with more coccoliths per cell than morphotype A cells. Growth rates of morphotypes A and B are significantly different (van Bleijswijk *et al.* 1991), while individual clones of the A morphotype of *E. huxleyi* taken from a single sampling bottle exhibit different acclimatized growth rates suggesting that the A morphotype is also genetically diverse (Brand 1982). Antisera raised against coccolith

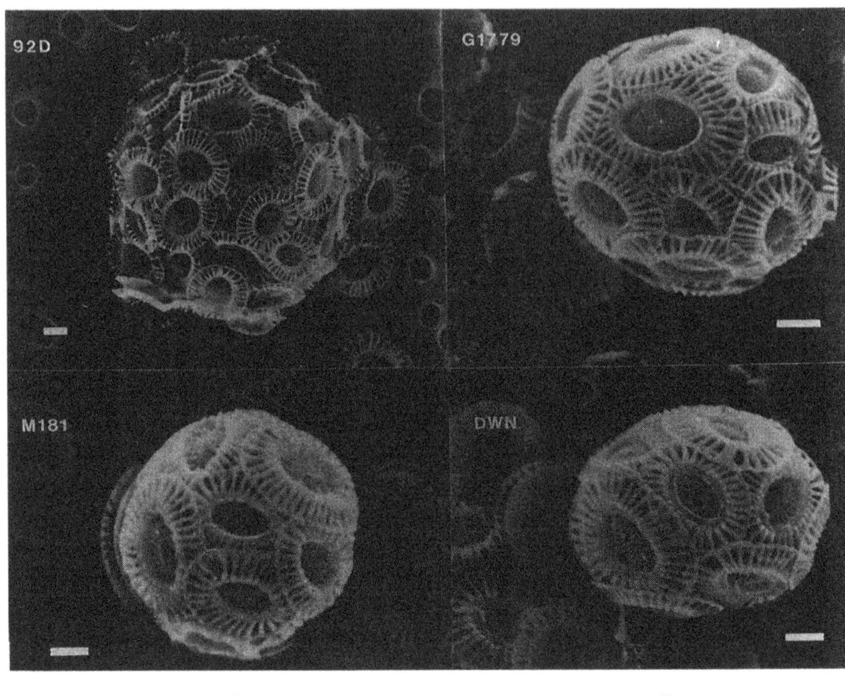

Fig. 21.4 *E. huxleyi.* Morphotypes A and B from different locations (see Table 21.4). Type A found in clones G1779Ga, CCMP1A1 (=M181) and DWN; type B in clone 92D. Note the larger size of the type B coccosphere compared with that of the type A clones. (Scale bar = 1 μm)

associated polysaccharide from either A or B morphotypes are morphotype specific, a feature retained by cells kept in culture for 20 years and implying a stable genetic basis (Young and Westbroek 1991). Coastal and oceanic strains also exhibit different key biomarker compounds and different amounts of fucoxanthin, but these isolates do not segregate into A and B morphotypes (Conte *et al.*, Chapter 19).

The molecular techniques initially used to investigate the genetic relatedness of geographically isolated clones of *E. huxleyi* involved a comparison of the nuclear-encoded ssu rRNA from three geographically isolated type A clones (Fig. 21.5) with that from 92D, a type B clone (Bhattacharya *et al.* 1992; Table 21.4). Only a single ambiguous nucleotide position separated clone 92D and CCMP1A1 (=M181) from the remaining clones. Because ambiguous positions are ignored in phylogenetic analyses, all clones are identical.

Our analysis then extended to the functionally non-conserved spacer region separating *rbc*L and *rbc*S (coding for the large and small subunits of

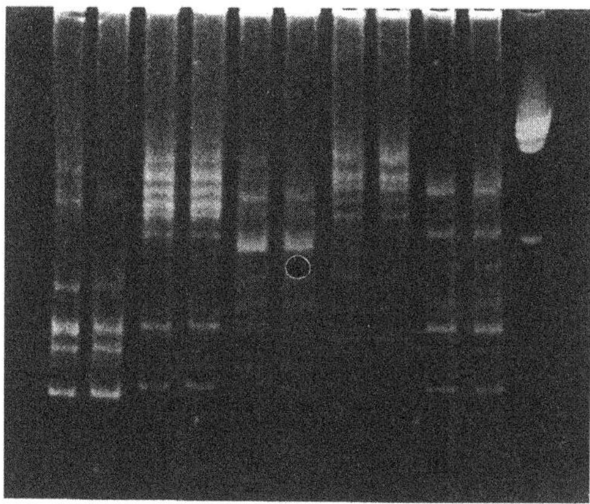

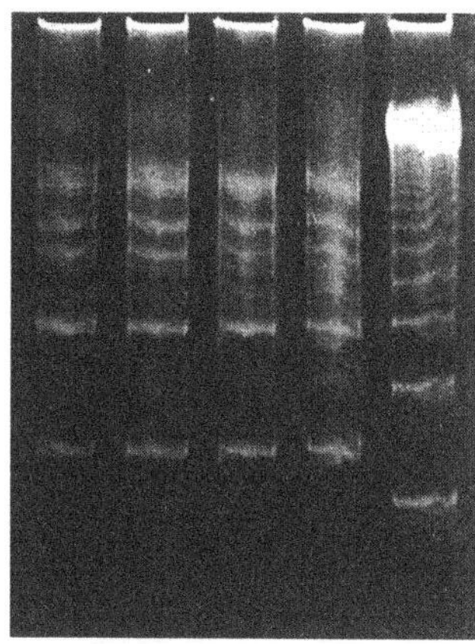

Fig. 21.5 Randomly Amplified Polymorphic DNA (RAPD) analysis of six *Emiliania huxleyi* clones (see Table 21.4 for locations). On the left-hand gel, the pairs of lanes were derived from (reading left to right) clones 92D, G1779Ga, CCMP1A1 (=PML M181), DWN 53/74/6, and L respectively. On the right-hand gel, the left pair of lanes represents clone G1779Ga, the pair on the right, clone G1779Ge. The extreme right-hand lanes on each gel represent the Lambda *Hin*D III and 100bp Ladder size markers respectively. Both gels were 5% acrylamide. Distinctive banding patterns were produced by PCR using the pUC18 sequencing primer under low stringency annealing conditions to amplify genomic DNA from each of the clones. Note the diversity of banding patterns among clones with identical ssu rRNA and RUBISCO spacer sequences (92D, G1779Ga, CCMP1A1, DWN53/74/6, L), and the identical banding patterns of the duplicate isolates G1779Ga and G1779Ge set up from single cells isolated from the same sample bottle. Each amplification was replicated to enable non-template differences to be detected.

ribulose 1, 5-bisphosphate carboxylase). Both subunits are plastid encoded in most chromophytes (Assali *et al.* 1991). This region appears to evolve rapidly; a 40% divergence within the spacer region was documented in what were initially believed to be conspecific clones of the red alga *Gymnocongrus devoniensis* from both sides of the Atlantic (Maggs *et al.* 1992).

This spacer region has been amplified from four type A clones and two type B clones of *E. huxleyi*, as well as from *Coccolithus pelagicus* (Table 21.4). All products are the same size. Preliminary sequence analysis from the 87 nucleotides of the spacer region and 60–70 nucleotides of coding sequence on either side indicates that all of the *E. huxleyi* clones are identical, with the exception of one possible ambiguous position in the 92D sequence.

Given the lack of variation in the geographic isolates of *E. huxleyi* from both coding and non-coding regions, the RAPD-PCR analysis was implemented as a means of distinguishing the clones (Table 21.4). All produce diverse banding patterns indicative of interclonal variation

Table 21.4 List of clones used in the analysis of *Emiliania huxleyi*

Clone	Location	ssu rRNA sequence	RAPD analysis	*rbc*L-*rbc*S spacer	
				Amplified	Sequenced
E. huxleyi					
G1779 Ga and Ge	60°N 40°W Iceland Basin	+	+	+	+
CCMP1A1 (=PML M181)	32°N 62°W Sargasso Sea	+	+	+	+
DWN 53/74/6	24°N 20°W Atlantic	+	+	+	+
VAN 55	Vancouver Pacific	−	−	+	+
92D	50°N 04°W English Channel	+	+	+	partial
L	Oslo Fjord	−	+	−	−
Ch25-90	Texel, Holland	−	−	+	+
C. pelagicus		+	−	+	−

(Fig. 21.5). Closely related clones, such as G1779Ga and G1779Ge, isolated from the same sampling bottle, have identical banding patterns, suggesting that they are descended from the same individual. This is the most sensitive of the techniques we have used; the production of additional fingerprints with other arbitrary primers will allow phylogenetic affiliations at the population level to be assessed.

The lack of genetic variation within *E. huxleyi*, as measured by ssu rRNA or Rubisco spacer sequence analysis, does not support the separation of A and B morphotypes at the species level. Our results suggest that *E. huxleyi* is a truly cosmopolitan species, able to exploit a wide range of habitats by means of phenotypic plasticity. At present, antibody reactions, physiological, and morphological differences appear to be the only methods of clone identification available to ecologists. We recommend that the presence of the A and B morphotypes in natural populations be recorded at the varietal level.

Acknowledgements

This work was supported in part by grants from the NERC (GR3/8139 and GR9/715), MAST-CT92-0038 and from the DFG (Schwerpunkt 'Antarktisforschung'). The support of T. Colbourn, M. Conte, J. C. Green, R. J. V. Green, R. Jordan, L. Madureira, T. Ruttkowski, A. Thompson, and U. Wellbrock is gratefully acknowledged. Isolates of *E. huxleyi* were obtained from CCMP (Bigelow Laboratory) and PML, and new clones were set up at PML from material collected during the BOFS cruises (NERC GST/02/386). Dr S. Giovannoni, Dr V. Huss, Dr U. Maid, and Professor K. Zetsche kindly provided access to unpublished sequences. This is Contribution No. 659 from the Alfred-Wegener-Institut.

References

Amann, R. I., Krumholz, L., and Stahl, D. A. (1990). Fluorescent-oligonucleotide probing of whole cells for determinative, phylogenetic, and environmental studies in microbiology. *Journal of Bacteriology*, **172**, 762–70.

Aiken, J., Moore, G. F., and Holligan, P. M. (1992). Remote sensing of oceanic biology in relation to global climate change. *Journal of Phycology*, **28**, 579–90.

Andersen, R. A. (1991). The cytoskeleton of chromophyte algae. *Protoplasma*, **164**, 143–59.

Assali, N.-E., Martin, W. F., Sommerville, C. C., and Loiseaux-de Goër, S. (1991). Evolution of the Rubisco operon from prokaryotes to algae: Structure and analysis of the rbcS gene of the brown alga *Phylaiella littoralis*. *Plant Molecular Biology*, **17**, 853–63.

Bakker, F. T., Olsen, J. L., Stam, W. T., and van den Hoek, C. (1992). Nuclear ribosomal DNA internal transcribed spacer regions (ITS1 and ITS2) define discrete biogeographic groups in *Cladophora albida* (Chlorophyta). *Journal of Phycology*, **28**, 839–45.

Baumann, M. and Jahnke, J. (1986). Marine Planktonalgen der Arktis. I. Die Haptophycee *Phaeocystis pouchetii. Mikrokosmos*, **75**, 262–5.

Baumann, M. E. M., Lancelot, C., Brandini, F. P., Sakshaug, E., and John, D. M. (1994). The taxonomic identity of the cosmopolitan prymnesiosphyte *Phaeocystis*, a morphological and ecophysiological approach. *Journal of Marine Sciences*, (In press).

Bhattacharya, D., Medlin, L., Wainwright, P. O., Arizitia, E. V., Bibeau, C., Stickel, S. K., and Sogin, M. L. (1992). Algae containing chlorophylls *a* + *c* are paraphyletic: Molecular evolutionary analysis of the Chromophyta. *Evolution*, **46**, 1801–17.

Bird, C. J., Rice, E. L., Murphy, C. A., and Ragan, M. A. (1992). Phylogenetic relationships in the Gracilariales (Rhodophyta) as determined by 18S rDNA sequences. *Phycologia*, **31**, 510–22.

Bleijswijk, J. van, van der Wal, P., Kempers, R., Veldhuis, M., Young, J. R., Muyzer, G., *et al.* (1991). Distribution of two types of *Emiliania huxleyi* (Prymnesiophyceae) in the northeast Atlantic region as determined by immunofluorescence and coccolith morphology. *Journal of Phycology*, **27**, 566–70.

Booth, B. C., Lewin, J., and Norris, R. E. (1982). Nanoplankton species predominant in the subarctic Pacific in May and June 1978. *Deep-Sea Research*, **29**, 185–200.

Brand, L. E. (1982). Genetic variability and spatial patterns of genetic differentiation in the reproductive rates of the marine coccolithophores *Emiliania huxleyi* and *Gephyrocapsa oceanica. Limnology and Oceanography*, **27**, 236–45.

Buma, A. G. J., Bano, N., Veldhuis, M. J. W., and Kraay, G. W. (1991). Comparison of the pigmentation of two strains of the prymnesiophyte *Phaeocystis* sp. *Netherlands Journal of Sea Research*, **27**, 173–82.

Cadée G. C. (1991). *Phaeocystis* colonies wintering in the water column. *Netherlands Journal of Sea Research*, **28**, 227–30.

Cadée G. C. and Hegeman, J. (1986). Seasonal and annual variation in *Phaeocystis pouchetii* (Haptophyceae) in the westernmost inlet of the Wadden Sea during the 1973 to 1985 period. *Netherlands Journal of Sea Research*, **20**, 29–36.

Cavalier-Smith, T. (1982). The origin of plastids. *Biological Journal of the Linnean Society*, **17**, 289–306.

Cavalier-Smith, T. (1986). The kingdom Chromista: Origin and systematics. In *Progress in phycological research*, Vol. 4, (ed. F. E. Round and D. J. Chapman), pp. 309–47. Biopress, Bristol.

Cavalier-Smith, T. (1987). Glaucophyceae and the origin of plants. *Evolutionary Trends in Plants*, **2**, 75–8.

Chapman, R. L. and Buchheim, M. A. (1991). Ribosomal RNA gene sequences: Analysis and significance in the phylogeny and taxonomy of green algae. *Critical Reviews in Plant Sciences*, **10**, 343–68.

Christensen, T., (1962). Alger. In *Botanik*, Bd. 2, *Systematik Botanik*, Nr. 2, (ed. T. W. Böcher, M. Lange, and T. Sørensen), pp. 1–178. Munksgaard, Copenhagen.

Davidson, A. T. and Marchant, H. (1990). The biology and ecology of *Phaeocystis* (Prymnesiophyceae). In *Progress in phycological research*, (ed. F. E. Round and D. J. Chapman), pp. 1–45, Biopress, Bristol.

Dodge, J. D. (1989). Phylogenetic relationship of dinoflagellates and their plastids. In *The chromophyte algae: problems and perspectives*, Systematics Association Special Volume No. 38, (ed. J. C. Green. B. S. C. Leadbeater, and W. L. Diver), pp. 207–27. Clarendon Press, Oxford.

Douglas, S. E., Durnford, D. G., and Morden, C. W. (1990). Nucleotide sequence of the gene for the large subunit of Ribulose-1, 5-bisphosphate carboxylase/oxygenase from *Cryptomonas* Φ: evidence supporting the polyphyletic origin of plastids. *Journal of Phycology*, **26**, 500–8.

Douglas, S. E., Murphy, C. A., Spencer, D. F., and Gray, M. W. (1991). Cryptomonad algae are evolutionary chimeras of two phylogenetically distinct unicellular eukaryotes. *Nature*, **350**, 148–51.

Douglas, S. E. and Turner, S. (1991). Molecular evidence for the origin of plastids from an cyanobacterial-like ancestor. *Journal of Molecular Evolution*, **33**, 267–73.

Fujiwara, S., Iwahashi, H., Someya, J., and Nishikawa, S. (1993). Structure and cotranscription of the plastid-encoded *rbc*L and *rbc*S genes of *Pleurochrysis carterae* (Prymnesiophyta). *Journal of Phycology*, *29*, 347–55.

Gibbs, S. P. (1981). Chloroplasts of some groups may have evolved from endosymbiotic eukaryotic algae. *Annals of the New York Academy of Science*, **36**, 193–207.

Goff, L. J. and Coleman, A. W. (1988). The use of plastid DNA restriction endonuclease patterns in delineating red algal species and populations. *Journal of Phycology*, **24**, 357–68.

Hallegraeff, G. (1983). Scale-bearing and loricate nanoplankton from the East Australian current. *Botanica Marina*, **26**, 493–515.

Hibberd, D. J. (1976). The ultrastructure and taxonomy of the Chrysophyceae and Prymnesiophyceae (Haptophyceae); a survey with some new observations on the ultrastructure of the Chrysophyceae. *Botanical Journal of the Linnean Society*, **72**, 55–80.

Huss, V. A. R. and Sogin, M. L., (1991). Phylogenetic position of some *Chlorella* species within the Chlorococcales based upon complete small-subunit ribosomal RNA sequences. *Journal of Molecular Evolution*, **31**, 432–42.

Jahnke, J. (1989). The light and temperature dependence of growth rate and elemental composition of *Phaeocystis globosa* Scherffel and *P. pouchetii* (Har.) Lagerh. in batch cultures. *Netherlands Journal of Sea Research*, **23**, 15–21.

Jahnke, J. and Baumann, M. (1987). Differentiation between *Phaeocystis pouchettii* (Har.) Lagerheim and *Phaeocystis globosa* Scherffel. I. Colony shapes and temperature tolerances. *Hydrobiological Bulletin*, **21**, 141–7.

Karsten, G. (1905). Das Phytoplankton des Antarktischen Meeres nach dem material der Deutschen Tiefsee-Expedition 1898–1899. *Wissenschaftliche Ergebnisse der Deutschen Tiefsee-Expedition auf dem Dampfer 'Valdivia'*. Band II, Teil 2, 1–136.

Kornmann, P. (1955). Beobachtungen an *Phaeocystis*-Kulturen. *Helgoländer Wissenschaftliche Meeresuntersuchungen*, **5**, 218–33.

Lancelot, C., Billen, G., Sournia, A., Weisse, T., Colijn, F. Veldhuis, M. J. W., *et al.* (1987). *Phaeocystis* blooms and nutrient enrichment in the continental coastal zones of the North Sea. *Ambio*, **16**, 38–46.

Lagerheim, G. (1893). *Phaeocystis*, nov. gen. grundadt på *Tetraspora poucheti* Har. *Botaniska Notiser*, **1**, 32–3.

Lynch, M. (1988). Estimation of relatedness by DNA fingerprinting. *Molecular Biology and Evolution*, **5**, 584–99.

Maggs, C. A., Douglas, S. E., Fenety, J., and Bird, C. J. (1992). A molecular and morphological analysis of the *Gymnogongrus devoniensis* (Rhodophyta) complex in the North Atlantic. *Journal of Phycology*, **28**, 214–32.

Markowicz, Y. and Loiseaux-de Goër, S. (1991). Plastid genomes of the Rhodophyta and Chromophyta constitute a distinct lineage which differs from that of the Chlorophyta and have a composite phylogenetic origin, perhaps like that of the Euglenophyta. *Current Genetics*, **20**, 427–30.

Martin, W., Somerville, C. C., and Loiseaux-de Goër, S. (1992). Molecular phylogenies of plastid origins and algal evolution. *Journal of Molecular Evolution*, **35**, 385–404.

Medlin, L. K., Elwood, H. J., Stickel, S., and Sogin, M. L. (1991). Morphological and genetic variation within the diatom *Skeletonema costatum* (Bacillariophyta): Evidence for a new species, *Skeletonema pseudocostatum*. *Journal of Phycology*, **27**, 514–24.

Medlin, L. K., Lange, M., and Baumann, M. E. M. (1994). Genetic differentiation among three colony-forming species of *Phaeocystis*; further evidence for the phylogeny of the Prymnesiophyta. *Phycologia*, **33**, (In press).

Mereschkowsky, C. (1905). Über Natur und Ursprung der Chromatophoren im Pflanzenreiche. *Biologische Centralblatt*, **25**, 593–604.

Mereschkowsky, C. (1910). Theories der zwei Plasmaarten als Grundlage der Symbiogenesis, einer neuen Lehre von der Entstehung der Organismen. *Biologische Centralblatt*, **30**, 278–303.

Moestrup, Ø. (1979). Identification by electron microscopy of marine nanoplankton from New Zealand, including the description of four new species. *New Zealand Journal of Botany*, **17**, 61–95.

Oppen, M. J. H. van, Olsen, J. L., Stam, W. T., van den Hoek, C., and Wiencke, C. (1993). Arctic-Antarctic disjunctions in the benthic seaweeds *Acrosiphonia arcta* (Chlorophyta) and *Desmarestia viridis/willii* (Phaeophyta) are of recent origin. *Marine Biology*, **115**, 381–6.

Parke, M., Green, J. C., and Manton, I. (1971). Observations on the fine structure of zoids of the genus *Phaeocystis* (Haptophyceae). *Journal of the Marine Biological Association of the United Kingdom*, **51**, 927–41.

Perasso, R., Baroin, A., Qu, L. H., Bachellerie, J. P., and Adoutte, A. (1989). Origin of the algae. *Nature*, **339**, 142–4.

Pouchet, G. (1892). Sur une algue pélagique nouvelle. *Compte Rendus de la Societé Biologique*, Series IX, **4**, 34–6.

Raven, P. H. (1970). A multiple origin for plastids and mitochondria. *Science*, **169**, 641–5.

Sagan, L. (1967). On the origin of mitosing cells. *Journal of Theoretical Biology*, **14**, 225–74.

Saunders, G. W. and Druehl, L. D. (1991). Nucleotide sequences of the small-subunit ribosomal RNA genes from selected Laminariales (Phaeophyta): Implications for kelp evolution. *Journal of Phycology*, **28**, 544–9.

Scherffel, A. (1900). *Phaeocystis globosa* nov. spec. nebst einigen Betrachtungen über die Phylogenie niederer, insbesonderer brauner Organismen. *Wissenschaftliche Meeresuntersuchungen. Abteilung Helgoland*, **4**, 1–129.

Schimper, A. F. W. (1883). Über die Entwicklung der Chlorophyllkorner und Farbkorner (1 Teil). *Botanische Zeitung*, **41**, 105–14.

Sogin, M. L., Ingold, A., Karlok, M., Nielsen, H., and Engberg, J. (1986). Phylogenetic evidence for the acquisition of ribosomal RNA introns subsequent to the divergence of some of the major *Tetrahymena* groups. *The EMBO Journal*, **5**, 3625–30.

Sournia, A. (1988). *Phaeocystis* (Prymnesiophyceae): How many species? *Nova Hedwigia*, **47**, 211–7.

Tomas, R. N. and Cox, E. R. (1973). Observations on the symbiosis of *Peridinium balticum* and its intracellular alga. I. Ultrastructure. *Journal of Phycology*, **9**, 304–23.

Valentin, K. and Zetsche, K. (1990). Rubisco genes indicate a close phylogenetic relation between the plastids of Chromophyta and Rhodophyta. *Plant Molecular Biology*, **15**, 575–84.

Welsh, J. and McClelland, M. (1990). Fingerprinting genomes using PCR with arbitrary primers. *Nucleic Acids Research*, **18**, 7213–18.

Woese, C. R. (1987). Bacterial evolution. *Microbacteriological Reviews*, **51**, 221–71.

Young, J. R. and Westbroek, P. (1991). Genotypic variation in the coccolithophorid species *Emiliania huxleyi*. *Marine Micropaleontology*, **18**, 5–23.

22. Origin and relationships of Haptophyta

T. CAVALIER-SMITH

Department of Botany, The University of British Columbia, Vancouver, Canada

Abstract

Haptophyta are placed, with Heterokonta, Chlorarachniophyta, and Cryptista, in the kingdom Chromista because in all four phyla the chloroplasts are surrounded by a periplastid membrane, located either inside the lumen of the rough endoplasmic reticulum (RER) or (*Chlorarachnion* only) inside a smooth endomembrane. This unique arrangement arose abut 600 millions years ago by incorporating a symbiotic eukaryotic alga into a biciliate protozoan host. Ancestral chromists were photophagotrophs, but phagotrophy and photosynthesis were independently lost several times in different diverging lineages. Haptophyta and Heterokonta (i.e. ochristan algae, Pseudofungi, Bicoecia, Labyrinthulea) are grouped as the infrakingdom Chromobiota because they share five derived characters: (1) chloroplasts with thylakoids stacked in threes, and chlorophylls c_1, c_2, c_3, and fucoxanthin, but no phycobilins; (2) chloroplasts located inside the RER and periplastid membrane, but (unlike cryptomonads and *Chlorarachnion*) with no nucleomorph; (3) an autofluorescent posterior cilium, (4) tubular mitochondrial cristae with intracristal filaments; (5) large body scales. Ultrastructural and molecular evidence favour a single symbiotic origin for all chromists, and a very early divergence between the four phyla. The symbiont was probably intermediate between red algae, green algae, and dinoflagellates, and closely related to their immediate common ancestor, while the host was an opalozoan with tubular surface hairs and probably body scales.

I argue that Haptophyta and Heterokonta diverged by adopting alternative adaptations for phagotrophy: the haptonema versus the feeding current generated by retronemes on the heterokont anterior cilium. I suggest that the haptonematal axoneme arose by duplication and modification of the nucleating centre for the R3 microtubular ciliary root, and that its origin made retronemes selectively disadvantageous; thus they were inevitably rapidly lost. The class Pavlovea and the scale-bearing haptophytes (class Patelliferea Cavalier-Smith 1993, containing Prymnesiales, Coccosphaerales, and Isochrysidales) diverged relatively early.

The Haptophyte Algae (ed. J. C. Green and B. S. C. Leadbeater), Systematics Association Special Volume No. 51, pp. 413–35. Clarendon Press, Oxford, 1994.

Introduction

Haptophyta have been recognized as a distinct phylum for over thirty years (Hibberd 1972). Marked differences from heterokont algae were first revealed by electron microscopy; discovery of the haptonema (Parke *et al.* 1955) led to their separation from chrysomonads, with which they were formerly grouped on account of their similarly pigmented chloroplasts, as the class Haptophyceae (Christensen 1962; unnecessarily renamed Prymnesiophyceae by Hibberd 1976). Unlike chrysomonads and other heterokont algae, haptophytes lack the rigid tubular hairs (retronemes; Cavalier-Smith 1989) on the anterior cilium that characterize Heterokonta (*sensu* Cavalier-Smith 1981*a*, 1986). Because of their uniqueness it has been suggested that Haptophyta form a separate kingdom (Leedale 1974; Cavalier-Smith 1978) or eukaryote lineage (Bhattacharya *et al.* 1992).

However, important similarities between Haptophyta and Heterokonta relate them more closely to each other than to other organisms, so the two phyla were grouped into a subkingdom, Chromophyta (Cavalier-Smith 1981*a*); more recently, because of changing and increasingly ambiguous meanings of 'Chromophyta', I renamed the joint taxon Chromobiota (Cavalier-Smith 1991*a*, 1993*a*, *b*). I chose this name because it was first used (but not validly published) by Jeffrey (1971) for a kingdom which innovatively grouped oomycetes with chromophyte algae; Chromophyta Cavalier-Smith 1981 included not only algae but also oomycetes, hyphochytrids, labyrinthulids, and thraustochytrids. My Chromobiota (now an infrakingdom: Cavalier-Smith 1993*b*) differs from Jeffrey's by excluding two groups of chromophyte algae: dinoflagellates and cryptomonads. Dinoflagellates differ so greatly from other chromophytes that they should not be in the same kingdom (Cavalier-Smith 1981*a*), still less the same division as advocated by Christensen (1989). Though cryptomonads differ from Chromobiota in many ways, the chloroplasts of both are surrounded by a periplastid membrane and located inside the RER lumen; moreover, like heterokonts, cryptomonads have retronemes, though usually on both cilia, not just the anterior one. As neither synapomorphy is likely to have evolved more than once, I used them to unite cryptomonads and Chromobiota in the kingdom Chromista (Cavalier-Smith 1981*a*, 1986, 1989).

Ribosomal RNA (rRNA) sequences support my separation of chromophyte algae into four phyla: (1) dinoflagellates as subphylum Dinoflagellata in phylum Dinozoa, (Cavalier-Smith 1991*a*, 1993*b*); (2) cryptomonads as class Cryptomonadea in phylum Cryptista (Cavalier-Smith 1989); (3) heterokont algae as subphylum Ochrista (Cavalier-Smith 1986) in the phylum Heterokonta (Cavalier-Smith 1989); and (4) the phylum Haptophyta. rRNA

Table 22.1 Classification of the kingdom Chromista Cavalier-Smith 1981

Subkingdom 1. CHLORARACHNIA Cavalier-Smith, 1993
- **Phylum 1. CHLORARACHNIOPHYTA Hibberd & Norris, 1984**
  - Class 1. Chlorarachniophycea Hibberd & Norris, 1984 orthogr. emend.

Subkingdom 2. EUCHROMISTA Cavalier-Smith, 1993
- **Infrakingdom 1. CRYPTISTA Cavalier-Smith, 1989 stat. nov.**
  - **Phylum 1. CRYPTISTA Cavalier-Smith, 1989 (syn. Cryptophyta)**
    - Class 1. Cryptomonadea Stein, 1878 stat. nov. Pascher ex Schoenichen, 1925
    - Class 2. Goniomonadea nom. nov. pro Cyathomonadea Cavalier-Smith, 1989
- **Infrakingdom 2. CHROMOBIOTA orthogr. emend. pro. CHROMOPHYTA Cavalier-Smith 1986**
  - **Phylum 1. HETEROKONTA Cavalier-Smith, 1986**
    - Subphylum 1. BICOECIA Cavalier-Smith, 1989
      - Class 1. Bicoecea orthogr. emend. pro Bicosoecea Cavalier-Smith, 1986
    - Subphylum 2. LABYRINTHISTA Cavalier-Smith, 1986 stat. nov., 1989
      - Class 1. Labyrinthulea Olive ex Cavalier-Smith, 1986
    - Subphylum 3. OCHRISTA Cavalier-Smith, 1986
      - Infraphylum 1. Raphidoista Cavalier-Smith, 1986 emend. stat. nov.
        - Superclass 1. Raphidomonadia Cavalier-Smith, 1993
          - Class 1. Raphidomonadea Chadefaud ex Silva, 1980 (syn. Chloromonadea)
        - Superclass 2. Dictyochia Haeckel, 1894 stat. nov. emend.
          - Class 1. Pedinellea Cavalier-Smith, 1986
          - Class 2. Silicoflagellatea Borgert, 1891 stat. nov.
          - Class 3. Oikomonadea Cavalier-Smith, 1993 (*Oikomonas*)
      - Infraphylum 2. Chrysista Cavalier-Smith, 1986 stat. nov.
        - Class 1. Chrysophycea Pascher ex Hibberd, 1976 orthogr. emend.
        - Class 2. Flavoretea Cavalier-Smith, 1993 (*Reticulosphaera*)
        - Class 3. Xanthophycea Allorge ex Fritsch, 1935 (syn. Tribophyceae Hibberd)
        - Class 4. Phaeophycea Kjellman, 1891 orthogr. emend.
      - Infraphylum 3. Eustigmista Cavalier–Smith, 1993
        - Class 1. Eustigmatophycea Hibberd et Leedale, 1971
      - Infraphylum 4. Diatomea Agardh, 1824 stat. nov. Cavalier-Smith, 1993
        - Class 1. Centricea Cavalier-Smith, 1986 (syn. Centrobacillariophycea Silva, 1962)
        - Class 2. Pennatea Cavalier-Smith, 1986 (syn. Pennatibacillariophyceae Silva, 1962)

Table 22.1 (*cont.*)

Subphylum 4.	PSEUDOFUNGI Cavalier-Smith, 1986 emend. 1989
Class 1.	Pythiistea Cavalier-Smith, 1986 stat. nov. 1989 (oomycetes and hyphochytrids)
Phylum 2.	**HAPTOPHYTA Hibberd ex Cavalier-Smith 1986**
Class 1.	Patelliferea Cavalier-Smith, 1993 (orders: Isochrysidales, Coccosphaerales, Prymnesiales)
Class 2.	Pavlovea Cavalier-Smith, 1986 stat. nov. 1993 (order: Pavlovales)

sequences for *Chlorarachnion* (McFadden *et al.* 1994) suggest that their nucleomorphs are related to those of cryptomonads, while the host cell is distantly related to Heterokonta, Cryptista, and Haptophyta (Fig. 22.1); for these and other reasons (Cavalier-Smith 1992*a*) I now include Chlorarachniophyta in Chromista (Cavalier-Smith 1993*b*) as a subkingdom (see Table 22.1). The three original chromist phyla become the subkingdom Euchromista, distinguished from Chlorarachnia by the absence of chlorophyll *b* and having ribosomes on the outermost membrane around the chloroplast.

The key question about the origin of Haptophyta concerns their relationship with Ochrista: did these taxa acquire chloroplasts by separate endosymbioses (Margulis 1970; Leedale 1974; Whatley *et al.* 1979; Gibbs 1993), or a single one (Cavalier-Smith 1981*a*, *b*, 1992*a*; Andersen 1991)? Here I review evidence for the monophyly of Chromobiota, and explain how Heterokonta and Haptophyta could have diverged from a photophagotrophic common ancestor. I argue that: (1) the haptonema and heterokont retronemes are mutually exclusive adaptations for phagotrophic feeding that caused the two phyla to diverge in earliest chromobiote evolution; (2) the ancestor of chromobiotes was a chimaeric cell derived from the merger of a biciliated protozoan of the phylum Opalozoa (Cavalier-Smith 1993*c*) with a photosynthetic endosymbiont; (3) the symbiont ancestral to the chromobiote chloroplast and periplastid membrane was not a member of a modern algal class, but a close relative of the first eukaryotic algae, having chlorophylls *a*, *b*, and *c* as well as phycobilins (Cavalier-Smith 1982, 1986, 1992*a*, 1993*a*); and (4) cryptomonad and *Chlorarachnion* chloroplasts probably originated by the same endosymbiosis.

Monophyly of Chromobiota

Heterokonts and haptophytes share three major characters which occur in no other organisms; all are very unlikely to have evolved independently. I therefore consider them shared derived characters that first evolved in the immediate common ancestor of the two phyla, i.e. the ancestral chromobiote. I discuss them and other possible shared derived characters in turn.

Chloroplast structure and chemistry

The thylakoids of Chromobiota are stacked regularly in threes, as in dinoflagellates, but unlike all other organisms. They lack both phycobilins and chlorophyll *b*, like peridinean dinoflagellates, but unlike all other eukaryotes. Except for eustigmatophytes, which have only chlorophyll *a*, they have at least one, more usually two, of the pigments chlorophyll c_1, c_2, and c_3. Chlorophyll c_3 is unique to haptophytes and heterokonts; in the latter it occurs only in chrysomonads and diatoms, though it is not found in all members of these three groups (Jeffrey 1989). Chlorophyll c_2 is also present in both dinoflagellates and cryptomonads (both very occasionally also have chlorophyll c_1). Haptophytes, heterokonts, and dinoflagellates share many carotenoids; fucoxanthin, however, is almost uniquely shared between the majority of haptophytes and the majority of heterokont algae (though it is totally absent from xanthophytes and eustigmatophytes). Typical peridinin-containing dinoflagellates never have fucoxanthin, but four other species do (Bjørnland and Liaaen-Jensen 1989); the plastids of two of these (*Glenodinium foliaceum* and *Peridinium balticum*) are not their own, but those of symbiotic chromobiotes (Schnepf 1993). If this is also true of the other two, then fucoxanthin also is restricted to chromobiotes. At least one haptophyte lacks fucoxanthin, while some have also 19′-acyloxy-fucoxanthin which has not been found in heterokonts; the latter is present in some of the fucoxanthin-containing dinoflagellates, suggesting that their endosymbionts are haptophytes. The simplest interpretation of this pigment distribution is that chlorophyll c_1 and c_2 were both present in the group ancestral to the chromobiote chloroplast, but that chlorophyll c_3 and fucoxanthin were not and first evolved only in the immediate common ancestor of chromobiotes; and that all four pigments were lost polyphyletically within both Heterokonta and Haptophyta following their divergence.

Location of plastids

As in cryptomonads, the chloroplast envelope is double and surrounded by a separate periplastid membrane (Cavalier-Smith 1989), the whole being located within the RER lumen; but in contrast to cryptomonads the periplastid space has no nucleomorph, ribosomes or starch. A minor difference between heterokont algae and haptophytes is that the haptophyte periplastid space is virtually empty, whereas that of heterokonts commonly contains a periplastid reticulum or at least a few obvious vesicles (Gibbs 1981*a*). This unique arrangement of membranes in euchromistan algae, as proposed by Greenwood (1974) and Greenwood *et al.* (1977), and developed in detail by Whatley *et al.* (1979), Whatley and Whatley (1981), Gibbs (1981*a*, *b*), and myself (Cavalier-Smith 1981*b*, 1982, 1986, 1989), arose by the symbiotic merger of a eukaryotic algal endosymbiont with a protozoan host. The nucleomorph is the former endosymbiont's nucleus, while the 80S periplastid ribosomes of cryptomonads are descended from its cytosolic ribosomes. The periplastid membrane is the relic of the plasmamembrane of the endosymbiont (Whatley and Whatley 1981; Cavalier-Smith 1982, 1986, 1989); contrary to earlier assumptions (Bouck 1965; Gibbs 1981*a*), it is not homologous with the host endoplasmic reticulum (ER); therefore the misleading name chloroplast ER, which embraces both the periplastid membrane and the periplastid RER membrane, should no longer be used (Cavalier-Smith 1989). It is still debated whether such a symbiosis took place once only, as Gibbs (1981*b*) and I (Cavalier-Smith 1981*a*, 1982, 1986, 1989) have argued, or separately in cryptomonads and chromobiotes as Whatley *et al.* (1979), Whatley (1989), and Gibbs (1993) suggested, or more than once within Chromobiota as Whatley *et al.* (1979), Margulis (1981), Patterson (1989), and Gibbs (1993) supposed.

Conversion of a eukaryotic endosymbiont into the chromist chloroplast and periplastid membrane required the evolution of a unique and novel targeting mechanism for importing proteins made in the cytosol across the four membranes that surround the chloroplast matrix (Cavalier-Smith 1982, 1986, 1989, 1993*a*). Because thousands of separate mutations must have been involved in this it is very improbable that the three euchromist phyla acquired their chloroplasts independently. This was one of the two major reasons for creating the kingdom Chromista for all three (Cavalier-Smith 1981*a*). My thesis predicts that the targeting mechanism should be homologous in all euchromists. A recent finding suggests that this may be so. Jenkins *et al.* (1990) have found that a cryptomonad α-phycoerythrin gene located in the nucleus (McFadden *et al.*, personal communication) has an exceptionally long NH_2-terminal leader sequence, strongly homologous with the leader of a diatom chlorophyll *c* binding protein; thus cryptomonads and chromobiotes may well share an homologous targeting mechanism.

These considerations of protein targeting strongly favour the monophyly of Euchromista; those who have advocated a polyphyletic origin have ignored them. In contrast, the common absence of the nucleomorph, periplastid ribosomes, and starch from both Heterokonta and Haptophyta, on superficial consideration, might be thought to support chromobiote monophyly only weakly. Since the presence of these three structures in cryptomonads is the ancestral state, because the symbiont must once have had a nucleus and ribosomes (Cavalier-Smith 1986, 1989), it might be argued that they could relatively easily have been lost independently in the ancestral heterokont and haptophyte. But this is not so, since the total loss of the symbiont nucleus required the prior transfer to the host nucleus of all its genes coding for essential chloroplast proteins, and the acquisition by all of them of topogenic sequences for the reimport of their proteins made in the cytosol into the chloroplast. Thus, we are here dealing not merely with potentially independent character losses, but with the transfer of genes from the nucleomorph into the nucleus, and with the potentially independent gain of many new topogenic sequences. Absence of the nucleomorph is not a purely negative character, but involves major positive molecular changes in the chromobiote nucleus compared with that of cryptomonads and *Chlorarachnion*.

Autofluorescent flavin in the rear cilium

The rear cilium autofluoresces (Kawai 1988; Kawai and Inouye 1989) in two haptophyte orders (Prymnesiales and Isochrysidales), and in most members of three heterokont algal classes (Chrysophycea, Xanthophycea, and Phaeophycea) that with Flavoretea (*Reticulosphaera*; Grell *et al.* 1990) constitute the infraphylum Chrysista (Cavalier-Smith 1986, 1989), but in no other organisms. This strongly supports one of three possibilities: (1) haptophytes evolved from a chrysist by losing retronemes and evolving a haptonema; (2) Chrysista evolved from a haptophyte by losing the haptonema and evolving retronemes; or (3) haptophytes and chrysomonads diverged from a common ancestor with a single autofluorescent cilium, while Pavlovea and the heterokont classes without an autofluorescent cilium secondarily lost either autofluorescence (Pavlovea, Raphidomonadea, Eustigmatophycea), or the whole cilium (Dictyochea, Diatomea). I favour the third one, because (1) autofluorescence clearly can be secondarily lost (it is absent from some species in each of the four classes that have it), (2) it does not require the loss of retronemes or haptonema, and (3) because other evidence (e.g. Fig. 22.1) favours a very early divergence between Heterokonta and Haptophyta rather than a recent derivation of one from the other.

Filaments within the tubular cristae of the mitochondria

Intracristal filaments appear as a central dot in cross-sectional views of cristae in all haptophytes and heterokonts, and apparently some dinoflagellates (e.g. Larsen 1988); but are absent from cryptomonads and *Chlorarachnion* (Cavalier-Smith 1993*b*). Intracristal filaments are absent from 12 of the 13 protozoan phyla that have tubular cristae (i.e. all except Dinozoa, Cavalier-Smith 1993*b*). This generalization requires further study since cristal filaments are sometimes hard to see when fixation is poor, which has often been so for studies of Opalozoa (Cavalier-Smith 1993*c*),

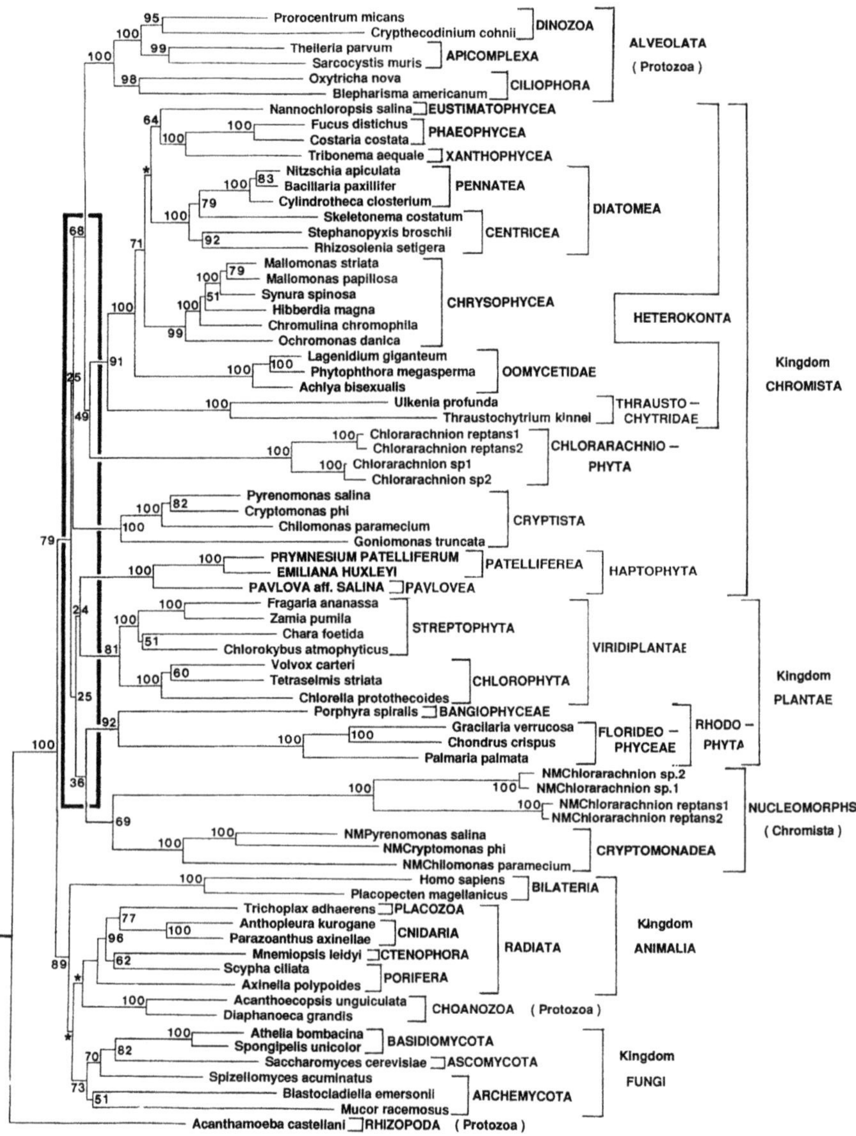

Fig. 22.1 Neighbour joining distance tree of 73 nuclear 18s rRNA sequences. The tree was calculated using Felsenstein's PHYLIP v. 3.5, with the Jukes-Cantor correction for genetic distances; jumbling the input order; the amoeba *Hartmanella vermiformis* as outgroup; and 2142 alignment positions (the complete alignment had 2862 positions, but insertions unique to one group were excluded from the analysis except for one nucleotide at each end). Bootstrap values for 100 replicates are at the nodes, except where three clades (*asterisks*) were absent from the consensus tree, in which diatoms and chrysophytes interchanged their positions, Choanozoa grouped (boot-

which were probably ancestral to chromists. Nonetheless intracristal filaments may prove to be a synapomorphy for Chromobiota plus at least some dinoflagellates. These are also the only taxa with chlorophyll c_2, suggesting that intracristal filaments and chlorophyll c_2 evolved in the same organism and were both retained by its descendants. As they were probably absent from the opalozoan ancestor of chromists, I suggest that the genes for the filament proteins were transferred into the ancestral chromobiote nucleus from the dinoflagellate-like endosymbiont that provided its chloroplast.

Other possible synapomorphies

A possible synapomorphy for Chromobiota is the large (1–15 μm) ribbed and fenestrated body scales found only in Haptophyta (specifically in Patelliferea, not Pavlovea), and Chrysophycea and Labyrinthulea in the Heterokonta, and formed in Golgi cisternae (see Leadbeater, Chapter 2). Possibly such scales evolved first in the ancestral chromobiote and were lost in classes that lack them. Though large open scales are not found in other protists, small scales are widespread (almost universal in Prasinophyceae; common in Opalozoa; rare in Dinozoa, Rhizopoda, and Cryptista). The major carbohydrate is glucose in haptophyte scales, and galactose in thraustochytrids, so they may not be homologous; to judge this we need to know more about their proteins.

Leucosin is a β-1,3 glucan stored as cytosolic liquid drops in five groups of heterokonts (chrysophytes, silicoflagellates, pedinellids, xanthophytes, diatoms). Often stated to be present in haptophytes, this has not been confirmed (Green *et al.* 1990). The β-1,3 laminarin of Phaeophycea is similar; oomycetes also store β-1,3 glucans. However, euglenoid paramylon is also a β-1,3 glucan, so storage β-glucans *per se* are not a chromobiote

strap value 47) with Fungi, and Bilateria (bootstrap value 49) with Radiata, making Animalia monophyletic. The branching pattern within the heavy boxed area is also highly unreliable, the close branches and low bootstrap values suggesting an almost simultaneous divergence of these taxa (probably about 600 million years ago), as predicted (Cavalier-Smith 1978). (Using the DNAPARS algorithm instead, Cryptista grouped with Chlorarachnia and thraustochytrids to form a sister clade to other heterokonts; Haptophyta grouped with these plus Alveolata rather than with green plants; Animalia were monophyletic and closer to Fungi than to Choanozoa. The FITCH algorithm grouped Haptophyta with *Chlorarachnion*, Heterokonta, and Alveolata; and monophyletic Animalia with Choanozoa.) Sequences were from Genbank or M. T. E. P. Allsopp in my laboratory (*Pavlova, Prymnesium, Axinella, Parazoanthus, Ulkenia, Thraustochytrium, Chilomonas*: Cavalier-Smith *et al.* in preparation), or kindly provided by G. McFadden (*Goniomonas, Chlorarachnion* spp., and *C. reptans* 2), U. Maier (*C. reptans* 1) or M. A. Ragan (*Porphyra*).

synapomorphy. The haptophyte storage material is commonly inside a membrane that caps the projecting pyrenoid (Manton 1966). The arrangement in *Chlorarachnion* is remarkably similar; its storage material is not starch, and might be β-glucan (Hibberd and Norris, 1984). In the symbiotic merger that created the Chromista, the host probably stored β-glucans, while the endosymbiont used α-1-4 glucans (starch). Thereafter only one need have been retained: periplastidal α-1-4 glucans in cryptomonads and cytosolic β 1-3 glucans in chromobiotes and chlorarachnians the other was lost.

The ancestors of Haptophyta and Heterokonta

The host, i.e. the cell excluding the chloroplast and periplastid membrane

The closest non-chromist relatives of chromobiotes must be free-living biciliate protozoa with tubular cristae; of the 18 protozoan phyla only Dinozoa and Opalozoa answer this description (Cavalier-Smith 1993*b*, *c*). Dinozoa can be ruled out because unlike chromobiotes they have cortical alveoli. Therefore, the only protozoa likely to have been ancestral to Haptophyta or Heterokonta are the Opalozoa. Six of the eight Opalozoan classes can be ruled out as likely ancestors for chromists because they have very different body plans. Only the classes Heteromitea and Cyathobodonea are reasonable candidates. Heteromitea are more likely to be ancestral to chromobiotes, because like all biciliate heterokonts, pavloveans, and some patelliferans they are anisokont, and because one heteromitean (*Proteromonas*) has tubular hairs (somatonemes) on its cell surface (not on its cilia) that closely resemble chromist retronemes. Like those of cryptomonads these somatonemes are bipartite, with a tubular shaft plus a single terminal filament. Therefore a *Proteromonas*-like heteromitean of the opalozoan subphylum Proterozoa is, as argued previously (Cavalier-Smith 1986; 1989), the best candidate for the ancestor of both heterokonts and cryptomonads. It is an equally good candidate for the ancestor of haptophytes and *Chlorarachnion.* The fact that many patelliferans are isokont, not anisokont, is not a good reason for preferring the isokont Cyathobodonea as their body form and/or extrusomes are always quite different from those of haptophytes. Moreover, since all Pavlovea and some isochrysidalean patelliferans are fundamentally anisokont, it is likely that anisokonty with an asymmetric cell shape is the ancestral state for all Haptophyta (Cavalier-Smith 1986, 1989; Green *et al.* 1990). Since heterokonts are fundamentally anisokont the ancestor of all chromobiotes was almost certainly anisokont: most probably a heteromitean with somatonemes like *Proteromonas* (but unlike *Proteromonas* was free-living, not a gut symbiont, and therefore had not lost its peroxisomes as has *Proteromonas*).

An anisokont proterozoan opalozoan would be a good ancestor not only for Cryptista and Chromobiota but also for *Chlorarachnion* (if the anterior flagellum was lost). The opalozoan *Kathablepharis* (Lee and Kugrens 1991) has scrolled ejectisomes similar (not identical) to those of Cryptista; therefore there may once have existed a proterozoan intermediate between *Proteromonas* and *Kathablepharis* in having both bipartite tubular somatonemes and scrolled ejectisomes. Such a proterozoan would have been the ideal host for the origin of Chromista, which then could have diversified as outlined in Fig. 22.2.

Source of the haptophyte chloroplast

The only organisms with chlorophyll c_1 and c_2 other than chromobiotes are photosynthetic dinoflagellates and cryptomonads. Photosynthetic dinoflagellates are divided into two major classes (Cavalier-Smith 1991, 1993*b*): Peridinea with chloroplasts with a three-membraned envelope, peridinin and chlorophyll *c*, but no phycobilins; and Bilidinea (sole order Dinophysida) with chloroplasts with a two-membraned envelope and phycobilin as well as chlorophyll *c*. If any dinoflagellates contain a fucoxanthin-containing chloroplast not secondarily acquired from a chromobiote, they would be the best candidate for the ancestor of the chromobiote chloroplast. If such dinoflagellates do not exist, the best extant organism would be a bilidinean with a *Dinophysis*-like chloroplast (Cavalier-Smith 1989, 1992*a*, 1993*a*), which could have evolved into a chromobiote chloroplast simply by losing the intrathylakoidal phycobilin (so long as they have fucoxanthin, which is unknown); a peridinean would have to have lost one of the three envelope membranes to produce the chromobiote chloroplast.

Since rRNA trees (Fig. 22.1) support the idea (Cavalier-Smith 1982, 1986) that Rhodophyta, Glaucophyta, Viridiplantae, Dinozoa, Chlorarachnia, Cryptophyta, Chromobiota, and chlorarachnian and cryptomonad nucleomorphs, originated at about the same time, it is most likely that all four chromist phyla acquired their chloroplasts in a single endosymbiotic event and that the symbiont was an early alga that, unlike any extant alga, had chlorophylls *a*, *b*, and *c* as well as phycobilins (Cavalier-Smith 1992*a*).

Ribosomal RNA sequences and chromist evolution

Molecular phylogenetic analysis of nuclear (18s and 28s) and chloroplast (16s) rRNA genes supports most of the major features of my first detailed treatment of the diversification of chloroplasts and algae (Cavalier-Smith 1982). Firstly, chloroplast 16s rRNA trees (Douglas and Turner, 1991; Medlin, Chapter 21) support the idea (Taylor 1978; Cavalier-Smith 1982) of

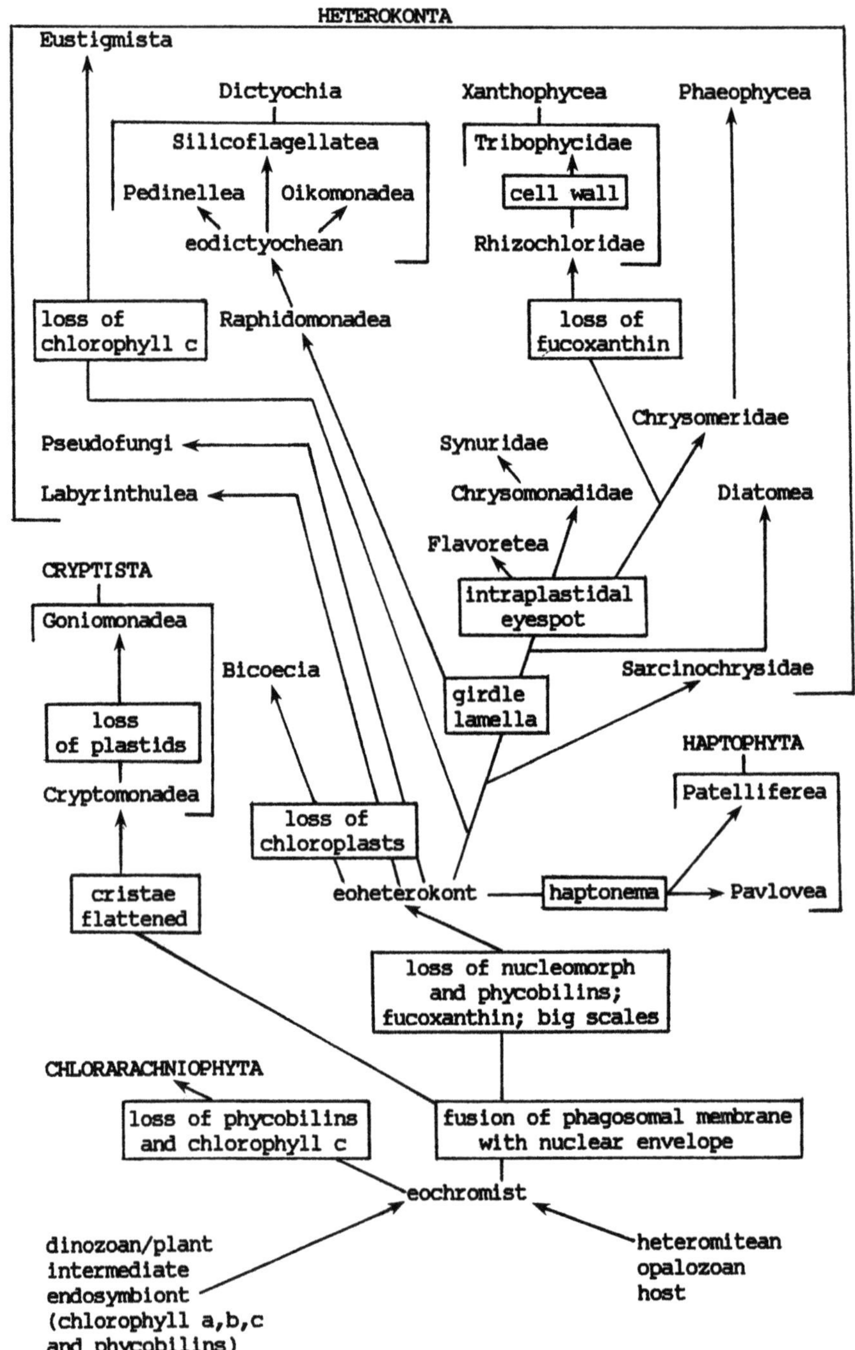

Fig. 22.2 Postulated relationships between Haptophyta, other chromists, and their two probable ancestors.

a single symbiotic origin for all chloroplasts from a cyanobacterium, rather than several (Mereschkowsky 1910; Raven 1970; Whatley *et al.* 1979) or numerous symbioses (Margulis 1970, 1981). They thus support my theses that green plant and euglenoid chloroplasts evolved from cyanobacteria (not prochlorophytes), and that chromophyte chloroplasts also ultimately evolved from cyanobacteria, rather than a mythical bacterium with chlorophyll c_1 and c_2 (Margulis 1970; Raven 1970; Whatley *et al.* 1979). Secondly, they also support my thesis that the chloroplast of *Cyanophora* (and other Glaucophyta) has retained peptidoglycan for 550–600 million years and diverged from the same cyanobacterium as other chloroplasts, and are not recent independently symbiotic cyanobacteria ('cyanelles') as commonly thought (Pascher 1929; Margulis 1970; 1981; Taylor 1978; Whatley *et al.* 1979). Thirdly, the 16s and 18s trees (Fig. 22.1) both contradict a common assumption that retention of cryptomonad and *Chlorarachnion* nucleomorphs implies a much more recent endosymbiotic origin for cryptomonads and *Chlorarachnion* than for chromobiotes. Instead nucleomorphs diverged from other nuclei (Fig. 22.1), and the cryptomonad chloroplasts from other chloroplasts, at about the same time as each diverged from chromobiotes, exactly as argued (Cavalier-Smith 1982, 1986). Fourthly the nuclear (Schlegel 1991; Sogin 1991) and chloroplast trees support my thesis that the major groups of algae diverged approximately simultaneously; the sole exception is euglenoids, which diverge on the nuclear 18s RNA tree (but not the chloroplast tree) much earlier than other algae. This mismatch between the two trees is perhaps explicable by the secondary acquisition by euglenoids of a preexisting eukaryotic chloroplast (Gibbs 1978), as discussed elsewhere (Cavalier-Smith 1992*a*, 1993*a*). Fifthly, nucleomorph sequences confirm the symbiotic origin of cryptomonads (Douglas *et al.* 1991; Maier *et al.* 1991). However, as explained elsewhere (Cavalier-Smith 1992*a*), since the chromist phyla diverged almost simultaneously with the origin of the group, as predicted (Cavalier-Smith 1982, 1986), 18s rRNA phylogenetics are unable conclusively to confirm or refute the monophyly of the Chromista.

Ciliary transformation and the origin of Haptophyta

Two key changes are needed to convert a heterokont into a haptophyte: origin of the haptonema and loss of ciliary retronemes. I argue that these two changes were intimately linked. How retronemes may have been lost has been clarified by the discovery of ciliary transformation in biciliate protists (Melkonian *et al.* 1987). Every cell cycle the anterior tinsel cilium of heterokonts retracts, loses its retronemes, and re-elongates after its centriole reorients to become the posterior whiplash cilium (Wetherbee *et al.* 1988; Beech and Wetherbee 1990*a*, *b*; Beech *et al.* 1991). Thus tinsel and whiplash cilia of heterokonts are not fundamentally distinct as formerly thought, but different developmental stages of the same thing. A ciliary developmental cycle with transformation of a younger cilium into a structurally different older cilium is also true for cryptomonads (Perasso *et al.* 1992) and anisokont biciliates in phyla as varied as green algae, Euglenozoa, and Glaucophyta (Beech *et al.* 1991); ciliary transformation occurs even in isokonts like *Chlamydomonas* (Holmes and Dutcher 1989), almost isokont species like the haptophyte *Pleurochrysis* (Beech *et al.* 1988), and organisms with two centrioles but only one cilium (e.g. pedinellids). In pedinellids the young centriole bearing the tinsel cilium transforms into the older barren centriole in the next generation. Thus ciliary and centriolar transformation over two cell generations, coupled with a semi-conservative segregation of new and old centrioles, is probably universal in eukaryotes with paired centrioles.

A heterokont could lose retronemes either by mutation of their structural genes or by modifying developmental controls over ciliary transformation to give the younger cilium characteristics formerly delayed by a generation: an organellar neoteny or heterochrony. This makes retroneme loss mechanistically easier than previously thought, but would not make it evolutionarily easier or frequent, because the great rarity of retroneme loss implies a very strong stabilizing selection against it. Loss of retronemes would alter the direction of thrust of a cilium to the detriment of swimming and taxes, and for phagotrophs, their feeding behaviour (Cavalier-Smith 1986, 1989). Therefore retroneme gain or loss could be successful only if associated with a drastic and advantageous change in these properties.

I propose that what enabled, nay stimulated, the loss of retronemes was evolution of the haptonema as an alternative feeding apparatus. This is suggested by the proof that the long coilable haptonema of *Chrysochromulina hirta* is a highly efficient food capturing device (Kawachi *et al.* 1991). Cells of *Chrysochromulina hirta* swim with their elongated haptonema sticking out ahead, which catches prey by adhesion, concentrates it in a particle aggregating centre, moves the aggregate to the haptonematal tip, and then bends to transfer the aggregate to the site of phagocytosis at the rear of the cell. Their isokont cilia do not have a planar oar-like beat like the anterior cilia of *Chlamydomonas*: instead they point backwards and undulate from base to tip like the rear-pointing whiplash cilium of heterokonts. Thus, they resemble the heterokont whiplash cilium both in the absence of retronemes and in orientation and beat. A heterochronic mutation simply by converting the 'young' anterior cilium into an 'old' posterior cilium could have caused all these changes simultaneously and converted the ciliary pattern of a heterokont into that of a prymnesialean haptophyte in one step. The importance of this 'macromutation' is that, had the loss and transformation been gradual with many steps, it is unlikely that intermediates would have been advantageous. Loss of retronemes was beneficial to the ancestral haptophyte only because it coincided with the origin of the haptonema and its novel use in feeding.

Phagotrophic heterokonts feed on bacteria propelled towards the cell by the reverse current generated by retronemes on the anterior cilium as it beats from base to tip. Retronemes not only increase ciliary thrust for more powerful swimming (Holwill 1982), but are also an adaptation for phagotrophy which directs prey towards the base of the cilium. The three phagotrophic heterokont classes have different mechanisms for trapping prey from this feeding current: (1) pedinellids use a symmetrical ring of axopodia, each supported by three microtubules in triangular array, surrounding the tinsel cilium; because this trap of sticky axopodia is most efficient when radially symmetrical, they lost the posterior whiplash cilium by converting it to a barren centriole; (2) chrysomonads use a temporary feeding basket (*Epipyxis*) or pouch (*Chrysosphaerella*) to catch their prey (Andersen and Wetherbee 1991; Moestrup and Andersen 1991; Wetherbee and Andersen 1992); this forms around a large loop created by the upwards sliding of a microtubule of the R3 root; (3) Bicoecea have a permanent lip-like peristome supported by a broad band of eight microtubules. The divergence between these three classes I interpret as an early functional radiation of the ancestral photophagotrophic heterokont.

Origin of the haptonema

I regard the haptonema as a fourth chromobiote specialization for phagotrophy; but one which, unlike the others, would have worked much better, not worse, if retronemes were lost. The fishing rod function of a long anteriorly pointing haptonema, like that of *Chrysochromulina hirta*, would be impaired by a closely parallel anterior tinsel cilium which would tend to sweep bacteria away from it. By losing retronemes and reflexing the former tinsel cilium like the whiplash cilium, the sticky, unimpeded haptonema would precede the swimming cell and trap prey. Thus retroneme loss would have been grossly harmful for a phagotrophic heterokont, without a haptonema, because base to tip undulation of the now posterior former tinsel cilium would thereafter propel prey away from the cell not towards it; but, immediately after the origin of a haptonema, loss of retronemes would instead have been strongly beneficial. Since, as argued above, such loss by a heterochronic mutation would have been very easy, retronemes would have been lost almost immediately after the origin of the haptonema. This is why haptophytes, alone among chromobiotes, lack retronemes. Thus their two key major differences from heterokonts, absence of retronemes and presence of haptonema, arose together as a striking case of functionally constrained organellar coevolution. The 18s rRNA tree (Fig. 22.1) shows that retroneme loss must have occurred in the very earliest stages of chromobiote evolution, as originally proposed (Cavalier-Smith 1986).

To discuss the origin of the haptonema in more detail we need to decide which haptophyte cilium is homologous with the tinsel and which with the whiplash cilium of heterokonts. I assume that ciliary transformation is fundamentally conserved in all eukaryotes with bikinetids. In the patelliferan *Pleurochrysis,* the younger first generation 'right' centriole bears the shorter cilium; the older (second or later generation) centriole bears the longer 'left' cilium (Beech and Wetherbee 1988; Beech *et al.* 1988; Beech *et al.* 1991), which makes it homologous to the centriole of the heterokont whiplash cilium. Since the anterior centriole of *Pavlova* is positionally homologous to the 'right' centriole of patelliferans it must be homologous with the younger centriole that bears the anterior tinsel cilium of heterokonts, as Beech *et al.* (1991) conclude. This suggests that the anterior cilium of the strongly anisokont pavloveans evolved from the anterior tinsel cilium of heterokonts by losing retronemes, thus supporting the thesis (Green 1980; Cavalier-Smith 1986) that the asymmetry and anisokonty of *Pavlova* is ancestral for haptophytes, and that the isokonty of Prymnesiales is secondary. Prymnesialean isokonty makes it difficult to determine which cilium is the fluorescent one; I predict that it will be the 'left' one, i.e. that homologous with the whiplash cilium: in *Prymnesium* it is the longer, in *Pleurochrysis* the left cilium; i.e. the long cilia of *Prymnesium* and *Pleurochrysis* are homologous.

Given these positional homologies let me attempt to homologize the haptonematal axoneme with one of the microtubular roots of heterokonts. The roots of Haptophyta are rather varied (Preisig 1989; Andersen 1991). Those of *Pleurochrysis* (Inouye and Pienaar 1985) seem simplest and ancestral, since they consist of three major roots arranged as in the archezoan retortamonads and Euglenozoa: i.e. an R_1 root that curves around the younger centriole and nucleates its own microtubules, and two dissimilar roots (R_3 and R_2) emanating from either side of the older centriole, which is probably the ancestral state for biciliate eukaryotes (Cavalier-Smith 1992*b*). The base of the 8-microtubule haptonematal axoneme starts near, and to the outer side of, the base of the R3 root. Its position, though not identical with that of any microtubular roots of heterokonts, is quite close to the R3 root, which has a similar number of microtubules, six in chrysomonads; seven in *Pleurochrysis*; seven or eight in the peristomial lip of bicoecids, which I considered homologous with the R3 root (Cavalier-Smith 1989) I therefore propose that the haptonematal axoneme arose by duplication of the nucleating centre for the R3 microtubular root, and the attachment of the duplicate to a different part of the older centriole in such an orientation that the microtubules grew towards the cell surface, and pushed the plasma membrane outwards to form a primitive haptonema. A simple long haptonema could thus have been formed in a single step. From the outset its surface could have carried glycoproteins able to stick to bacterial prey, since these would have been on the plasma membrane in the ciliary base region of the ancestral photo-phagotrophic heterokont (because phagocytosis occurs there). It also could have had the molecular motors for moving these glycoproteins with adhering prey. Initially, I suggest, these mechano-chemical ATPases (dynein for basipetal and kinesin for acropetal movement) moved them all the way to the cell surface for engulfment, as in pedinellid axopodia or choanoflagellate microvilli. Such ability to move particles is widespread in protists; not only phagotrophic ones, but also on the ciliary surfaces of autotrophs like *Chlamydomonas* (Bloodgood 1991).

The ability to bend and coil a long haptonema, I suggest, evolved later by controlling the calcium levels in it by calcium pumps and gatable calcium channels located in the smooth reticulum now present in haptonemata, which I suggest was added after the original trapping function of the haptonema so as to achieve such controls. Eventually, in haptophytes that gave up phagotrophy and concentrated on photosynthesis, the haptonema was greatly shortened (Pavlovea and coccolithophorids) or even lost (e.g. *Dicrateria*).

Haptophyte diversification

No modern haptophyte fits the picture of the ancestral haptophyte that I have painted: an asymmetric body shape like *Pavlova*, a long predatory haptonema as in *Chrysochromulina hirta*. Different lineages have probably lost different characters at various times; probably more than one reduced the haptonema or made the body more symmetrical and isodiametric. The latter could have occurred for two different reasons: to perfect the *Chrysochromulina*-like feeding mechanism or to expand and simplify the cell in non-motile photosynthetic plankton. If the large scales of chrysomonads and patelliferans are homologous, as I suspect but cannot demonstrate, they must have been lost by the ancestral pavlovean; if they are not, they could have originated in the patelliferan ancestor after it diverged from Pavlovea. The knobs on pavlovean cilia are usually called knob-scales, but I do not think them homologous with the scales of other haptophytes. They seem more like modified hairs, so I call them 'knobbed hairs'.

Since Pavlovales differ from other haptophytes not only in having ciliary hairs and knobbed hairs, and lacking scales (Cavalier-Smith 1986), but also in having very different ciliary roots (Preisig 1989) and mitosis (Green

1989), as well as significant chemical differences (Green *et al.* 1990), and since the 18s rRNA tree shows that the divergence between *Prymnesium* and coccolithophorids is significantly less deep than between them and *Pavlova*, I have raised the subclass Pavlovidae Cavalier-Smith 1986 to the class Pavlovea (Cavalier-Smith 1993*b*), and grouped scale-bearing haptophytes (orders Isochrysidales, Coccosphaerales, Prymnesiales) in the new class Patelliferea (Cavalier-Smith 1993*b*). It has long been supposed that Patelliferea and Pavlovea diverged rather early (Green 1980; Cavalier-Smith 1986, 1989; Green *et al.* 1990). The 18s rRNA tree dates this divergence before that between Phaeophycea and Xanthophycea, but not as early as the likely time of origin of Haptophyta. Though I oppose the idea that divergence time is a sufficient basis for assigning ranks (Woese *et al.* 1990), it can be a useful supplementary consideration when it corroborates morphological evidence. The fact that haptophyte 18s rRNA is about as different from that of heterokonts as from those of dinoflagellates, cryptomonads, and ciliates, whereas that of the different heterokont classes is much more similar, supports the ranking of Heterokonta and Haptophyta as phyla (divisions).

The recent discovery of many non-photosynthetic haptophytes (Marchant and Thomsen, Chapter 11) makes it important to determine whether they have residual leucoplasts. If they do, then haptophytes probably became obligately dependent on plastids for a benefit other than photosynthesis very early in their evolution. If they do not, then either the ancestral haptophyte was not photosynthetic, contrary to what I have argued, or else haptophytes can dispense altogether with their plastids. It would be valuable to learn more about the partition of non-photosynthetic functions between the chloroplast and the rest of the cell, and how the distribution of phagotrophy in the phylum relates to the degree of development of the haptonema and the presence or absence of photosynthesis.

Acknowledgements

I thank T. Chappell for typing; E. Chao for help with sequence alignment and figures; NSERC for a research grant; F. E. Round for comments on the manuscript; and the Canadian Institute for Advanced Research for Fellowship support.

References

Andersen, R. A. (1991). The cytoskeleton of chromophyte algae. *Protoplasma*, **164**, 143–59.

Andersen, R. A. and Wetherbee, R. (1991). Microtubules of the flagellar apparatus are active during prey capture in the chrysophycean alga *Epipyxis pulchra*. *Protoplasma*, **166**, 8–20.

Beech, P. L. and Wetherbee, R. (1988). Observations on the flagellar apparatus and peripheral endoplasmic reticulum of the coccolithophorid, *Pleurochrysis carterae* (Prymnesiophyceae). *Phycologia*, **27**, 142–58.

Beech, P. L. and Wetherbee, R. (1990*a*). Direct observations on flagellar transformation in *Mallomonas splendens* (Synurophyceae). *Journal of Phycology*, **26**, 90–5.

Beech, P. L. and Wetherbee, R. (1990*b*). The flagellar apparatus of *Mallomonas splendens* (Synurophyceae) at interphase and its development during the cell cycle. *Journal of Phycology*, **26**, 95–111.

Beech, P. L., Wetherbee, R., and Pickett-Heaps, J. D. (1988). Transformation of the flagella and associated flagellar components during cell division in the coccolithophorid *Pleurochrysis carterae*. *Protoplasma*, **145**, 37–46.

Beech, P. L., Heimann, K., and Melkonian, M. (1991). Development of the flagellar apparatus during the cell cycle in unicellular algae. *Protoplasma*, **164**, 23–37.

Bhattacharya, D., Medlin, L., Wainright, P. O., Ariztia, E. V., Bibeau, C., Stickel, S. K., and Sogin, M. L. (1992). Algae containing chlorophylls *a* + *c* are paraphyletic: molecular evolutionary analysis of the Chromophyta. *Evolution*, **46**, 1801–17.

Bjørnland, T. and Liaaen-Jensen, S. (1989). Distribution patterns of carotenoids in relation to chromophyte phylogeny and systematics. In *The chromophyte algae: problems and perspectives*, (ed. J. C. Green, B. S. C. Leadbeater, and W. L. Diver), pp. 37–61. Clarendon Press, Oxford.

Bloodgood, R. A. (1991). Regulation of flagellar glycoprotein movement by protein phosphorylation. *European Journal of Cell Biology*, **54**, 85–9.

Bouck, G. B. (1965). Fine structure and organelle associations in brown algae. *Journal of Cell Science*, **26**, 523–37.

Cavalier-Smith, T. (1978). The evolutionary origin and phylogeny of microtubules, mitotic spindles and eukaryote flagella. *BioSystems*, **10**, 93–114.

Cavalier-Smith, T. (1981*a*). Eukaryote kingdoms: seven or nine? *BioSystems*, **14**, 461–81.

Cavalier-Smith, T. (1981*b*). The origin and early evolution of the eukaryote cell. In *Molecular and cellular aspects of microbial evolution*, (ed. M. J. Carlile, J. F. Collins, and B. E. B. Moseley), pp. 33–84. Cambridge University Press.

Cavalier-Smith, T. (1982). The origins of plastids. *Biological Journal of the Linnean Society*, **17**, 289–306.

Cavalier-Smith, T. (1986). The kingdom Chromista: origin and systematics. In *Progress in phycological research*, Vol. 4, (ed. F. E. Round and D. J. Chapman), pp. 309–47. Biopress Ltd., Bristol.

Cavalier-Smith, T. (1989). The kingdom Chromista. In *The chromophyte algae: problems and perspectives*, (ed. J. C. Green, B. S. C. Leadbeater, and W. L. Diver), pp. 379–405. Clarendon Press, Oxford.

Cavalier-Smith, T. (1991). Cell diversification in heterotrophic flagellates. In *The biology of free-living heterotrophic flagellates*, (ed. D. J. Patterson and J. Larsen), pp. 113–31. Clarendon Press, Oxford.

Cavalier-Smith, T. (1992*a*). The number of symbiotic origins of organelles. *BioSystems*, **28**, 91–106.

Cavalier-Smith, T. (1992*b*). Origin of the cytoskeleton. In *The origin and evolution of the cell*, (ed. H. Hartman and K. Matsuno), pp. 79–106. World Scientific Publishing, Singapore.

Cavalier-Smith, T. (1993*a*). The origin, losses and gains of chloroplasts. In *Origins of plastids: symbiogenesis, prochlorophytes, and the origins of chloroplasts*, (ed. R. A. Lewin), pp. 291–349. Chapman & Hall, New York.

Cavalier-Smith, T. (1993*b*). Kingdom Protozoa and its 18 phyla. *Microbiological Reviews*, **57**, 953–94.

Cavalier-Smith, T. (1993*c*). The protozoan phylum Opalozoa. *Journal of Eukaryotic Microbiology*, **40**, 609–15.

Christensen, T. (1962). Alger. In *Botanik*, Bd-2, *Systematisk Botanik*, Nr. 2, (ed. T. W. Böcher, M. Lange, and T. Sørensen), pp. 1–178. Munksgaard. København.

Christensen, T. (1989). The Chromophyta, past and present. In *The chromophyte algae: problems and perspectives*, (ed. J. C. Green, B. S. C. Leadbeater, and W. L. Diver), Systematics Association Special Volume No. 38, pp. 1–12. Clarendon Press, Oxford.

Douglas, S. E. and Turner, S. (1991). Molecular evidence for the origin of plastids from a cyanobacterium-like ancestor. *Journal of Molecular Evolution*, **33**, 267–73.

Douglas, S. E., Murphy, C. A., Spencer, D. F., and Gray, M. W. (1991). Cryptomonad algae are evolutionary chimaeras of two phylogenetically distinct unicellular eukaryotes. *Nature*, **350**, 148–51.

Gibbs, S. P. (1978). The chloroplasts of *Euglena* may have evolved from symbiotic green algae. *Canadian Journal of Botany* **56**, 2883–9.

Gibbs, S. P. (1981*a*). The chloroplast endoplasmic reticulum structure, function and evolutionary significance. *International Review of Cytology*, **72**, 49–99.

Gibbs, S. P. (1981*b*). The chloroplasts of some algal groups may have evolved from endosymbiotic eukaryotic algae. *Annals of the New York Academy of Sciences*, **361**, 193–208.

Gibbs, S. P. (1993). The evolution of algal chloroplasts. In *Origins of plastids: symbiogenesis, prochlorophytes, and the origins of chloroplasts*, (ed. R. A. Lewin), pp. 107–21. Chapman & Hall, New York.

Green, J. C. (1980). The fine structure of *Pavlova pinguis* Green and a preliminary survey of the order Pavlovales (Prymnesiophyceae). *British Phycological Journal*, **15**, 151–91.

Green, J. C. (1989). Relationships between the chromophyte algae: the evidence from studies of mitosis. In *The chromophyte algae: problems and perspectives*, (ed. J. C. Green, B. S. C. Leadbeater, and W. L. Diver), Systematics Association Special Volume No. 38, pp. 189–206. Clarendon Press, Oxford.

Green, J. C., Perch-Nielsen, K., and Westbroek, P. (1990). Phylum Prymnesiophyta. In *Handbook of Protoctista*, (ed. L. Margulis, J. O. Corliss, M. Melkonian, and D. J. Chapman), pp. 293–317. Jones and Bartlett, Boston.

Greenwood, A. D. (1974). The Cryptophyta in relation to phylogeny and photosynthesis. In *8th International Congress of Electron Microscopy*, pp. 566–7. Canberra.

Greenwood, A. D., Griffiths, H. B., and Santore, U. J. (1977). Chloroplasts and cell compartments in Cryptophyceae. *British Phycological Journal*, **12**, 119.

Grell, K. G., Heini, A., and Schüller, S. (1990). The ultrastructure of *Reticulosphaera socialis* Grell (Heterokontophyta). *European Journal of Protistology*, **26**, 37–54.

Hibberd, D. J. (1972). Chrysophyta: definition and interpretation. *British Phycological Journal*, **7**, 281.

Hibberd, D. J. (1976). The ultrastructure and taxonomy of the Chrysophyceae and Prymnesiophyceae (Haptophyceae): a summary with some new observations on the ultrastructure of the Chrysophyceae. *Botanical Journal of the Linnean Society*, **72**, 55–80.

Hibberd, D. J. and Norris, R. E. (1984). Cytology and ultrastructure of *Chlorarachnion reptans* (Chlorarachniophyta divisio nova, Chlorarachniophyceae classis nova). *Journal of Phycology*, **20**, 310–30.

Hollwill, M. E. J. (1982). Dynamics of eukaryotic flagellar movement. In *Prokaryotic and eukaryotic flagella*, (ed. W. B. Amos and J. G. Duckett), pp. 289–312. Cambridge University Press.

Holmes, J. A. and Dutcher, S. K. (1989). Cellular asymmetry in *Chlamydomonas reinhardii*. *Journal of Cell Science*, **94**, 275–85.

Inouye, I. and Pienaar, R. N. (1985). Ultrastructure of the flagellar apparatus in *Pleurochrysis* (Class *Prymnesiophyceae*). *Protoplasma* **125**, 24–35.

Jeffrey, C. (1971). Thallophytes and kingdoms — a critique. *Kew Bulletin*, **25**, 291–9.

Jeffrey, S. W. (1989). Chlorophyll *c* pigments and their distribution in the chromophyte algae. In *The chromophyte algae: problems and perspectives*, (ed. J. C. Green, B. S. C. Leadbeater, and W. L. Diver), Systematics Association Special Volume No. 38, pp. 13–36. Clarendon Press, Oxford.

Jenkins, J., Hiller, R. G., Speirs, J., and Godovac-Zimmerman, J. (1990). A genomic clone encoding a cryptophyte phycoerythrin α-subunit. Evidence for three α-subunits and an N-terminal membrane transit sequence. *FEBS Letters*, **273**, 191–4.

Kawachi, M., Inouye, I., Maeda, O., and Chihara, M. (1991). The haptonema as a food-capturing device: observations on *Chrysochromulina hirta* (Prymnesiophyceae). *Phycologia*, **30**, 563–73.

Kawai, H. (1988). A flavin-like autofluorescent substance in the posterior flagellum of golden and brown algae. *Journal of Phycology*, **24**, 114–17.

Kawai, H. and Inouye, I. (1989). Flagellar autofluorescence in forty-four chlorophyll *c*-containing algae. *Phycologia*, **28**, 222–7.

Larsen, J., (1988). An ultrastructural study of *Amphidinium poecilochroum* (Dinophyceae), a phagotrophic dinoflagellate feeding on small species of cryptophytes. *Phycologia*, **27**, 366–77.

Lee, R. and Kugrens, P. (1991). *Katablepharis ovalis*, a colourless flagellate with interesting cytological characteristics. *Journal of Phycology* **27**, 505–13.

Leedale, G.F. (1974). How many are the kingdoms of organisms? *Taxon*, **32**, 261–70.

Maier, U.-G., Hofmann, C. J. B., Eschbach, S., Wolters, J., and Igloi, G. (1991) Demonstration of nucleomorph-encoded eukaryotic small subunit ribosomal RNA in cryptomonads. *Molecular and General Genetics*, **230**, 155–60.

Manton, I. (1966). Further observations on the fine structure of *Chrysochromulina chiton*, with special reference to the pyrenoid. *Journal of Cell Science* **1**, 187–92.

Margulis, L. (1970). *Origin of eukaryotic cells*. Yale University Press, New Haven.

Margulis, L. (1981). *Symbiosis in cell evolution*. Freeman, San Francisco.

McFadden, G., Gilson, P. R., Hofmann, C. J. B., Adcock, G. J., and Maier, U.-G. (1994). Evidence that an amoeba acquired a chloroplast by retaining part of an engulfed eukaryotic alga. *Proceedings of the National Academy of Sciences of the USA*, (In press).

Mereschkowsky, C. (1910). Theorie der zwei Plasmaarten als Grundlage der Symbiogenesis, einer neuen Lehre von der Entstehung der Organismen. *Biologisches Centralblatt*, **30**, 278–303, 321–47, 353–67.

Melkonian, M., Reize, I. B., and Preisig, H. R. (1987). Maturation of a flagellum/basal body requires more than one cell cycle in algal flagellates: studies on *Nephroselmis olivacea (Prasinophyceae)*. In *Algal development: molecular and cellular aspects*, (ed. W. Wiessner, D. G. Robinson, and R. C. Starr), pp. 102–13. Springer, Berlin.

Moestrup, Ø. and Andersen, R. A. (1991). Organization of heterotrophic heterokonts. In *The biology of free-living heterotrophic flagellates*, (ed. D. J. Patterson and J. Larsen), Systematics Association Special Volume No. 45, pp. 333–60. Clarendon Press, Oxford.

Parke, M., Manton, I., and Clarke, B. (1955). Studies on marine flagellates. II. Three new species of *Chrysochromulina*. *Journal of the Marine Biological Association of the United Kingdom*, **34**, 579–609.

Pascher, A. (1929). Studien über Symbiosen. I. Über einige Endosymbiosen von Blaualgen in Einzellern. *Jahrbücher für Wissenshaftliche Botanik*, **71**, 386–462.

Patterson, D. J. (1989). Stramenopiles: chromophytes from a protistan perspective. In *The chromophyte algae: problems and perspectives*, (ed. J. C. Green, B. S. C. Leadbeater, and W. L. Diver), Systematics Association Special Volume No. 38, pp. 357–79. Clarendon Press, Oxford.

Perasso, R., Hill, D. R. A., and Wetherbee, R. (1992). Transformation and development of the flagellar apparatus of *Cryptomonas ovata* (Cryptophyceae) during cell division. *Protoplasma* **170**, 53–67.

Preisig, H. R. (1989). The flagellar base ultrastructure and phylogeny of chromophytes. In *The chromophyte algae: problems and perspectives*, (ed. J. C. Green, B. S. C. Leadbeater, and W. L. Diver), Systematics Association Special Volume No. 38, pp. 167–87. Clarendon Press, Oxford.

Raven, P. H. (1970). A multiple origin for plastids and mitochondria. *Science*, **169**, 641–6.

Schnepf, E. (1993). From prey via endosymbiont to plastid: comparative studies in dinoflagellates. In *Origins of plastids: symbiogenesis, prochlorophytes, and the origins of chloroplasts*, (ed. R. A. Lewin), pp. 53–76. Chapman & Hall, New York.

Schlegel, M. (1991). Protist evolution and phylogeny as discerned from small subunit ribosomal RNA sequence comparisons. *European Journal of Protistology*, **27**, 207–19.

Sogin, M. L. (1991). Early evolution and the origin of eukaryotes. *Current Opinion in Genetics and Development*, **1**, 457–63.

Taylor, F. J. R. (1978). Problems in the development of an explicit phylogeny of the lower eukaryotes. *BioSystems*, **10**, 67–89.

Wetherbee, R. and Andersen, R. A. (1992). Flagella of chrysophycean algae play on active role in prey capture and selection: direct observations on *Epipyxis pulchra* and *Ochromonas danica* using image enhanced video microscopy. *Protoplasma*, **166**, 1–7.

Wetherbee, R., Platt, S. J., Beech, P. L., and Pickett-Heaps, J. D. (1988). Flagellar transformation in the heterokont *Epipyxis pulchra (Chrysophyceae)*: direct observations using image enhanced light microscopy. *Protoplasma* **145**, 47–54.

Whatley, J. M. (1989). Chromophyte chloroplasts — a polyphyletic origin? In *The chromophyte algae: problems and perspectives*, (ed. J. C. Green, B. S. C. Leadbeater, and W. L. Diver), Systematics Association Special Volume No. 38, pp. 125–144. Clarendon Press, Oxford.

Whatley, J. M. and Whatley, F. R. (1981). Chloroplast evolution. *New Phytologist*, **87**, 233–47.

Whatley, J. M., John, P., and F. R. Whatley (1979). From extracellular to intracellular: the establishment of mitochondria and chloroplasts. *Proceedings of the Royal Society of London*, Series B, **204**, 165–87.

Woese, C. R., Kandler, O., and Wheelis, M. L. (1990). Towards a natural system of organisms: proposal for the domains Archaea, Bacteria, and Eucarya. *Proceedings of the National Academy of Sciences of the USA*, **87**, 4576–9.

Index

Page numbers in *italics* are references to Figures.

Systematics Association Publications

1. Bibliography of key works for the identification of the British fauna and flora, *3rd edition* (1967)†
   Edited by G. J. Kerrich, R. D. Meikle, and N. Tebble
2. Function and taxonomic importance (1959)†
   Edited by A. J. Cain
3. The species concept in palaeontology (1956)†
   Edited by P. C. Sylvester-Bradley
4. Taxonomy and geography (1962)†
   Edited by D. Nichols
5. Speciation in the sea (1963)†
   Edited by J. P. Harding and N. Tebble
6. Phenetic and phylogenetic classification (1964)†
   Edited by V. H. Heywood and J. McNeill
7. Aspects of Tethyan biogeography (1967)†
   Edited by C. G. Adams and D. V. Ager
8. The soil ecosystem (1969)†
   Edited by H. Sheals
9. Organisms and continents through time (1973)†
   Edited by N. F. Hughes
10. Cladistics: a practical course in systematics (1992)
    P. L. Forey, C. J. Humphries, I. J. Kitching, R. W. Scotland, D. J. Siebert, and D. M. Williams.

†Published by the Association (out of print)

Systematics Association Special Volumes

1. The new systematics (1940)
   Edited by J. S. Huxley (Reprinted 1971)
2. Chemotaxonomy and serotaxonomy (1968)*
   Edited by J. G. Hawkes
3. Data processing in biology and geology (1971)*
   Edited by J. L. Cutbill
4. Scanning electron microscopy (1971)*
   Edited by V. H. Heywood
   Out of print

5. Taxonomy and ecology (1973)*
   Edited by V. H. Heywood
6. The changing flora and fauna of Britain (1974)*
   Edited by D. L. Hawksworth
   Out of print
7. Biological identification with computers (1975)*
   Edited by R. J. Pankhurst
8. Lichenology: progress and problems (1976)*
   Edited by D. H. Brown, D. L. Hawksworth, and R. H. Bailey
9. Key works to the fauna and flora of the British Isles and north-western Europe, 4th edition (1978)*
   Edited by G. J. Kerrich, D. L. Hawksworth, and R. W. Sims
10. Modern approaches to the taxonomy of red and brown algae (1978)
    Edited by D. E. G. Irvine and J. H. Price
11. Biology and systematics of colonial organisms (1979)*
    Edited by G. Larwood and B. R. Rosen
12. The origin of major invertebrate groups (1979)*
    Edited by M. R. House
13. Advances in bryozoology (1979)*
    Edited by G. P. Larwood and M. B. Abbot
14. Bryophyte systematics (1979)*
    Edited by G. C. S. Clarke and J. G. Duckett
15. The terrestrial environment and the origin of land vertebrates (1980)
    Edited by A. L. Panchen
16. Chemosystematics: principles and practice (1980)*
    Edited by F. A. Bisby, J. G. Vaughan, and C. A. Wright
17. The shore environment: methods and ecosystems (2 Volumes) (1980)*
    Edited by J. H. Price, D. E. G. Irvine, and W. F. Farnham
18. The Ammonoidea (1981)*
    Edited by M. R. House and J. R. Senior
19. Biosystematics of social insects (1981)*
    Edited by P. E. Howse and J. -L. Clément
20. Genome evolution (1982)*
    Edited by G. A. Dover and R. B. Flavell
21. Problems of phylogenetic reconstruction (1982)*
    Edited by K. A. Joysey and A. E. Friday
22. Concepts in nematode systematics (1983)*
    Edited by A. R. Stone, H. M. Platt, and L. F. Khalil
23. Evolution, time and space: the emergence of the biosphere (1983)*
    Edited by R. W. Sims, J. H. Price, and P. E. S. Whalley
24. Protein polymorphism: adaptive and taxonomic significance (1983)*
    Edited by G. S. Oxford and D. Rollinson
25. Current concepts in plant taxonomy (1983)*
    Edited by V. H. Heywood and D. M. Moore
26. Databases in systematics (1984)*
    Edited by R. Allkin and F. A. Bisby

27. Systematics of the green algae (1984)*
    Edited by D. E. G. Irvine and D. M. John
28. The origins and relationships of lower invertebrates (1985)‡
    Edited by S. Conway Morris, J. D. George, R. Gibson, and H. M. Platt
29. Infraspecific classification of wild and cultivated plants (1986)‡
    Edited by B. T. Styles
30. Biomineralization in lower plants and animals (1986)‡
    Edited by B. S. C. Leadbeater and R. Riding
31. Systematic and taxonomic approaches in palaeobotany (1986)‡
    Edited by R. A. Spicer and B. A. Thomas
32. Coevolution and systematics (1986)‡
    Edited by A. R. Stone and D. L. Hawksworth
33. Key works to the fauna and flora of the British Isles and north-western Europe, 5th edition (1988)‡
    Edited by R. W. Sims, P. Freeman, and D. L. Hawksworth
34. Extinction and survival in the fossil record (1988)‡
    Edited by G. P. Larwood
35. The phylogeny and classification of the tetrapods (2 Volumes) (1988)‡
    Edited by M. J. Benton
36. Prospects in systematics (1988)‡
    Edited by D. L. Hawksworth
37. Biosystematics of haematophagous insects (1988)‡
    Edited by M. W. Service
38. The chromophyte algae: problems and perspective (1989)‡
    Edited by J. C. Green, B. S. C. Leadbeater, and W. L. Diver
39. Electrophoretic studies on agricultural pests (1989)‡
    Edited by Hugh D. Loxdale and J. den Hollander
40. Evolution, systematics, and fossil history of the Hamamelidae (2 Volumes) (1989)‡
    Edited by Peter R. Crane and Stephen Blackmore
41. Scanning electron microscopy in taxonomy and functional morphology (1990)‡
    Edited by D. Claugher
42. Major evolutionary radiations (1990)‡
    Edited by P. D. Taylor and G. P. Larwood
43. Tropical lichens: their systematics, conservation, and ecology (1991)‡
    Edited by D. J. Galloway
44. Pollen and spores: patterns of diversification (1991)‡
    Edited by S. Blackmore and S. H. Barnes
45. The biology of free-living heterotrophic flagellates (1991)‡
    Edited by D. J. Patterson and J. Larsen
46. Plant-animal interactions in the marine benthos (1992)‡
    Edited by D. M. John, S. J. Hawkins, and J. H. Price
47. The Ammonoidea: environment, ecology, and evolutionary change (1993)‡
    Edited by M. R. House

48. Designs for a global plant species information system‡
Edited by F. A. Bisby, G. F. Russell, and R. J. Pankhurst
49. Plant galls: organisms, interactions, populations‡
Edited by Michèle A. J. Williams
50. Systematics and conservation evaluation‡
Edited by P. L. Forey, C. J. Humphries, and R. I. Vane-Wright
51. The Haptophyte algae‡
Edited by J. C. Green and B. S. C. Leadbeater

*Published by Academic Press for the Systematics Association
†Published by the Palaeontological Association in conjunction with Systematics Association
‡Published by the Oxford University Press for the Systematics Association

The manufacturer's authorised representative in the EU for product safety is Oxford University Press España S.A. of el Parque Empresarial San Fernando de Henares, Avenida de Castilla, 2 – 28830 Madrid (www.oup.es/en or product.safety@oup.com). OUP España S.A. also acts as importer into Spain of products made by the manufacturer.

www.ingramcontent.com/pod-product-compliance
Ingram Content Group UK Ltd.
Pitfield, Milton Keynes, MK11 3LW, UK
UKHW021439280726
14060UKWH00001BA/148